Geological Evolution of Ocean Basins: Results from the Ocean Drilling Program

Robert Benjamin Kidd

1947–1996

Robert Benjamin Kidd will be long remembered for his enthusiastic support of marine geosciences world-wide, his innovative thinking, and his organizational skills.

Rob gained his BSc from Kingston, and then went on to complete a PhD from Southampton University (Oceanography) in 1973. As a research student, he made a comprehensive study of the then poorly understood sedimentary processes in the Tyrrhenian Sea, an area of research which he continued to pursue for much of his career. With his thesis safely under his belt, Rob took up a position at the Institute of Oceanographic Sciences (IOS) in Wormley, UK. At IOS, Rob worked on a wide range of projects which included initial work on interpreting GLORIA sidescan sonar images of the sea floor, organic-rich sediments or 'sapropels' in the Mediterranean, and evaluating the potential for the disposal of radioactive waste at sea. The impact of his many papers in these fields came from his multidisciplinary approach, one of which proved of immense importance in solving major geological problems. In 1973 Rob moved to America, becoming a visiting scientist at the Scripps Institute of Oceanography in La Jolla. It was during this period that Rob's interest in deep sea drilling developed. He became closely associated with the Deep Sea Drilling Project (DSDP), participating on five cruises, three as co-chief scientist. Research during this period led to numerous publications on ocean circulation, sediment drifts and high-resolution stratigraphy, many of which received wide acclaim from his peers.

After a short return to IOS, Rob moved to the Ocean Drilling Program (ODP) at Texas A&M University, College Station, where he was appointed Professor of Oceanography and Head of ODP Science Operations. He quickly initiated many new working procedures which still live on today at ODP, including pre- and post-cruise science operations and publication procedure. He was also very proud of the state-of-the-art drillship laboratory facilities which he helped design.

In 1986 Rob moved back to Wales where he had been born 39 years before, to take up the appointment of Professor of Geology and Head of Department at University College Swansea. Soon after, he presided over the integration of the Geology and Oceanography Departments into the Department of Earth Sciences. Through uncertain times during the first universities research assessment exercise, Rob continued to maintain a cheerful and supportive attitude to his staff and students, despite prolonged periods of illness. Rationalization led to a move to Cardiff where he took on the role of Professor of Marine Geology and head what became a vibrant Marine Geosciences Research Group.

Rob continued to be a major guiding force within ODP, chairing the Site Survey Panel from 1989 and later becoming Chair of the Scientific Planning Committee (PCOM). Through his determined efforts a major coup was achieved in 1994 when the Joint Oceanographic Institute Deep Earth Sampling (JOIDES) office was relocated to Cardiff, the first time it had been located outside America. For two years the scientific planning of the ODP was co-ordinated and organized by Rob and his staff.

During the later part of his career Rob developed a passionate link with the UNESCO-funded Training Through Research (TREDMAR) programme. The project involved training a wide variety of international students at sea using the Russian research vessel *Gelendzhik* as the floating classroom. Rob and other international colleagues enthused about this programme and enjoyed the trips in the Mediterranean. The reputation and continued success of the TREDMAR programme owes much to Rob's efforts.

In June of 1996 the geological community under the banner of the Geological Society of London recognized the international scientific efforts of Rob and awarded him the prestigious Major John Sachervell A'Deane Coke Medal.

In the time we knew him, Rob was friend to us all, a diligent, well-liked man who was respected. His warmth and sense of fun were limitless. Most importantly he loved his family and his friends. We all miss him.

Rob Benjamin Kidd was born in Milford Haven on 17 September 1947, and died in London on 9 June 1996. He is survived by his wife Rosalie and four sons.

GEOLOGICAL SOCIETY SPECIAL PUBLICATION NO. 131

Geological Evolution of Ocean Basins: Results from the Ocean Drilling Program

EDITED BY

A. CRAMP
Department of Earth Sciences, University of Wales Cardiff, UK

C. J. MacLEOD
Department of Earth Sciences, University of Wales Cardiff, UK

S. V. LEE
Department of Earth Sciences, University of Wales Cardiff, UK

and

E. J. W. JONES
Department of Geological Sciences, University College, London, UK

1998
Published by
The Geological Society
London

Geological Society Special Publications
Series Editor A. J. FLEET

THE GEOLOGICAL SOCIETY

The Society was founded in 1807 as The Geological Society of London and is the oldest geological society in the world. It received its Royal Charter in 1825 for the purpose of `investigating the mineral structure of the Earth'. The Society is Britain's national society for geology with a membership of around 8500. It has countrywide coverage and approximately 1500 members reside overseas. The Society is responsible for all aspects of the geological sciences including professional matters. The Society has its own publishing house, which produces the Society's international journals, books and maps, and which acts as the European distributor for publications of the American Association of Petroleum Geologists, SEPM and the Geological Society of America.

Fellowship is open to those holding a recognized honours degree in geology or cognate subject and who have at least two years' relevant postgraduate experience, or who have not less than six years' relevant experience in geology or a cognate subject. A Fellow who has not less than five years' relevant postgraduate experience in the practice of geology may apply for validation and, subject to approval, may be able to use the designatory letters C Geol (Chartered Geologist).

Further information about the Society is available from the Membership Manager, The Geological Society, Burlington House, Piccadilly, London W1V 0JU, UK. The Society is a Registered Charity, No. 210161.

Published by The Geological Society from:
The Geological Society Publishing House
Unit 7, Brassmill Enterprise Centre
Brassmill Lane
Bath BA1 3JN
UK
(*Orders*:   Tel. 01225 445046
            Fax 01225 442836)

First published 1998

The publishers make no representation, express or implied, with regard to the accuracy of the information contained in this book and cannot accept any legal responsibility for any errors or omissions that may be made.

British Library Cataloguing in Publication Data
A catalogue record for this book is available from the British Library.
ISBN 1-86239-003-7
ISSN 0305-8719

Distributors
USA
    AAPG Bookstore
    PO Box 979
    Tulsa
    OK 74101-0979
    USA
(*Orders*: Tel. (918) 584-2555
           Fax (918) 560-2652)

Australia
    Australian Mineral Foundation
    63 Conyngham Street
    Glenside
    South Australia 5065
    Australia
(*Orders*: Tel. (08) 379-0444
           Fax (08) 379-4634)

India
    Affiliated East-West Press PVT Ltd
    G-1/16 Ansari Road
    New Delhi 110 002
    India
(*Orders*: Tel. (11) 327-9113
           Fax (11) 326-0538)

Japan
    Kanda Book Trading Co.
    Tanikawa Building
    3-2 Kanda Surugadai
    Chiyoda-Ku
    Tokyo 101
    Japan
(*Orders*: Tel. (03) 3255-3497
           Fax (03) 3255-3495)

Typeset by E & M Graphics, Midsomer Norton, Bath, UK.

Printed by The Alden Press, Osney Mead, Oxford, UK.

Contents

CRAMP, A., MACLEOD, C. J., LEE, S. V. & JONES, E. J. W. Introduction: recent results from the Ocean Drilling Program   vii

Palaeoceanographic Issues

SYKES, T. J. S., ROYER, J.-Y., RAMSAY, A. T. S. & KIDD, R. B. Southern hemisphere palaeobathymetry   3

SYKES, T. J. S., RAMSAY, A. T. S. & KIDD, R. B. Southern hemisphere Miocene bottom- water circulation: a palaeobathymetric analysis   43

RAMSAY, A. T. S., SMART, C. W. & ZACHOS, J. C. A model of early to middle Miocene deep ocean circulation for the Atlantic and Indian Oceans   55

ERBACHER, J. & THUROW, J. Mid-Cretaceous radiolarian zonation for the North Atlantic: an example of oceanographically controlled evolutionary processes in the marine biosphere?   71

HASLETT, S. K. & FUNNELL, B. M. Low-latitude Plio-Pleistocene temporal abundance variations in the radiolarian *Cycladophora davisiana* Ehrenberg: stratigraphic and palaeoceanographic significance   83

MASLIN, M., SARNTHEIN, M., KNAACK, J.-J., GROOTES, P. & TZEDAKIS, C. Intra-interglacial cold events: an Eemian–Holocene comparison   91

SCHAAF, M. & THUROW, J. Two 30 000 year high-resolution greyvalue time series from the Santa Barbara Basin and the Guaymas Basin   101

MASLIN, M. Equatorial western Atlantic Ocean circulation changes linked to the Heinrich events: deep-sea sediment evidence from the Amazon Fan   111

MASLIN, M. & MIKKELSEN, N. Timing of the late Quaternary Amazon Fan Complex mass-transport deposits   129

CROWLEY, S. F., STOW, D. A. V. & CROUDACE, I. W. Mineralogy and geochemistry of Bay of Bengal deep-sea fan sediments, ODP Leg 116: evidence for an Indian subcontinent contribution to distal fan sedimentation   151

Structural, Tectonic and Sedimentary Issues

DILEK ,Y. Structure and tectonics of intermediate-spread oceanic crust drilled at DSDP/ODP Holes 504B and 896A, Costa Rica Rift   177

MITCHELL, N. C. Sediment accumulation rates from Deep Tow profiler records and DSDP Leg 70 cores over the Galapagos spreading centre   199

ROTHWELL, R. G. Sedimentary evidence relating to the tectonic evolution of the Lau Basin, SW Pacific, from ODP Sites 834–839 (ODP Leg 135)   211

HODKINSON, R. A. & CRONAN, D. S. Hydrothermal inputs at ODP Sites 836, 837, 838 and 839 in relation to Eastern Lau Spreading Centre propagation in the Lau Basin, southwest Pacific   231

ROBERTSON, A. H. F., EMEIS, K.-C., RICHTER, C. *ET AL.* Collision-related break-up of a carbonate platform (Eratosthenes Seamount) and mud volcanism on the Mediterranean Ridge: preliminary synthesis and implications of tectonic results of ODP Leg 160 in the Eastern Mediterranean Sea   243

ALEXANDER, J. L. Rare earth element anomalies in the Nankai accretionary prism, Japan   273

LOVELL, M. A., HARVEY, P. K., BREWER, T. S., WILLIAMS, C., JACKSON, P. D. & WILLIAMSON, G.
Application of FMS images in the Ocean Drilling Program: an overview   287

LEE, YIR-DER E. & FRANCIS, T. J. G. A statistical study of hydraulic piston coring, ODP Legs
101–149   305

Index   317

References to this volume

It is recommended that reference to all or part of this book should be made in one of the following ways:

CRAMP, A., MACLEOD, C. J., LEE, S. V. & JONES, E. J. W. (eds) 1998. *Geological Evolution of Ocean Basins: Results from the Ocean Drilling Program*. Geological Society, London, Special Publications, **131**.

ALEXANDER, J. L. 1998. Rare earth element anomalies in the Nankai accretionary prism, Japan. *In:* CRAMP, A., MACLEOD, C. J., LEE, S. V. & JONES, E. J. W. (eds) *Geological Evolution of Ocean Basins: Results from the Ocean Drilling Program*. Geological Society, London, Special Publications, **131**, 273–286.

SYKES, T. J. S., RAMSAY, A. T. S. & KIDD, R. B. 1998. Southern hemisphere Miocene bottom-water circulation: a palaeobathymetric analysis. *In:* CRAMP, A., MACLEOD, C. J., LEE, S. V. & JONES, E. J. W. (eds) *Geological Evolution of Ocean Basins: Results from the Ocean Drilling Program*. Geological Society, London, Special Publications, **131**, 43–54.

Introduction: recent results from the Ocean Drilling Program

A. CRAMP[1], C. J. MacLEOD[1], S. V. LEE[1] & E. J. W. JONES[2]

[1] Marine Geosciences Research Group, Department of Earth Sciences, University of Wales Cardiff, UK

[2] Department of Geological Sciences, University College London, London, UK

Advances in the field of marine geoscience through the medium of deep-ocean drilling have been rapid and continue to be so. No one volume can cover all aspects of drilling-related science, so this publication attempts to provide a snapshot, circa 1995/1996, of recent scientific output under the aegis of the international Ocean Drilling Program (ODP). In late 1995 the drillship *JOIDES Resolution* had successfully completed major programmes in the Mediterranean and Arctic Seas and was about to embark on a leg dedicated to the intriguing and topical subject of gas hydrates.

Part of this volume reflects the excitement of findings from recent legs; other sections provide invaluable syntheses of the voluminous drilling information collected over a period of more than 20 years, which provide a detailed picture of how the oceans have evolved since the late Mesozoic.

Most of the research was carried out when Robert Kidd was closely involved with both the planning and implementation of ODP strategy. It is therefore entirely appropriate that we dedicate this volume to his memory.

In the first chapter, **Southern hemisphere palaeobathymetry,** Sykes, Royer, Ramsay & Kidd present a much-needed synthesis of the Southern Ocean palaeobathymetry which forms the basis for interpreting late Mesozoic and Cenozoic sedimentary patterns. They have synthesized 20 plate reconstruction maps with inferred palaeobathymetry which is constrained by no fewer than 39 drill sites. They show that the difference between the expected 'true bathymetry' that takes account of sediment decompaction and loading, and the sediment-free palaeobathymetry is less than 600 m for 85% of the region. The greatest differences occur close to continental margins where sediment thicknesses are greatest.

In the second paper, **Southern hemisphere Miocene bottom-water circulation: a palaeobathymetric analysis,** Sykes, Ramsay & Kidd, use palaeobathymetry derived in the previous paper to examine the path of bottom water flow in the Southern Ocean during the Miocene. The paper draws comparisons between the Miocene and the present-day circulation patterns, and demonstrates that the palaeobathymetry and northward movement of Antarctic Bottom Water (AABW) during

the early Miocene are broadly consistent with today's water mass migrations. They conclude that by the middle Miocene (approximately 15 Ma), the waning of proto-AABW flow paths led to the restriction of this water mass to the Antarctic Basin. Waxing of proto-AABW during the late Miocene (approximately 10 Ma) allowed AABW to re-enter many of the deep basins, though not to the same extent as at 20 Ma.

In the third paper Ramsay, Smart & Zachos present **A model of early to middle Miocene deep-ocean circulation for the Atlantic and Indian Oceans.** They focus on stable isotope (carbon and oxygen) data and micropalaeontological information to investigate a unique event, namely the occurrence of high abundances of the smooth-walled bolivinid in the Atlantic and northwest Indian Oceans in the early and middle Miocene. They suggest that Tethyan outflow water formed a component of deep water masses in the Atlantic from approximately 13 Ma to 20 Ma ago, and in the Indian Ocean from about 20 Ma to 14.5 Ma. They argue that the termination of the Tethyan outflow in the Atlantic, and to a lesser extent in the Indian Ocean, contributed to global cooling and to the expansion of the Antarctic ice sheet by limiting the meridional heat transport.

The fourth and fifth papers by Erbacher & Thurow and by Haslett & Funnell respectively discuss palaeoceanographic evolution in the light of radiolarian studies. In **Mid-Cretaceous radiolarian zonation for the North Atlantic – an example of oceanographically controlled evolutionary processes in the marine biosphere?,** Erbacher & Thurow provide an innovative zonation for mid-Cretaceous North Atlantic radiolaria. In addition, they present a calibration based upon planktonic foraminiferal zonations obtained from calcareous sections in central Italy. They go on to explain the critical interactive links between sea-level fluctuations, nutrient supply and dissolved oxygen which appear to have influenced the evolution of mid-Cretaceous radiolaria. The paper by Haslett & Funnell, **Low-latitude Plio-Pleistocene temporal abundance variations in the radiolarian *Cycladophora davisiana* Ehrenberg: stratigraphic and palaeoceanographic significance,** summarizes a study of the abundance of

Cycladophora davisiana from ODP sites in the Pacific and Indian Oceans. They recognize 12 peak abundance events at different sites, most of which appear to be coeval. At Site 709, however, the results diverge and are approximately 15 ka older than those recorded in the Pacific. They suggest that this difference may be due to diachronous *C. davisiana* events or an inaccuracy in the existing erected timescales for ODP Site 709 compared with sites in the eastern Pacific. They conclude that the recognition of correlatable *C. davisiana* events in the equatorial Pacific and Indian Oceans during the Late Quaternary permits further fine-tuning of high-resolution timescales, enabling more accurate evaluation of sedimentary sequences on the regional, oceanic and possibly interoceanic scales.

In **Intra-Interglacial cold events: an Eemian–Holocene comparison,** Maslin, Sarnthein, Knaack, Grootes & Tzedakis provide a high-resolution (30–500 year) oxygen isotope comparison between the GRIP and GISP ice cores and similar records obtained from the marine and lacustrine environments, concentrating on Site 658 located on the low-latitude east Atlantic margin. They show that there are rapid correlatable oscillations between warm and cold climates recorded in the GRIP ice core and the Eemian/Marine oxygen isotope Stage 5e, but observed no such correlation between the GISP core and the Atlantic deep-sea record. They suggest that the isotope record from the eastern Atlantic indicates close similarities in climatic signatures between the Eemian and the Holocene. Further studies indicate that there was a cold period (<400 years) in the Eemian which correlates with a reduction in the upper North Atlantic Deep Water ventilation. They hypothesize that the cold period could be correlated with a similar interval detected in terrestrial records from Europe. The authors conclude that both the marine and terrestrial records appear to be incompatible with those recorded in the GRIP core, adding weight to the argument that the area from where the GRIP core was recovered has been subject to isostatic readjustment.

Ultra-high-resolution stratigraphy has evolved rapidly over the last ten years. Advances in analytical techniques coupled with developments in ODP coring technology have, in some instances, provided us with decadal or better resolution. This is particularly the case when sediments are laminated and contain enhanced levels of organic carbon. **Two 30 000 year high-resolution grey-value time series from the Santa Barbara Basin and the Guaymas Basin**, by Schaaf & Thurow, provides such an example. The authors identify substantial variability on all timescales ranging from millennial to annual. They note a strong 1500 year cyclicity in both records. Using spectral analysis, they identify the existence of Gleissberg

(86 year), sunspot (11 year, 22 year) and El Niño (3–9 years) events. They conclude that both basins have responded in a similar manner to external forcing mechanisms.

In his paper **Equatorial western Atlantic circulation changes linked to the Heinrich events: deep-sea sediment evidence from the Amazon Fan**, Maslin provides isotopic, palaeo-magnetic and AMS ^{14}C evidence to propose that the influence of Heinrich events has been preserved in the sediments of the Amazon submarine fan. Heinrich events, defined as intense, quasi-periodic ice-rafting pulses, originating from the Laurentide ice sheet, have been identified within North Atlantic sediments. These events appear to occur every 7–13 ka and have a duration of between 50 and 1000 years, though some investigators shorten their duration to between 100 and 500 years. Heinrich events have been correlated with other major shifts in climate records, including the record preserved in Greenland ice cores, and vegetation sequences in Florida, Europe and the North Pacific. At present, there are two views on the origins of these events. Some believe that the events are caused by internal ice sheet dynamics; others favour periodic climate changes resulting from variations in orbital parameters. The geographical limit of the influence of Heinrich events is the subject of considerable investigation. Are these events limited to the North Atlantic and surrounding hinterlands or do they have a much wider, possibly global influence? Maslin concludes that enhanced ice-rafting in the North Atlantic during Heinrich events increased the latitudinal thermal gradient and thus the zonal component of the wind system. The reduced penetration of the Intertropical Convergence Zone curtailed the cross-equatorial export of the North Brazilian Coastal Current and resulted in permanent North Brazilian Coastal Current retroflection. An increase in surface water salinity above the Amazon Fan led to positive oxygen isotope deviations.

Submarine fans form the largest accumulations of sediment found on continental margins. In most cases, deep-sea fans located in the modern oceans are associated with large fluvial systems. Fan deposits are thought to accumulate most rapidly during low sea-level stands when fluvial systems issued directly to the continental slope. Deposition rates tend to be very high (up to 30 m ka^{-1}); sediments accumulate in the form of channels and levees which 'stack' upon each other sequentially through time. These rapidly developing sedimentary bodies contain 'amplified' signatures of climate change and continental denudation. They also record valuable information on sedimentary processes, architecture and stratigraphy. Mikkelsen & Maslin and Crowley, Stow & Croudace discuss major submarine fan systems in their two papers. Mikkelsen & Maslin address the problem of mass-transport deposition on the Amazon Fan in **Timing of the late Quaternary Amazon Fan Complex mass-transport deposits**. They use high-resolution oxygen isotope stratigraphy to define the timing of mass-transport deposition on the fan. They demonstrate that the two mass-transport deposits on the fan, named the Deep Eastern Unit and Unit R, were last active 33 ka and 45 ka ago respectively. The interglacial deposits located beneath these units were laid down during Oxygen Isotope Stages 7, 9, and 11. They speculate that the mass-transport deposits were emplaced by climatically induced changes in sea level, and that the interglacial deposits may have acted as slip planes for these vast sedimentary units.

In the following paper, **Mineralogy and geochemistry of Bay of Bengal deep sea fan sediments, ODP Leg 116: evidence for an Indian subcontinent contribution to distal fan sedimentation**, Crowley, Stow & Croudace use mineralogical and geochemical data to identify the provenance of sediments in the world's largest deep-sea fan system. They identify three distinct groups of sediment. Group I is made up of quartz–mica-rich turbidites derived from a Himalayan sedimentary/granitic source; Group II consists of smectite–kaolinite, organic-rich turbidites probably originating from the Deccan Trap basalts of central India; and Group III is composed of carbonate-rich sediments originating from southern India and/or Sri Lanka which contain low-latitude marine bioclastic faunas. They outline the relative contributions of the Himalayas and Indian subcontinent to distal fan sediments with time, evaluating the influence of uplift, weathering and erosion rates, eustatic sea-level changes and the switching of major distributary fan channels. The paper concludes by warning of the unreliability of Himalayan tectonic and climate reconstructions based on the record of distal fan sediments, since the supply of sedimentary material to the distal fan was discontinuous through time.

The next section of the volume is dedicated to structural, tectonic and sedimentary issues. Dilek, in his paper **Structure and tectonics of inter-mediate-spread ocean crust drilled at DSDP/ODP Holes 504B and 896A, Costa Rica Rift,** uses information obtained from Hole 504B where a 2.1 km section into 5.9 Ma old crust is the deepest penetration yet made into oceanic lithosphere, and forms the *de facto* reference section for the upper crust. The extensive and comprehensive database reviewed in the paper provides an insight into the physical and chemical processes of sea-floor spreading. The upper oceanic crustal rocks recovered include massive and pillow lava flows with

interlayered breccias. The bathymetry in the vicinity of the sites is defined by linear hills and intervening troughs and is interpreted to reflect a basement topography defined by fault-bounded asymmetric hills that developed within the crustal accretion zone in the spreading environment; it is inferred that the volcanic section drilled at Hole 896A represents one such abyssal hill.

In **Sediment accumulation rates from Deep Tow profiler records and DSDP Leg 70 cores over the Galapagos spreading centre,** Mitchell adopts a novel approach to determine sedimentation rates over a spreading centre. He reports a method of calculating sedimentation rates from a statistical analysis of variations in sediment thickness with distance from the spreading centre. Regression analysis tends to average out the changes in sediment thickness in spreading areas – a result of, in most instances, sediment reworking and redistribution. In an attempt to 'groundtruth' the method, Mitchell uses physical properties determined at DSDP sites to calculate mass accumulation rates for the area. High accumulation rates are attributed to enhanced equatorial productivity of pelagic organisms. In addition, Mitchell reports on the scaling of sediment thickness variability and examines the possibility of utilizing sediment thickness to calculate the age of the sea floor.

The Lau backarc basin is the subject of Rothwell's paper, **Sedimentary evidence relating to the tectonic evolution of the Lau Basin, SW Pacific, from ODP Sites 834–839 (ODP Leg 135).** Rothwell examines Miocene to Holocene sedimentary sequences, in particular gravity-flow deposits, recovered from six sites located within a horst and graben terrain in the Lau Basin to shed light on the regional tectonics. The lithologies he identifies are characterized by a lower succession of volcaniclastic gravity flows interbedded with hemipelagic clayey nannofossil oozes and nanno-fossil clays, overlain by a distinctive succession of hydrothermally stained hemipelagic, and locally re-deposited, clayey nannofossil oozes. The volcani-clastic deposits tend to be massive, proximal, vitric gravels, sands and silts that appear to be locally derived from adjacent basement ridges and intrabasin seamount volcanoes. Rothwell identifies a suite of muddy debris-flow deposits together with a number of coherent rafted blocks of older hemipelagic sediment. He postulates that these allochthonous deposits reveal evidence of several episodes of instability in the sub-basin which may be related to large-scale tectonic activity.

Hydrothermal activity and sedimentation in the Lau Basin is the subject of Hodkinson & Cronan's paper **Hydrothermal inputs at ODP Sites 836, 837, 838 and 839 in relation to Eastern Lau**

Spreading Centre propagation in the Lau Basin, southwest Pacific. This contribution describes an investigation of a sequence of late Pliocene to Pleistocene clayey nannofossil oozes with sparse calcareous turbidites, which are found overlying a thick sequence of redeposited volcaniclastic sediments interbedded with hemipelagic clayey nannofossil oozes. All calcareous oozes contain hydrothermal ferromanganese oxides. Hodkinson & Cronan assess the hydrothermal flux to the sediments by determining the non-detrital Mn+Fe accumulation rates for the hemipelagic sediment intervals which had not been subjected to reworking. They conclude that hydrothermal fluxes vary across the Basin, and that hydrothermal plume fallout associated with the southern, more recently generated portion of the Eastern Lau Spreading Centre is lower than that of the older portion in the north.

Drilling in the Mediterranean Sea has revealed many fascinating facts relating to the evolution of a semi-enclosed marine basin. This highly complex tectonic region has been the research focus of many structural geologists and palaeoceanographers in recent years. The paper by Robertson, Emeis, Richter & the Scientific Party of ODP Leg 160, **Collision-related break-up of a carbonate platform (Eratosthenes Seamount) and mud volcanism on the Mediterranean Ridge: preliminary synthesis and implications of tectonic results of ODP Leg 160 in the Eastern Mediterranean Sea,** provides an insight into the evolution and subsequent destruction of the Eratosthenes Seamount, a carbonate platform located in the eastern basin. Prior to drilling, the evolution of this region of collision between the African and Eurasian plates was controversial. During recent drilling, the oldest sediments recovered from the seamount were identified as ?mid-Cretaceous shallow-water limestones, overlain by upper Cretaceous to lower Oligocene pelagic carbonates, with several hiatuses. Following uplift, it appears that a carbonate platform was established in the Miocene and that the platform remained submerged during the Messinian salinity crisis. The platform then subsided to bathyal depths during the Lower Pliocene. Subsidence continued in Late Pliocene and Quaternary time and was associated with strong surface uplift in southern Cyprus. Subsidence and break-up of the Eratosthenes Seamount was achieved by a combination of flexural loading and normal faulting. In addition, the authors present some exciting new data relating to the evolution and the processes associated with submarine mud volcanoes. For the first time, active deep-water mud volcanoes were drilled. It is clear that the Milano and Napoli mud volcanoes on the northern flank of the Mediterranean Ridge accretionary complex are mainly extrusive sedimentary features composed of multiple debris-flow deposits containing both sandstone and limestone clasts of Miocene age. Both mud volcanoes began to evolve more than 1 Ma ago and now appear to be episodically active. Hydrocarbon gas was detected on both features; methane hydrates (clathrates) exist locally at Milano. It appears that the driving force of mud volcanism is overpressuring caused by incipient plate collision at plate boundaries. It is postulated that Messinian evaporites may have acted as a seal with material escaping through a zone of backthrusting against rigid Cretan crust located to the north.

Alexander provides an insight into rare element anomalies in an accretionary prism. In her paper entitled **Rare earth element anomalies in the Nankai accretionary prism, Japan,** she questions the use of rare earth elements as provenance indicators since the provenance signal may be distorted by rare earth element mobility. Most of the sequences recovered from the Nankai accretionary prism have typical shale rare earth element signatures: four units identified appear to be hydrothermal deposits; two are enriched in zircon and flourencite, with high rare earth concentrations; and one sample recovered from the décollement zone has a heavy rare earth enrichment. Alexander concludes that this enrichment does not appear to have a mineralogical control.

The final two papers in this volume emphasize the importance of advances in ODP technology to our understanding of sedimentary sequences. Lovell, Harvey, Brewer, Williams, Jackson & Williamson from the Leicester University Borehole Research Group, provide a thorough review of the application of Formation Microscanner (FMS) downhole measurements in deep-water drilling. Their paper, **Application of FMS images in the Ocean Drilling Program: an overview,** illustrates how FMS borehole wall images can be used for a wide range of geological applications including structural and stratigraphic reconstructions. The determination of core orientation and the mapping of core intervals where recovery is poor is discussed. The FMS instrument provides the geologist with a means of carrying out field studies based on borehole and core observations which were previously unthinkable. The final paper, entitled **A statistical study of hydraulic piston coring, ODP Legs 101–149,** by Lee & Francis, examines the operational performance of the ODP Hydraulic Piston Coring (HPC) system. The performance of the Advanced Piston Coring (APC) system over the first 49 legs of the ODP is compared to the physical properties of the cores recovered. It appears that there is a direct correlation between pullout force and the shear strength of the sediments. The authors argue that information derived from deep-water hydraulic piston coring could be used to determine the tensional load-bearing capabilities of sediment where structures need to be tethered.

This volume arose from a meeting held at the Geological Society in October 1995. The editors would like to thank all those who contributed to the success of the meeting, and to the production of this Special Publication. In particular, thanks go to Sid and Norma Barton and their staff, together with Heidie Gould at Burlington House and Angharad Hills and Jo Cooke at the Publishing House in Bath.

Valuable and essential financial support to stage the meeting was provided by the Natural Environment Research Council (NERC), the Challenger Society for Marine Science, the Marine Studies Group of the Geological Society, Shell, Texaco and Badley Ashton and Associates.

Adrian Cramp thanks Tony Ramsay for practical advice on the production of this volume.

Palaeoceanographic Issues

Southern hemisphere palaeobathymetry

T. J. S. SYKES[1,2], J.-Y. ROYER[3], A. T. S. RAMSAY[1] & R. B. KIDD[1]

[1] *Marine Geosciences Research Group, Department of Earth Sciences, UWCC, PO Box 914, Cardiff, CF1 3YE, UK*

[2] *Present address: Landmark EAME Ltd, 4 Albert Street, Aberdeen, AB25 1XQ, UK*

[3] *Laboratoire de Geodynamique sous Marine, BP 48, 06239 Villefranche-sur-Mer, France*

Abstract: Digital grids, with a spatial resolution of 0.5°, were compiled using bathymetry, sediment thickness and oceanic crustal ages for the southern hemisphere to calculate sediment-free palaeobathymetric charts for late Cretaceous (110 Ma) to the present. These reconstructions allow the definition and quantification of the bathymetric evolution of the seaways and gateways within this region. Pre-60 Ma the palaeobathymetry and the interconnection of the deeper ocean basin was dominated by the formation of hotspot-related ridges and plateaux. Between 40 Ma and the present the palaeobathymetry is characterized by the opening of numerous gateways as the Tasman Rise separated from Antarctica, the Mascarene Plateau split from the Chagos–Laccadive Ridge, Broken Ridge rifted from the northern Kerguelen Plateau, the Crozet Plateau separated from the Madagascar Ridge, and the Scotia Sea opened.

The validity of these palaeobathymetric reconstructions is demonstrated by comparing the digital gridded data sets, from which they were calculated, with data derived from 39 Deep Sea Drilling Project and Ocean Drilling Program drill sites. Comparison of drill site data with the gridded data revealed positive correlations between crustal ages and bathymetry, whereas the sediment thickness data showed no correlation. The lack of any correlation for the sediment thickness data, however, did not preclude a positive correlation between sediment-free bathymetries.

The difference between expected 'true' palaeobathymetry, that is with sediments older than the time of reconstruction decompacted and reloaded back onto oceanic crust, and the sediment-free palaeobathymetry was shown to be less than 600 m for 87.5% of the data. The greatest differences occur where sediment thicknesses are greatest, i.e. within the deep ocean basins which are in close proximity to the continental margin.

The Palaeoceanographic Indian Ocean Synthesis (PALIOS) (1989–1992) and its successor the Palaeoceanographic Southern Ocean Synthesis (PALSOS) (1993–1995) set out to synthesize data from the Deep Sea Drilling Project (DSDP) and Ocean Drilling Program (ODP). One of the aims of these projects was to establish the palaeobathymetric evolution of the southern hemisphere to allow the four-dimensional (latitude, longitude, depth and age) interpretation of DSDP/ODP drill site data within a single integrated time framework (Kidd *et al.* 1992). In particular it was deemed necessary to be able to constrain the boundaries of sediment facies or water mass in a less subjective way than was previously possible. For example, the attempt by Ramsay *et al.* (1994) to determine the spatial extent of bottom waters within the Eocene to Recent Indian Ocean, based on a synthesis of biostratigraphic data and the distribution of hiatuses, lacked any palaeodepth information away from drill sites. This paper sets out to establish a palaeobathymetric framework upon which such

interpretations can be better undertaken. The effectiveness and use of these palaeobathymetric reconstructions is illustrated in Sykes *et al.* (1998). The northern extent of the area considered within this work is illustrated in Fig. 1.

Background

Once the kinematic history of the Earth's continents and oceans was constructed, allowing the past position of the continents to be defined (Sclater & Fisher 1974; Royer *et al.* 1992*a*), palaeoceanographers soon realized that this could be combined with the age–depth relationship of ocean crust to produce palaeobathymetric charts (e.g. Sclater & McKenzie 1973; Sclater *et al.* 1977*a, b*, 1985). Earlier workers used the following methods to reconstruct palaeobathymetry.

(i) Rotation of modern bathymetry and isochrons. The position of the mid-ocean ridge in these reconstructions was frequently defined by the

SYKES, T. J. S., ROYER, J.-Y., RAMSAY, A. T. S. & KIDD, R. B. 1998. Southern hemisphere palaeobathymetry. *In*: CRAMP, A., MACLEOD, C. J., LEE, S. V. & JONES, E. J. W. (eds) *Geological Evolution of Ocean Basins: Results from the Ocean Drilling Program*. Geological Society, London, Special Publications, **131**, 3–42.

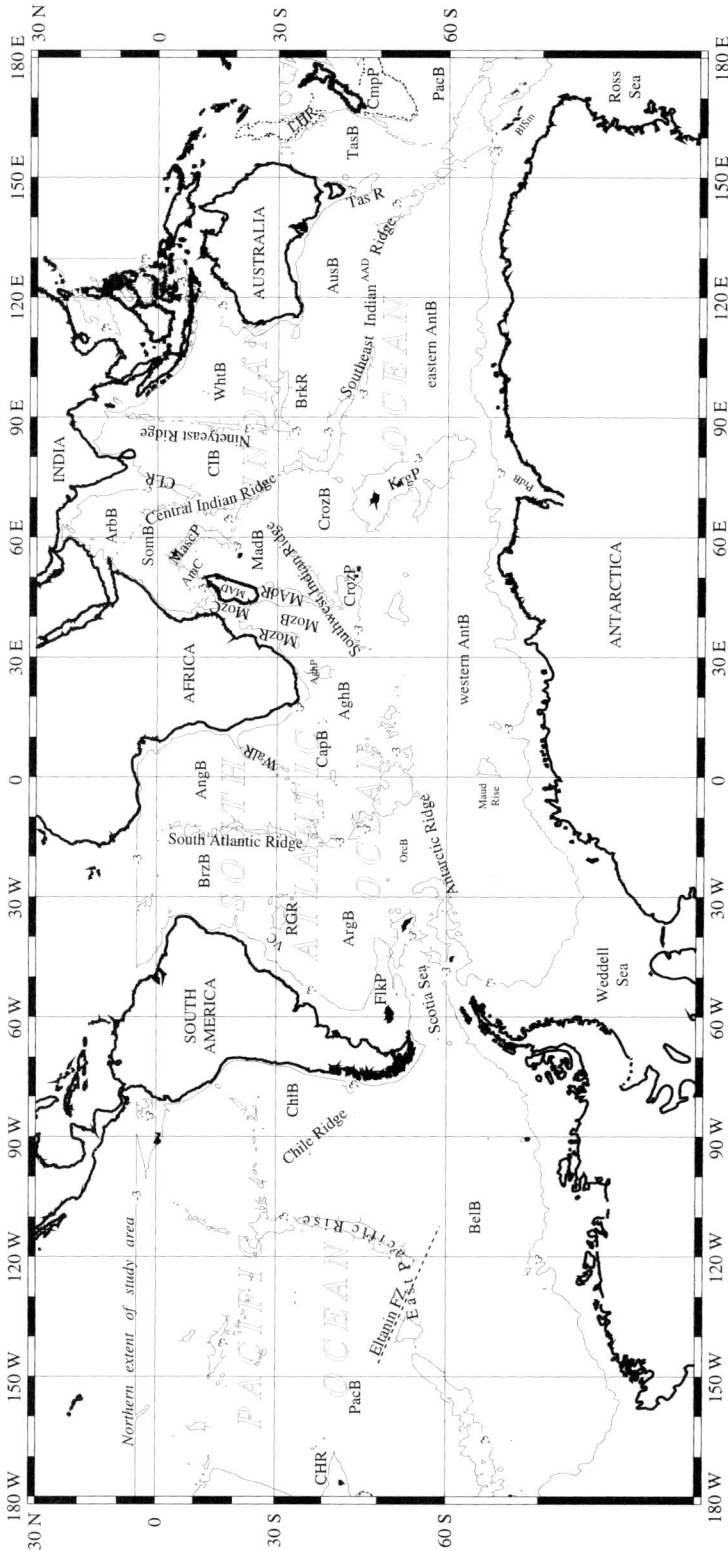

Fig. 1. Location of features mentioned in text; 3000 m isobath shown. Abbreviations: AAD, Antarctic-Australian Discordance; AghB, Agulhas Basin; AghP, Agulhas Plateau; AmC, Amirante Channel; AngB, Angola Basin; AntB, Antarctic Basin; ArbB, Arabian Basin; ArgB, Argentine Basin; AusB, Australian Basin; BelB, Bellinghausen Basin; BlSm, Bellany Seamounts; BrB, Brasil Basin; BrkR, Broken Ridge; CmpP, Campbell Plateau; CapB, Cape basin; ChlB, Chile Basin; CIB, Central Indian Basin; CLR, Chagos–Laccadive Ridge; CrozB, Crozet Basin; CrozP, Crozet Plateau; FlkP, Falkland Plateau; KrgP, Kerguelen Plateau; LHR, Lord Howe Rise;MAD, Madagascar; MadB, Madagascar Basin; MadR, Madagascar Ridge;MascP, Mascarene Plateau; MozB, Mozambique Basin; MozC, Mozambique channel; MozR, Mozambique Ridge;OrcB, Orcadas Basin; PacB, Pacific Basin; PrdB, Prydz Bay; RGR, Rio Grande Rise; SomB, Somali Basin; TasB, Tasman Basin; TasR, Tasman Rise; VC, Vema Channel; WalR, Walvis Ridge; WhtB, Wharton Basin.

position of rotated isochrons (Ramsay 1977; Lawver *et al.* 1992). Frequently modern bathymetric features are rotated, along with isochrons, to provide quasi-palaeobathymetric reconstructions (e.g. Royer *et al.* 1992*b*). These studies generally took no account of the sediment load. This 'simple' method is probably the most frequently used to assess the timing of the opening of oceanic gateways.

(ii) Palaeobathymetry was defined by assigning depths to oceanic crust (isochrons) of a particular age, for example, crust of 2, 20 and 50 Ma occurs at depths of −3000, −4000 and −5000 m, respectively (Sclater & McKenzie 1973; Sclater *et al.* 1977*a*, *b*; 1985) using the age–depth functions of Parsons & Sclater (1977). This empirical age–depth relationship was validated by comparing the depth to basement, after correcting for sediment load, from DSDP drill sites. Thus these charts represent smoothed sediment-free palaeobathymetry and consider, to a limited extent, residual depth anomalies (i.e. the difference between the theoretical and observed water depths).

(iii) Palaeobathymetry was modelled (Wold 1992) by mass balancing subsiding oceanic crust according to the age–depth relationship of Parsons & Sclater (1977). Wold's (1992) study was restricted to the northern North Atlantic Ocean and the charts produced took into account the sediment load, the reloading and decompacting of sediment older than the time of reconstruction and the residual depth anomalies.

Methods (i) and (ii) have been used to model ocean and global palaeobathymetry. Methods (ii) and (iii) both utilized a single subsidence curve for all ocean crust which clearly misrepresents the along-ridge variations in the depths and subsidence characteristics of the world's mid-ocean ridges (Hayes 1988, 1992; Marty & Cazenave 1989; Kane & Hayes 1992, 1994; Calcagno & Cazenave 1994; Hayes & Kane 1994; Marks & Stock 1994; Sykes 1995) and hence reduces the accuracy of the reconstructions.

Despite limitations which are related to the effect of the sediment load, the residual depth anomalies and the use of a single subsidence curve, these reconstructions allowed the timing of opening of palaeoseaways to be assessed. In addition, drill site data from the ODP and DSDP are often plotted on these reconstructions in order to elicit the palaeoceanographic history of various oceanic basins recorded by the accumulation of sediments (e.g. van Andel *et al.* 1977; Davies & Kidd 1977; Kidd & Davies 1978; Davies *et al.* 1995).

Methods

A digital database, with a spatial resolution of $0.5° \times 0.5°$, of ocean crust ages, sediment thickness and bathymetry was compiled for the southern hemisphere. The oceanic crust digital data were based on a global isochron data set, which includes data for poles of rotation. These data were compiled for the Paleoceanographic Mapping Project (POMP), based at the Institute for Geophysics, University of Texas at Austin (Royer *et al.* 1992*a*). (At the time of writing, the references/data used in this technical report, along with diagrams, can be seen on the internet at: http://omphacite.es.su. oz.au/Staff Profiles/dietmar/Agegrid/utig_report. html or via http://www.ig.utexas.edu/research/ projects/plates.html). The continents were placed in their palaeolatitudes using the central African palaeomagnetic reference frame of Ziegler *et al.* (1983). The ages of the isochrons were updated to the Cande & Kent (1992) timescale for the Cenozoic; the Kent & Gradstein (1986) timescale was used for the Mesozoic. The Cande & Kent (1992) timescale was chosen because it was the latest, widely accepted, Cenozoic timescale when work within the PALSOS project commenced (1993). The Kent & Gradstein (1986) timescale was originally adopted by the PALIOS project team (1989) and continued to be used within the PALSOS project. The latest Mesozoic timescale at commencement of the PALSOS project, that of Harland *et al.* (1989), provided ages for stage boundaries between the Albian and Maestrichtian which differed from Kent & Gradstein (1986) by less than 0.5 Ma. A difference of less than 0.5 Ma is considered by the authors to be insignificant when considering the uncertainties involved within the plate model for the Late Cretaceous (65 Ma to 110 Ma) (see below). Therefore we do not consider that our continued use of the Kent & Gradstein (1986) timescale has had a significant impact on the palaeobathymetric results.

Additionally, it was necessary to model the tectonic evolution of the southern hemisphere through the Cretaceous Quiet Zone (118.7 to 83 Ma) so that crust formed during this interval could be assigned an age. Synthetic isochrons were calculated for 115, 105, 97 and 94 Ma, assuming symmetrical mid-ocean ridge spreading, for the South Atlantic Ocean and the seaway between Antarctica, Africa/Madagascar/India and Australia during this interval. Within the plate model used it is assumed that there was a major plate reorganization within the developing Indian Ocean at 94 Ma, associated with a northward ridge jump. The position of the mid-oceanic ridge, after the ridge jump, was calculated by assuming that the spreading rates between chron 34 and the synthetic isochron at 94 Ma were the same as those between chrons 34 and 33. This ridge jump effectively transferred large areas of the Indian and Madagascan Plates onto the Antarctic Plate within the vicinity of the Kerguelen Plateau. Similarly a portion of the South American Plate was transferred onto the African plate at chron 30 times (66.6 Ma), after a simple westward ridge jump. The remnants of the later abandoned ridge are clearly visible within gravity data (Smith & Sandwell 1994), whilst evidence for the former is lacking. A more expansive description of the plate modelling, including numerous illustrations, is presented in Sykes (1995).

The isochrons and synthetic isochrons, in addition to continental–oceanic crust boundaries from Royer *et al.* (1992*b*), were subsampled every 0.2° to provided to *c.* 55 000 raw data points used in computing the digital age grids. Fracture zones defined from the isochron data set were used as faults through the age data. The digital age grid was computed using the Integrated Surface Mapping (ISM, Dynamic Graphics, Inc.) package which was the only software package capable of handling the large volume of fault data (*c.* 5000 fault pairs) which were essential for the calculation of a meaningful age grid. ISM used a minimum tension gridding technique requiring the input of additional synthetic isochrons to control the gridding process, i.e. preventing 'overshoots' in areas of relatively sparse data (see Smith 1993). The digital grid was iteratively examined and regridded to determine the existence of overshoots and 'leakage' of data across fractures until both these problems were eliminated.

Sediment thickness data were derived by digitizing, gridding and merging of the sediment isopach maps for the Indian Ocean (Mathias *et al.* 1988), the Southern Ocean (Hayes & LaBrecque 1991) and the South Atlantic (Divins & Rabinowitz 1991). A digital isopach map for the south Pacific Ocean was provided by NGDC, which was based on the isopach map of Houtz *et al.* (1973). The ETOPO5 digital bathymetry database (NGDC 1988) was averaged into a half-degree grid using a weighted average. The limitations of this bathymetric data set have been identified and discussed by Smith (1993). In the absence of an alternative bathymetric data set with the same spatial coverage, ETOPO5 was used.

For each grid node with a value for sediment thickness and bathymetry, the isostatically corrected water depth, i.e. after the sediment load was removed, was calculated using the method of Sykes (1996). The method for calculating the isostatic correction used a sediment thickness–density function to calculate the cumulative load of 10 m layers. Sediment density was held constant at a sediment thickness greater than 5 km. The isostatic correction is subtracted from the sediment–water interface to produce the depth of the deeper sediment-free oceanic crust–water interface. (In this paper and Sykes (1995, 1996), bathymetric values are negative and the isostatic correction is positive, therefore this correction is subtracted from the sediment–water interface.) The southern hemisphere was divided into 86 tectonic corridors and into six older ocean basin areas; the latter were not directly linked to a currently active mid-ocean ridge segment. A tectonic corridor is defined, within this study, as the area between two adjacent fracture zones which is assumed to represent a section of the ocean floor characterized by long-lived distinctive geophysical characteristics. The fracture zones used to delimit the tectonic corridors were defined from the isochron data set. They can be considered to be flowline-parallel and accommodate any crustal age asymmetry. Corrected water depths and crustal ages were extracted from the digital database using the tectonic corridors to allow the age–depth relationships to be determined for the conjugate halves of the mid-ocean ridges and for the older basinal areas. Details of these agedepth analyses are beyond the scope of this paper and are fully presented in Sykes (1995) and Sykes & Royer (*under review*).

Palaeobathymetric charts were constructed at 5 Ma intervals for the Southern Ocean by gridding rotated data points for the following: (i) actual palaeobathymetry, defined as grid nodes where the calculation of the isostatic correction was possible, i.e. values for bathymetry, sediment thickness and basement age were defined, and the residual depth anomaly for each of these nodes was calculated; (ii) synthetic palaeobathymetry, defined as grid nodes where age data alone existed, e.g. subducted crust;(iii) control bathymetry, in the form of coast lines (0 m depth), shelf edges (–200 m) and plateaux bathymetry for the Falklands Plateau, the plateaux along the western margin of Australia, plateaux from the Scotia Sea and the Tasman Rise. The Generic Mapping Tools (GMT) package of Wessel & Smith (1991) was used to grid and illustrate the palaeobathymetric charts. Gridding the rotated palaeobathymetry was undertaken using the adjustable tension continuous curvature surface gridding algorithm of the GMT package (with the tension factor set to 0.35) which has been shown to be suitable for bathymetric data (Smith & Wessel, 1990).

Residual depth anomalies

The subsidence rate associated with the large positive residual depth anomalies, generally related to hotspot volcanism, e.g. the Ninetyeast Ridge, was assumed to be the same as the subsidence rate for surrounding oceanic crust. Differential subsidence of these anomalously shallow ridges was not identified either within this study or in the earlier work of Detrick *et al.* (1977). Residual depth anomalies are, therefore, maintained until they are back-tracked to the ridge crest at which point they disappear. This results in variations in ridge crest depth through time as succeeding positive and negative anomalies are back-tracked to the ridge crest, causing the ridge to shoal or deepen respectively.

The presence of younger hotspot-generated crustal material emplaced upon older oceanic crust results in anomalously shallow palaeobathymetry, which is highlighted where possible. Many of the smaller seamounts of unknown age are assumed to have formed at the ridge crest which is consistent with observations of seamounts near the crests of currently active mid-ocean ridges (e.g. Scheirer & Macdonald 1995). These smaller seamounts were important and controlled the depth of gateways during the early evolution of southern hemisphere palaeobathymetry (see below). Recognition of the true mode of formation of smaller seamounts, i.e. ridge crest or intraplate volcanism, and their ages in the southern South Atlantic sector may subsequently modify our view of the timing of opening and configuration of gateways in this area.

Palaeobathymetry equations

The equation for calculating palaeobathymetry is defined as:

$$d(t) = rd + \left(R_{0i} + (K_i \times \sqrt{age - t})\right) \quad (1)$$

where $d(t)$ is the depth (m) of a grid node at time t (Ma), rd is the residual depth anomaly (m), R_{0i} and K_i are the zero-age depth (m) and subsidence rate (m/$\sqrt{my}$) for subsidence zone i, respectively, and *age* is the age of the ocean crust (Ma), i.e. if *age* = 30 Ma and $t = 10$ Ma then the square-root subsidence is calculated for 20 Ma old crust. Where flattening of bathymetry with increasing age occurs, palaeobathymetry was calculated:

$$d(t) = rd + \left(\Phi_i + 22473e^{(-0.0278(age - t))}\right) \quad (2)$$

where Φ_i is the maximum depth reached in subsidence zone i. This equation was modified from Stein & Stein (1992) so that Φ_i replaced their value of 5651 m for maximum ocean depths. The shape of the subsidence curve defined by Stein & Stein (1992) is preserved but the subsidence curve was moved either up or down to allow a smooth transition from Equation 1 to Equation 2 and allowed the maximum depth (Φ_i) to be defined. The

age at which subsidence changed from Equation 1 to Equation 2 varied for each subsidence zone between 33 Ma and 80 Ma.

For grid nodes lacking sediment thickness data (e.g. subducted crust), synthetic palaeobathymetry was calculated, using Equations 1 and 2 but with rd = 0.

Palaeobathymetric reconstructions

The results embodied within the palaeobathymetric reconstructions are illustrated in Figs 2 to 22 and are accompanied by DSDP and ODP drill sites; named locations mentioned in the text can be located in Fig. 1. (To aid interpretation of the palaeobathymetric reconstructions, full-colour digital files are available; an example of one of these colour figures can be found on the internet at http://servant.geol.cf.ac.uk/marine/Tim_Sykes. html. All digital files are in Adobe System's Portable Document Format (PDF).) Before commenting on the evolving palaeobathymetry of the southern hemisphere we feel it is necessary to reiterate that the palaeobathymetric maps are based on a sediment-free ocean. The palaeobathymetric evolution can be subdivided into three time intervals: fragmentation of Gondwana (110–95 Ma), late Cretaceous to late Eocene (95–40 Ma) and late Eocene to Recent (40–0 Ma). The discussion of these intervals is by no means exhaustive and is intended to highlight the main developments in the evolving palaeobathymetry. All depths are negative.

Fragmentation of Gondwana (110–95 Ma) (Figs 2–4)

The fragmentation of Gondwana at approximately 160 Ma and subsequent sea-floor spreading resulted in the development of ocean basins with depths deeper than –4000 m by 110 Ma. Mid-ocean ridges and numerous shallow plateaux with depths ranging from –2000 m to above sea level formed barriers to oceanic circulation and may have contributed to the formation of deep restricted basins

A major arcuate seaway was formed between Antarctica, Africa, India and Australia (Figs 2–4), ranging in depth from –2800 m, along the poorly defined mid-ocean ridge, to deeper than –4000 m in the flanking basins. Palaeogeographic reconstructions show that continental barriers in the west and east, between South America and Antarctica and Australia and Antarctica, precluded the development of deep oceanic gateways between the southern seaway and the Pacific Ocean. The as yet

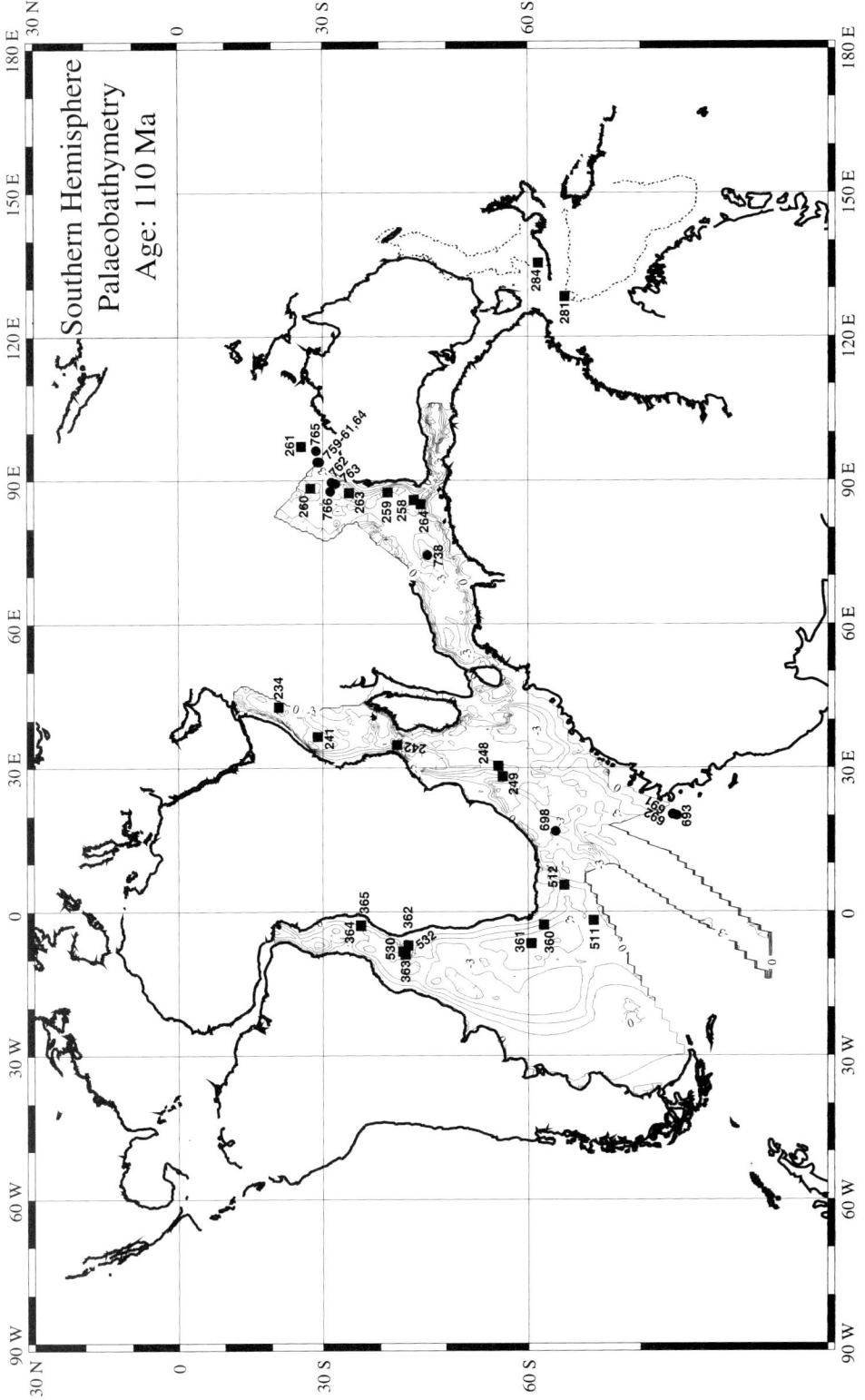

Figs 2 to 22. Palaeobathymetric contour charts, with drill site locations, for the southern hemisphere. Contour interval is 1 km. Dashed thick lines represent the outlines of the Campbell Plateau and Lord Howe Rise. DSDP drill sites shown as squares and ODP drill sites shown as circles.

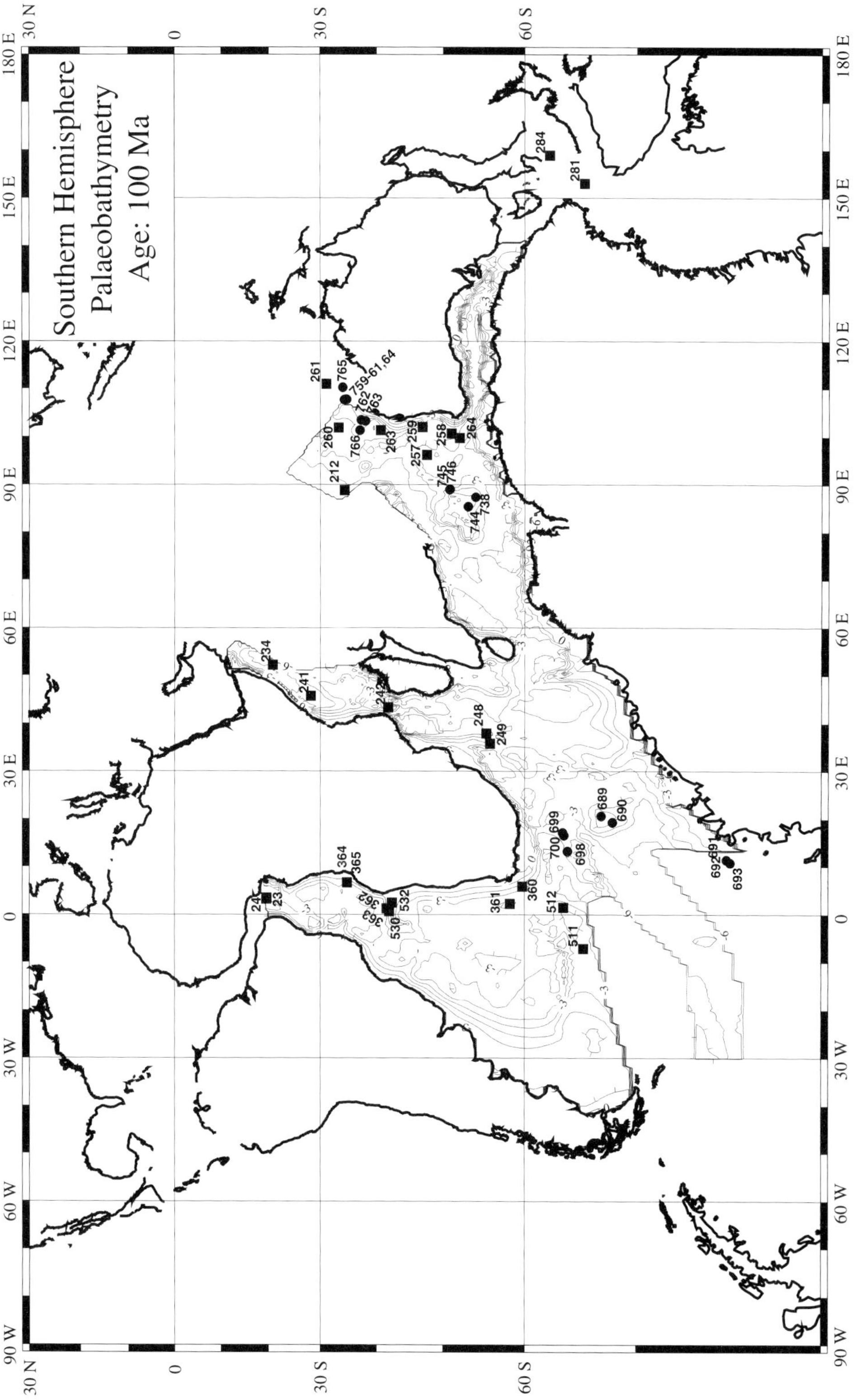

Southern Hemisphere
Palaeobathymetry
Age: 100 Ma

Fig. 3.

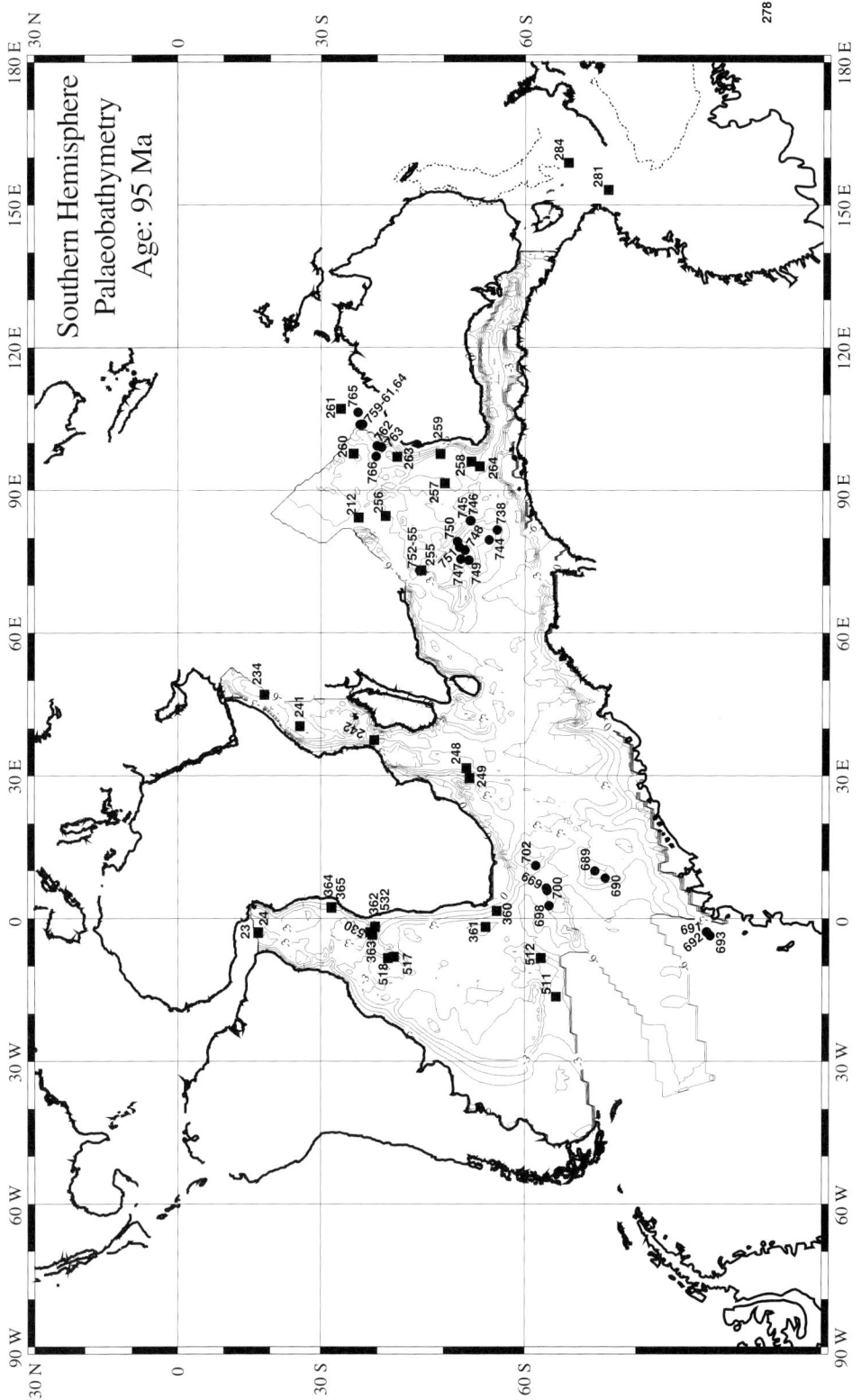

Southern Hemisphere
Palaeobathymetry
Age: 95 Ma

Fig. 4.

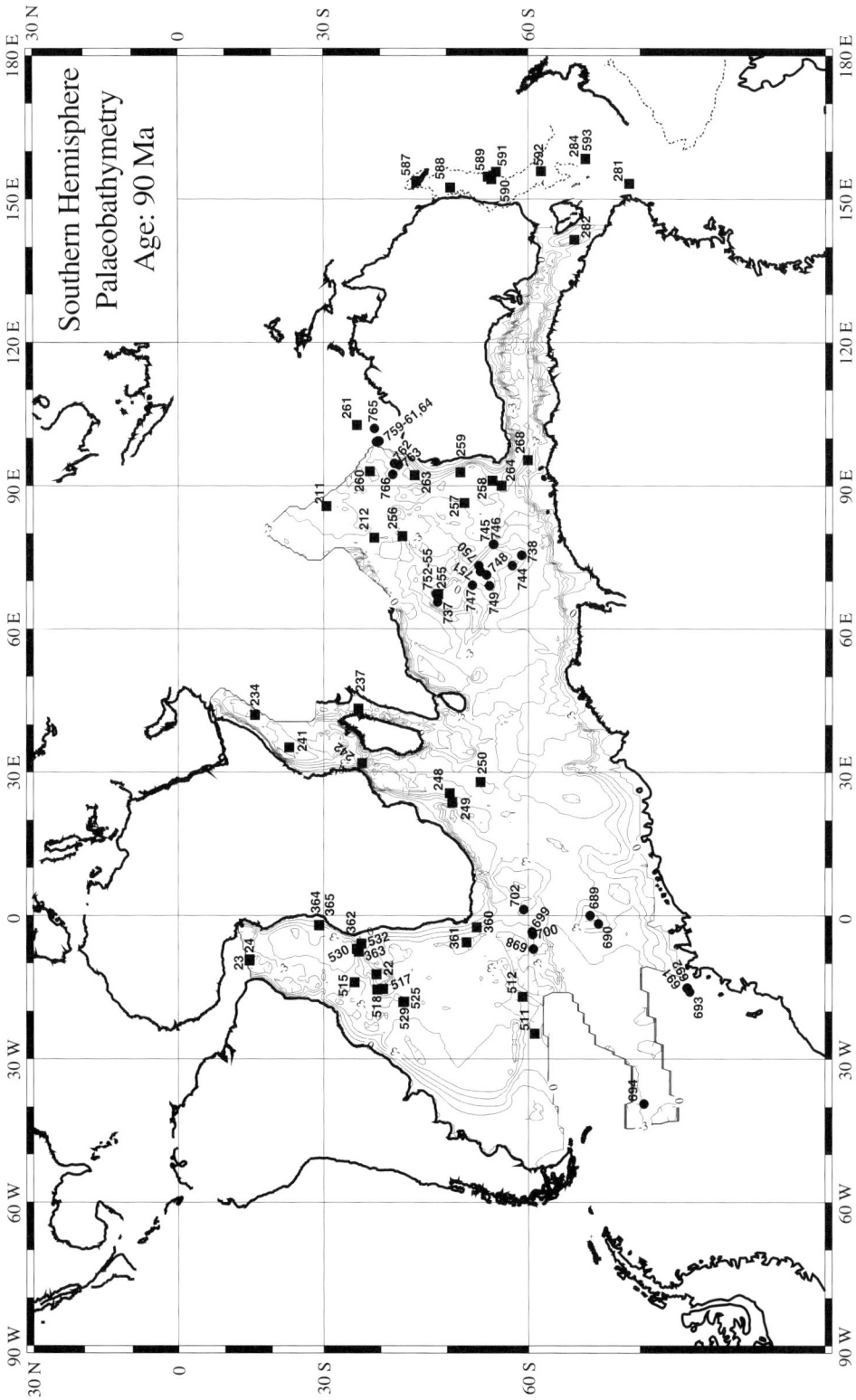

Southern Hemisphere
Palaeobathymetry
Age: 90 Ma

Fig. 5.

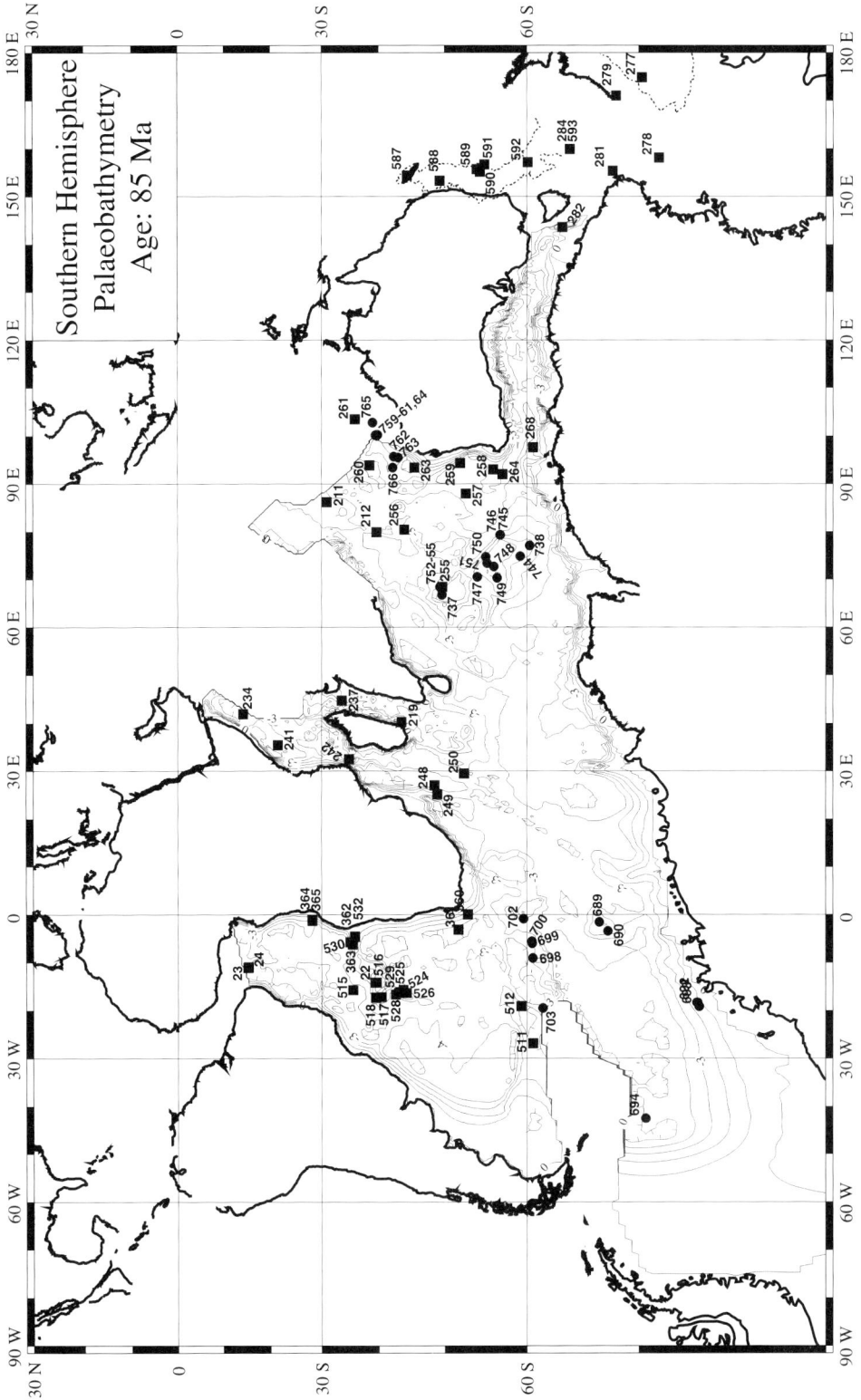

Fig. 6.

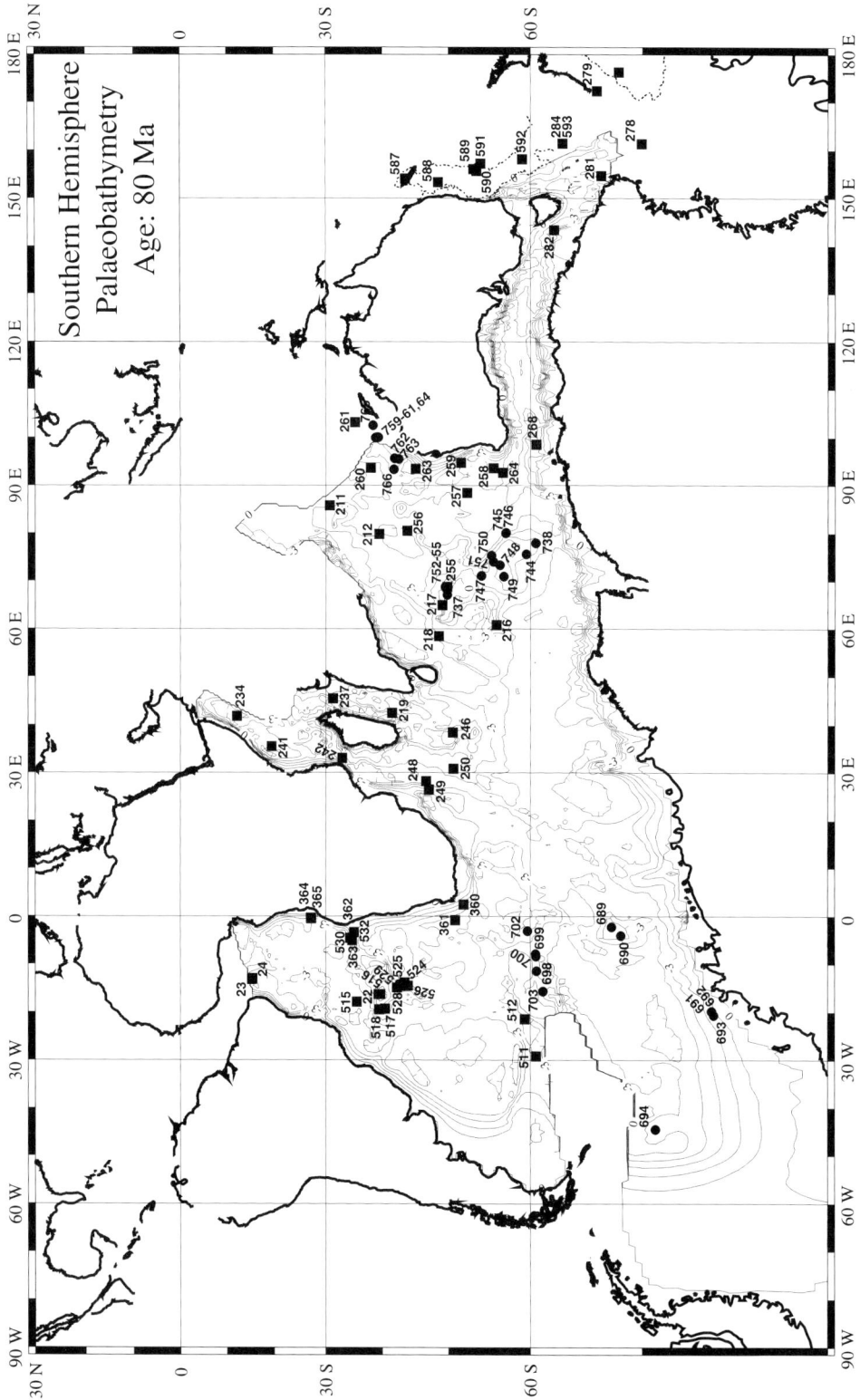

Southern Hemisphere
Palaeobathymetry
Age: 80 Ma

Fig. 7.

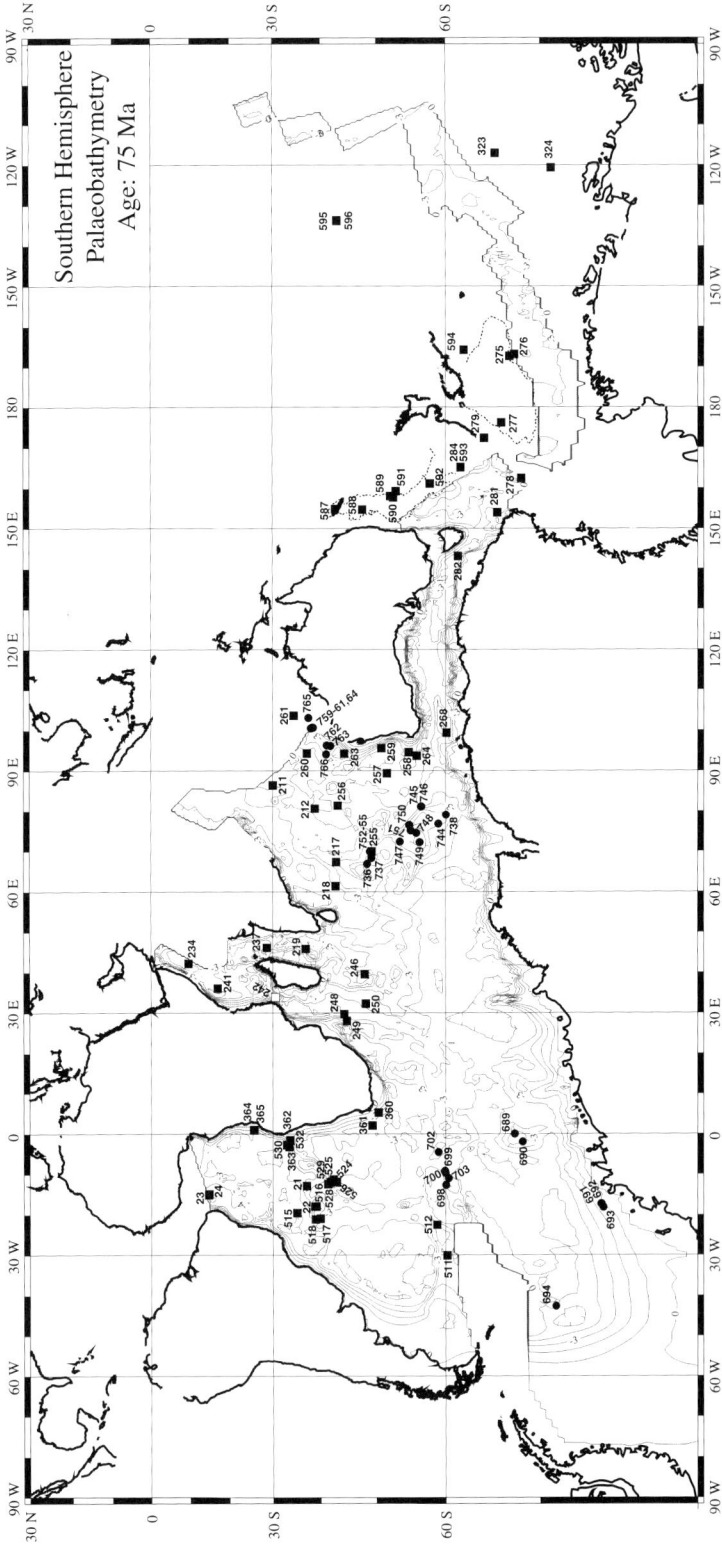

Fig. 8.

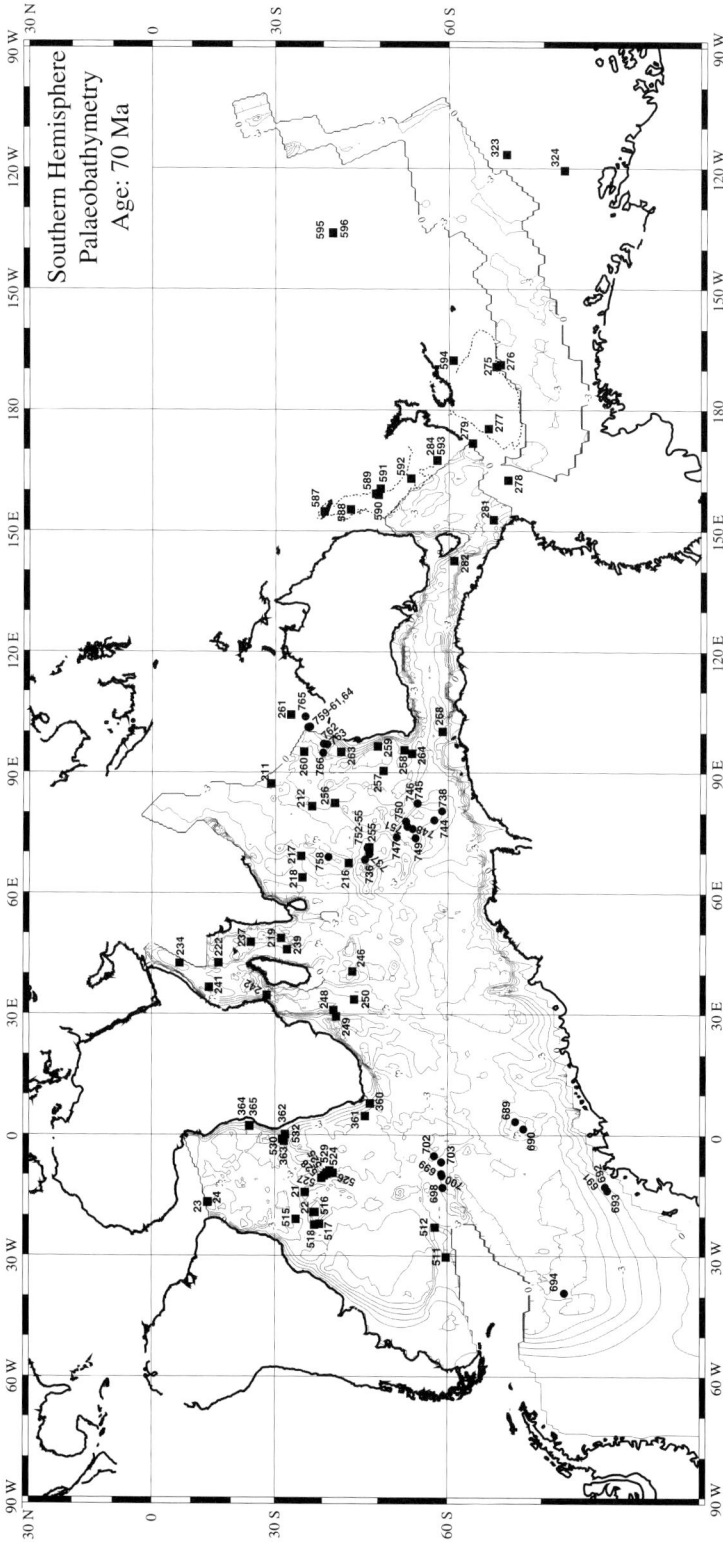

Southern Hemisphere
Palaeobathymetry
Age: 70 Ma

Fig. 9.

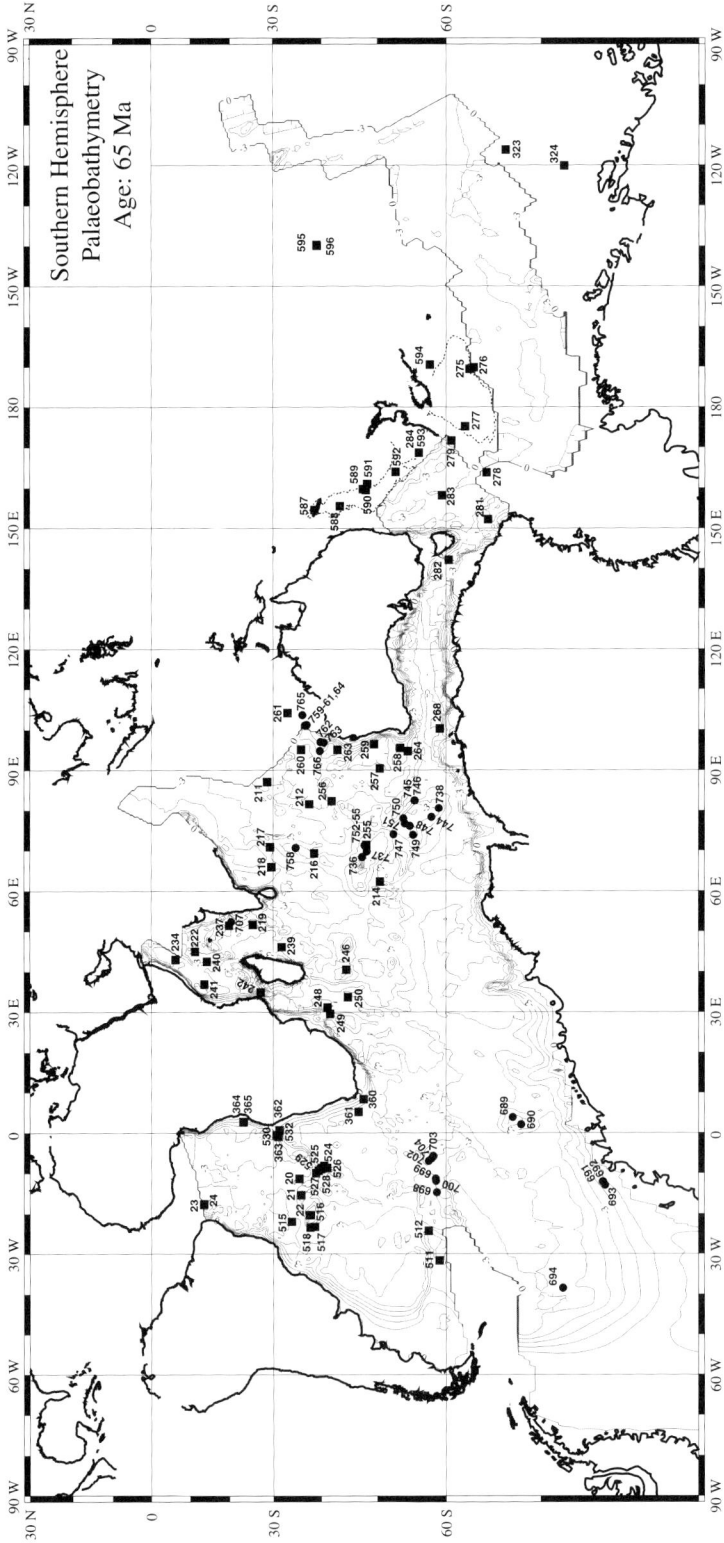

Fig. 10.

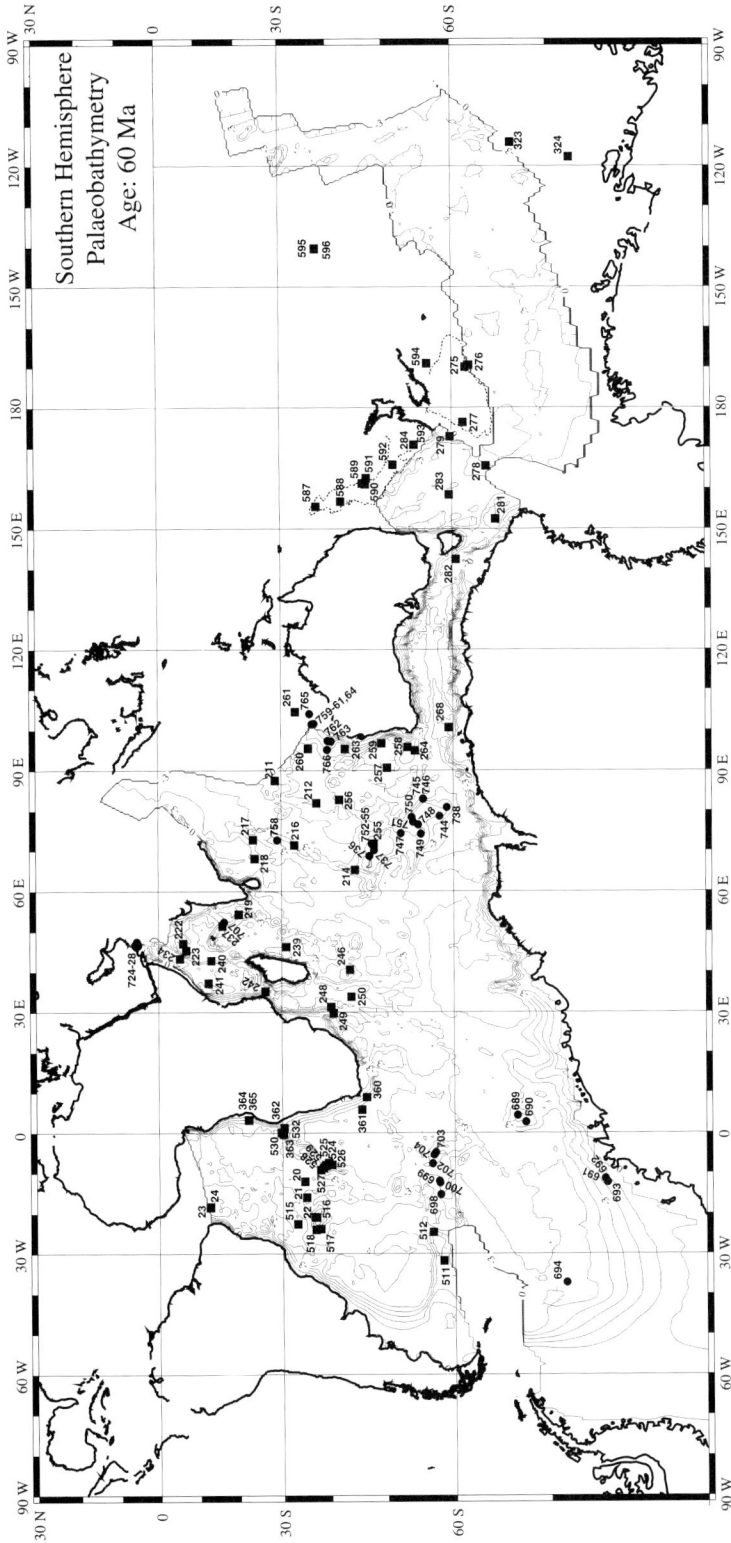

Southern Hemisphere Palaeobathymetry Age: 60 Ma

Fig. 11.

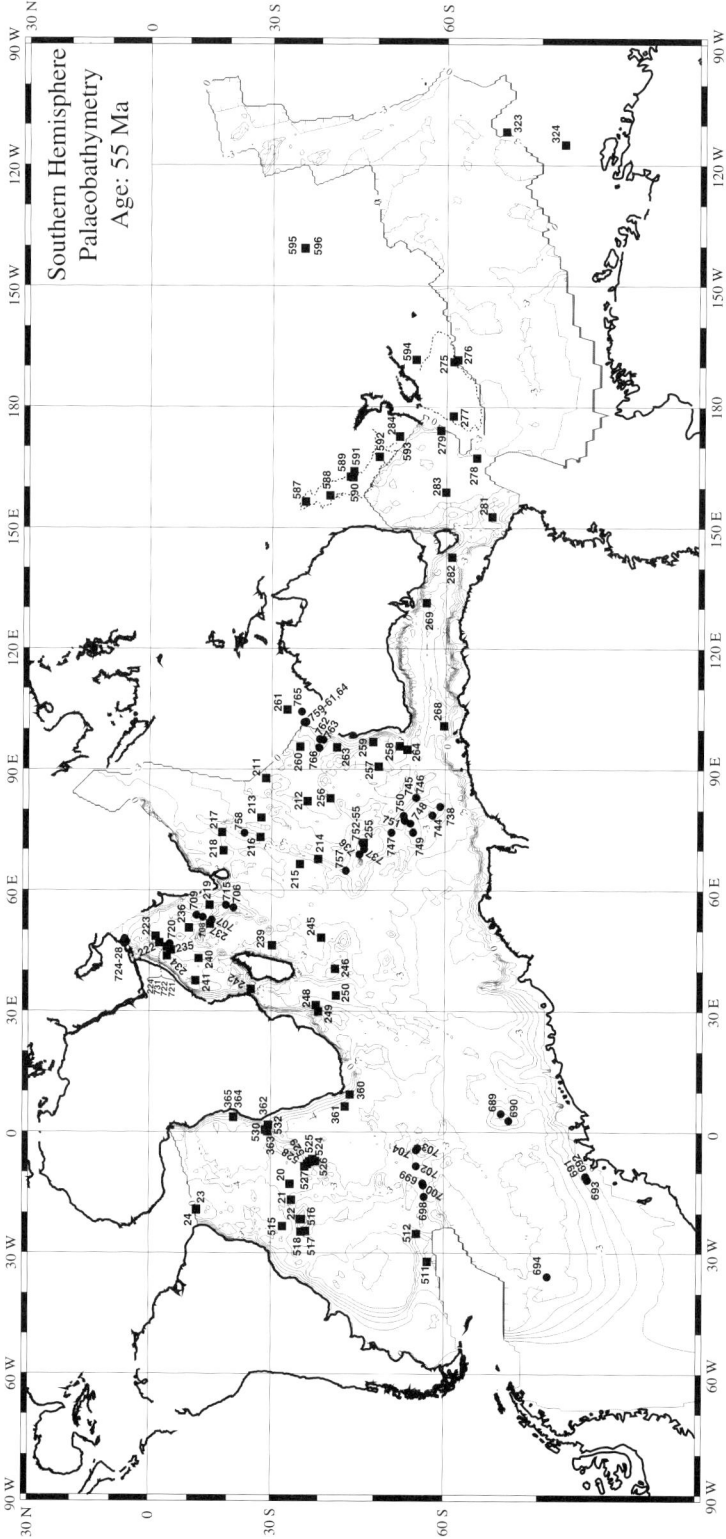

Fig. 12.

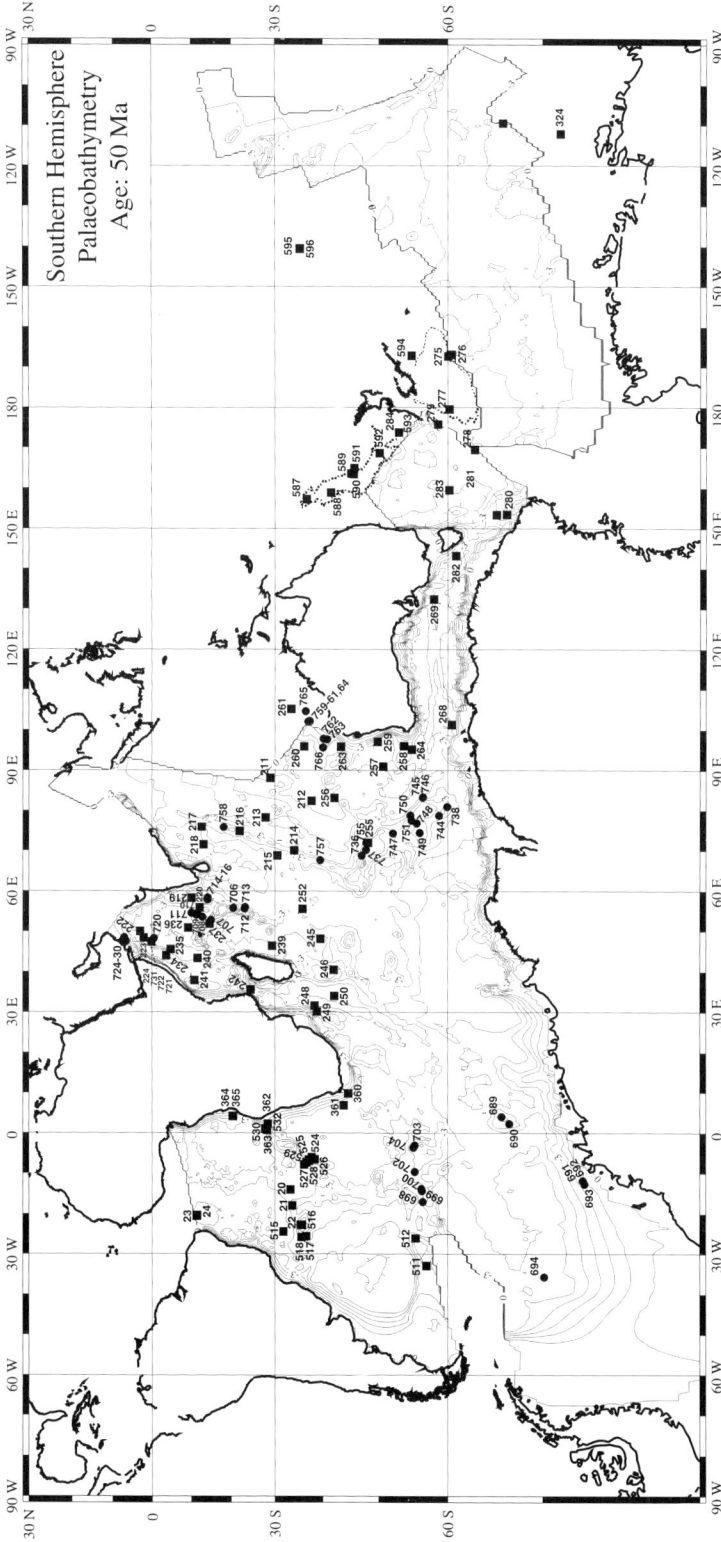

Southern Hemisphere
Palaeobathymetry
Age: 50 Ma

Fig. 13.

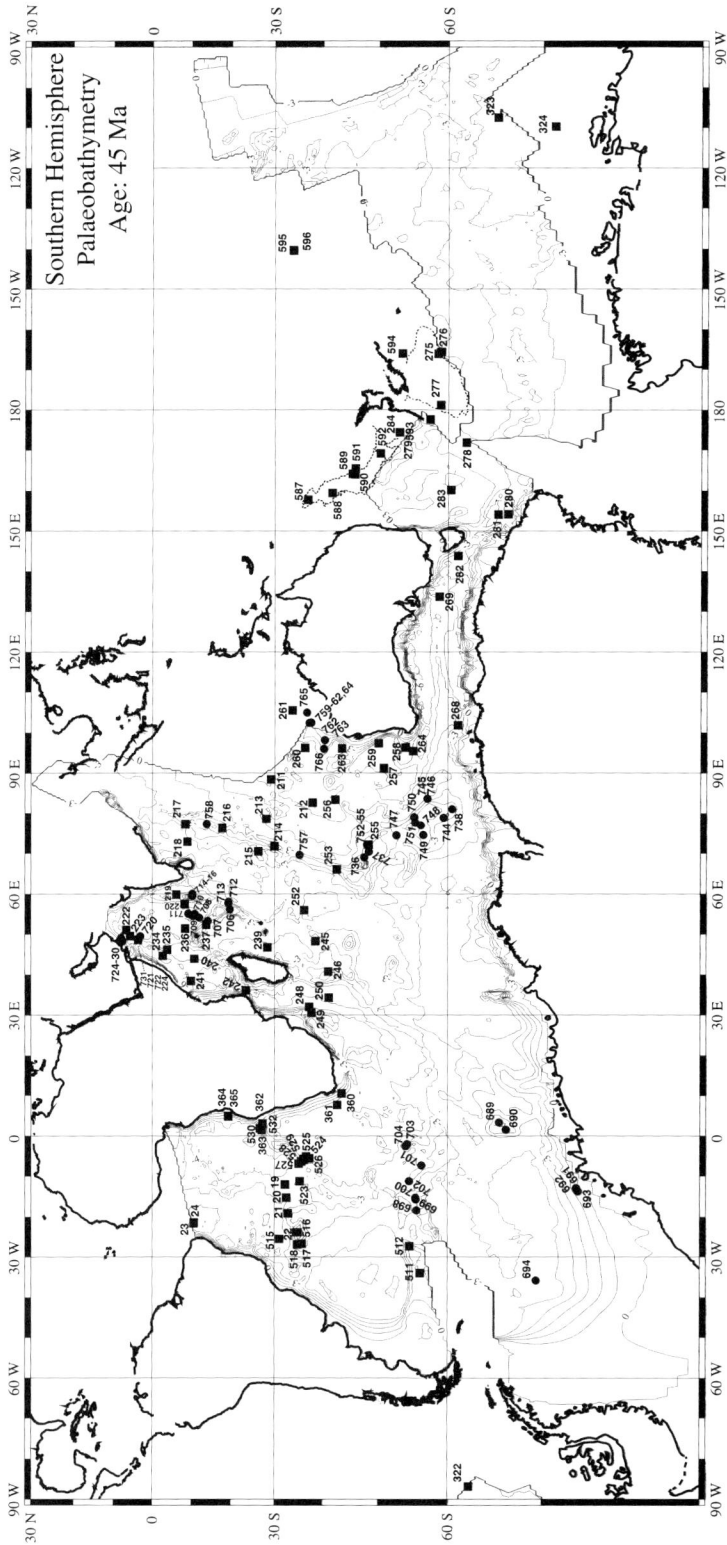

Southern Hemisphere
Palaeobathymetry
Age: 45 Ma

Fig. 14.

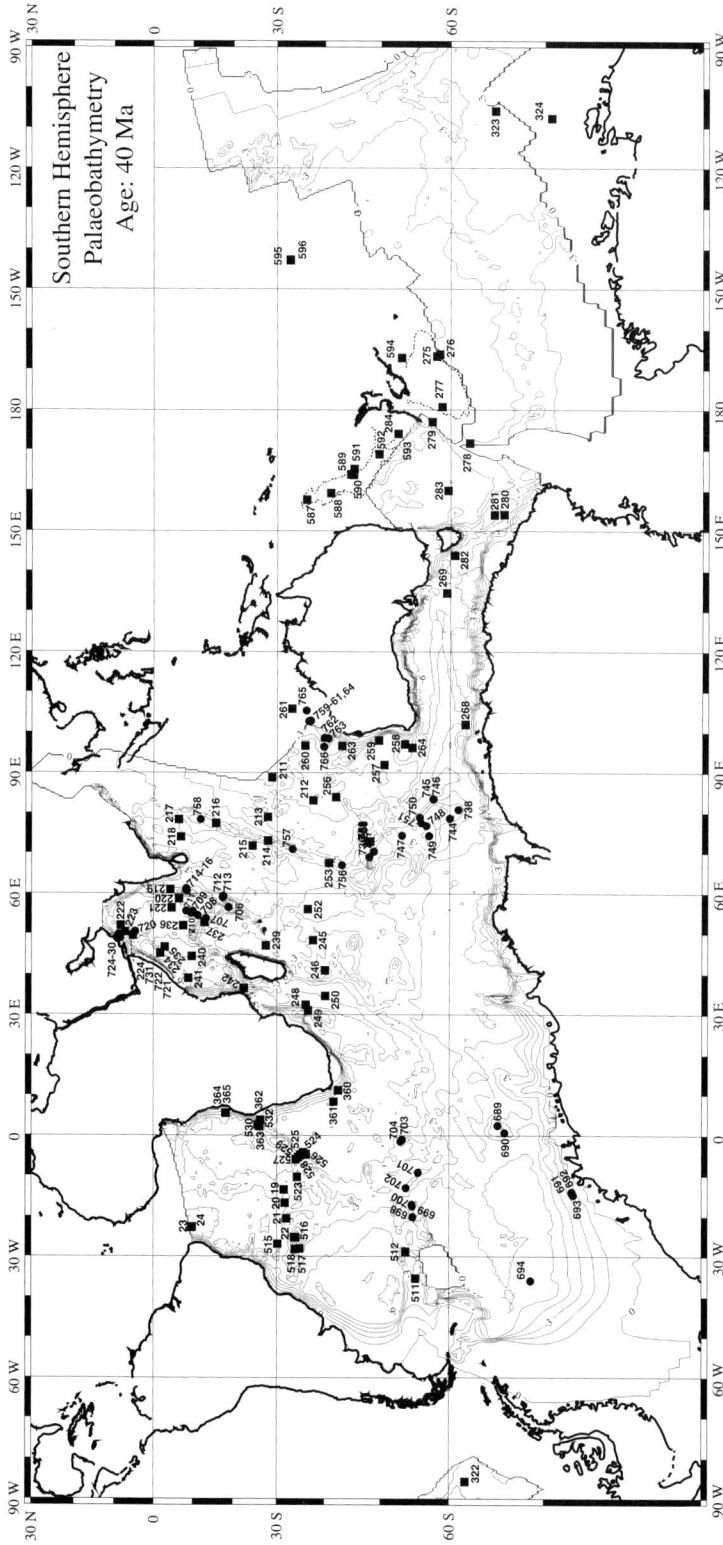

Fig. 15.

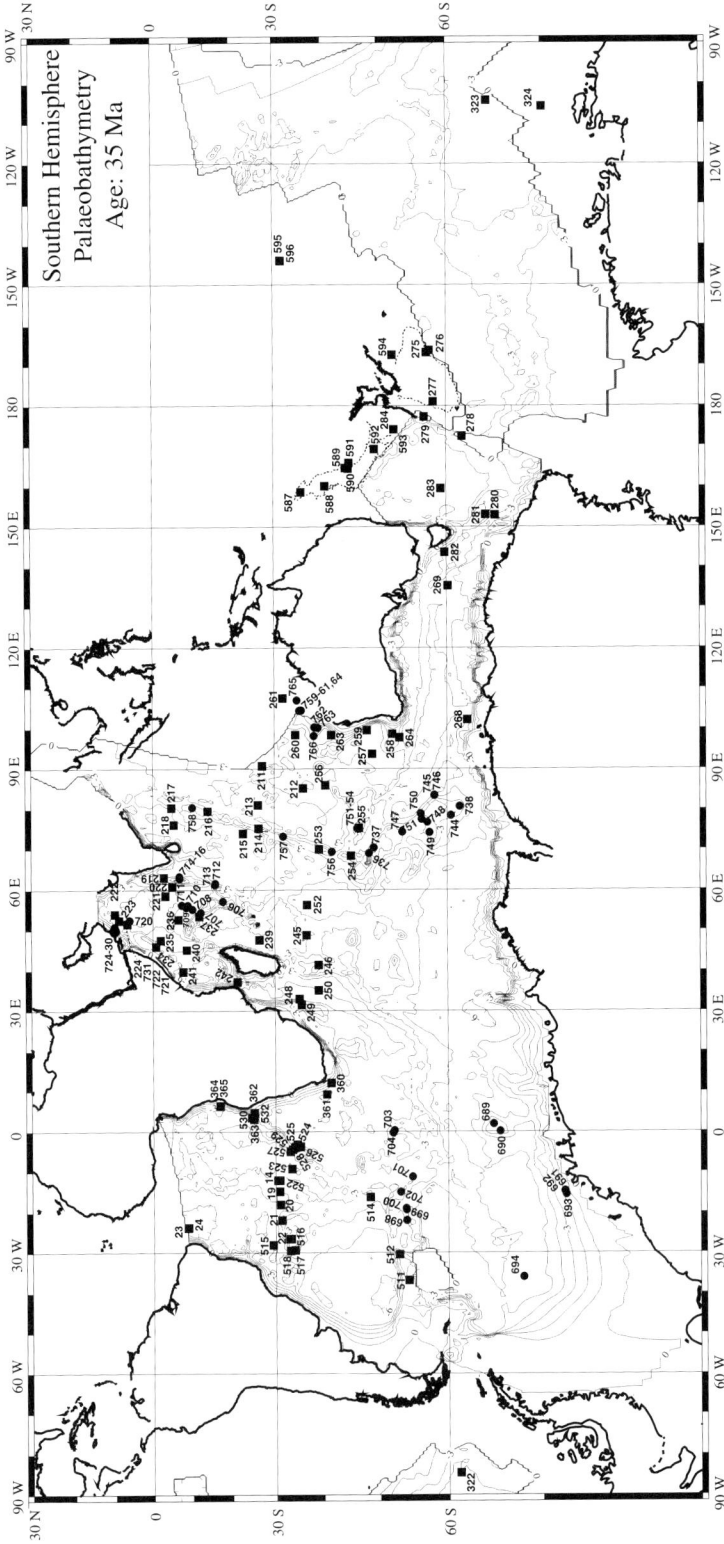

Fig. 16.

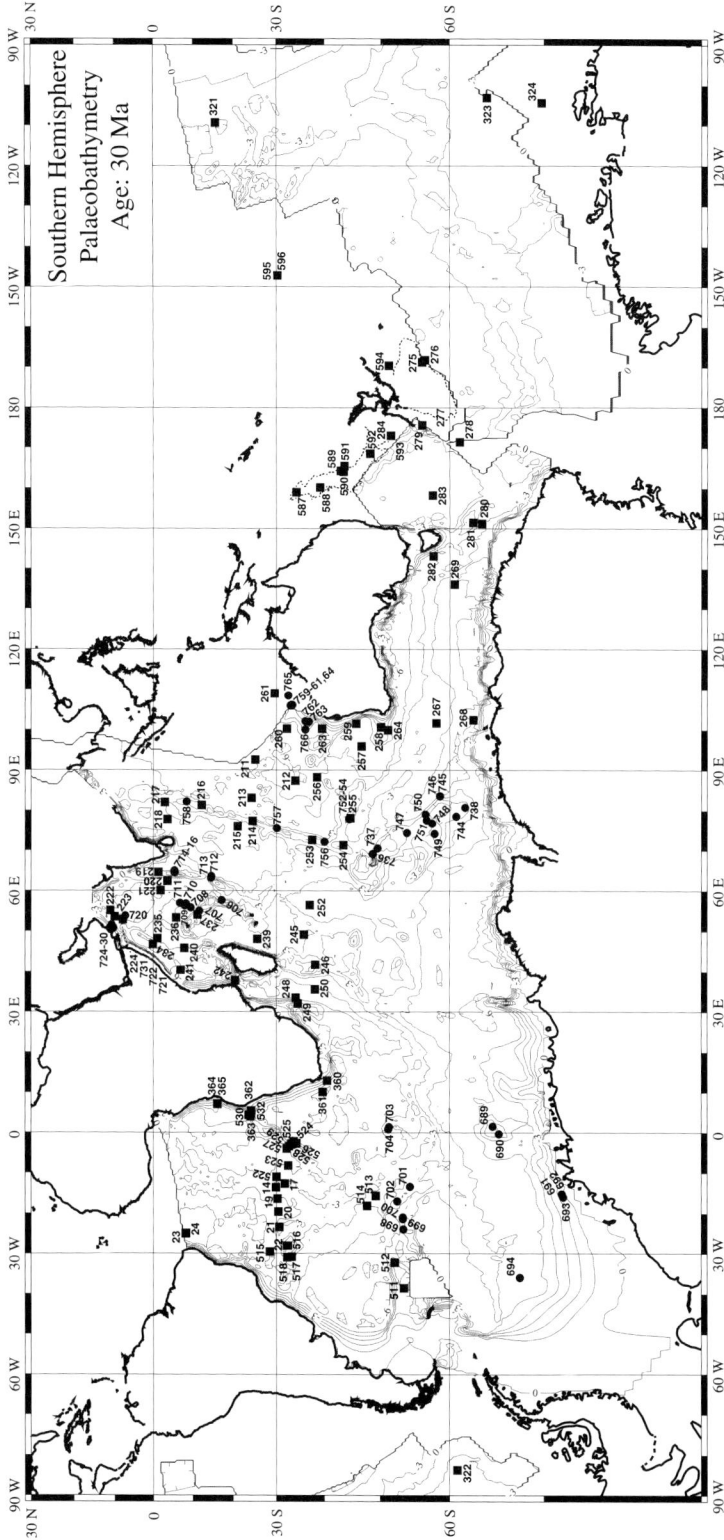

Southern Hemisphere
Palaeobathymetry
Age: 30 Ma

Fig. 17.

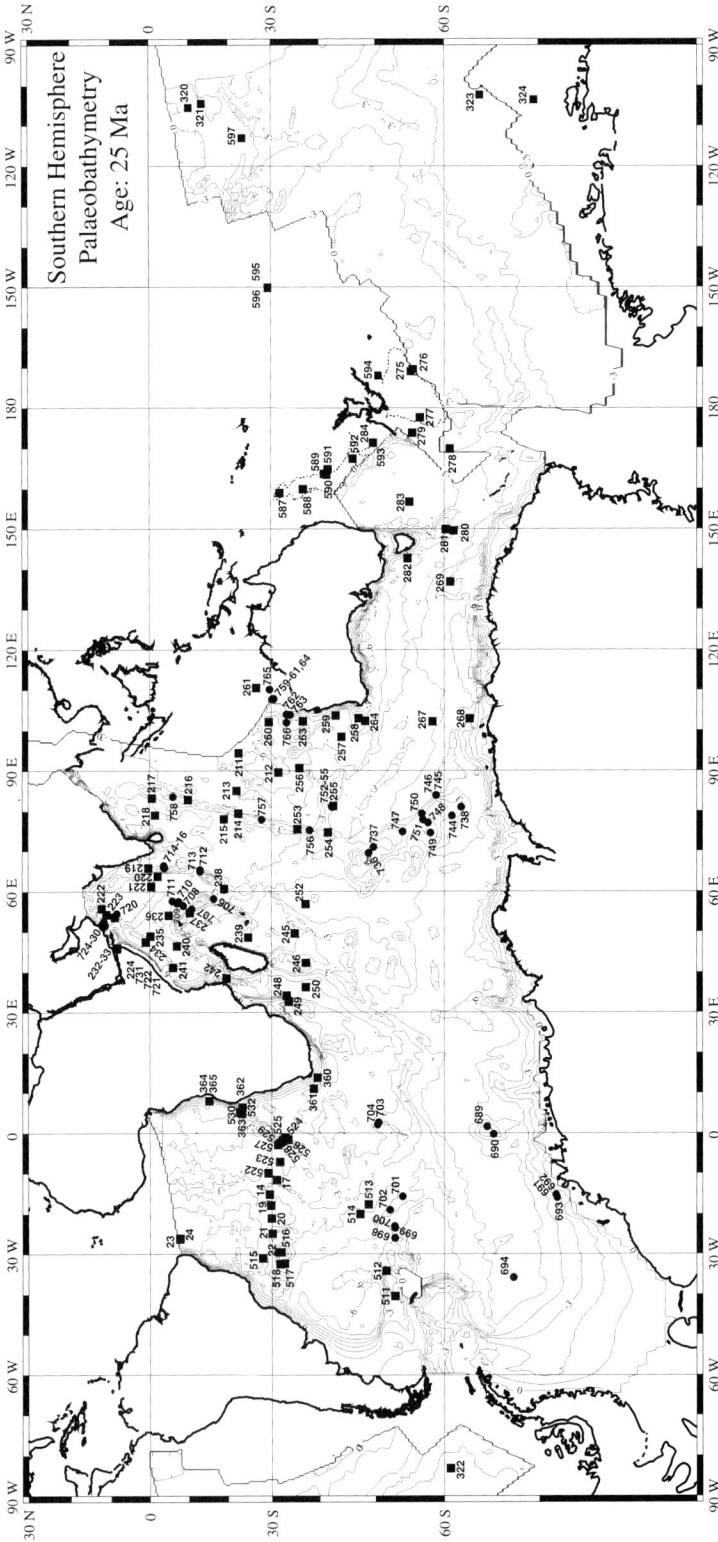

Fig. 18.

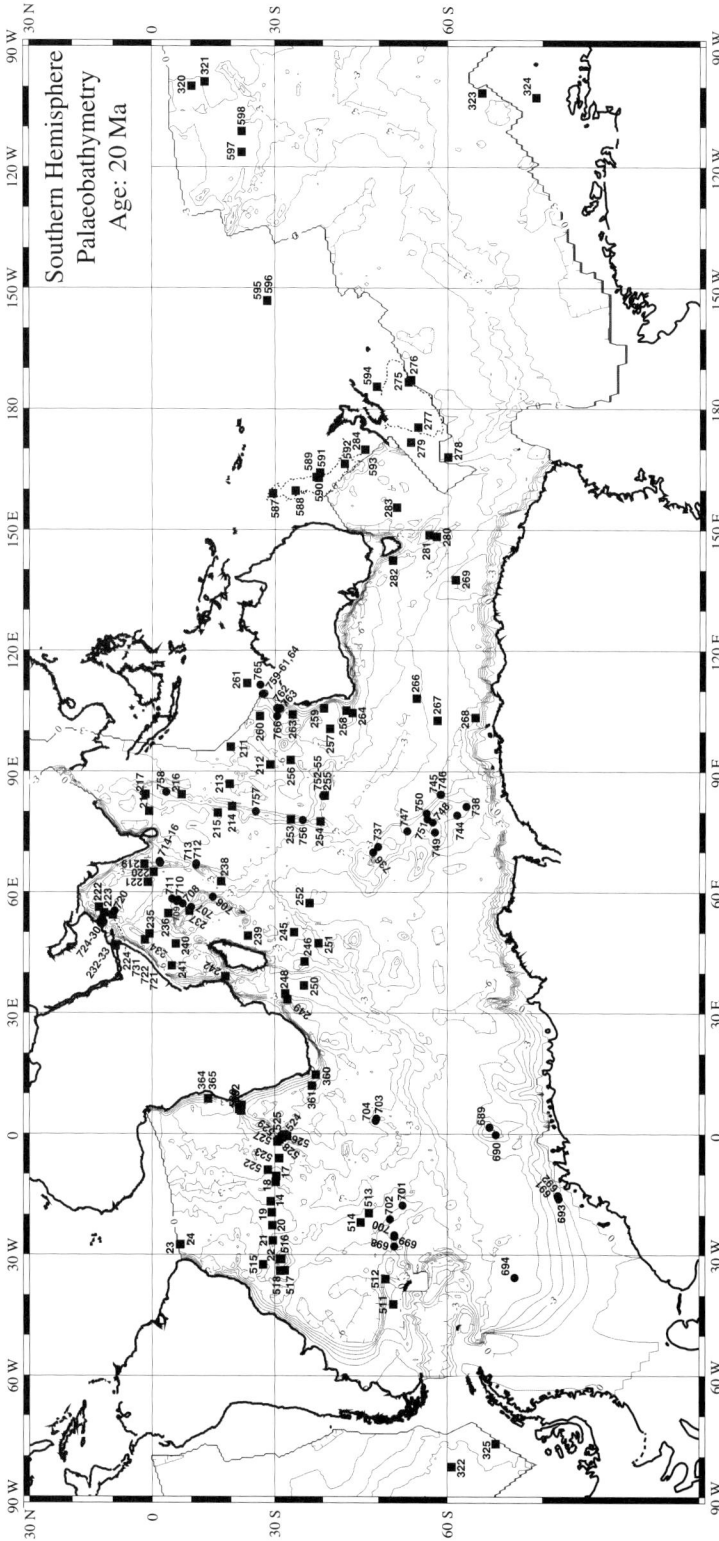

Southern Hemisphere
Palaeobathymetry
Age: 20 Ma

Fig. 19.

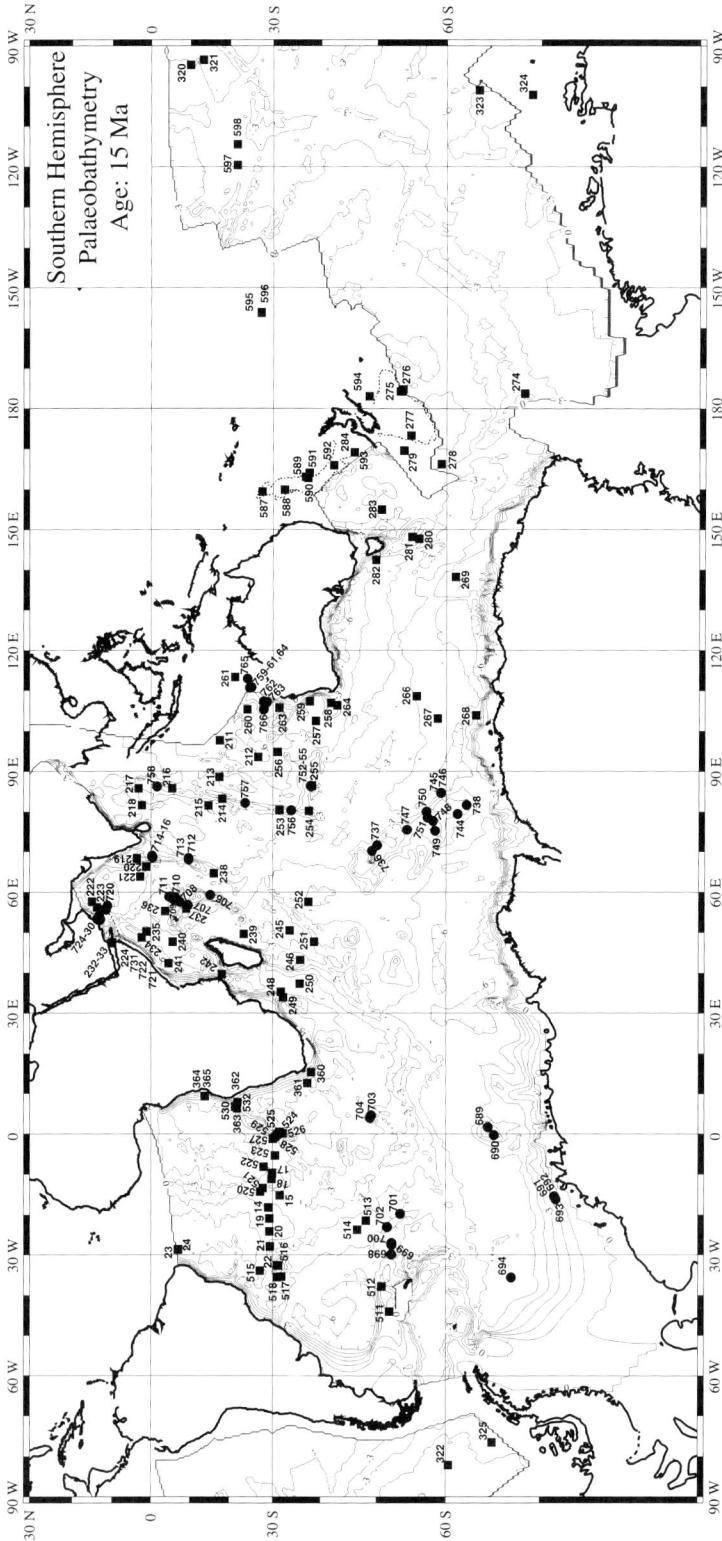

Southern Hemisphere
Palaeobathymetry
Age: 15 Ma

Fig. 20.

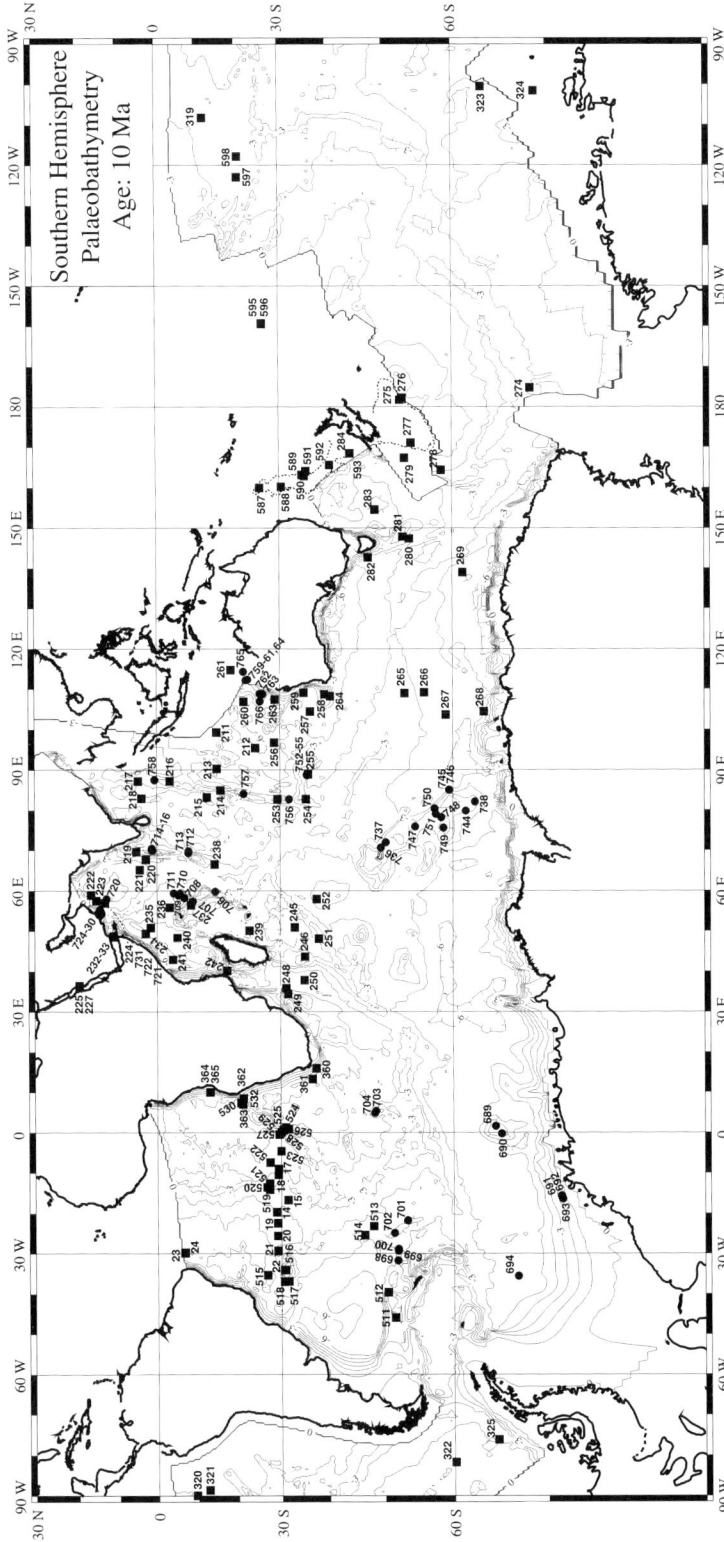

Southern Hemisphere
Palaeobathymetry
Age: 10 Ma

Fig. 21.

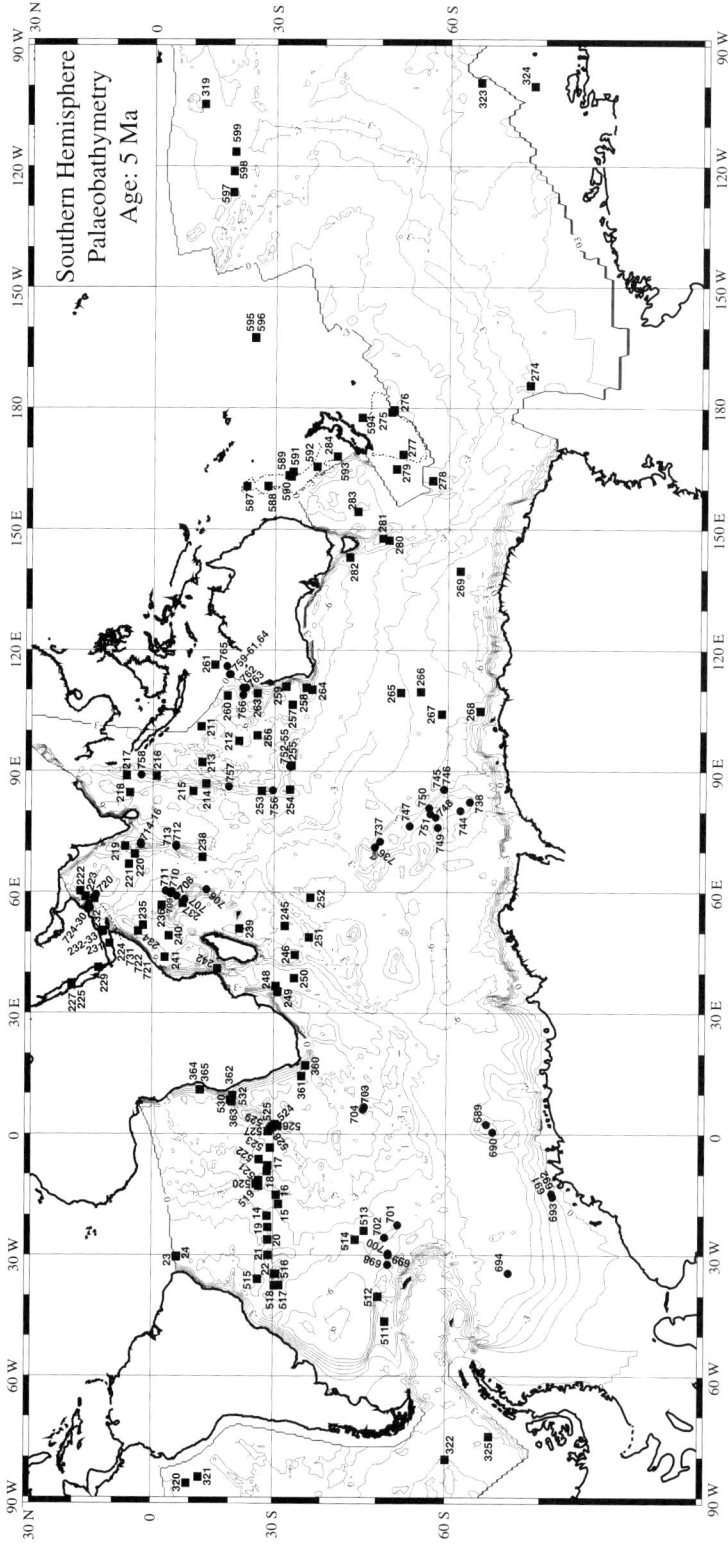

Fig. 22.

unknown tectonic history of the Weddell Sea prevents an analysis of its palaeobathymetry. It is likely that this basin was land-locked (Lawver *et al.* 1992). Connections between the southern seaway and Tethys occurred via the Mozambique Channel between East Africa and Madagascar, a shallow, narrow, shelf sea between India and Madagascar and through the Wharton Basin which separated India from Australia. The deep nature (*c.* −4200 m) of the Mozambique Channel in the palaeobathymetric reconstructions differs little from its present-day bathymetric configuration (Simpson *et al.* 1974). The depths shown here, which are an artefact of the gridding process and data input into the gridding process, can be justified on the grounds that the channel has a sediment fill in excess of 1 km (Mougenot *et al.* 1987). These workers also showed that seismic reflector 'A', the Eocene/Oligocene boundary, occurred at a palaeodepth of between −3500 m and −4500 m. These values were not isostatically corrected and are therefore too shallow by approximately 200 to 300 m.

Major changes during the evolution of the Southern Ocean between 110 and 95 Ma involve the opening of the following gateways within the southern hemisphere: (i) South Atlantic basins by continued separation between South America and Africa; (ii) a South Atlantic deep gateway, −3600 m to −4000 m deep and 750 km to 800 km wide by 100 Ma formed by the separation of the Falklands Plateau from southwest Africa at approximately 108 Ma; (iii) a gateway between India and Antarctica which developed as a result of continued rifting between Sri Lanka and Antarctica. This gateway was approximately 700 km wide and deeper than −4000 m by 95 Ma.

Access between the South Atlantic and the Indian Ocean was restricted at depths below −2400 m by the formation of the Agulhas Platform, Maud Rise, Mozambique Ridge and seamounts in the Orcadas Basin. The South Atlantic was bisected by the Walvis Ridge–Rio Grande Rise hotspot complex which occurred at depths shallower than −1000 m. The basins to the north of this complex were small and ranged in depth from −2080 to −4000 m. Prior to 110 Ma some of these basins were extremely restricted and formed centres of salt deposition (Cande & Rabinowitz 1978). Evidence for the restricted circulation within the basins to the north and south of the Walvis Ridge–Rio Grande Rise complex between 110 and 95 Ma is equivocal. The sapropels recorded on the western margin of the Angola Basin and Sao Paulo Rise (van Andel *et al.* 1977; Weissert 1981; Thurow *et al.* 1992) either represent phases of anoxia that extended through the whole water column or alternatively were associated with expanded oxygen minimum zones

and oxic conditions at greater depth. The large depth range of the Angolan margin sapropels indicates that the Angola Basin may, like the NE Atlantic (Thurow *et al.* 1992) have been characterized by oxygen depletion from approximately −200 m to the ocean floor. Thurow *et al.* (1992) document the occurrence and propose a model for the development of an expanded oxygen minimum zone, from the NW Australian margin. The NW Australian example is in part coeval (Cenomanian/Turonian) with the Angolan margin sapropels. The preservation of phytodetritus in DSDP Site 361 on the South African margin (Ryan 1978) is possibly related to the influx of turbidites.

The continued separation of India and Antarctica was associated with the growth of the Kerguelen Plateau. The elevation of this feature evolved from a submarine position at 110 Ma to a subaerial position by 100 Ma. Significant areas of the Kerguelen Plateau (>125 000 km^2) were above sea level at 95 Ma with elevations of 1229 m (Coffin 1992), which compare with values of 1100 m in this study. The difference in these elevations is a function of the use of different subsidence rates.

The slow northward drift of Australia away from Antarctica, commencing at 130 Ma, allowed the development of an east–west elongated basin which rapidly attained a depth greater than −4000 m. Australia remained attached to Antarctica in the eastern part of this basin, precluding an oceanic connection with the southwest Pacific Ocean. Indications of marine incursions on the Australian margin, however, are lacking (Hegarty *et al.* 1988).

Late Cretaceous to late Eocene (95–40 Ma) (Figs 4 to 15)

The Walvis Ridge–Rio Grande Rise hotspot complex dominated the palaeobathymetry of the mid-latitudes of the South Atlantic between 95 and 40 Ma. During its activity between 90 and 85 Ma this plateau formed a single island with an estimated area >10 000 km^2 and reached heights in excess of 1000 m above sea level. A sinuous deep-water connection (at depths greater than −3600 m) through the northern part of the South Atlantic had developed by 80 Ma. This meandering feature crossed the South Atlantic mid-ocean ridge repeatedly. The asymmetrical nature of subsidence within the South Atlantic is clearly shown by the development of deeper basins on the South American Plate.

The southern oceanic gateway between the South Atlantic and Weddell Sea through to the Indian Ocean, i.e. between the Falklands Plateau and Africa, was congested by numerous deep-sea plateaux. Continued subsidence and spreading along the mid-oceanic ridge allowed the gradual development of narrow oceanic gateways through this area by 70 Ma. These gateways remained shallower than −3000 m until 60 Ma when palaeodepths exceeded −4000 m.

The configuration of the seaway between India, Australia and Antarctica changed significantly post-80 Ma when the Wharton, Madagascar and Central Indian Basins started to open. The development of the Ninetyeast Ridge is clearly visible in the bathymetric reconstructions post-85 Ma (Figs 7–22). Apart from at its northern end, the Ninetyeast Ridge restricted the bathymetric connection between the Central Indian and Wharton Basins at depths greater than −2000 m. The opening of the Central Indian and Madagascar Basins was associated with the development of many shallow submarine ridges, at depths shallower than −1600 m. The majority probably formed at the Crozet and Réunion hotspots. Thus the centre of the Indian Ocean was bisected by numerous shallow ridges and plateaux between 75 and 60 Ma. These were generally shallower than crust formed at nearby mid-oceanic ridges of comparable age. The Wharton Basin and, to a lesser degree, the Central Indian Basin attained their present configuration by 40 Ma. In contrast the Madagascar Basin attained its present-day configuration by 60 Ma. The trace of the Réunion Hotspot across the Madagascar Basin is erroneous. It is an artefact of the original bathymetry data set (i.e. ETOPO5) and represents a younger feature situated on considerably older crust. Throughout the bathymetric evolution of the Arabian Basin the combined Chagos–Laccadive–Mascarene Plateau formed a barrier to oceanic circulation at palaeodepths below −1200 m until the complete separation of these features by 25 Ma.

The continued opening of the seaway between Australia and Antarctica resulted in the development of an enclosed basin by 90 Ma, which was bounded to the east by Tasmania. This basin remained enclosed but increased in depth between 90 and 45 Ma. It is thought to have been characterized by sluggish circulation which resulted in the accumulation of poorly sorted detrital sediments with a high organic content but poor in biogenic components (Andrews *et al.* 1975). The palaeobathymetric charts indicate that oceanic connections between this seaway and the Tasman Sea occurred at depths shallower than −3000 m. This connection may be an artefact of the tectonic model used because no account was taken of the crustal stretching and subsidence of the Tasman Rise. The oceanic connection across the saddle on this rise may not have been fully developed until the rise separated from Antarctica. Examination of palaeobathymetric reconstructions reveals that the

southern and shallower part of the Tasman Rise is distinguishable from the Antarctic margin post-65 Ma. The presence of this saddle may therefore be valid from 65 Ma but its presence before 65 Ma is questionable. The formation of this oceanic gateway is considered a major component in the development of Circum-Antarctic circulation and was thought to have evolved post-40 Ma (Kennett *et al.* 1972, 1975*a,b*). The Tasman Sea was completely formed between 85 and 50 Ma and during this time the basin was clearly divided by the then active mid-ocean ridge. Following the cessation of active crust formation, continued thermal relaxation resulted in the basin-wide development of depths in excess of −4000 m by 30 Ma.

Palaeodepths along the East Pacific Rise are not illustrated prior to 75 Ma as they provide little palaeobathymetric insight. The separation of the shallow (−1600 m) Campbell Plateau from Antarctica, however, resulted in the formation of a narrow (< 300 km wide) and deep ocean gateway with depths below −3000 m between the Tasman Sea and the Pacific Ocean by 75 Ma. This gateway was initiated in response to spreading along the East Pacific Rise which commenced at 83 Ma. The early history of this ridge is characterized by short ridge segments offset by fracture zones and it is unlikely to have acted as a significant barrier to deep ocean circulation. Between 75 and 55 Ma the ridge crest segments at the northern end of the East Pacific Rise are organized into triangular segments within which water depths shoal from the southern apices by as much as 800 m to depths of between −2400 and −2800 m at the base of the triangle. At present we have no explanation for the origin of these triangular features. These ridge segments become rectilinear, narrower and generally deeper at depths between −2800 and −3200 m by 50 Ma.

Late Eocene to Recent (40–0 Ma)
(Figs 15 to 22)

The plates within the southern hemisphere reached their present-day configuration just prior to 40 Ma (Royer *et al.* 1992*a*). This was achieved by the abandonment of the Wharton Ridge combined with a southward ridge jump which resulted in the rifting of Broken Ridge from the northern margin of the Kerguelen Plateau. The following discussion focuses on the opening of major oceanic gateways and the general palaeobathymetric configuration of the southern hemisphere.

The separation of Broken Ridge from the northern Kerguelen Plateau is thought to have been accompanied by the rapid uplift of Broken Ridge, between 42 and 38 Ma, by *c.* 2500 m which effectively reset the subsidence history of this feature (Rea *et al.* 1990). The palaeobathymetry shows that this ridge was approximately −2400 m deep (Fig. 15), whereas palaeodepths deduced from benthic foraminfers indicate a depth of 200–400 m (Driscoll *et al.* 1989). This very localized phenomenon has not been specifically accounted for.

Major oceanic gateways

Pre-40 Ma the Chagos–Laccadive Ridge was connected to the Mascarene Plateau forming a shallow (shallower than −2000 m), partly emergent oceanic barrier between the Arabian and Central Indian Basins. However, the formation of an intermediate depth (shallower than −2400 m), narrow (*c.* 150 km wide) gateway at *c.* 40 Ma between the Maldive Bank and the Chagos Bank presumably marked a period of volcanic quiescence for the Réunion Hotspot. The southwestern extension of the Mascarene Plateau from the Saya de Malha Bank (ODP Site 706, Fig. 15) is anomalous and must be disregarded when interpreting the palaeobathymetry. This extension was formed by activity on the Réunion Hotspot post-40 Ma. By 30 Ma the southern part of the Chagos–Laccadive Ridge had formed and separated from the Saya de Malha Bank resulting in a narrow (*c.* 300 km) passage between these features. This was generally of intermediate depth (shallower than −2800 m) but was choked (shallower than −1600 m) at its southern end by a combination of hotspot and normal oceanic crust formation. A deep gateway between the Chagos Bank and the Mascarene Plateau was firmly established between 25 and 20 Ma, reaching depths of between −2600 and −2800 m. This gateway did not exceed depths of −3200 m before 15–10 Ma when the Central Indian Ridge crest was separated by *c.* 500 km from these submarine plateaux.

The rifting apart of the Kerguelen Plateau and Broken Ridge resulted in the development of a narrow (< 100 km wide), relatively shallow (shallower than −2600 m) passage at 40 Ma. This event was contemporaneous with the separation of the Chagos–Laccadive Ridge from the Mascarene Plateau. As this passage grew the palaeobathymetry was initially dominated by the Southeast Indian Ridge crest at depths shallower than −2400 m, e.g. at 35 Ma. By 30 Ma this gateway had attained intermediate depths (deeper than −3200 m) and was significantly wider (*c.* 350 km). The influence of the Kerguelen Hotspot can be clearly seen within the development of the southern flank of the Southeast Indian Ridge. This ridge flank was shallower (−2800 m deep) by 25 Ma, despite the *c.* 400 km separation between the Southeast Indian Ridge crest and the Kerguelen Plateau. The fact that the southern flank of the Southeast Indian Ridge

has remained at an intermediate depth, between −3200 and −2800 m, through to the Recent probably reflects the continued interaction of the Kerguelen Hotspot with this ridge. The gateway between the Central Indian Basin and the area between the Southeast Indian Ridge crest and Broken Ridge never exceeded −3200 m due to the development of a small seamount at the ridge crest (just to the west of Broken Ridge) at 20 Ma. This seamount has been influential in maintaining the intermediate depth of this gateway.

One of the two most influential deep oceanic gateways to develop over the last 40 Ma was between the Tasman Rise and Antarctica (Kennett *et al.* 1972, 1975*a,b*) leading to profound changes in global circulation and climate (Kennett 1982). Prior to 40 Ma a surface to intermediate depth connection existed across the Tasman Rise (see above). Between 40 and 35 Ma a narrow (<150 km) intermediate depth (−3600 to −4000 m) passage began to open. Post-15 Ma a deep (deeper than −4000 m) and narrow (*c.* 600 km) gateway had formed between the Campbell Plateau and the relatively shallow (−2600 to −2200 m) ridge crest of the western end of the East Pacific Rise. It is possible that a deep (shallower than −3200 m), narrow (<100 km) connection existed to the south of the shallow mid-ocean ridge crest between the Antarctic Basin and the Ross Sea. Its existence, prior to 15 Ma, cannot be confirmed due to limitations of the tectonic model. However, this gateway through the Tasman Fracture Zone had become restricted by 5 Ma as the ridge crest along the eastern Southeast Indian Ridge shoaled. Since 5 Ma the most direct deep (deeper than −2800 m) oceanic connection between the Indian and Pacific Oceans occurred south of the Southeast Indian Ridge and East Pacific Rise crests between the Balleny Seamounts and the Southeast Indian Ridge crest.

The opening of Scotia Sea post-30 Ma also represents the formation of a major oceanic gateway. This allowed the development of a circumpolar deep water mass and the thermal isolation of Antarctica, ultimately leading to the development of permanent Antarctic ice sheets (Kennett 1982). The western Scotia Sea Basin was well developed by 25 Ma and covered an area of *c.* 120 000 km^2, predominantly at water depths shallower than −2800 m. Bounding shallow (−1800 to −1400 m) plateaux restricted the gateway to shallow/intermediate depths. It is possible that deep (deeper than −3000 m) narrow gateways existed between these plateaux, which were too small to have been identified within the resolution of this study. The palaeobathymetric configuration of the Scotia Sea by 20 Ma indicates the existence of intermediate depth connections between this sea and the South

Atlantic together with narrow deep passages comparable to present-day passages. These connections have not changed noticeably during the continued evolution of the Scotia Sea between 20 Ma and the Recent.

General palaeobathymetry

Between 40 Ma and the Recent the ocean basins of the southern hemisphere continued to widen and deepen. The east and west sections of the Antarctic Basin had generally subsided below depths of –4800 m by 65–60 Ma, but were separated by a shallower (–4000 m) sill to the north of the Maud Rise. Depths in excess of –4800 m were attained across the entire basin by 25 Ma. The Antarctic Basin was connected to the Crozet Basin via a narrow channel between the Kerguelen Plateau and the Crozet Plateau at water depths shallower than –3600 m by 40 Ma. This gradually deepened to between –4400 and –4800 m by 15 Ma. The Antarctic Basin was connected at depth (deeper than –4400 m) to the western South Atlantic via the Orcadas Basin throughout the last 40 Ma. The palaeobathymetric reconstructions show that the deep (deeper than –4000 m) connection between the Agulhas Basin and the Antarctic Basin, across the Southwest Indian Ridge, was not established pre-25 Ma. However, it is possible that very narrow, deep connections may have existed prior to this time which were too small to be identified with the resolution used in this study. The Agulhas Basin is connected at depth (shallower than –4000 m) to the Cape Basin but the Walvis Ridge restricted deep connections between the Angola and Cape Basins at water depths below –2800 m. However, deep (deeper than –4000 m), narrow connections between the Cape and Angola Basins are visible in the palaeobathymetric reconstructions between 40 Ma and the Recent.

The extinction of the Wharton Ridge pre-40 Ma resulted in the thermal relaxation of its ridge crest and the almost complete removal of its bathymetrical expression above –3600 m by 20 Ma.

The Crozet Plateau/Madagascar Ridge have remained at water depths above –2800 m, thus precluding the development of a deep oceanic connection between these plateaux. The Southwest Indian Ridge can be clearly differentiated from the northern margins of the Crozet Plateau by 20 Ma though these features never separated. The continued shallow oceanic connection between these plateaux is related to the low spreading rates on the Southwest Indian Ridge and the presence of the Crozet Hotspot which probably interacts with the nearby Southwest Indian Ridge to maintain these shallow bathymetric features.

The deep basins of the Pacific Ocean at water depths below –4400 m were geographically limited pre-40 Ma, covered extensive areas by 30 Ma and attained depths in excess of –5000 m between 15 and 10 Ma. Deep connections across the East Pacific Rise between the Bellinghausen and Campbell Basins were restricted to the large offset fracture zones, e.g. Eltanin FZ system, throughout this time interval. The Chile Ridge remained a shallow (shallower than –2400 m) feature until the triple junction between this ridge and the East Pacific Rise changed from a ridge–ridge–ridge to ridge–ridge–fracture zone configuration at *c.* 20 Ma. After this triple junction reconfiguration the northern section of the Chile Ridge developed a shallower bathymetric profile than the southern section, which was being subducted beneath South America by 20 Ma.

Discussion

The palaeobathymetry presented here was constructed using compilations of geophysical data. In order to determine the accuracy of these geophysically derived grids, and hence the accuracy of the sediment-free palaeobathymetry, these data are compared with DSDP and ODP bathymetry, sediment thickness, crustal age and corrected water depth data from 39 southern hemisphere drill sites with penetration to oceanic crust (Fig. 23). The corrected water depths, for each drill site, were calculated using the method of Sykes (1996). The position of each of these 39 drill sites was rounded to the nearest half degree to allow comparison with equivalent values extracted from the digital database (Table 1). Figure 24 shows the comparison between drill sites versus gridded data for (A) bathymetry, (B) crustal age, (C) sediment thickness and (D) corrected bathymetry. In this discussion we consider the degree of correspondence between unique point data at the drill sites and the trend represented by the gridded data.

Bathymetry at individual drill sites correlates positively with the half-degree averaged ETOPO5 bathymetry, clustering along the 1:1 relationship (Fig. 24A). The differences between the drill site data and gridded data probably arise from: (i) the averaging process which has probably smoothed out any short-wavelength features (<20 km); such features may have also been poorly represented in the original ETOPO5 data; (ii) if the half-degree grid in which the drill site fell was dominated by deeper bathymetries, the averaged bathymetry will reflect this dominance, i.e. sites located on the edge of a shallower bathymetric feature on the edge of a half-degree grid cell will have a deeper gridded bathymetry and *vice versa*; (iii) many DSDP/ODP drill sites were deliberately drilled on locally upstanding features (Kidd 1995, pers. comm.). The

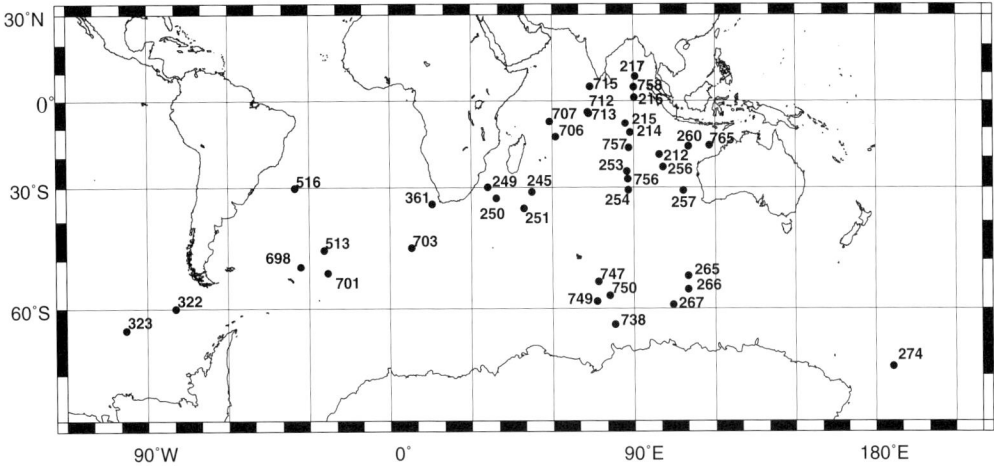

Fig. 23. Location of DSDP and ODP drill sites, penetrating to oceanic crust. The bathymetries, total sediment thickness, crustal age and sediment-corrected bathymetries for thes sites are compared to similar data extracted from the gridded data sets.

majority of these data show a difference of <500 m, suggesting that within the deep ocean ETOPO5 is not as inaccurate as portrayed by Smith (1993), although this conclusion is based on a limited analysis.

Crustal ages for the drill sites were derived either from published radiometric age determinations or from the age of the oldest sediment (Table 1). The age of the oldest sediment was usually defined to the subepoch level, e.g. early Palaeocene (60.4–66.0 Ma), therefore the age ascribed to such a sample would be 63.2 ± 2.8 Ma and would follow the timescalcs adopted here. A positive correlation occurs between the gridded and the drill site ages, although the latter are generally 10% younger (Fig. 24B). The largest difference occurred when the ages of the samples were dated on the basis of the age of the oldest sediment, e.g. ODP Site 703. However, the difference between the age of ODP Site 703 and the gridded crustal age may also be a function of its location, which is close to the site of a ridge jump (Sykes 1995). In addition it is also possible that the drill site ages record the ages of the final stages of volcanism and therefore do not accurately reflect the age of the formation of the bulk of the crust. A time lag between the end of crustal formation and the onset of sedimentation will also yield younger ages for the drill sites dated by the age of the oldest sediment.

The gridded sediment thicknesses vary by a factor of ± 0.5 and ± 2.0 of the sediment thicknesses recorded at the DSDP/ODP drill sites and can be said to show no correlation (Fig. 24C). This clearly raises the question: What exactly do the isopach maps record? Short-wavelength (<10 km (Kennett

1982)) variations in sediment thickness could arise if the topography of the oceanic crust is rugged, with a relief of between 500 and 1000 m, which would allow relatively thin and thick sediment sequences to co-exist in close proximity. Examination of single-channel seismic profiles presented by Heezen & Ewing (1963), Ewing & Ewing (1970), Mathias *et al.* (1988) and Hayes & LaBrecque (1991) show that the range of basement relief for normal ocean crust can vary between 0.5 and 1.0 s of two-way travel time, i.e. between 375 and 850 m assuming a range in sediment velocity of between 1500 and 1700 m s^{-1}. Clearly sediment isopach maps represent sediment thicknesses above an assumed generally smooth basement topography and do not consider small scale variations in relief. It is therefore not surprising that drill site sediment thickness data show such variability when compared with the isopach data. Hayes & LaBrecque (1991) suggested that their Southern Ocean sediment isopach data were accurate to approximately 10% in planform and accurate to better than 10% in the vertical. The validity of the latter statement is questionable because a constant sediment velocity was used to compute sediment thickness from seismic sections.

The sediment-corrected bathymetries for the drill sites and gridded data correlate positively. The difference between the two data sets generally ranges between ± 500 m and ± 700 m (Fig. 24D), irrespective of the difference in the bathymetry (Fig. 24A) or sediment thickness (Fig. 24C). From this analysis it is considered that the sediment-corrected water depths for the gridded database accurately portray the trends of the drill site data.

Table 1. *Comparison of drill site data to the gridded data sets*

Leg	Site	Drill site data							Gridded data				Data source
		Latitude	Longitude	Age (Ma)	Age (Ma) error	WD (m)	Ts (km)	CWD (m)	Age (Ma)	WD (m)	Ts (km)	CWD (m)	
22	212	−19.19	99.3	*70*	*10*	−6243	0.516	−6464	79.4	−5999	0.130	−6096	von der Borch et al. (1974)
22	214	−11.34	88.72	**59**		−1665	0.390	−1832	66.4	−1992	0.378	−2243	von der Borch et al. (1974), Duncan (1978)
22	215	−8.12	86.79	*60*	*4*	−5319	0.151	−5384	59.6	−4990	0.130	5087	von der Borch et al. (1974)
22	216	1.46	90.21	**81**		−2247	0.422	−2428	80.5	−2407	0.363	−2650	von der Borch et al. (1974), Duncan (1978)
22	217	8.93	90.54	*82*	*4*	−3020	0.664	−3303	80.2	−3023	0.572	−3384	von der Borch et al. (1974)
25	245	−31.53	52.3	*63*	*4*	−5857	0.389	−6024	70.0	−4646	0.393	−4899	Simpson et al. (1974)
25	249	−29.95	36.08	*134*	*10*	−2088	0.408	−2263		−2023	1.008	−2621	Simpson et al. (1974)
26	250	−33.46	39.37	*88*	*4*	−5119	0.710	−5422	85.8	−5004	0.786	−5457	Davies et al. (1974)
26	251	−36.5	49.48	*39*	*2*	−3489	0.486	−3697	24.2	−2957	0.026	−3103	Davies et al. (1974)
26	253	−24.88	87.37	*42.5*	*11*	−1962	0.549	−2197	46.1	−2180	0.605	−2560	Davies et al. (1974)
26	254	−30.97	87.9	**38**		−1253	0.176	−1329	39.0	−1455	0.554	−1807	Davies et al. (1974), Duncan (1978)
26	256	−23.46	100.77	**92**	*4*	−5361	0.251	−5469	103.5	−4681	0.568	−5024	Davies et al. (1974), Rudle et al. (1974)
26	257	−30.99	108.35	**97**		−5278	0.262	−5391	106.1	−5131	0.366	−5370	Davies et al. (1974), Rudle et al. (1974)
27	260	−16.14	110.3	*105*	*5*	−5702	0.323	−5841	130.8	−5545	0.380	−5791	Veevers et al. (1974)
28	265	−53.34	109.95	*13*	*16*	−3582	0.444	−3772	13.7	−3660	0.277	−3850	Hayes et al. (1975)
28	266	−56.4	110.11	*20*	*5*	−4173	0.370	−4332	23.1	−4601	0.308	−4808	Hayes et al. (1975)
28	267	−59.26	104.49	*28.5*	*6*	−4564	0.205	−4652	34.1	−4446	0.431	−4719	Hayes et al. (1975)
28	274	−69	173.43	*35.3*	*4*	−3326	0.415	−3504	44.1	−3354	0.830	−3847	Hayes et al. (1975)
35	322	−60.02	−79.42	*29*	*1.5*	−5026	0.513	−5246	48.1	−4874	0.236	−5039	Hollister et al. (1976)
35	323	−63.68	−97.99	*70*	*8*	−5004	0.701	−5303	82.7	−4905	0.548	−5238	Hollister et al. (1976)
40	361	−35.07	15.45	*121.5*	*5*	−4549	1.314	−5101	125.4	−4558	2.030	−5597	Bolli et al. (1978)
71	513	−47.58	−24.64	*31*	*2.5*	−4373	0.380	−4536	34.1	−4599	0.380	−4845	Ludwig et al. (1983)

72	516	-30.28	-35.28	**86**	**4**	-1313	1.250	-1839	89.7	-1017	0.630	-1411	Barker et al. (1983), Mussett & Barker (1983)
114	698	-51.46	-33.1	*90*	*20*	-2138	0.210	-2228	110.1	-3491	0.630	3875	Ciesielski et al. (1988)
114	701	-51.98	-23.21	*43*	*12*	-4636	0.481	-4842	50.6	-4621	0.510	-4935	Ciesielski et al. (1988)
114	703	-47.05	7.89	*43*	*12*	-1796	0.364	-1952	87.3	-2925	0.400	-3189	Ciesielski et al. (1988)
115	706	-13.17	61.37	**33**		-2504	0.047	-2524	56.9	-2933	0.401	-3198	Backman et al. (1988), Duncan & Hargraves (1990)
115	707	-7.55	59.05			-1541	0.375	-1702	66.8	-1544	0.263	-1726	Backman et al. (1988), Duncan & Hargraves (1990)
115	712	-4.22	73.04	**64**	*49*	-2904	0.115	-2953	52.2	-2563	0.214	-2714	Backman et al. (1988)
115	713	-4.19	73.04			-2915	0.155	-2982	52.2	-2563	0.214	-2714	Backman et al. (1988), Duncan & Hargraves (1990)
115	715	-4.19	73.83	**49**	*57*	-2266	0.287	-2389	58.0	-2596	0.389	-2854	Backman (1988), Duncan & Hargraves (1990)
119	738	-62.71	82.79	*91*	*105*	-2252	0.485	-2460	112.6	-2491	0.262	-2672	Barron et al. (1989)
120	747	-54.81	76.79			-1695	0.295	-1822	97.3	-1874	0.975	-2454	Schlich et al. (1989), Whitchurch et al. (1992)
120	749	-58.72	76.41	**110**		-1069	0.197	-1154	95.7	-941	0.685	-1365	Schlich et al. (1989), Whitchurch et al. (1992)
120	750	-57.59	81.24	**101**		-2030	0.671	-2316	95.9	-1955	0.809	-2447	Schlich et al. (1989), Whitchurch et al. (1992)
121	756	-27.53	87.6	**43**		-1515	0.150	-1579	42.1	-3048	0.195	-3187	Peirce et al. (1989), Duncan (1991)
121	757	-17.02	88.18	**58**		-1643	0.372	-1803	55.5	-1927	1.281	-2665	Peirce et al. (1989), Duncan (1991)
121	758	5.38	90.36	**82**		-2923	0.527	-3148	73.3	-3371	0.371	-3616	Peirce et al. (1989), Duncan (1991)
123	765	-15.98	117.57	**156**		-6919	0.936	-7316		-5719	0.691	-6124	Gradstein et al. (1990), Ludden (1992)

Bold ages refer to radiometric age determinations. Italicized ages determined from the age of the oldest sediment (see text for explanation)

WD = Water Depth.

Ts = Thickness of sedimentary cover.

CWD = Isostatically corrected water depth.

A

B

C

D

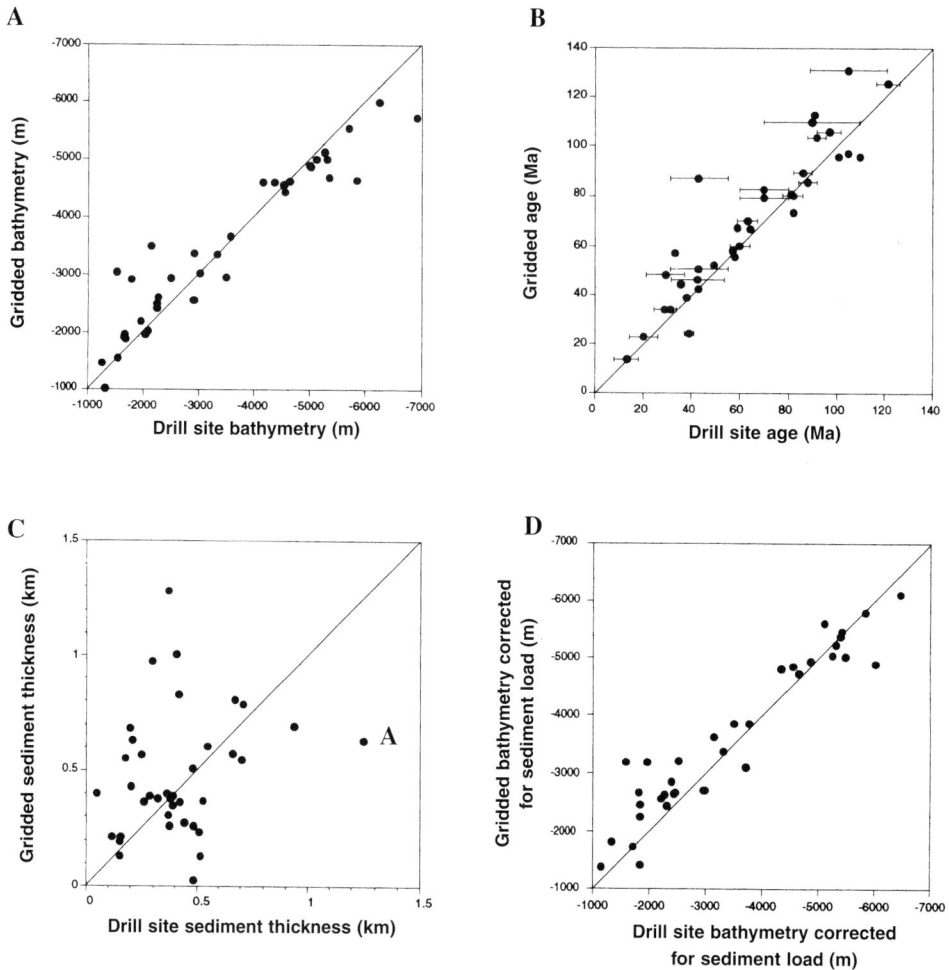

Fig. 24. Comparison of drill site versus gridded data for (A) bathymetry, (B) crustal age, (C) sediment thickness and (D) sediment-corrected bathymetry.

The largest depth differences between the sedimented (ETOPO5) and the sediment-free bathymetry occur at 0 Ma. The effect of these differences through geological time, within the reconstructions, has not been addressed because of the lack of stratigraphic control within the sediment thickness data. It was considered impractical to correlate between drill sites with an average spatial distribution of one drill site every 650 000 km² (Davies *et al.* 1995) or to calculate an assumed accumulation rate by dividing the total sediment thickness by the crustal age. The latter calculation is not applicable for deep-sea fans, e.g. Bengal and Indus Fans, in which the bulk of the sediments were deposited since the Himalayas started to rise in the early to middle Miocene (Kolla & Kidd 1982; Emmal & Curray 1985; Kolla & Coumes 1985).

Clearly these sediments accumulated long after the formation of the oceanic crust upon which they rest, thus the concept of average accumulation rates is invalid in this situation.

A statistical analysis of the sediment thickness data reveals that 12.5% of the sediment sequences are >1 km in thickness (Fig. 25) and that 0.5% of these are thicker than 4 km, and occur predominantly in the deeper ocean basins marginal to land. The maximum difference between true and sediment-free palaeobathymetry for the remaining 87.5% of the data is <600 m. This difference would decrease with the increasing age of the reconstruction and decreasing thickness of sediment which predates the age of reconstruction. The greatest discrepancies (>600 m) between true and sediment-free palaeobathymetry are therefore

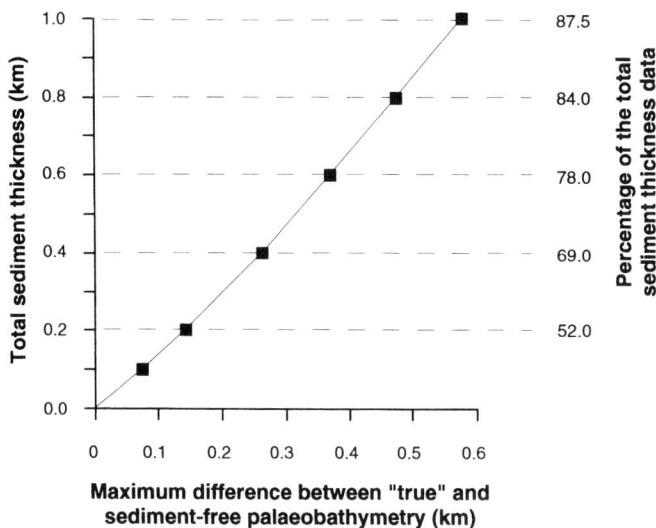

Fig. 25. Maximum difference between 'true' and sediment-free palaeobathymetry for sediment thickness less than 1 km. The difference is taken to be equivalent to the isostatic correction factor, calculated using the method of Sykes (1996).

associated with the thicker sediment sequences which rest on older ocean crust. The geographical distribution of the difference between modern and sediment-free modern bathymetry is shown in Fig. 26, i.e. equivalent to the maximum error between sediment-free palaeobathymetry and 'true' palaeobathymetry. The restoration of this crust to positions at or near a palaeomid-ocean ridge during palaeobathymetric reconstruction entails a considerable reduction in sediment thickness.

Finally, true palaeobathymetry implies that sediments older than the time of reconstruction are reloaded back onto the oceanic crust and decompacted (Wold 1992). Decompaction calculations require an understanding of the solidity, i.e. the inverse of porosity, through the sediment thickness data, in addition to stratigraphic control. Estimations of porosity, and hence solidity, could be attempted through use of a depth–density function (e.g. Sykes 1996); however, in the absence of stratigraphic control (see above) it is not considered worthwhile.

Conclusions

Palaeobathymetric reconstructions provide a useful tool for quantifying the depths of evolving oceanic seaways and gateways within the southern hemisphere, particularly for the palaeoceanographic interpretations of drill site data. Development of major oceanic gateways was related to either the fragmentation of Gondwana or the development

of hotspot-related submarine plateaux/ridges. Between 110 and 60 Ma the interconnection between the various deep basins of the South Atlantic and Indian Oceans was controlled by hotspot-related features.

The accuracy of these palaeobathymetric reconstructions was assessed by comparing the gridded data sets with similar data from 39 DSDP/ODP drill sites. Drill site bathymetry, crustal age and sediment-corrected bathymetry correlated positively with the gridded data extracted from the equivalent geographic positions of the drill sites, despite the lack of correlation between the two data sets for sediment thickness. The use of sediment-free palaeobathymetry does not limit the usefulness of the reconstructions because the largest differences (>600 m) with 'true' palaeobathymetry occur in the deeper basins associated with thicker sediments (>1 km, which accounted for only 12.5% of the data). These differences represent maxima at 0 Ma, which decrease with the increasing age of reconstruction.

One of the drawbacks in using DSDP/ODP drill site data as a control arises from the wide spacing of these sites. Nevertheless, the palaeobathymetric reconstructions provide a tool for interpreting these data. In addition, the palaeobathymetry data can be used to identify spatial and temporal histories of oceanic gateways and document the evolution of specific basins. They therefore provide a valuable aid in the location of drill sites for future drilling campaigns in the southern hemisphere.

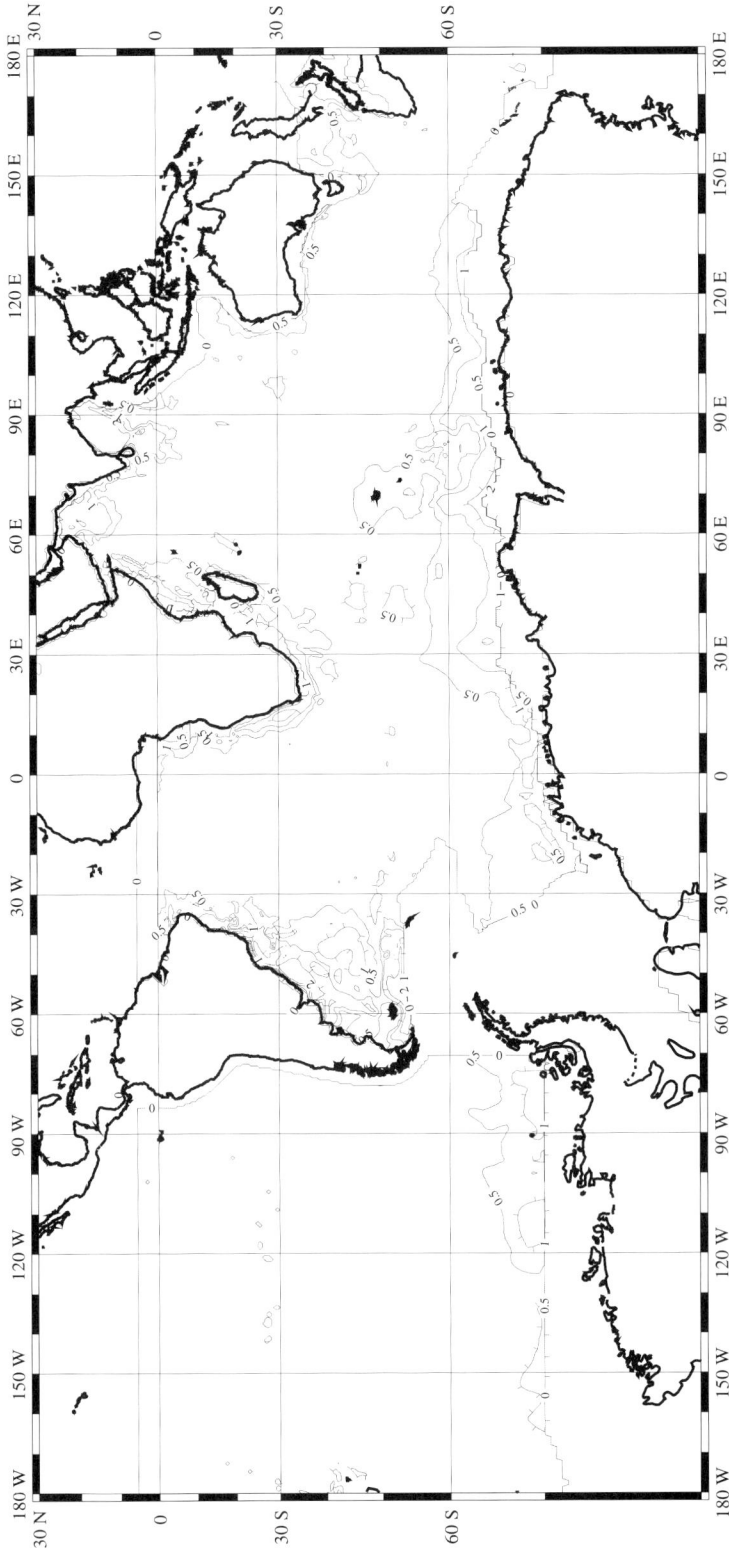

Fig. 26. Geographical distribution of the maximum difference between modern bathymetry and modern sediment-free bathymetry. Contours are in kilometres.

The authors were funded by Natural Environment Research Council ODP Special Topic grants GST-02434 and GST-02698 (Principal Investigators R. B. Kidd and A. T. S. Ramsay). The authors wish to thank N. Hamilton (Southampton Oceanography Centre) and an anonymous reviewer for their constructive criticism which greatly improved this manuscript. PALSOS contribution No. 2.

This paper is dedicated to the memory of Rob Kidd, who died 9th June 1996. This contribution is a consequence of his enthusiasm for all aspects of the Ocean Drilling Program and particularly the need to synthesize the information obtained through ocean drilling. Sykes, Ramsay and Royer would like to express their appreciation to Rob who was a great colleague and a good friend.

References

ANDREWS, P. B., GOSTIN, P. B., HAMPTON, M., MARGOLIS, S. V. & OVENSHINE, A. T. 1975. Synthesis - sediments of the southwest Pacific Ocean, southeast Indian Ocean and Tasman Sea. *In*: KENNETT, J. P., HOUTZ, R. E. *et al.* (eds) *Initial Reports of the Deep Sea Drilling Program*, **29**. US Govt Printing Office, Washington, 1147–1154.

BACKMAN, J., DUNCAN , R. A. *ET AL.* 1988. *Proceedings of the Ocean Drilling Program, Initial Reports*, **115**. Ocean Drilling Program, College Station, TX.

BARKER, P. F., CARLSON, R. C., JOHNSON, R L *ET AL.* 1983. *Initial Reports of the Deep Sea Drilling Program*, **72**. US Govt Printing Office, Washington.

BARRON, J., LARSEN, B. *ET AL.* 1989. *Proceedings of the Ocean Drilling Program, Initial Reports*, **119**. Ocean Drilling Program, College Station, TX.

BOLLI, H. M., RYAN, W. B. F. *ET AL.* 1978. *Initial Reports of the Deep Sea Drilling Program*, **40**. US Govt Printing Office, Washington.

CALCAGNO, P. & CAZENAVE, A. 1994. Subsidence of the seafloor in the Atlantic and Pacific Oceans: Regional and large scale variations. *Earth and Planetary Science Letters,* **126**, 473–492.

CANDE, S. C. & KENT, D. V. 1992. A new geomagnetic polarity timescale for the late Cretaceous and Cenozoic. *Journal of Geophysical Research,* **97** (B10), 13 917–13 951.

—— & RABINOWITZ, P. D. 1978. Mesozoic seafloor spreading bordering conjugate continental margins of Angola and Brazil. *Proceedings of Offshore Technology Conference*. Houston, TX, 1869–1874.

CIESIELSKI, P. F., KRISTOFFERSEN, Y. *ET AL.* 1988. *Proceedings of the Ocean Drilling Program, Initial Reports*, **114**. Ocean Drilling Program, College Station, TX.

COFFIN, M. F. 1992. Emplacement and subsidence of Indian Ocean plateaus and submarine ridges. *In*: DUNCAN, R. A., REA, D. K., KIDD, R. B., VON RAD, U. & WEISSEL, J. K. (eds) *Synthesis of Results from Scientific Drilling in the Indian Ocean*. Geophysical Monograph **70**, American Geophysical Union, Washington, DC, 115–126.

DAVIES, T. A. & KIDD, R. B. 1977. Sedimentation in the Indian Ocean through time. *In*: HEIRTZLER, J. R., BOLLI, H. M., DAVIES, T. A., SAUNDERS, J. B. & SCLATER, J. G. (eds) *Indian Ocean Geology and Biostratigraphy*. American Geophysical Union, Washington, DC, 61–86.

——, LUYENDYK, B. P. *ET AL.* 1974. *Initial Reports of the Deep Sea Drilling Program*, **26**. US Govt Printing Office, Washington.

——, KIDD, R. B. & RAMSAY, A. T. S. 1995. A time-slice approach to the history of Cenozoic sedimentation in the Indian Ocean. *Sedimentary Geology,* **96**, 157–179.

DETRICK, R. S., SCLATER, J. G. & THIEDE, J. 1977. The subsidence of aseismic ridges. *Earth and Planetary Science Letters,* **34**, 185–196.

DIVINS, D. L. & RABINOWITZ, P. D. 1991. Total sediment thickness map of the South Atlantic Ocean. *In*: *International Geological and Geophysical Atlas of the Atlantic and Pacific Oceans (GAPA)*. Intergovernmental Oceanographic Committee, 43–44.

DRISCOLL, N. W., KARNER, G. D., WEISSEL, J. K. & SHIPBOARD SCIENTIFIC PARTY. 1989. Stratigraphic and tectonic evolution of Broken Ridge from seismic stratigraphy and Leg 121 drilling. *In*: PEIRCE, J., WEISSEL, J. K. *ET AL.* (eds) *Proceedings of the Ocean Drilling Program, Initial Reports*, **121**. Ocean Drilling Program, College Station, TX, 71–91.

DUNCAN, R. A. 1978. Geochronology of basalts from the Ninetyeast Ridge and continental dispersion in the eastern Indian Ocean. *Journal of Volcanology and Geothermal Research,* **4**, 283–305.

—— 1991. Age distribution of volcanism along aseismic ridges in the eastern Indian Ocean. *In*: WEISSEL, J., PEIRCE, J., TAYLOR, E., ALT, J. *ET AL.* (eds) *Proceedings of the Ocean Drilling Program, Scientific Results*, **121**. Ocean Drilling Program, College Station, TX, 507–517.

—— & HARGRAVES, R. B. 1990. 40Ar/39Ar geochronology of basement rocks from the Mascarene Plateau, the Chagos Bank, and the Maldives Ridge. *In*: DUNCAN, R. A., BACKMAN, J., PETERSON, L. C. *ET AL.* (eds) *Proceedings of the Ocean Drilling Program, Scientific Results*, **115**. Ocean Drilling Program, College Station, TX, 43–51.

EMMAL, F. J. & CURRAY, J. R. 1985. Bengal Fan, Indian Ocean. *In*: BOUMA, A. H., NORMARK, W. R. & BARNES, N. E. (eds) *Submarine Fans and Related Turbidite Systems*. Springer-Verlag, New York, 107–112.

EWING, J. & EWING, M. 1970. Seismic reflections. *In*: MAXWELL, A. E. (ed.) *The Sea (Volume 4): Ideas and observations on progress in the study of the sea*. Wiley-Interscience, New York, 1–52.

GRADSTEIN, F. M., LUDDEN, J. N. *ET AL.* 1990. *Proceedings of the Ocean Drilling Program, Initial Reports*, **123**. Ocean Drilling Program, College Station, TX.

HARLAND, W. B., ARMSTRONG, R. L., COX, A. V., CRAIG, L. E., SMITH, A. G. & SMITH, D. G. 1989. *A Geological Timescale*. Cambridge University Press, New York.

HAYES, D. E. 1988. Age–depth relationships and depth anomalies in the southeast Indian Ocean and South Atlantic Ocean. *Journal of Geophysical Research,* **93** (B4), 2937–2954.

—— 1992. Tectonics and age of the oceanic crust: Circum-Antarctic to 30°S. *In*: HAYES, D. E. (ed.) *Marine Geological and Geophysical Atlas of the Circum-Antarctic to 30°S,* **54**. American Geophysical Union, Washington, DC, 47-56.

—— & KANE, K. A. 1994. Long-lived mid-ocean ridge segmentation of the Pacific–Antarctic ridge and the Southeast Indian Ridge. *Journal of Geophysical Research,* **99**(B10), 19 679–19 692.

—— & LABRECQUE, J. L. 1991. Sediment isopachs: Circum-Antarctic to 30°S. *In*: HAYES, D. E. (ed.) *Marine Geological and Geophysical Atlas of the Circum-Antarctic to 30°S.* American Geophysical Union, Washington, DC, 29–36.

——, FRAKES, L. A. *ET AL.* 1975. *Initial Reports of the Deep Sea Drilling Program,* **28**. US Govt Printing Office, Washington.

HEEZEN, B. C. & EWING, M. 1963. The mid-oceanic ridge. *In*: HILL, M. N. (ed.) *The Sea (Volume 3).* Wiley-Interscience, New York, 388–410.

HEGARTY, K. A., WEISSEL, J. K. & MUTTER, J. C. 1988. Subsidence history of Australia's southern margin. *American Association of Petroleum Geologists' Bulletin,* **74**, 615–633.

HOLLISTER, C. D., CRADDOCK, C. *ET AL.* 1976. *Initial Reports of the Deep Sea Drilling Program,* **35**. US Govt Printing Office, Washington.

HOUTZ, R. E., EWING, M., HAYES, D. E. & NAINI, B. 1973. *Sediments, Antarctic Map Folio Series, Folio 17, Plate 5.* American Geographical Society, Washington, DC.

KANE, K. A. & HAYES, D. E. 1992. Tectonic corridors in the South Atlantic: Evidence for long-lived mid-ocean ridge segmentation. *Journal of Geophysical Research,* **97**(B12), 17 317–17 330.

—— & HAYES, D. E. 1994. A new relationship between subsidence rate and zero-age depth. *Journal of Geophysical Research,* **99**(B11), 21 759–21 777.

KENNETT, J. P. 1982. *Marine Geology.* Prentice-Hall, Englewood Cliffs, NJ.

——, BURNS, R. E., ANDREW, J. E. *ET AL.* 1972. Australian–Antarctic continental drift, paleo-circulation changes and Oligocene deep-sea erosion. *Nature,* **239**, 51–55.

——, HOUTZ, R. E. *ET AL.* 1975a. *Initial Reports of the Deep Sea Drilling Program,* **29**. US Govt Printing Office, Washington.

——, Houtz, R. E., Andrews, P. B.*ET AL.* 1975b. Cenozoic paleoceanography in the southwest Pacific Ocean, Antarctic glaciation and the development of the Cirucm-Antarctic current. *In*: KENNETT, J. P., HOUTZ, R. E. *ET AL.* (eds) *Initial Reports of the Deep Sea Drilling Program,* **29**. US Govt Printing Office, Washington, 1155–1170.

KENT, D. V. & GRADSTEIN, F. M. 1986. A Jurassic to Recent timescale. *In*: VOGT, P. R. & TUCHOLKE, B. E. (eds) *The Geology of North America, Volume M: The western Atlantic region.* Geological Society of America, Boulder, CO, 45–58.

KIDD, R. B. & DAVIES, T. A. 1978. Indian Ocean sediment distribution through since the Late Jurassic. *Marine Geology,* **26**, 49–70.

——, RAMSAY, A. T. S., SYKES, T. J. S., BALDAUF, J. G., DAVIES, T. A., JENKINS, D. G. & WISE, S. W., JR. 1992. An Indian Ocean framework for paleoceanographic synthesis based on DSDP and ODP results. *In*: DUNCAN, R. A., REA, D. K., KIDD, R. B., VON RAD, U. & WEISSEL, J. K. (eds) *Synthesis of Results from Scientific Drilling in the Indian Ocean.* Geophysical Monograph **70**, American Geophysical Union, Washington, DC, 403–422.

KOLLA, V. & COUMES, F. 1985. Indus Fan, Indian Ocean. *In*: BOUMA, A. H., NORMARK, W. R. & BARNES, N. E. (eds) *Submarine Fans and Related Turbidite Systems.* Springer-Verlag, New York, 129–136.

—— & KIDD, R. B. 1982. Sedimentation and sedimentary processes in the Indian Ocean. *In*: NAIRN, A. E. M. & STELHI, F. G. (eds) *The Indian Ocean,* **6**. Plenum Press, New York, 1–50.

LAWVER, L. A., GAHAGAN, L. M. & COFFIN, M. F. 1992. The development of paleoseaways around Antarctica. *In*: KENNETT, J. P. & WARNKE, D. A. (eds) *The Antarctic Paleoenvironment: a perspective on global change.* Antarctic Research Series **56**, American Geophysical Union, Washington, DC, 7–30.

LUDDEN, J. N. 1992. Radiometric age determinations for basement from Sites 765 and 766, Argo Abyssal Plain and northwestern Australian Margin. *In*: GRADSTEIN, F. M., LUDDEN, N. J. *ET AL.* (eds) *Proceedings of the Ocean Drilling Program, Scientific Results,* **123**. Ocean Drilling Program, College Station, TX, 557–559.

LUDWIG, W. J., KRASHENENINOV, V. A. *ET AL.* 1983. *Initial Reports of the Deep Sea Drilling Program,* **71**. US Govt Printing Office, Washington.

MARKS, K. M. & STOCK, J. M. 1994. Variations in ridge morphology and depth–age relationships on the Pacific–Antarctic ridge. *Journal of Geophysical Research,* **99**(B1), 531–541.

MARTY, J. C. & CAZENAVE, A. 1989. Regional variations in subsidence rate of oceanic plates: A global analysis. *Earth and Planetary Science Letters,* **94**, 301–315.

MATHIAS, P. K., RABINOWITZ, P. D. & DIPIAZZA, N. 1988. *Sediment Thickness Map of the Indian Ocean.* The American Association of Petroleum Geologists, Tulsa, OK.

MOUGENOT, D., RAILLARD, S. & VIRLOGEUX, P. 1987. Structure de la marge et du canal de Mozambique. *Actes du Colloque sur la Recherche Francaise dans les Terre Australes,* 323–336.

MUSSETT, A. E. & BARKER, P. F. 1983. 40Ar/39Ar age spectra of basalts, Deep Sea Drilling Project Site 516. *In*: BARKER, P. F., CARLSON, R. C., JOHNSON, R. L. *ET AL.* (eds) *Initial Reports of the Deep Sea Drilling Program,* **72**. US Govt Printing Office, Washington, 467–470.

NGDC. 1988. ETOPO-5 bathymetry/topography data.

PARSONS, B. & SCLATER, J. G. 1977. An analysis of the variation of ocean floor bathymetry and heat flow with age. *Journal of Geophysical Research,* **82**(5), 803–827.

PEIRCE, J., WEISSEL, J. *ET AL.* 1989. *Proceedings of the*

Ocean Drilling Program, Initial Reports, **121**. Ocean Drilling Program, College Station, TX.

RAMSAY, A. T. S. 1977. The distribution of calcium carbonate in deep sea sediments. *In*: HAY, W. A. (ed.) *Studies in palaeo-oceanography*. Special Publication **20**, Society of Economic Paleontologists and Mineralogists, Tulsa, OK, 58–76.

——, SYKES, T. J. S. & KIDD, R. B. 1994. Waxing (and waning) lyrical on hiatuses: Eocene–Quaternary Indian Ocean hiatuses as proxy indicators of water mass production. *Paleoceanography,* **9**(6), 857–977.

REA, D. K., DEHN, J., DRISCOLL, N. W. *ET AL.* 1990. Paleoceanography of the eastern Indian Ocean from ODP Leg 121 drilling on Broken Ridge. *Geological Society of America Bulletin,* **102**, 679–690.

ROYER, J.-Y., MÜLLER, R. D., GAHAGAN, L. M., LAWVER, L. A., MAYES, C. L., NÜRNBERG, D. & SCLATER, J. G. 1992*a*. *A global isochron chart*. University of Texas Institute for Geophysics, Technical Report No. **117**.

——, SCLATER, J. G., SANDWELL, D. T. *ET AL.* 1992*b*. Indian Ocean plate reconstructions since the late Jurassic. *In*: DUNCAN, R. A., REA, D. K., KIDD, R. B., VON RAD, U. & WEISSEL, J. K. (eds) *Synthesis of Results from Scientific Drilling in the Indian Ocean*. Geophysical Monograph **70**, American Geophysical Union, Washington DC, 471–475.

RUDLE, C. C., BROOKS, M., SNELLING, N. J., REYNOLDS, P. H. & BARR, S. M. 1974. Radiometric age determinations. *In*: DAVIES, T. A., LUYENDYK, B. P. *ET AL.* (eds) *Initial Reports of the Deep Sea Drilling Program*, **26**. US Govt Printing Office, Washington, 513–516.

RYAN, W. B. F. 1978. Objectives, principal results, operations, and explanatory notes of Leg 40, South Atlantic. *In*: BOLLI, H. M., RYAN, W. B. F. *ET AL.* (eds) *Initial Reports of the Deep Sea Drilling Program*, **40**. US Govt Printing Office, Washington, 5–28.

SCHEIRER, D. S. & MACDONALD, K. C. 1995. Near-axis seamounts on the flanks of the East Pacific Rise, 8°N to 17°N. *Journal of Geophysical Research,* **100** (B2), pp. 2239-2259.

SCHLICH, R., WISE, S. W., JR. *ET AL.* 1989. *Proceedings of the Ocean Drilling Program, Initial Reports*, **120**. Ocean Drilling Program, College Station, TX.

SCLATER, J. G. & FISHER, R. L. 1974. The evolution of the east central Indian Ocean. *Bulletin of the Geological Society of America,* **85**, 683–702.

—— & MCKENZIE, D. P. 1973. Paleobathymetry of the South Atlantic. *Geological Society of America Bulletin,* **84**, 3203–3216.

——, ABBOTT, D. & THIEDE, J. 1977*a*. Paleobathymetry and sediments of the Indian Ocean. *In*: HEIRTZLER, J. R., BOLLI, H. M., DAVIES, T. A., SAUNDERS, J. B. & SCLATER, J. G. (eds) *Indian Ocean Geology and Biostratigraphy*. American Geophysical Union, Washington, DC, 25–60.

——, HELLINGER, S. & TAPSCOTT, C. 1977*b*. The paleobathymetry of the Atlantic Ocean from the Jurassic to the Present. *The Journal of Geology,* **85**, 509–552.

——, MEINKE, L., BENNETT, A. & MURPHY, C. 1985. The depth of the ocean through the Neogene. *In*: KENNETT, J. P. (ed.) *The Miocene Ocean*. Geological Society of America, 1–19.

SIMPSON, E. S. W., SCHLICH, R. *ET AL.* 1974. *Initial Reports of the Deep Sea Drilling Program*, **25**. US Govt Printing Office, Washington.

SMITH, W. H. F. 1993. On the accuracy of digital bathymetry data. *Journal of Geophysical Research,* **98**, 9591–9603.

—— & SANDWELL, D. T. 1994. Bathymetric prediction from dense satellite altimetry and sparse shipboard bathymetry. *Journal of Geophysical Research,* **99**(B11),21 803–21 824.

—— & WESSEL, P. 1990. Gridding with continuous curvature splines in tension. *Geophysics,* **55**(3), 293–305.

STEIN, C. A. & STEIN, S. 1992. A model for the global variation in oceanic depth and heat flow with lithospheric age. *Nature,* **359**, 123–129.

SYKES, T. J. S. 1995. *Palaeobathymetry of the southern hemisphere*. PhD thesis, University of Wales College Cardiff.

—— 1996. A correction for sediment load upon the ocean floor: Fixed versus varying sediment density estimations – implications for isostatic correction. *Marine Geology*.

—— & ROYER, J.-Y. *under review*. Continuous along ridge age-depth analyses: A southern hemisphere perspective. *Marine Geology*.

——, RAMSAY, A. T. S. & KIDD, R. B. 1998. Southern hemisphere Miocene bottom-water circulation: a palaeobathymetric analysis. *This volume*.

THUROW, J., BRUMSACK, H.-J., RULLKOTTER, J., LITTKE, R. & MEYERS, P. 1992. The Cenomanian/Turonian boundary event in the Indian Ocean - a key to understanding the global picture. *In*: DUNCAN, R. A., REA, D. K., KIDD, R. B., VON RAD, U. & WEISSEL, J. K. (eds) *Synthesis of Results from Scientific Drilling in the Indian Ocean*. Geophysical Monograph **70**, American Geophysical Union, Washington, DC, 253–273.

VAN ANDEL, T., THIEDE, J., SCLATER, J. G. & HAY, W. W. 1977. Depositional history of the South Atlantic Ocean during the last 125 million years. *Journal of Geology,* **85**, 651–698.

VEEVERS, J. J., HIERTZLER, J. R. *ET AL.* 1974. *Initial Reports of the Deep Sea Drilling Program*, **27**. US Govt Printing Office, Washington.

VON DER BORCH, C. C., SCLATER, J. G. *ET AL.* 1974. *Initial Reports of the Deep Sea Drilling Program*, **22**. US Govt Printing Office, Washington.

WEISSERT, H. 1981. The environment of deposition of black shales in the early Cretaceous: An ongoing controversy. *In*: WARME, J. E., DOUGLAS, R. G. & WINTERER, E. L. (eds) *The Deep Sea Drilling Project: A decade of progress*. Special Publication **32**, Society of Economic Paleontologists and Mineralogists, Tulsa, OK, 547–560.

WESSEL, P. & SMITH, W. H. F. 1991. Free software helps map and display data. *EOS (Transactions, American Geophysical Union),* **72** (441), 445–446.

WHITCHURCH, H., MONTIGNY, R., SEVIGNY, R., STOREY, M. & SALTERS, V. 1992. K-Ar and 40Ar/39Ar ages of central Kerguelen Plateau basalts. *In*: WISE, S. W. J.,

SCHLICH, R. *ET AL.* (eds) *Proceedings of the Ocean Drilling Program, Scientific Results*, **120**. Ocean Drilling Program, College Station, TX, 71–77.

WOLD, C. N. 1992. *Paleobathymetry and sediment accumulation in the northern North Atlantic and southern Greenland–Iceland–Norwegian Sea.* PhD thesis, GEOMAR.

ZIEGLER, A., SCOTESE, C. R. & BARRET, S. F. 1983. Mesozoic and Cenozoic paleogeographic maps. *In*: BROCHE & SUNDERMANN (eds) *Tidal Friction and the Earth's Rotation II*. Springer-Verlag, Berlin, 240–252.

Southern hemisphere Miocene bottom-water circulation: a palaeobathymetric analysis

T. J. S. SYKES[1,2], A. T. S. RAMSAY[1] & R. B. KIDD[1]

[1] *Marine Geosciences Research Group, Department of Earth Sciences, UWCC, PO Box 914, Cardiff, CF1 3YE, UK*

[2] *Present address: Landmark EAME Ltd, 4 Albert Street, Aberdeen, AB25 1XQ, UK*

Abstract: Palaeobathymetric reconstructions of the southern hemisphere are used to assess the potential flow paths and extent of proto-Antarctic Bottom Water (proto-AABW) for three time slices during the Miocene: 20 Ma (early), 15 Ma (middle) and 10 Ma (late). The depth ranges of fluctuations in this water mass were derived from its waxing/waning curve, based on an analysis of hiatuses within drill sites from the Indian Ocean. Proto-AABW was likely to be expressed at depths below –4400 m, –5500 m and –4700 m at 20 Ma, 15 Ma and 10 Ma respectively. These depth ranges were extended into the South Atlantic and South Pacific Oceans to allow a first approximation of potential proto-AABW flow paths. An understanding of modern oceanography and the distribution of suspended sediment particle concentrations within the nepheloid layer and sediment drifts were used to constrain flow paths. This interpretation was made on the basis that proto-Antarctic Bottom Water was an analogue of it modern equivalent and was likely to have followed similar flow paths.

The palaeobathymetric analysis reveals that at 20 Ma (early Miocene) the deep basins of the southern hemisphere were well connected and proto-AABW flow paths were likely to have been widespread. By 15 Ma (middle Miocene), however, the waning of proto-AABW led to the disconnection of the deep ocean basins and this water mass was probably restricted to the Antarctic Basin. Waxing of proto-AABW during the late Miocene allowed the reconnection of the deep basins but not to the same extent as at 20 Ma (early Miocene).

The palaeobathymetric reconstruction of the southern hemisphere (Sykes *et al.* 1998) provides a useful tool for palaeoceanographic investigations. Here we use palaeobathymetric time slices at 20 Ma, 15 Ma and 10 Ma (corresponding to early, middle and late Miocene respectively) to assess the impact of fluctuations in the production and vertical extent of proto-Antarctic Bottom Water (proto-AABW) in the Indian, South Atlantic and Pacific Oceans. These fluctuations were inferred from the palaeodepth and chronostratigraphical distribution of hiatuses from Ocean Drilling Program (ODP) and Deep Sea Drilling Project (DSDP) cores from the Indian Ocean (Ramsay *et al.* 1994). These authors defined 'areas of potential hiatus formation' (APHiDs) determined from palaeogeographical reconstructions, subsidence curves for individual drill sites and an understanding of modern bathymetry and deep-water oceanography. The APHiD concept is complex since it embraces erosion, transport and deposition of sediment within the flow path of a bottom water mass. It also embraces the possibility of enhanced dissolution of calcium carbonate through increased proto-AABW activity.

The vertical stratification of the oceans and water-mass movements through time have been determined by examination of the depth–geographical variation of isotopes of carbon ($\partial^{13}C$) derived from benthic foraminifers (e.g. Woodruff & Savin 1989; Wright *et al.* 1992). The dissolution of calcareous foraminifers below the CCD, however, precluded an analysis of bottom-water stratification at palaeodepths below 3.0–3.5 km during the Miocene in the southern hemisphere. In addition an understanding of the modern vertical structure of water masses (Tchernia 1980; Mantylla & Reid 1983), contemporary measurements of suspended sediment particles (SSPs) within the nepheloid layer (Kolla *et al.* 1976; Biscaye & Eittreim 1977) and the distribution of sediment drifts (Johnson & Damuth 1979; Kolla & Kidd 1982; Masson *et al.* 1982; Johnson *et al.* 1983; Kidd & Hill 1986; McCave & Tucholke 1986; Wold 1992; Carter &

SYKES, T. J. S., RAMSAY, A. T. S. & KIDD, R. B. 1998. Southern hemisphere Miocene bottom-water circulation: a palaeobathymetric analysis. *In*: CRAMP, A., MACLEOD, C. J., LEE, S. V. & JONES, E. J. W. (eds) *Geological Evolution of Ocean Basins: Results from the Ocean Drilling Program.* Geological Society, London, Special Publications, **131**, 43–54.

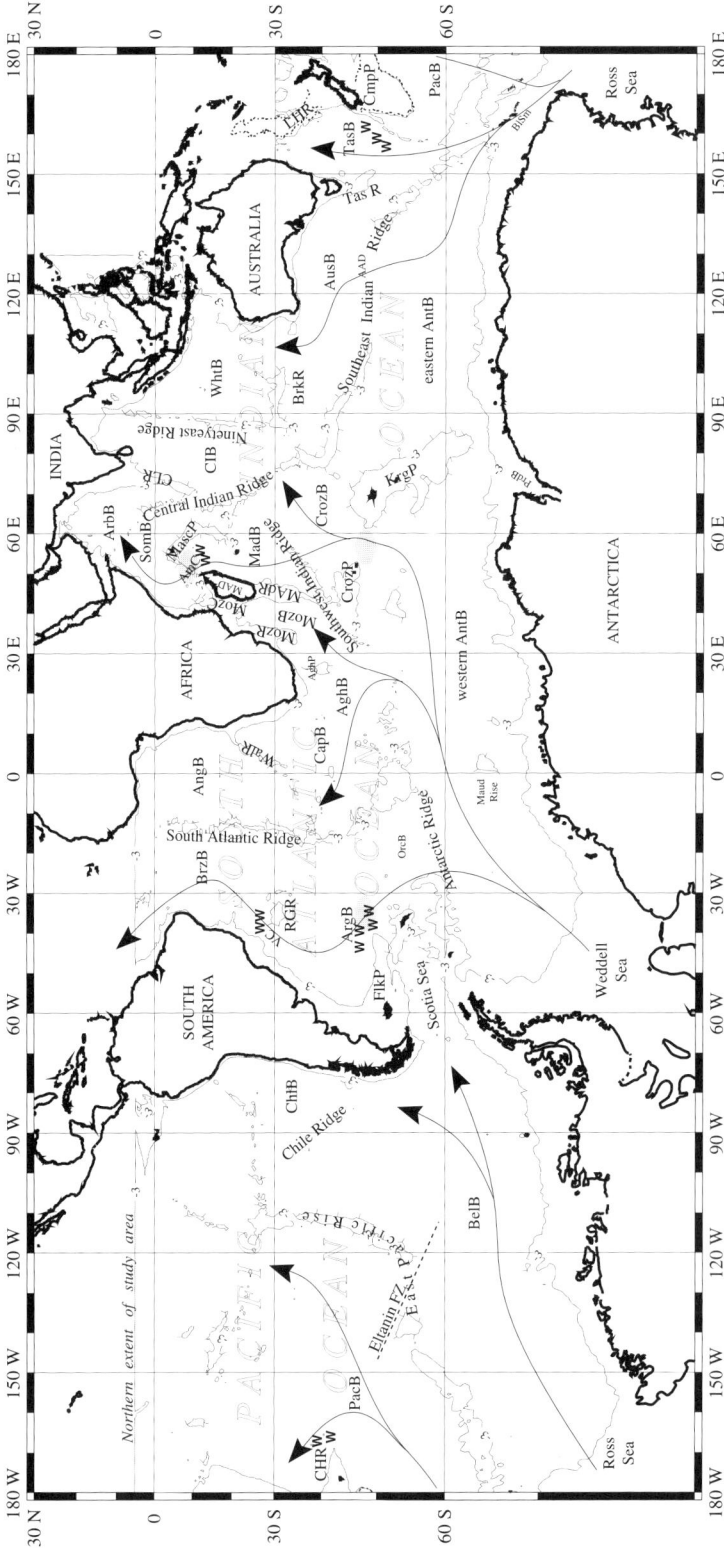

Fig. 1. Location of features mentioned in text: −3 km isobath is shown. W indicates locations of suspended sediment particle measurement. Arrows indicate the flow paths of modern AABW. Abbreviations: AAD, Antarctic–Australian Discordance; AghB, Agulhas Basin; AghP, Agulhas Plateau; AmC, Amirante Channel; AngB, Angola Basin; AntB, Antarctic Basin; ArbB, Arabian Basin; ArgB, Argentine Basin; AusB, Australian Basin; BelB, Bellinghausen Basin; BlSm, Bellany Seamounts; BrB, Brazil Basin; BrkR, Broken Ridge; CmpP, Campbell Plateau; CapB, Cape Basin; ChlB, Chile Basin; CHR, Chattam Rise; CIB, Central Indian Basin; CLR, Chagos–Laccadive Ridge; CrozB, Crozet Basin; CrozP, Crozet Plateau; FlkP, Falkland Plateau; KrgP, Kerguelen Plateau; LHR, Lord Howe Rise; MAD, Madagascar; MadB, Madagascar Basin; MadR, Madagascar Ridge; MascP, Mascarene Plateau; MozB, Mozambique Basin; MozC, Mozambique Channel; MozR, Mozambique Ridge; OrcB, Orcadas Basin; PacB, Pacific Basin; PrdB, Prydz Bay; RGR, Rio Grande Rise; SomB, Somali Basin; TasB, Tasman Basin; TasR, Tasman Rise; VC, Vema Channel; WalR, Walvis Ridge; WhtB, Wharton Basin.

McCave 1994; Jones 1994; Masse *et al.* 1994) within the southern hemisphere (Fig. 1) are used to constrain the interpretations.

Method

The geographical extent of proto-AABW was inferred by utilizing depth ranges derived from the waxing/waning proto-AABW curve (Fig. 2) in conjunction with palaeobathymetric reconstructions. Based on this curve it was assumed that proto-AABW was expressed at depths below –4400 m, –5500 m and –4700 m at 20 Ma, 15 Ma and 10 Ma respectively. Depths exceeding these values are shaded on palaeobathymetric maps for these time interval, which also show 1 km palaeoisobaths.

The method used here, i.e. defining the depth of proto-AABW from a single depth range, is simplistic. The depth range of Miocene proto-AABW, like its modern analogue AABW, probably decreased with increasing distance from Antarctica. In the modern Antarctic Basin, south of *c.* 55–60°S, AABW is found at depths below –3000 m and rapidly attains depths below –4000 m by 20°N within the Indian and South Atlantic Oceans (Tchernia, 1980; Mantylla & Reid 1983). The waxing/waning curve presented in Fig. 2 represents a whole-ocean curve, as it was derived from deep ocean drill sites spread across the Indian Ocean between the equator and 45°S. Despite its simplicity the method provides a first approximation of the maximum vertical extent of proto-AABW for the individual time slices.

Modern AABW formation and flow paths

Modern AABW is formed on the shelf areas of the Weddell and Ross Seas by the super cooling of the water under the floating ice shelves (Foldvik &

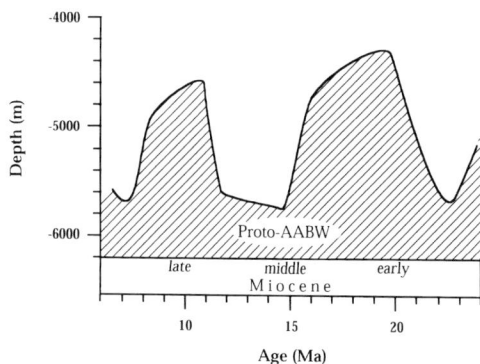

Fig. 2. Depth range of proto-AABW during the Miocene. Modified from Ramsay *et al.* (1994*b*).

Gammelsrod 1988). This becomes more dense and flows northward over the shelf edge and enters the deep basins around Antarctica (Tchernia 1980; Mantylla & Reid 1983).

Weddell Sea AABW flows northwards from the western Antarctic Basin and follows three distinct flow routes. Firstly, Weddell Sea AABW flows through the western South Atlantic and is relatively unimpeded except to the west of the Rio Grande Rise where flow is directed through the Vema Channel. The southern parts of the Argentine and Brazil Basins, which are in close proximity to the points of exit of AABW from confined channels, are associated with the formation of some of the largest sediment waves identified (Jones 1994; Masse *et al.* 1994) and with high levels of SSPs in the nepheloid layer (Biscaye & Eittreim 1977). These observations indicate long-term sediment transport and accumulation under relatively strong currents. Secondly, Weddell Sea AABW crosses the Southwest Indian Ridge, via fracture zones, into the Agulhas Basin where it spreads to the northeast and northwest. The shallow nature of the Mozambique Channel, between Madagascar and Africa, and the Walvis Ridge define the northward limit of the spread of AABW in the deep basins around the southern tip of Africa. However, Shannon & Chapman (1991) have detected the slow leakage of AABW through the Walvis Ridge. Finally, Weddell Sea AABW enters the western Indian Ocean via the Crozet Basin and across fracture zones in the Southwest Indian Ridge and flows into the Madagascar and Somali Basins, entering the latter via Amirante Channel. High concentrations of SSPs are found in the southern Crozet Basin (Kolla *et al.* 1976) and Amirante Channel at depths below –3800 m (Johnson & Damuth 1979; Masson *et al.* 1982; Johnson *et al.* 1983).

Sluggish easterly directed currents within the deep (below –4000 m) passage between the southern Kerguelen Plateau and Antarctica (Rodman & Gordon 1982) and the fact that bottom water to the west of the Kerguelen Plateau is colder and less saline than those to the east (Warren 1978) indicate little intermixing of Weddell and Ross Sea derived AABW.

Ross Sea AABW flows directly northwards through the Tasman Sea, into the Bellinghausen and Pacific Basins and westwards into the eastern Antarctic Basin (as far as the Kerguelen Plateau) (Rodman & Gordon 1982; Mantylla & Reid 1983). From the eastern Antarctic Basin this flow crosses the Southeast Indian Ridge via the Antarctic–Australian Discordance through the Australian Basin and finally into the Wharton Basin. Analyses of the age of surficial sediments at the entrance to the Wharton Basin indicate that the removal of sediments (Kennett *et al.* 1975) was most likely associated with erosion and transport by AABW. The modern southeastern Tasman Basin is characterized by sediment drifts, which are interpreted to have formed since the onset of circum-Antarctic circulation in the late Eocene/early Oligocene (Jenkins 1992). Sediment drifts which have been recognized along the eastern base of the Campbell Plateau are interpreted to have been formed by the flow of AABW during the Miocene (Carter & McCave 1994).

Results and discussion

The palaeobathymetric reconstructions show that all the major gateways utilized by the modern AABW were available for flow of proto-AABW. Thus the modern AABW flow routes, the distribution of sediment drifts and areas of high SSPs combined with the waxing/waning history of proto-AABW during the Miocene can be used to delimit the extent of proto-AABW. Within the deep ocean basins sediment drifts obviously define areas subject to long-term sediment deposition within the bottom-water flow.

20 Ma time slice (early Miocene) (Fig. 3)

At 20 Ma the production of proto-AABW was at a peak with flow expressed at depths greater than -4400 m within the Indian Ocean (Fig. 2). The time interval between 22.5 and 18.2 Ma is thought to be characterized by repeated advances/retreats of ice across the Antarctic shelves, inferred from seismic studies in the Ross Sea (Anderson & Bartek 1992). The fluctuation of $\partial^{18}O$ values during this time interval are also interpreted as evidence for changes in the growth of Antarctic ice sheets (Miller *et al.* 1991). The accumulation of subglacial deposits within the Ross Sea (Anderson & Bartek 1992) clearly indicates that this interval was not subject to continuous ice-grounding across the shelf. Consequently the production of proto-AABW could have been enhanced beneath the ice sheets. The waxed state of this water-mass is also indicated by the shallow depth (–3800 to –4000 m) of the CCD in the equatorial Indian Ocean (Peterson & Backman 1990; Peterson *et al.* 1992), which may be due in part to the upward mixing of cold corrosive bottom waters. During the early Miocene large areas of ocean floor in the southern hemisphere were at depths greater than –4400 m and therefore potentially under the influence of proto-AABW (Figure 3).

Western Indian Ocean. Ramsay *et al.* (1994) inferred that bottom water in the western Indian Ocean was derived via the Agulhas Basin (their

figures 6 and 7) across the southern Madagascar Ridge into the Madagascar Basin. The palaeobathymetric analysis, however, indicates that although the Mozambique Basin could have been flooded by proto-AABW, the presence of the Madagascar Ridge in the east would have acted as a barrier to bottom-water flow. It is unlikely that the Mozambique Channel acted as a deep-water conduit during the early Miocene because it never exceeded –4000 m in depth (Mougenot *et al.* 1987). Bottom waters within the Madagascar Basin were more likely to have flowed via the Crozet Basin through the Southwest Indian Ridge as at the present (Fig. 1). Proto-AABW could also clearly have flowed through the Madagascar Basin into the Somali and even into the Arabian Basin. The connection of the Crozet Basin with the Antarctic Basin at 20 Ma was relatively narrow (400–450 km) and at palaeodepths of between –4300 and –4500 m. Waning of the proto-AABW by as little as 500 m would have isolated these two basins from the flow of bottom water.

The development of hiatuses at basinal DSDP Sites 240 and 245, which encompass the Miocene and the early Miocene at Site 235, indicate that eroding bottom waters may have utilized this flow path. The interpretation of bottom-water flow based solely on the presence of hiatuses, however, is dubious, as indicated by Ramsay *et al.* (1994).

Johnson *et al.* (1983) suggested that the presence of extensive sand waves within the Amirante Channel indicates the occurrence of active pre-early Miocene bottom-water flow. This view concurs with the interpretation made here for the flow of proto-AABW in the western Indian Ocean during the early Miocene. The development of sediment waves within the Madagascar and Somali Basins is not widespread. Seismic lines near DSDP Sites 239, 240, 241 and 245 (Simpson *et al.* 1974) do not show the occurrence of large bedforms indicative of bottom-water flow. Examination of the sediment isopach map for the Indian Ocean at 45°S and 55°E (Mathias *et al.* 1988) reveals a tongue of sediments extending northwards into the southern part of the Crozet Basin, which may have formed under the influence of proto-AABW. The timing of this sediment accumulation is unknown but occurred post-63 Ma, the time of initial formation of this basin (Sykes 1995). At present the high levels of SSPs recorded in the southern entrance of the Crozet Basin (Fig. 1) probably contribute to the development of sediment drifts. We propose that this AABW gateway has a long history of sediment drift accumulation.

Eastern Indian Ocean. The palaeobathymetry shows that the mean depth of the Australian–Antarctic Discordance (AAD) at 20 Ma lay

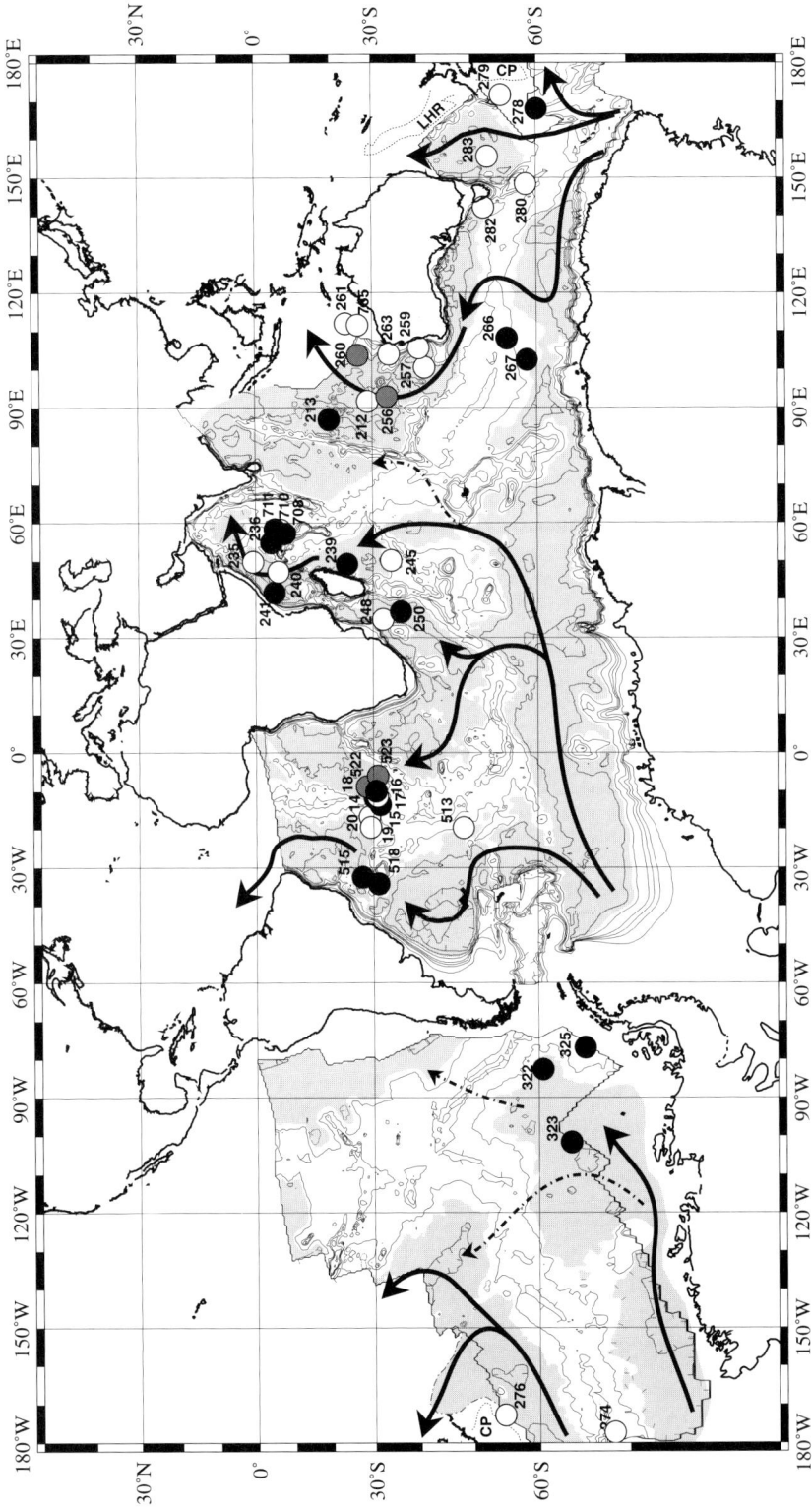

Fig. 3. Extent and flow paths of proto-AABW during the early Miocene (20 Ma). ODP and DSDP drill sites are shown, where solid, shaded and open circles indicate complete sedimentary sequence, condensed sedimentary sequence and hiatus respectively. Proto-AABW flow paths are indicated by solid arrows when constrained, or dashed when speculative (see text for discussion). The shaded area represents the maximum extent of proto-AABW.

between −4000 m and −4200 m. Although this is shallower than the depth predicted by the waxing/waning curve for proto-AABW, it is probable that fracture zones that are too small to be represented at the scale of the reconstruction formed conduits for bottom water. The palaeobathymetric reconstruction also shows that the deep connection (below −4000 m) between the Australian and Wharton Basins was less than *c.* 350 km wide.

The development of hiatuses and condensed sequences at drill sites within the Wharton Basin during the early Miocene and the fact that the eastern Wharton Basin is characterized by sediment sequences up to 800 m thick (most noticeably between the Naturaliste Plateau and Broken Ridge) suggest that this region occurred within an APHiD, i.e. an area characterized by bottom-water flow in which erosion (including dissolution), sedimentation and non-deposition can occur. The western portion of the Wharton Basin is characterized by thin sediment sequences (<100 m; Mathias *et al.* 1988) which does not necessarily infer bottom-water flow but merely signifies the lack of sediment supply to create sediment drifts. Sediment supply to the northwestern Wharton Basin was likely to have increased in response to the development of the Nicobar Fan. (The Nicobar Fan is the part of the Bengal Fan that has spilt down the eastern side of the Ninetyeast Ridge (Kolla & Kidd 1982).) It is not possible, however, to determine if these sediments were reworked by proto-AABW. The development of the Indonesian Arc would presumably have restricted the northward extent of proto-AABW.

DSDP Sites 266 and 267 (Hayes *et al.* 1975), which were situated close to the Southeast Indian Ridge crest, recovered complete hemipelagic sedimentary sections and were probably not influenced by Ross Sea proto-AABW.

Central Indian Basin. The palaeobathymetric analysis shows that the Central Indian Basin floor occurred at a palaeodepth which was potentially subject to the influence of proto-AABW flow at 20 Ma. Unfortunately no drill sites exist in the Central Indian Basin, therefore we can only postulate the proto-AABW flow patterns within this basin using the palaeobathymetric analysis and knowledge of modern AABW within this basin. Presumably proto-AABW would have utilized fracture zones to flow through the Southwest Indian Ridge, as at present. The northward extension of proto-AABW within this basin would have been increasingly constrained, however, as the Bengal Fan expanded, both in area and volume, during the Miocene (Kolla & Kidd 1982; Emmal & Curray 1985; Kolla & Coumes 1985).

Antarctic Basin. The palaeobathymetric reconstruction at 20 Ma shows that the passage between the Kerguelen Plateau and Antarctica could have been utilized by proto-AABW. Evidence for flow through this passage is lacking, particularly at the present (see above). Asymmetrical build-up of ice in either west or east Antarctica could, however, have led to the increased flow through this passage. The thick sediment sequences which occur on both sides of this passage (Mathias *et al.* 1988) are likely to be derived from glacial activity in the Prydz Bay area (Barron *et al.* 1989) and may preserve a record of proto-AABW flow within their seismic architecture.

South Atlantic Ocean. The palaeobathymetric analysis indicates that proto-AABW could have flowed up through the basins of the western South Atlantic, utilizing narrow channels to negotiate the shallower sills, e.g. the Vema Channel to the west of the Rio Grande Rise. In contrast the Walvis Ridge formed a barrier between the Angola and Cape Basins at depths below −3800 m. The palaeobathymetry of the Scotia Sea was shallower than the depth range at which proto-AABW was expressed, therefore bottom-water flow through this opening seaway was unlikely during the early Miocene. Well developed intermediate depth passages (shallower than −3600 m) existed at this time which would have allowed the northward incursion of circum-polar deep-water flow.

Deep drill sites within the South Atlantic Ocean (Maxwell *et al.* 1970; Barker *et al.* 1983) are limited within a latitudinal band (*c.* 30°S). These drill sites indicate that the southern Brazil Basin, i.e. western South Atlantic, was characterized by sedimentation, probably within sediment drifts, whilst those in the Angola Basin recorded hiatuses or condensed sequences during the early Miocene. The isolation of the Angola Basin from the influence of southward-derived bottom water may have occurred since the late Cretaceous (95 Ma; Sykes 1995; Sykes *et al.* 1998). The accumulation of dissolution facies at DSDP Sites 522 and 523 during the Miocene (Hsu *et al.* 1983) suggests that leakage of cold corrosive bottom water across the Walvis Ridge may have also occurred in the Miocene, in a similar manner to the present leakage of AABW into this basin demonstrated by Shannon & Chapman (1991). At DSDP Sites 17 and 18, approximately 300–400 m shallower than DSDP Sites 522 and 523, nanofossil oozes were cored (Maxwell *et al.* 1970) suggesting that these sites were not influenced by dissolution associated with proto-AABW during this time interval.

Wright *et al.* (1992) suggested, on the basis of $\partial^{13}C$ data, that between 23.7 and 18.3 Ma a northward-flowing water mass at palaeodepths of between −3000 m and −4000 m ceased to flow in the Atlantic. Although the cessation of this water mass in the Atlantic is earlier than the postulated time of waning of the proto-AABW (17–15 Ma) it is possible that Wright *et al.* (1992) interpreted data which record the behaviour of an upper component of proto-AABW rather than the behaviour of the whole water mass which is obscured by a lack of data from deeper palaeodepths. Similarly, the absence of data within the Indian and Pacific Oceans prevented the recognition of a deep northward flowing water mass by these workers.

Tasman Sea and Pacific Ocean. The Tasman, Pacific and Bellinghausen Basins could have been flooded by proto-AABW. It is not possible to constrain northward extensions of proto-AABW owing to a lack of drill site data and the northern extent of the palaeobathymetric reconstructions presented here. Proto-AABW was likely to have flowed eastwards, south of the Balleny Seamounts, into the Bellinghausen Basin. However, due to the shallow nature of the Scotia Sea (see above) any bottom water that escaped from the Bellinghausen Basin must have exited either via the Eltanin Fracture Zone system into the Pacific Basin, or northeastwards into the Chile Basin.

DSDP Sites 322, 323 and 325, in the southeast Pacific Ocean (Bellinghausen Basin), recovered Miocene clay and silt sediments (Hollister *et al.* 1976) which were considered to be ice-derived detritus from Antarctica. The origin of these sediments, therefore, does not provide evidence for the flow path of the proto-AABW in the southeast Pacific.

The DSDP drill sites within the Tasman Sea area (Sites 276, 279, 280, 282 and 283) have no sediment preserved for this time interval (Kennett *et al.* 1975). The occurrence of sediment drifts within the southeast Tasman Sea (Jenkins 1992) and along the foot of the eastern flank of Chatham Rise (Carter & McCave 1994) indicates the possibility of long-lived bottom-water flow in these areas. The drifts in the Tasman Sea were interpreted, from seismic sections (Jenkins 1992), as pre-late Eocene to early Oligocene hemipelagic sediments which were reworked by bottom flow. The palaeobathymetric analysis clearly indicates that these drifts were potentially active during the early Miocene; however, the absence of drill site data precludes an age determination.

15 Ma time slice (middle Miocene) (Fig. 4)

Oxygen isotope studies show that the middle Miocene was a time of significant ice accumulation on Antarctica (Shackleton & Kennett 1974; Miller *et al.* 1991), associated with a lower sea level (Haq *et al.* 1987) and the development of unconformities on the continental shelves of Antarctica as the ice-shelves grounded (Anderson & Bartek 1992). This increased ice cover on Antarctica resulted in the almost total cessation of the shelf component of proto-AABW production. In the Ramsay *et al.* (1994) model of proto-AABW (Fig. 2), bottom-water activity was restricted to water depths below −5500 m at 15 Ma.

The immediate effect of this postulated waning of the proto-AABW is the disconnection of the majority of the deep ocean basins at water depths below −5000 m. Therefore, many of the gateways that may have been utilized, both previously (Fig. 3) and subsequently (Fig. 5), were now too shallow to be utilized by proto-AABW. Ross Sea derived proto-AABW production either shut down or was severely reduced in response to ice-grounding across the shelf during the mid-Miocene (Anderson & Bartek 1992). Presumably the production of Weddell Sea derived proto-AABW was similarly affected. Despite the potential shutdown of shelf-derived proto-AABW production it is probable that a cold, oxygenated, deep water mass was still produced within the Southern Ocean but the area of production was restricted in its geographical extent. The development of a manganese pavement within the Amirante Channel between 17 and 13 Ma (Johnson & Damuth 1979) suggests the continued use of this gateway by a bottom-impinging water mass.

The northern Indian Ocean was probably influenced at depth by a southward-flowing highly saline water mass to depths of at least −3.1 km derived from Tethys between 24 and *c.* 14.5 Ma (Woodruff & Savin 1989; Wright *et al.* 1992). An intermediate to deep saline(?) oxygen-depleted southerly flowing water mass has been identified within the Atlantic Ocean and western Indian Ocean from the relative abundance of bolivinid (infaunal benthic) foraminifera between *c.* 19.5 and 16.5 Ma (Smart & Ramsay 1995) and it is suggested that the production of this water mass continued to approximately 14.5 Ma (see Ramsay *et al.* 1998). It is possible that around 15 Ma bottom water flow reversed, being dominated by the saline oxygen-depleted Tethyan outflow water. The development of a manganese pavement between 17 and 13 Ma within the Amirante Channel (see above) is indicative of bottom-water flow but gives no clue as to the direction of sediment transport. The recognition of such an event, however, may be precluded by the poor preservation or lack of microfossils at such depths.

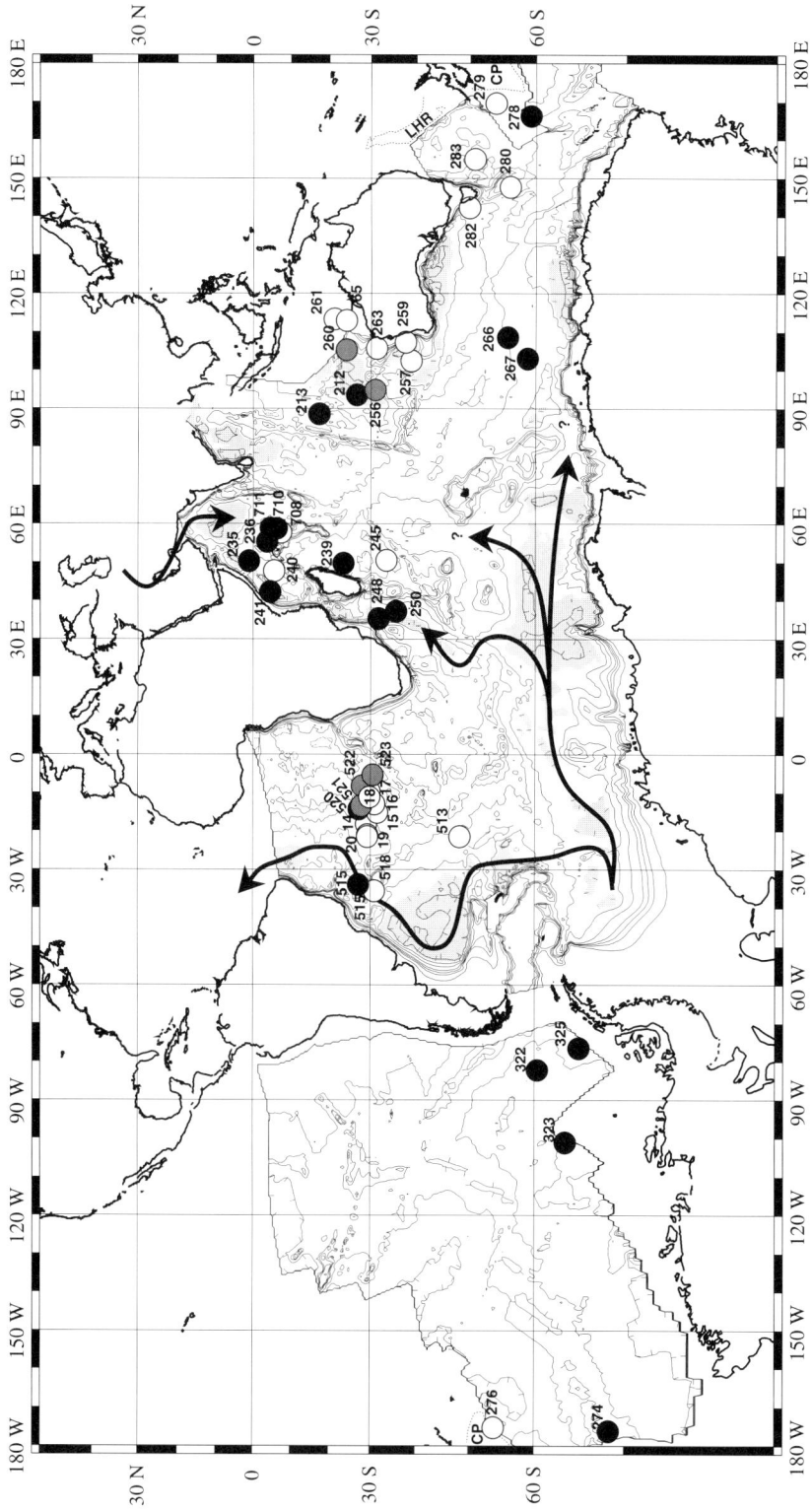

Fig. 4. Extent and flow paths of proto-AABW during the middle Miocene (15 Ma). See Fig. 3 for explanation of symbols.

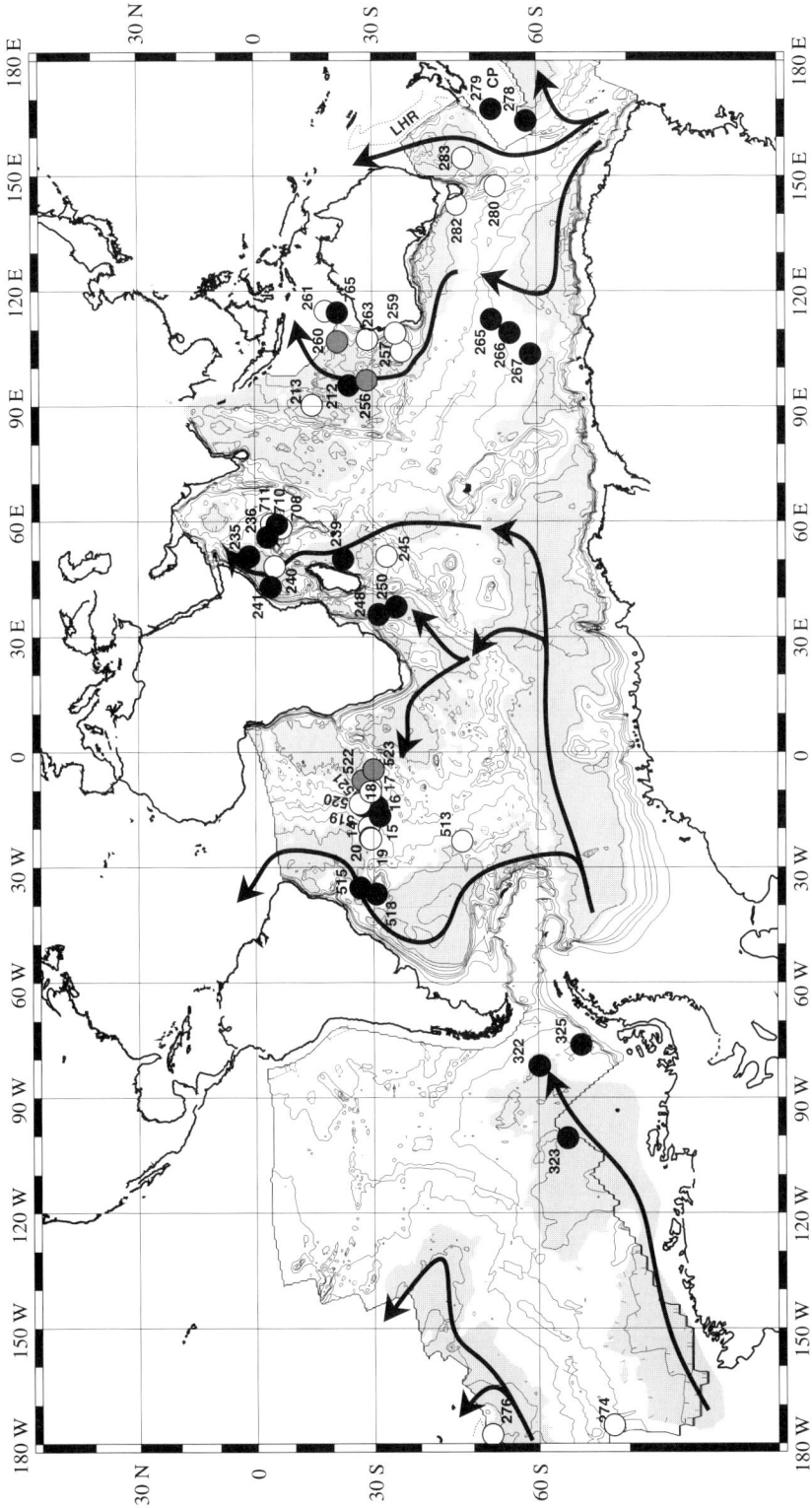

Fig. 5. Extent and flow paths of proto-AABW during the late Miocene (10 Ma). See Fig. 3 for explanation of symbols.

10 Ma time slice (late Miocene) (Fig. 5)

Proto-AABW began to wax at *c.* 12 Ma so that by 10 Ma proto-AABW was likely to occur at water depths below −4700 m, some 300 m deeper than at 20 Ma (Fig. 2). The slight deepening of proto-AABW at this time, compared to 20 Ma, had a significant effect on interbasin connectivity (cf. Fig. 3).

Flow of proto-AABW through the Agulhas Basin was likely to have been severely restricted due to the presence of two sills, one dividing the southern Agulhas Basin and the other separating the Agulhas and Cape Basins. Both of these sills would have acted as barriers to proto-AABW flow. The flow path of proto-AABW through the Crozet, Madagascar and Somali Basins was potentially restricted but may have been maintained through deep small fracture zones. The northern extension of this water mass flow path into the Arabian Sea was restricted by the northern end of the Central Indian Ridge.

In the Somali Basin the expansion of the proto-AABW coincides with a significant coeval increase in the relative abundance of *Nuttalides umboniferous* at deeper ODP Site 710 and shallower DSDP Site 237 (Ramsay *et al.*, 1998). This increase also post-dates the cessation of major Tethyan outflow and the initiation of production of a major southerly flowing analogue of North Atlantic Deep Water in the North Atlantic Ocean, and represents a faunal response to a major change in the deep-water mixing pattern of the Indian Ocean (Ramsay *et al.* 1998).

The proportion of the South East Indian Ridge shallower than the depth of proto-AABW within the vicinity of the Australian–Antarctic Discordance is greater than during the early Miocene, thus for bottom water to cross this ridge, it must have utilized small deep fracture zones. Once bottom water enters the Australian Basin there are no sills to inhibit the flow of proto-AABW through into the Wharton Basin.

The waxing of proto-AABW commenced at *c.* 12 Ma which may be related to the development of regional hiatuses recorded within the western Pacific (Carter & McCave 1994). The timing of drift build-up in the Tasman Sea was argued to be late Eocene–early Oligocene by Jenkins (1992), but it is possible that these drifts were being built during the late Miocene, as were the drifts defined by Carter & McCave (1994). Therefore, both of these drifts were subjected to bottom water generated from the same source and may have a similar history.

Conclusions

The definition of potential proto-AABW flow paths for three time slices during the Miocene (20, 15 and 10 Ma) can be inferred from a combination of: firstly, an understanding of modern factors (bottom-water circulation, concentration of suspended sediment particles and the distribution of large sediment drifts); secondly, the glacial history of Antarctica and hiatus terminations to define the production history of proto-AABW; and thirdly, the palaeobathymetric charts. The palaeobathymetry is useful in defining the dimensions of potential gateways and in assessing the likelihood of their utilization by a bottom water mass. In the modern ocean, however, deep narrow multiple fracture zones frequently act as conduits through which bottom water can flow (Kennett 1982). Conduits of this kind are not represented within the coarser resolution palaeobathymetry and so the 'gateways' identified here are not exclusive.

From this study it would appear that further drilling is needed to quantify bottom circulation through time. This drilling should concentrate on coring sediment drifts either side of gateways (to detect flow reversal) and ridge-to-flank transects. New surveys and further review of seismic data are necessary to precisely identify sediment drifts and locate future drill sites. Mid-ocean ridge-to-flank transects will allow standard palaeoceanographic methods (isotope, hiatus, micropalaeontological studies etc.) to be extended down to the CCD and combined with the results of the drift drilling. Flow paths that merit further investigation would include: (i) the southern Argentine Basin to the Southern Brazil Basin; (ii) from south of the Crozet Basin through to the Somali Basin via the Crozet and Madagascar Basins; (iii) the western Antarctic Basin through the Australian–Antarctic Discordance into the Australian and Wharton Basins; and (iv) the Tasman Basin and around the Campbell Plateau/Chatham Rise.

This work clearly highlights the importance of future deep drilling in the southern hemisphere. We speculate that the history of the various drifts will ultimately be similar, controlled by the production of Weddell Sea and Ross Sea proto-AABW. The timing of events within these drifts may vary dependent on whether they are formed through action of proto-AABW derived from the Weddell Sea or Ross Sea due to the asymmetrical development and grounding of ice in the respective shelf areas of Antarctica.

The authors were funded by Natural Environment Research Council ODP Special Topic grants GST-02434 and GST-02698 (Principal Investigator R. B. Kidd). The authors gratefully acknowledge the constructive criticism provided by N. Hamilton and P. Weaver (Southampton Oceanography Centre) which greatly improved this manuscript. PALSOS contribution No. 3.

References

ANDERSON, J. B. & BARTEK, L. R. 1992. Cenozoic glacial history of the Ross Sea revealed by intermediate resolution seismic reflection data combined with drill site information. *In*: KENNETT, J. P. & WARNKE, D. A. (eds) *The Antarctic paleoenvironment: a perspective on global change*. Antarctic Research Series **56**, American Geophysical Union, Washington, DC, 231–263.

BARKER, P. F., CARLSON, R. C., JOHNSON, R. L. *ET AL.* 1983. *Initial Reports of the Deep Sea Drilling Program*, **72**. US Govt Printing Office, Washington.

BARRON, J., LARSEN, B. *ET AL.* 1989. *Proceedings of the Ocean Drilling Program, Initial Reports*, **119**. Ocean Drilling Program, College Station, TX.

BISCAYE, P. E. & EITTREIM, S. L. 1977. Suspended particulate loads and transports in the nepheloid layer of the abyssal Atlantic Ocean. *Marine Geology*, **23**, 155–172.

CARTER, L. & McCAVE, N. 1994. Development of sediment drifts approaching an active plate margin under the SW Pacific Deep Boundary Current. *Paleoceanography*, **9**(6), 1061–1085.

EMMAL, F. J. & CURRAY, J. R. 1985. Bengal Fan, Indian Ocean. *In*: BOUMA, A. H., NORMARK, W. R. & BARNES, N. E. (eds) *Submarine Fans and Related Turbidite Systems*. Springer-Verlag, New York, 107–112.

FOLDVIK, A. & GAMMELSROD, T. 1988. Notes on Southern Ocean hydrography, sea-ice and bottom water formation. *Palaeogeography, Palaeoclimatology, Palaeoecology*, **67**, 3–17.

HAQ, B. L., HARDENBOL, J. & VAIL, P. R. 1987. Chronology of fluctuating sea levels since the Triassic. *Science*, **235**, 1156–1167.

HAYES, D. E., FRAKES, L. A. *ET AL.* 1975. *Initial Reports of the Deep Sea Drilling Program*, **28**. US Gov. Printing Office, Washington.

HOLLISTER, C. D., CRADDOCK, C. *ET AL.* 1976. *Initial Reports of the Deep Sea Drilling Program*, **35**. US Govt Printing Office, Washington.

HSU, K. J., LaBRECQUE, J. L. *ET AL.* 1983. *Initial Reports of the Deep Sea Drilling Program*, **73**. US Govt Printing Office, Washington.

JENKINS, C. J. 1992. Abyssal sediment drifts, erosion and history of bottom water flow in the Tasman Sea southwest of New Zealand. *Australian Journal of Earth Sciences*, **39**, 195–210.

JOHNSON, D. A. & DAMUTH, J. E. 1979. Deep thermohaline flow and current-controlled sedimentation in the Amirante Passage: Western Indian Ocean. *Marine Geology*, **33**, 1–44.

——, LEDBETTER, M. T. & DAMUTH, J. E. 1983. Neogene sedimentation and erosion in the Amirante Passage, western Indian Ocean. *Marine Geology*, **30**, 195–219.

JONES, G. A. 1994. Holocene climate and deep ocean circulation changes: Evidence from accelerator mass spectrometer radiocarbon dated Argentine Basin (SW Atlantic) mudwaves. *Paleoceanography*, **9**(6), 1001–1016.

KENNETT, J. P. 1982. *Marine Geology*. Prentice-Hall, Englewood Cliffs, NJ.

——, HOUTZ, R. E. *ET AL.* 1975. *Initial Reports of the Deep Sea Drilling Program*, **29**. US Govt Printing Office, Washington.

KIDD, R. B. & HILL, P. R. 1986. Sedimentation of mid-ocean sediment drifts. *In*: SUMMERHAYES, C. P. & SHACKLETON, N. J. (eds) *North Atlantic Palaeoceanography*. Geological Society of London, Special Publication **21**, Blackwell Scientific, Boston, MA, 87–102.

KOLLA, V. & COUMES, F. 1985. Indus Fan, Indian Ocean. *In*: BOUMA, A. H., NORMARK, W. R. & BARNES, N. E. (eds) *Submarine Fans and Related Turbidite Systems*. Springer-Verlag, New York, 129–136.

—— & KIDD, R. B. 1982. Sedimentation and sedimentary processes in the Indian Ocean. *In*: NAIRN, A. E. M. & STELHI, F. G. (eds) *The Indian Ocean*, **6**. Plenum Press, New York, 1–50.

——, SULLIVAN, L., STREETER, S. S. & LANGSETH, M. G. 1976. Spreading of bottom water and its effects on the Indian Ocean inferred from bottom-water temperature, turbidity and sea-floor topography. *Marine Geology*, **21**, 171–189.

MANTYLLA, A. W. & REID, J. L. 1983. Abyssal characteristics of the world ocean waters. *Deep-Sea Research*, **30**(8A), 805–833.

MASSE, L., FAUGERES, J.-C., BERNAT, M., PUJOS, A. & MEZERAIS, M.-L. 1994. A 600,000-year record of Antarctic Bottom Water activity inferred from sediment textures and structures in a sediment core from the Southern Brazil Basin. *Paleoceanography*, **9**(6), 1017–1026.

MASSON, D. G., KIDD, R. B. & ROBERTS, D. G. 1982. Late Cretaceous sample from the Amirante Passage, western Indian Ocean. *Geology*, **10**, 264–266.

MATHIAS, P. K., RABINOWITZ, P. D. & DIPIAZZA, N. 1988. *Sediment Thickness Map of the Indian Ocean*. The American Association of Petroleum Geologists, Tulsa, OK.

MAXWELL, A. *ET AL.* 1970. *Initial Reports of the Deep Sea Drilling Program*, **3**. US Govt Printing Office, Washington.

McCAVE, I. N. & TUCHOLKE, B. E. 1986. Deep current controlled sedimentation in the western North Atlantic. *In*: VOGT, P. R. & TUCHOLKE, B. E. (eds) *The Western North Atlantic Region*. Geological Society of America, Boulder, CO, 451–468.

MILLER, K. G., WRIGHT, J. D. & FAIRBANKS, R. G. 1991. Unlocking the ice house: Oligocene–Miocene oxygen isotopes, eustasy and margin erosion. *Journal of Geophysical Research*, **96**, 6829–6848.

MOUGENOT, D., RAILLARD, S. & VIRLOGEUX, P. 1987. Structure de la marge et du canal de Mozambique.

Actes du Colloque sur la Recherche Francaise dans les Terre Australes, 323–336.

PETERSON, L. C. & BACKMAN, J. 1990. Late Cenozoic calcium carbonate accumulation and history of the carbonate compensation depth in the western equatorial Indian Ocean. *In*: DUNCAN, R. A., BACKMAN, J., PETERSON, L. C. *ET AL.* (eds) *Proceedings of the Ocean Drilling Program, Scientific Results,* **115**. Ocean Drilling Program, College Station, TX, 467–507.

——, MURRAY, D. W., EHRMANN, W. U. & HEMPEL, P. 1992. Cenozoic carbonate accumulation and compensation depth changes in the Indian Ocean. *In*: DUNCAN, R. A., REA, D. K., KIDD, R. B., VON RAD, U. & WEISSEL, J. K. (eds) *Synthesis of Results from Scientific Drilling in the Indian Ocean.* Geophysical Monograph **70**, American Geophysical Union, Washington, DC, 311–333.

RAMSAY, A. T. S., SYKES, T. J. S. & KIDD, R. B. 1994. Waxing (and waning) lyrical on hiatuses: Eocene–Quaternary Indian Ocean hiatuses as proxy indicators of water mass production. *Paleoceanography,* **9**(6), 857–977.

——, SMART, C. W. & ZACHOS, J. C. 1998. A model of early to middle Miocene deep-ocean circulation for the Atlantic and Indian Oceans. *This volume.*

RODMAN, M. R. & GORDON, A. L. 1982. Southern Ocean bottom water of the Australian–New Zealand Sector. *Journal of Geophysical Research,* **87**, 5771–5778.

SHACKLETON, N. J. & KENNETT, J. P. 1974. Paleo-temperature history of the Cenozoic and the initiation of Antarctic glaciation: oxygen and carbon isotope analyses in DSDP Sites 277, 279 and 281. *In*: KENNETT, J. P., HOUTZ, R. E. *ET AL.* (eds) *Initial Reports of the Deep Sea Drilling Program,* **29**. US Govt Printing Office, Washington, 743–755.

SHANNON, L. V. & CHAPMAN, P. 1991. Evidence of Antarctic Bottom Water in the Angola Basin at 32°S. *Deep-Sea Research,* **38**(10),1299–1304.

SIMPSON, E. S. W., SCHLICH, R. *ET AL.* 1974. *Initial Reports of the Deep Sea Drilling Program,* **25**. US Govt Printing Office, Washington..

SMART, C. W. 1992a. Early to middle Miocene benthic foraminiferal faunas from DSDP Site 518 and 529, South Atlantic: Primary investigations. *In*: TAKAYANAGI, Y. & SAITO, T. (eds) *Studies in Benthic Foraminifera.* Proceedings of the Fourth International Symposium on Benthic Foraminifera (Benthos '90), Tokai University Press, Sendai, 245–248.

—— 1992b. *Ecological controls on patterns of speciation and extinction in deep-sea benthic foraminifera.* PhD thesis, Southampton University.

—— & RAMSAY, A. T. S. 1995. Benthic foraminiferal evidence for the existence of an early Miocene oxygen-depleted water mass? *Journal of the Geological Society, London,* **152**, 735–738.

SYKES, T. J. S. 1995. *Palaeobathymetry of the southern hemisphere.* PhD thesis, University of Wales College Cardiff.

——, ROYER, J.-Y., RAMSAY, A. T. S. & KIDD, R. B. 1998. Southern hemisphere palaeobathymetry. *This volume.*

TCHERNIA, P. 1980. *Descriptive Regional Oceanography.* Pergamon Press, New York.

WARREN, B. A. 1978. Bottom water transport through the Southwest Indian Ridge. *Deep-Sea Research,* **25**, 315–321.

WOLD, C. N. 1992. *Paleobathymetry and sediment accumulation in the northern North Atlantic and southern Greenland–Iceland–Norwegian Sea.* PhD thesis, GEOMAR.

WOODRUFF, F. & SAVIN, S. M. 1989. Miocene deep-water oceanography. *Paleoceanography,* **4**, 87–140.

WRIGHT, J. D., MILLER, K. G. & FAIRBANKS, R. G. 1992. Early and middle Miocene stable isotopes: Implications for deep water circulation. *Paleoceanography,* **7**, 357–389.

A model of early to middle Miocene deep ocean circulation for the Atlantic and Indian Oceans

ANTHONY T. S. RAMSAY[1], CHRISTOPHER W. SMART[2], JAMES C. ZACHOS[3]

[1] *Marine Geosciences Research Group, Department of Earth Sciences, University of Wales College of Cardiff, PO Box 914, Cardiff CF1 3YE, UK*

[2] *Department of Geological Sciences, University of Plymouth, Drake Circus, Plymouth, Devon PL4 8AA, UK*

[3] *Department of Earth Sciences, University of California Santa Cruz, Santa Cruz, California, CA 95064, USA*

Abstract: Tethyan Outflow Water (TOW), characterized by high benthic $\delta^{13}C$ values, was a component of the early Miocene deep water mass of the Atlantic (*c.* 20 and *c.* 13 Ma) and Indian (*c.* 20 and *c.* 14.5 Ma) Oceans. Fluctuations in temperature, salinity and quantity of dissolved oxygen coincided with high abundances of infaunal, smooth-walled bolivinids within the Atlantic and Indian Ocean TOW between 19.5 and 16.5 Ma.

High $\delta^{18}O$ values recorded in the Atlantic water mass may have originated, in part, from evaporation and the formation of warm, saline water within the Tethys. The southward flowing Atlantic TOW at depths of 2–4.5 km enhanced meridional heat transport. It was overlain by Northern Component Water (NCW) with a lower $\delta^{13}C$ signal. TOW and NCW were replaced by southward flowing North Atlantic Deep Water at *c.* 13 Ma when cold dense Norwegian–Greenland Sea water flowed across the Greenland–Scotland Ridge.

Indian Ocean TOW was restricted laterally by the Chagos–Lacadive Ridge, and vertically (1.3 to 3 km) by overlying intermediate water, characterized by the occurrence of coarse-walled bolivinids, and underlying deep water characterized by high abundances (>10%) of *Nuttallides umboniferus*. The coarse-walled bolivinid and '*N. umboniferus*' faunas were associated with lower temperature, salinity and higher oxygen concentrations. Tethyan outflow to the northwest Indian Ocean ended at *c.* 14.5 Ma, probably in response to the closure of a gateway, and resulted in a two-component deep water mass structure. The coeval increase in the relative abundances of *N. umboniferus* at Indian Ocean Sites 237 (*c.* 2 km) and 710 (*c.* 3.5 km), at *c.* 13 Ma, probably records the impact of this change on the nature of Indian Ocean deep water at depths between 2 and 4 km.

The termination of the Atlantic and, to a lesser extent, Indian Ocean TOW contributed to global cooling and to the expansion of the Antarctic Ice Sheet by limiting the meridional heat transport.

The Miocene was an important period in the evolution of the ocean–atmosphere system. During this time there was a significant change from the relatively warm early Miocene global climate to the cooler climate of the middle Miocene (Shackleton & Kennett 1975; Savin *et al.* 1981; Miller *et al.* 1987, 1991; Woodruff & Savin 1989, 1991; Flower & Kennett 1993, 1994).

Our understanding of early to middle Miocene deep ocean circulation has come from studies of benthic foraminifera (e.g.Woodruff 1985; Thomas 1986*a,b*, 1992; Murray 1988) stable isotope records (e.g. Wright *et al.* 1992; Zachos et al. 1992; Seto 1995) and from the integration of these records

(Woodruff & Savin 1989). The benthic $\delta^{13}C$ stable isotope signal provides evidence for the source, direction of ageing and/or mixing of water masses away from their source regions (e.g. Kroopnick 1985; Woodruff & Savin 1989; Wright *et al.* 1992). Changes in the early to middle Miocene benthic $\delta^{18}O$ stable isotope signal have been used to interpret changes in deep water mass temperature (e.g. Savin *et al.* 1975; Kennett & Shackleton 1976) and ice volume (e.g. Shackleton & Opdyke 1973; Miller *et al.* 1991; Wright *et al.* 1992). In addition, Wise *et al.* (1992) discuss the potential for high rates of precipitation (^{18}O depletion) or evaporation (^{18}O enrichment) to generate localized

RAMSAY, A. T. S., SMART, C. W. & ZACHOS, J. C. 1998. A model of early middle Miocene deep ocean circulation for the Atlantic and Indian Oceans. *In*: CRAMP, A., MACLEOD, C. J., LEE, S. V. & JONES, E. J. W. (eds) *Geological Evolution of Ocean Basins: Results from the Ocean Drilling Program.* Geological Society, London, Special Publications, **131**, 55–70.

variation in the isotopic composition of sea water.

Comparisons between modern and fossil assemblages of deep-sea benthic foraminifera have been made and used either to infer the existence of similar water masses in the past (e.g. Schnitker 1980; Weston & Murray 1984) or to broadly define shallow, intermediate and deep water mass stratification. Recently, the view that organic input exerts a strong influence on at least some benthic foraminifera has been used to cast doubt on the exclusive relationionship between benthic foraminifera and water masses (e.g. Gooday 1994), and food input as well as water mass properties may play a role (Mackensen *et al.* 1993; Schnitker 1994).

In this study, we investigate changes in the pattern of deep water circulation and in water mass composition for the Atlantic and Indian Oceans, during the early to middle Miocene, using integrated benthic foraminiferal and stable isotope records.

Models for early to middle Miocene deep ocean circulation have been proposed on the basis of investigations of changes in benthic foraminiferal faunas (Thomas 1986a,b; Murray *et al.* 1986; Smart & Murray 1994; Smart & Ramsay 1995), stable isotope records (Seto 1995) and from integrated benthic faunal and stable isotope studies (Woodruff & Savin 1989). Woodruff & Savin (1989) proposed that the early Miocene deep ocean circulation pattern was strongly influenced by the influx of warm, Tethyan Indian Saline Water (TISW) into the northern Indian Ocean which mixed and became entrained with circum-Antarctic circulation and contributed to the early Miocene equivalent of Antarctic Bottom Water (AABW). Wright *et al.* (1992), Zachos *et al.* (1992) and Seto (1995) also recognized the occurrence of TISW in the early to middle Miocene Indian Ocean. The existence of a water mass of Southern Ocean origin within the Indian Ocean was also confirmed by Zachos *et al.* (1992), Nomura (1995) and Seto (1995). Seto (1995) argued that this southern component i.e. equivalent to Antarctic Intermediate Water, formed the main component of deep water flow.

Woodruff & Savin concluded that no significant formation of North Atlantic Deep Water (NADW) occurred prior to 14.5 Ma. In contrast Wright *et al.* (1991, 1992) and Wright & Miller (1996) proposed that Northern Component Water (NCW), characterized by high $\delta^{13}C$ values, formed in the North Atlantic between approximately 20 and 15 Ma and from 12.5 Ma into the Pliocene. Reductions or interruptions in flow occurred between *c.* 15 and *c.* 12.5 Ma and between *c.* 9 and *c.* 7 Ma . They suggest that Southern Component Water with a high $\delta^{13}C$ signature was the only source of deep water between 24 and 20 Ma and 15 and 12.5 Ma.

Thomas (1986a,b) and Smart & Murray (1994) recognized the occurrence of abundant smooth-walled bolivinids in the early Miocene Atlantic Ocean which they attributed to a period of sluggish circulation. More recently, Smart and Ramsay (1995) recognized that high abundances of smooth-walled bolivinids in the Indian Ocean between 19.5 and 16.5 Ma were coeval with their high abundances in the Atlantic Ocean and they attributed this phenomenon to oxygen depletion within NCW. They suggested that this Atlantic water mass contributed, via Tethys, to Tethyan outflow in the Indian Ocean.

Methods

Benthic stable isotope data were obtained from analyses of Cibicidoides spp. (including *Cibicides wuellerstorfi*) which secrete tests consistently offset from $\delta^{18}O$ equilibrium and which accurately record $\delta^{13}C$ variations in sea water (e.g. Duplessy *et al.* 1970; Woodruff *et al.* 1980; Belanger *et al.* 1981; Graham *et al.* 1981). We include new data for the relative abundances of *Nuttallides umboniferus* in the Indian Ocean. The benthic stable isotope data for Indian Ocean DSDP Site 237 are from Woodruff & Savin (1989). The benthic stable isotope data for Site 710 are new (Table 1).

Table 1. *Benthic stable isotope data for Site 710*

Core	Section	Interval	Depth (m)	Age (Ma)	$\delta^{13}C$
10H	3	73–75	89.93	10.99	0.66
10H	3	73–75	89.93	10.99	0.62
10H	5	74–76	92.94	12.22	1.31
10H	5	74–76	92.94	12.22	1.41
11H	1	73–75	96.53	12.85	1.60
11H	4	77–79	101.07	14.03	1.64
11H	5	72–74	102.52	14.41	1.86
11H	6	68–70	103.98	14.79	1.66
11H	7	29–31	105.09	15.07	1.61
12H	1	71–73	106.21	15.36	1.25
12H	2	71–73	107.78	15.77	1.62
12H	3	73–75	109.23	16.22	1.84
12H	4	78–80	110.78	16.92	1.37
12H	5	73–75	112.23	17.58	0.91
12H	6	77–79	113.77	18.27	1.42
12H	7	34–36	114.84	18.73	0.78
13H	1	72–74	115,92	19.40	0.83
13H	3	73–75	118.93	20.18	1.05
13H	6	78–80	123.48	21.29	0.89
14X	2	97–99	127.37	22.24	1.33
14X	5	76–78	131.66	23.28	1.31
15X	2	78–80	134.68	23.90	1.24

For the stable isotope analyses, foraminifer samples were loaded into stainless steel boats and roasted under vacuum at 390°C for one hour. Isotopic measurements were performed on 5–15 specimens. All samples were analysed using an Autocarb device coupled to either a Fisons Prism or Optima Mass Spectrometer (University of California, Santa Cruz). In this system, each sample is reacted in 100% H_3PO_4 in a common acid bath at 90°C. All values are reported in the delta (per mil) notation relative to the Pee Dee Belemnite (PDB) standard. External analytical precision based on replicate analyses of the NBS-19 standard was better than 0.1‰ for both oxygen and carbon isotopes. Unless otherwise stated, all isotope values discussed in this paper are relative to the PDB standard.

In order to achieve a reliable comparison between the benthic foraminferal faunas recorded by different workers, we have only included studies which examined the >63 µm size fraction so that small specimens, including the smooth-walled bolivinids, were included (e.g. Schrœder et al. 1987). These morphotypes do not occur in the >125 µm and 149 µm size fractions. A list of the data sources for the benthic foraminifera is given in Table 2.

Relative abundance values for benthic foraminifera have been used because we recognize a strong positive correlation between the relative and absolute abundance records of both smooth-walled bolivinids and N. umboniferus in the Indian Ocean during the Miocene. We assume that this observation is valid for the Atlantic data where information concerning the absolute abundances of benthic foraminifera are not available. Changes in the relative abundances of smooth-walled

bolivinids (published records) and N. umboniferus were used in conjunction with time series analyses of the benthic $\delta^{13}C$ and $\delta^{18}O$ records to infer changes in the deep water mass circulation patterns of the Atlantic and Indian Oceans.

The age–depth reconstructions were derived from subsidence curves which were generated by using the standard methods for calculating oceanic subsidence (Parsons & Sclater 1977; Hayes 1988). The age data were derived from magnetic anomaly data (where available) and micropalaeontological data documented by Berggren et al. (1985) and Fornaciari et al. (1990). Constant sedimentation rates were assumed between magnetic and micropalaeontological datums at each site and all age–depth data are calibrated to the Berggren et al. (1985) timescale.

Results

Benthic foraminifera

We have considered the abundances of smooth-walled bolivinids and N. umboniferus (Figs 1–4) from Atlantic, Mediterranean Sea and Indian Ocean sites. High abundances of smooth-walled bolivinids were coeval in the Atlantic, Mediterranean Sea and Indian Ocean sites between 19.5 and 16.5 Ma (Figs 1–3).

In the Atlantic Ocean, faunas with >10% smooth-walled bolivinids displaced an 'intermediate to deep water fauna' in which no single species was dominant between 19.5 and 16.5 Ma. Faunas characterized by smooth-walled bolivinids were geographically widespread (Fig. 1) at palaeo-depths ranging from 1.2 km (Site 548) to >4.5 km

Table 2. *Benthic foraminiferal faunal sources used in this study*

Site	Latitude	Longitude	Depth (m)	Size fraction	Reference
216	012°7′N	90°12′E	2247	>63µm	Smart & Ramsay (1995)
237	07°05′S	58°08′E	1623	>63µm	Smart & Ramsay (1995)
372	40°01′N	04°47′E	2699	>63µm	Smart & Ramsay (1995)
400	47°22′N	09°11′W	4399	>63µm	Smart & Ramsay (1995)
525	29°04′S	02°59′E	2467	>73µm	Boltovskoy & Boltovskoy (1989)
529	28°55′S	02°46′E	3035	>63µm	Smart & Murray (1994)
548	48°54′N	12°09′W	1251	>74µm	Poag & Low (1985)
555	56°40′N	20°47′W	1669	>63µm	This study
563	33°38′N	43°46′W	3786	>63µm	Smart & Murray (1994)
608	42°50′N	23°05′W	3534	>63µm	Thomas (1986 a,b)
610	53°13′N	18°53′W	2427	>63µm	Thomas (1986 a,b)
667	04°34′N	21°54′W	3529	>63µm	Smart & Ramsay (1995)
707	07°32′S	59°01′E	1552	>63µm	Smart & Ramsay (1995)
709	03°54′S	60°33′E	3040	>63µm	Smart & Ramsay (1995)
710	04°18′S	60°58′E	3824	>63µm	Smart & Ramsay (1995)
758	05°23′N	90°21′E	2923	>63µm	Smart & Ramsay (1995)

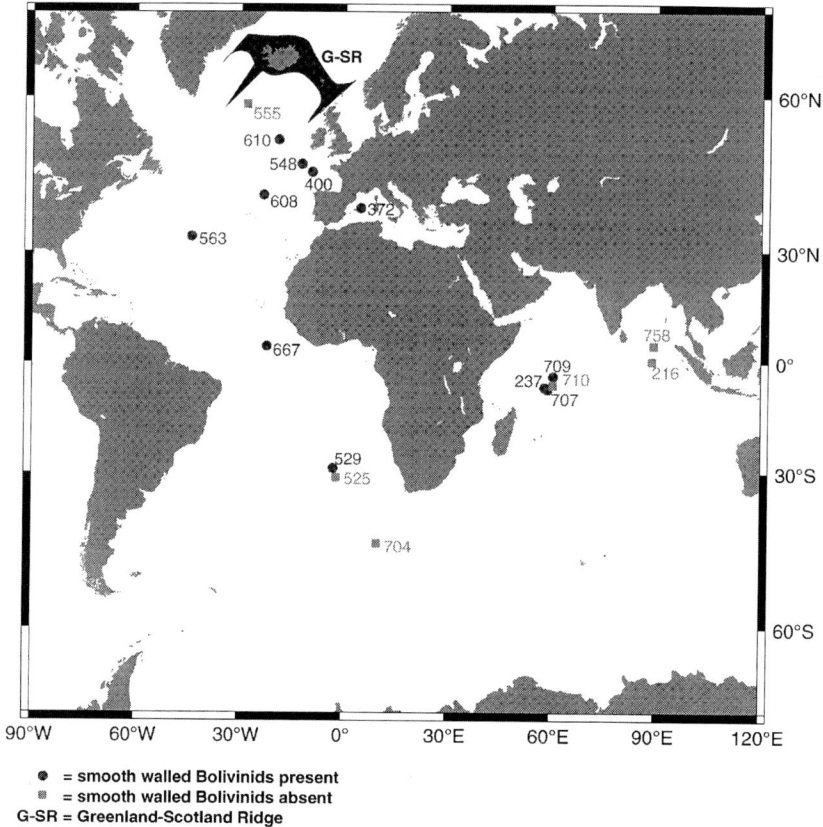

Fig. 1. Distribution of Atlantic and Indian Ocean sites used in this study.

(Site 400) (Fig. 2). They are absent at Site 525 (Boltovskoy & Boltovskoy 1989) and were not recorded in the few samples obtained from the poorly recovered lower Miocene section from Site 555 (this study). Smooth-walled bolivinids have been recorded at Site 548 by Poag & Low (1985) but no quantitative data are available. A time-slice analysis of the contoured relative abundances of smooth-walled bolivinids on age-depth plots (Fig. 2) show a vertical and latitudinal decrease in abundances away from a fauna with higher values. The depth and latitudinal distribution of this fauna varied temporally and latitudinally as shown by the 10% smooth-walled bolivinid contour (Fig. 2). The 'intermediate to deep water fauna' replaced the smooth-walled bolivinid fauna at 16.5 Ma. *N. umboniferus* is rare at the North Atlantic sites (Smart 1992).

In the Indian Ocean, the highest relative abundances of smooth-walled bolivinids are recorded at Site 237. In Somali Basin Sites 237, 707 and 709 they occurred between 1.3 and 3 km,

and are absent from Site 710 (Smart & Ramsay 1995). No smooth-walled bolivinids were found at eastern Indian Ocean Sites 216 and 758 (Smart & Ramsay 1995). The age–depth plot for this ocean (Fig. 3) shows evidence for vertical expansions of faunas with high values of smooth-walled bolivinids at 19 Ma (Site 707) and 17 Ma (Site 709). Between these times the depth distribution of these forms is poorly constrained. At Site 237 the 'intermediate to deep water fauna' was displaced by the smooth-walled bolivinid fauna at 19.5 Ma and was re-established between 16.5 and *c.* 13 Ma.

The distribution of benthic foraminifera, during the high-abundance phase of smooth-walled bolivinids, was depth-stratified. Shallow to inter-mediate sites were characterized by coarse-walled bolivinids (e.g. Site 707), intermediate to deep water Sites (237, 709) were characterized by smooth-walled bolivinids, and the deep water Site 710 contained higher abundances of *N. umboniferus* (Fig. 4). Temporal fluctuations in the relative abundances of *N. umboniferus* at Sites 237

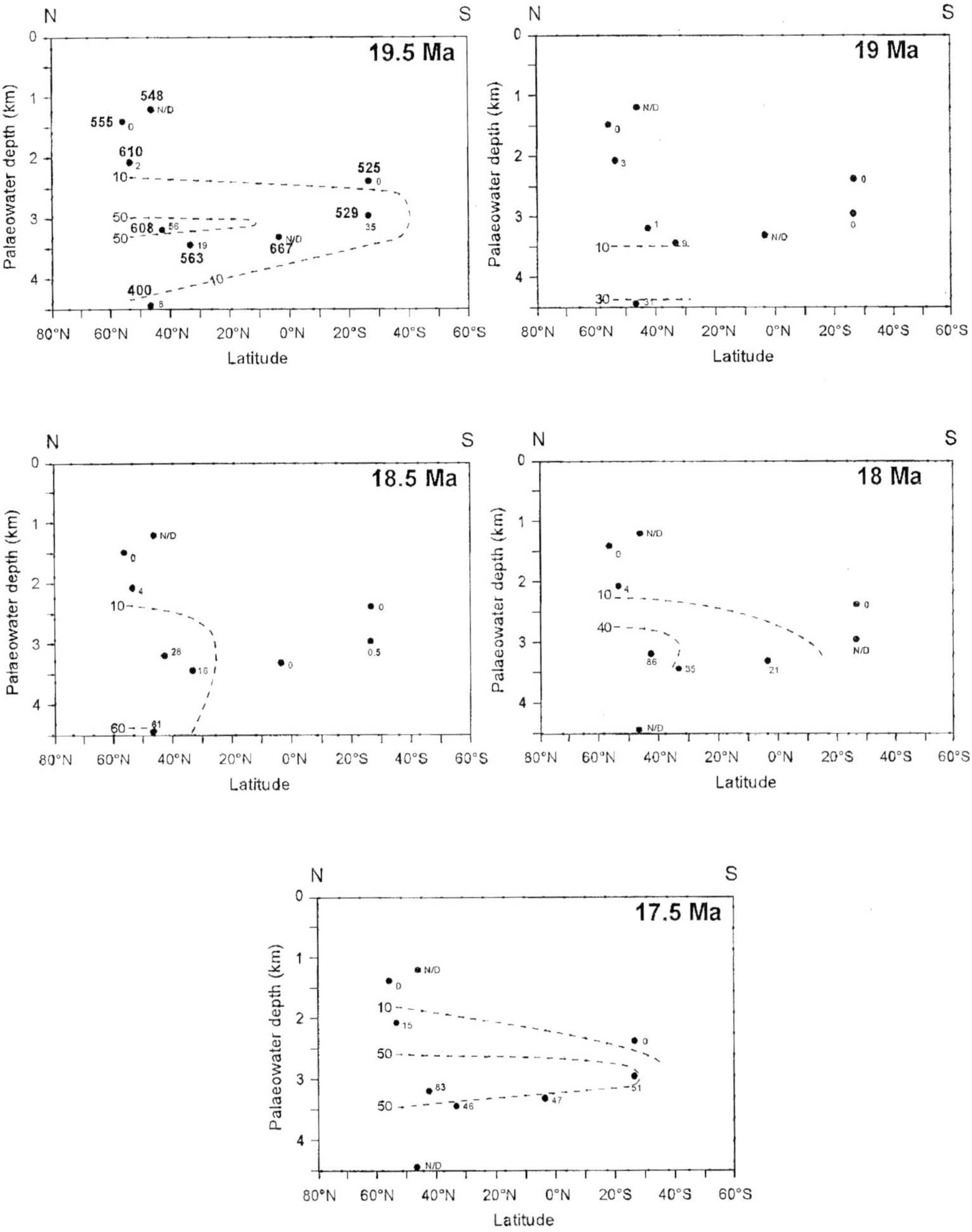

Fig. 2. Atlantic Ocean 0.5 Ma time slices calculated using the Berggren *et al.* (1985) timescale showing age/depth relationships of contoured relative abundance fluctuations of smooth-walled bolivinids. Data are derived from published records and values are expressed in percentages calculated for 0.2 Ma intervals about 0.5 Ma mid-points. N/D = no data. Data sources: Site 548, Poag & Low, (1985); Sites 608 and 610, Thomas (1986*a*,*b*); Site 525, Boltovskoy & Boltovskoy (1989); Sites 400, 529, 563 and 667, Smart & Murray (1994). Abundances of >10% are considered significant.

and 710 (Fig. 4) reveal a coeval increase in the abundance of this species at both sites beginning at approximately 13 Ma. *N. umboniferus* is rare in North Atlantic Site 563 which occurs at a

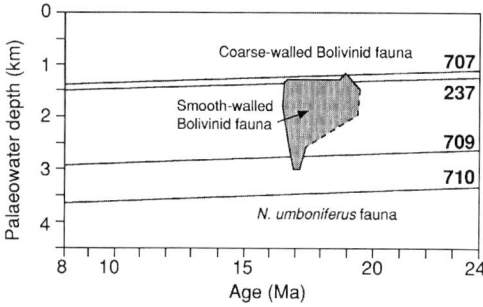

Fig. 3. Age–depth distribution of benthic foraminiferal faunas within the Indian Ocean.

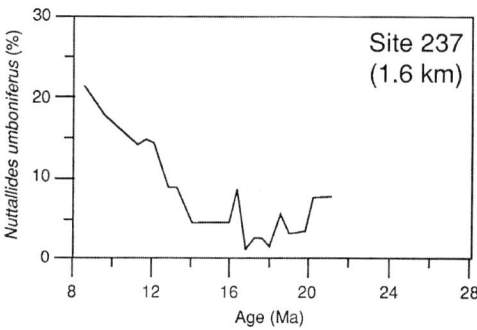

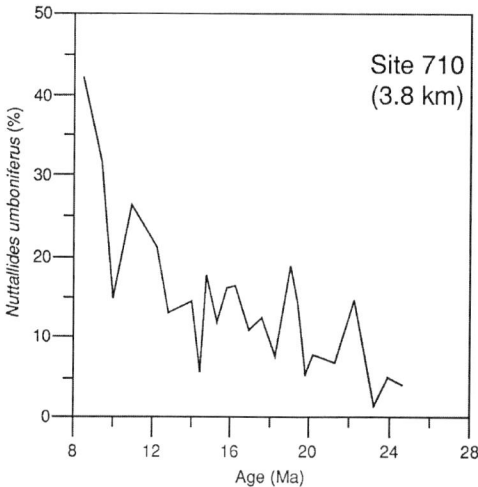

Fig. 4. Fluctuations in the relative abundance of *Nuttallides umboniferus* at Indian Ocean Sites 237 and 710.

comparable palaeodepth to Indian Ocean Site 710 (*c.* 3.4 km).

In the Mediterranean Sea, few sites have recovered sediments of lower Miocene age. Smooth-walled bolivinids were a significant component (>30%) of the lower Miocene sediments recovered at Site 372 (Fig. 1).

Benthic stable isotopes

Sites 563 and 608 have similar $\delta^{13}C$ records which diverge and converge with the record for Site 555 at 18 Ma and *c.* 13 Ma respectively (Figs 5a and b). Between 18 and *c.* 13 Ma Sites 563 and 608 have more positive $\delta^{13}C$ values than the shallower Site 555. Prior to 18 Ma the $\delta^{13}C$ values at Site 608 are either more positive and not so strongly divergent or equal to the values recorded at Site 555. The benthic $\delta^{13}C$ records for North Atlantic Sites 563 and 608 and South Atlantic Site 704 are compared in Figs 6a and b. The generally higher $\delta^{13}C$ values at the North Atlantic sites either reflect water mass ageing within the Atlantic or the mixing of water masses with different isotopic signals.

The benthic $\delta^{13}C$ records for North Atlantic Sites 563 and 608 and northwest Indian Ocean Site 237, compared in Figs 7a and b, show a succession of convergences and divergences in $\delta^{13}C$ values. The incomplete nature of these records prior to *c.* 18 Ma precludes any meaningful comparison. Between *c.* 18 Ma and *c.* 17 Ma, when the records diverge, the North Atlantic and northwest Indian Ocean sites have almost identical $\delta^{13}C$ values. From *c.* 17 Ma to *c.* 16 Ma, when the records converge, the North Atlantic sites have consistently higher $\delta^{13}C$ values than Site 237. The record of higher $\delta^{13}C$ values at Site 237, post *c.* 16 Ma , is replaced at 14.5 Ma by consistently higher values in the North Atlantic records.

The benthic $\delta^{13}C$ records for northwest Indian Ocean Sites 237 and 710 are compared in Fig. 8. This figure shows that, for time intervals where direct comparisons are meaningful, the $d^{13}C$ values are similar at both sites between 21.3 and 14.5 Ma. From 14.5 Ma the $\delta^{13}C$ values are higher at the deeper Site 710.

Time series analyses of benthic $\delta^{18}O$ for northern North Atlantic Site 555 and North Atlantic Sites 563 and 608 are compared in Figs 9a and b respectively. Between 19 and *c.* 17.3 Ma the record for Site 555 has more negative $\delta^{18}O$ values than Sites 563 and 608. The records are inconsistent between *c.* 14.3 and 13 Ma and are characterized by temporal inversions in higher values of $\delta^{18}O$ between the sites. From 13 Ma the $\delta^{18}O$ values are similar at the three sites, but generally higher at Site 555.

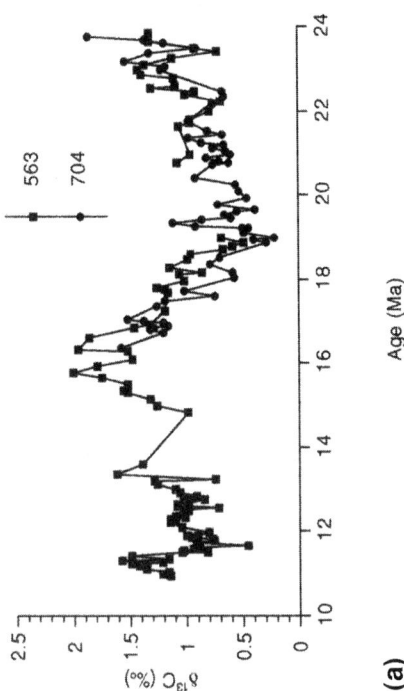

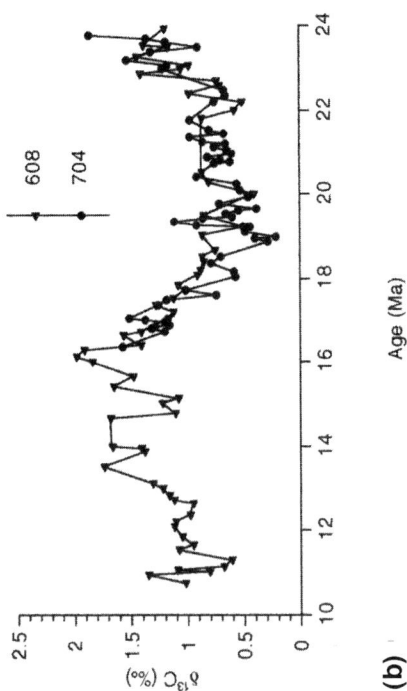

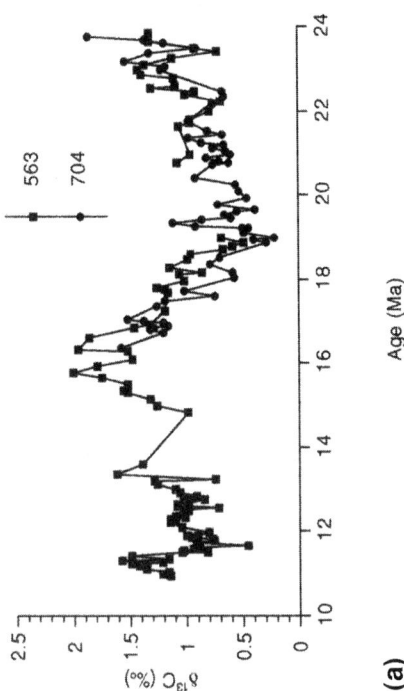

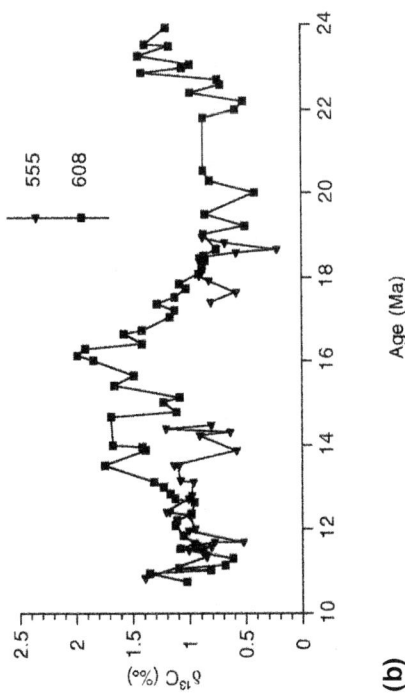

Fig. 5. Comparison between time-series analyses of $\delta^{13}C$ records for northern North Atlantic Site 555 and North Atlantic Sites 563 (a) and 608 (b). Data from Wright *et al.* (1992). Breaks in the records for Sites 555 and 563 are due to unconformities.

Fig. 6. Comparison between time-series analyses of $\delta^{13}C$ records for South Atlantic Site 704 and North Atlantic Sites 563 (a) and 608 (b). Data from Wright *et al.* (1992).

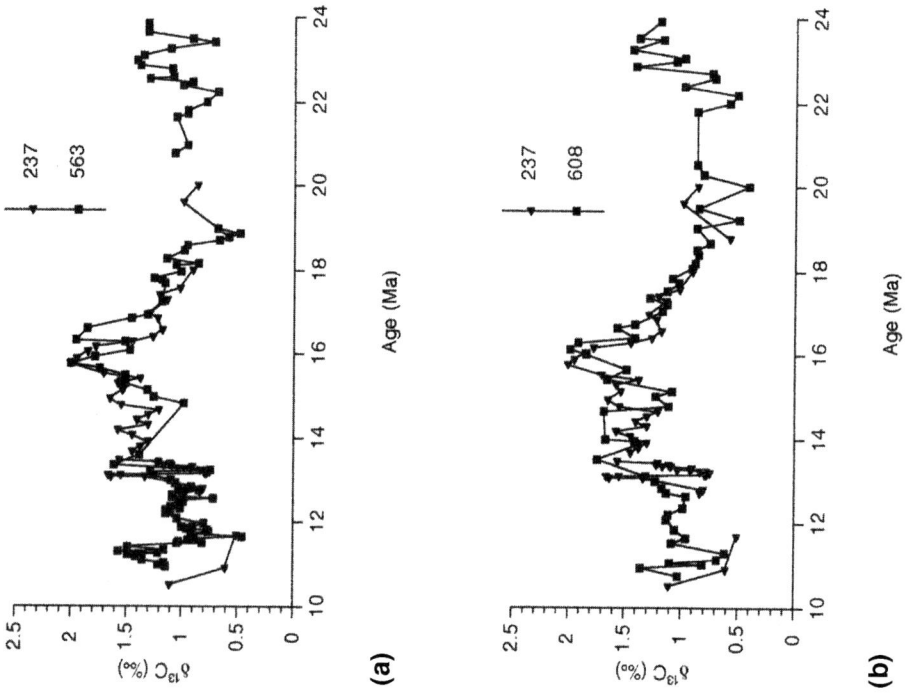

Fig. 8. Comparison between time-series analyses of δ^{13}C records for Indian Ocean Sites 237 and 710. Data for Site 237 are from Woodruff & Savin (1989).

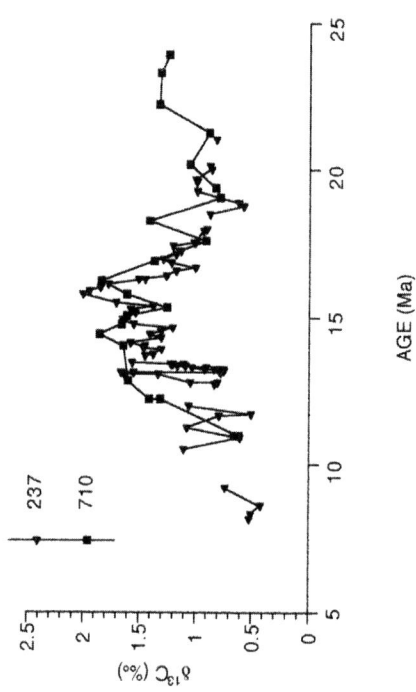

Fig. 7. Comparison between time-series analyses of δ^{13}C records for Indian Ocean Site 237 and North Atlantic Sites 563 (a) and 608 (b). Data from Woodruff & Savin (1989) and Wright *et al.* (1992).

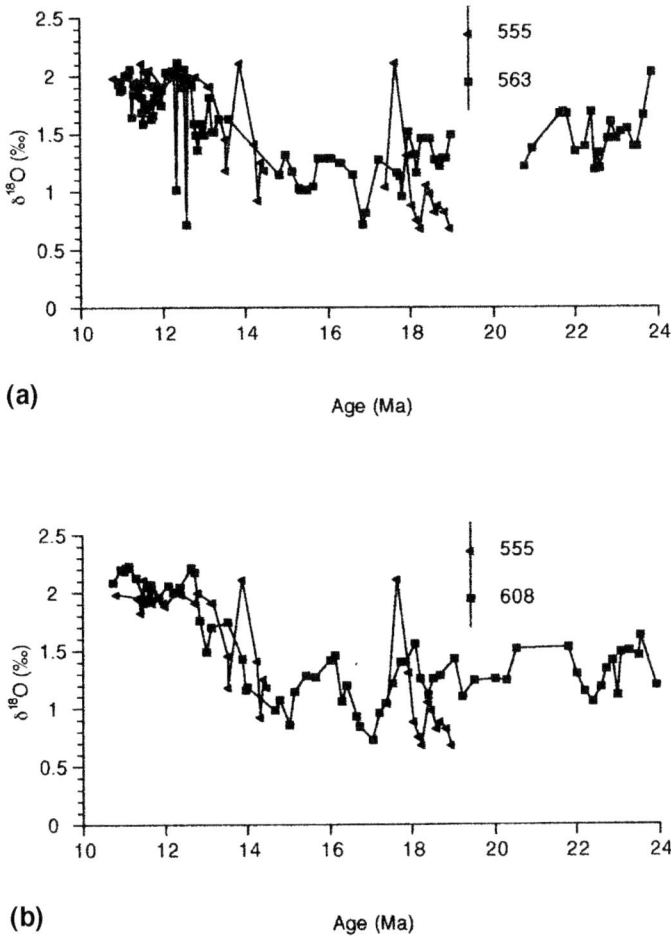

Fig. 9. Comparison between time-series analyses of $\delta^{18}O$ records for northern North Atlantic Site 555 and North Atlantic Sites 563 (a) and 608 (b). Data from Wright *et al.* (1992).

Discussion

Significance of the benthic foraminiferal faunas

In the modern ocean, smooth-walled bolivinids are a component of an infauna of thin-walled, elongate, flattened forms which characterizes low oxygen (dysoxic) conditions (Phleger & Soutar 1973; Douglas 1981; Kaiho 1994). They are recorded in restricted basins with sluggish circulation, and where oxygen minimum zones impinge on shelf and slope areas. In these settings the sediments are rich in organic matter (Boltovskoy & Wright 1976; Sen Gupta & Machain-Castillo 1993).

The high abundances of smooth-walled bolivinids (>10%) in the open-ocean Atlantic, northwest Indian Ocean and Mediterranean sites (Fig. 1) between 19.5 and 16.5 Ma are not associated either with organic-rich sediments or with evidence of increased surface water productivity (Thomas 1986a,b; Smart & Murray 1994; Smart & Ramsay 1995). These high abundances, therefore, do not define depositional environments that are strict analogues of the conditions described above. We propose that open-ocean high abundances of smooth-walled bolivinids coincided with phases of reduced oxygen concentration within the Tethyan Outflow Water (TOW), which for reasons which are unclear did not affect the abundance of other benthic foraminiferal taxa. At this time we are unable to account for the restricted infaunal benthic foraminiferal response to changes in the physico-chemical properties of a water mass.

The smooth-walled bolivinid phase of Smart & Ramsay (1995) formed a distinctive intermediate to deep component in the Atlantic and Indian Oceans between 19.5 and 16.5 Ma (Figs 2 and 3). In the Atlantic Ocean, between 19.5 and 16.5 Ma, there was an apparent depth stratification with an overlying 'intermediate to deep water fauna' (in which no single species was dominant) and the smooth-walled bolivinid fauna. After 16.5 Ma the 'intermediate to deep water fauna' was re-established. In the Indian Ocean, between 19.5 and 16.5 Ma, there was a clear stratification of the benthic foraminiferal faunas (Fig. 3). We attribute the stratification in both oceans to variations in the oxygen concentration within the water columns of these oceans. Thus the 'coarse-walled bolivinid', 'intermediate to deep' and '*N. umboniferus*' faunas (Figs 2 and 3) were associated with higher oxygen concentrations within the water mass. These observations imply that the early Miocene stratification of benthic foraminifera was related to specific deep water masses with different oxygen concentrations.

In the modern ocean, *N. umboniferus* is commonly associated with mixed water masses which contain a component of southerly derived, cold, corrosive, nutrient-depleted and highly oxygenated AABW (Bremer & Lohmann 1982; Mackensen *et al.* 1990, 1993, 1995; Murray 1991). In the Atlantic Ocean, the absence of an '*N. umboniferus* fauna' at depths greater than *c*. 3.4 km (Fig. 2) is attributed to an insignificant influx of Deep Southern Component Water (DSCW) (Fig. 10), an analogue of AABW. The high abundance of *N. umboniferus* at *c*. 3.4 km in Indian Ocean Site 710 (Figs 3 and 4) indicates the presence of a mixed water mass with a significant DSCW component. This deep water mass was probably an analogue of the modern Indian Ocean Common Water which occurs at depths of between 2 and 4 km (Corliss 1979, 1983; Tchernia 1980) and has a significant AABW component (Mantylla & Reid 1983).

Benthic foraminiferal distribution and the benthic stable isotope record

The coeval occurrence of high abundances of smooth-walled bolivinids in the Atlantic and Indian Oceans between 19.5 and 16.5 Ma coincides with high benthic $\delta^{13}C$ signals in the North Atlantic (Sites 563 and 608) and northwest Indian Oceans (Site 237). The high $\delta^{13}C$ values in Sites 563 and 608 (Fig. 5), between *c*. 20 and 16.5 Ma, have been previously attributed to the production of deep, southward flowing NCW, an analogue of NADW, within the North Atlantic (Wright *et al.* 1992; Wright & Miller 1996). The high to low $\delta^{13}C$ gradients recorded between the northern sites (563 and 608) and the southern site 704 (Fig. 6) were used to confirm this interpretation (Wright *et al.* 1992). It is our opinion that these gradients were probably a product of water mass mixing as opposed to water mass ageing. The high $\delta^{13}C$ signal at Site 237 is attributed to the outflow of warm, saline water from Tethys into the northwest Indian Ocean between *c*. 20 and 14.5 Ma (Woodruff & Savin 1989; Wright *et al.* 1992; Zachos *et al.* 1992). We propose that the young Atlantic (NCW of Wright *et al.* (1992) and Wright & Miller (1996)) and Indian Ocean water masses, which were characterized by high abundances of smooth-walled bolivinids, originated within the Tethys Ocean (Figs 10 and 11) and represent an oxygen depleted phase of TOW. The TOW probably formed in a similar manner to the process documented for the origin of modern Mediterranean Outflow Water, i.e. as a warm, dense, saline outflow generated by the intense evaporation of inflowing Atlantic near-surface water (Tchernia 1980). It is possible that during the Mioocene Indian Ocean near-surface water also contributed to TOW production. $\delta^{13}C$ values were almost identical within the Atlantic and Indian Ocean components of the TOW, between *c*. 18 and 17 Ma (Smart & Ramsay 1995). Variations between the

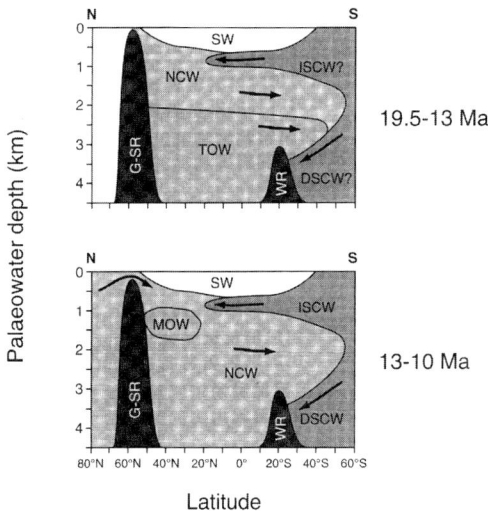

Fig. 10. Schematic diagram of deep water mass circulation in the Atlantic Ocean for 19.5 to 13 Ma and *c*. 13 Ma to 10 Ma. G-SR = Greenland–Scotland Ridge, WR = Walvis Ridge, SW = Surface Water, NCW = Northern Component Water, TOW = Tethyan Outflow Water, ISCW = Intermediate Southern Component Water, DSCW = Deep Southern Component Water. The existence of the ISCW is interpreted from the benthic $\delta^{13}C$ data of Wright *et al.* (1992).

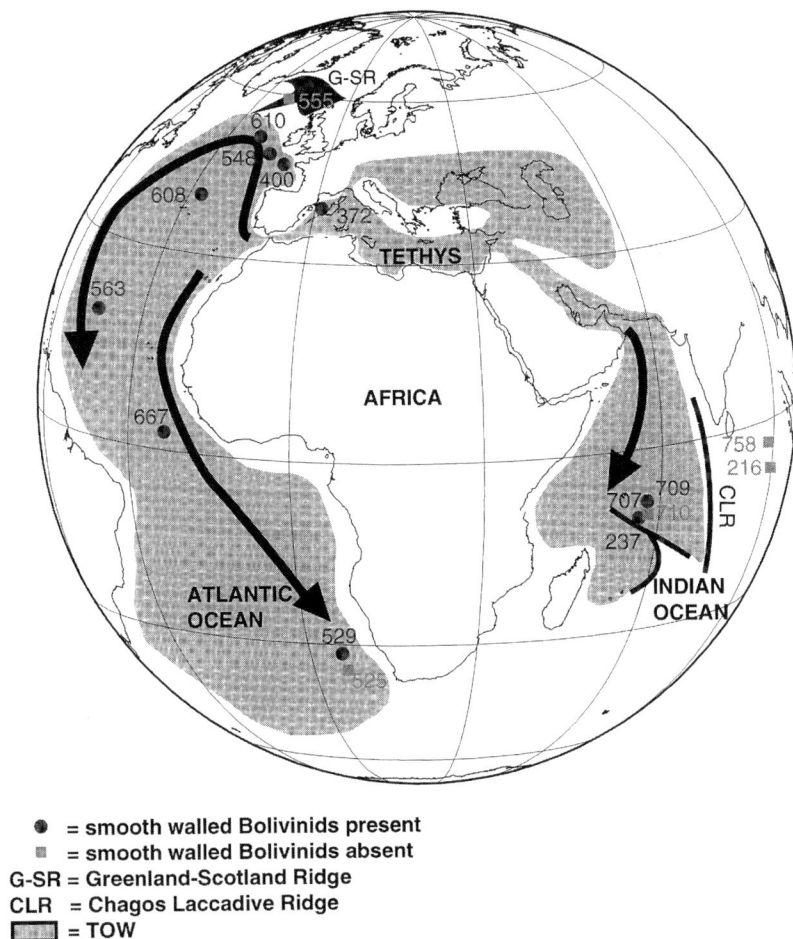

Fig. 11. Distribution and circulation pattern of Tethyan Outflow Water at 18 Ma.

$\delta^{13}C$ signals for the two components of the TOW would have arisen through the mixing of this water mass with the ambient water masses within the Atlantic and Indian Oceans. The discrepancy in the depth ranges of the TOW between the two oceans (Figs 2 and 3) was probably related to the density contrast between the TOW and the ambient water masses within these oceans. In the Atlantic Ocean the dense oxygen-depleted TOW formed a deep, southward outflow with a depth range of between 2 and c. 4.5 km (Fig. 2). The temporal fluctuations in the vertical and lateral extent of the smooth-walled bolivinids (Figs 2 and 3) record the product of mixing between ambient Atlantic water and Tethyan outflow which was characterized by variations in one or all of the following factors: (1) production rates, (2) oxygen concentrations, (3) temperature, salinity and density. Production rates

would have been sensitive to tectonic events which either influenced the rate of inflow and therefore outflow of Atlantic and/or Indian Ocean near-surface waters, or changed the size of the Tethys basin. Stille *et al.* (1996) suggest that the decoupling of the Nd isotopic composition of Atlantic and Tethyan sea water from 20 to 17 Ma (based on analyses of marine phosphorites) records an early stage in the isolation of Tethys. Changes in oxygen concentration were probably related to changes in temperature, salinity and density in response to climatic events within the Tethys region. In the Indian Ocean, the more restricted depth range (c. 1.4 to c. 3. 5 km) of the oxygen depleted TOW is compatible with the presence of a cold, denser underlying water mass inferred from the high abundance of *N.umboniferus* at Site 710 (see above discussion).

Palaeoceanographic significance of the TOW

The cessation of oxygen-depleted Tethyan outflow at 16.5 Ma, which is indicated by low abundances of smooth-walled bolivinids in the Atlantic and Indian oceans, probably did not coincide with the termination of TOW production. In the northwest Indian Ocean the age of this event coincided with the closure of the eastern end of Tethys, although there is a debate as to when this occurred. Hypotheses concerning the timing of closure of the Tethys to the Indian Ocean include the following. Adams *et al.* (1983) proposed that a definite barrier to the dispersal of marine organisms existed between the Mediterranean and Indian Ocean by the mid-Burdigalian (*c.* 19 Ma) based on the geographical distribution of age-diagnostic larger foraminifera. Woodruff & Savin (1989) suggest an age of *c.* 14.5 Ma for the reduction or termination of Tethyan outflow into the Indian Ocean. based on stable isotope analyses. Drooger (1979, 1993) suggested, on the basis of morphometric studies of larger foraminifera, that the the Mediterranean and Indian Ocean were disconnected by the end of the Oligocene. Rögl & Steininger (1984) proposed that a marine connection between the Mediterranean and Indian Ocean existed throughout the Oligocene, was disrupted at some time during the early Miocene and was briefly restored again in the early Middle Miocene. We suggest, on the basis of our data (see above), that the marine connection between Tethys and the Indian Ocean existed until 14.5 Ma. This view supports the conclusions of Woodruff & Savin (1989). The closure of this connection at 14.5 Ma coincides with a reduction in $\delta^{13}C$ values for Site 237 relative to Sites 608 (North Atlantic) and 710 (Indian Ocean). Wright *et al.* (1992) and Wright & Miller (1996) have used divergences and convergences in the magnitude of N Atlantic–N Pacific $\delta^{13}C$ values as indicators for fluctuations in the production rate of NCW. Intervals characterized by divergence and high production are identified between 20 to *c.* 15 Ma (= TOW, this paper) and 12.5–5 Ma (= NADW, this paper). Intervals of convergence and small differences in the magnitude of N Atlantic–N Pacific $\delta^{13}C$ values show that NCW production was either reduced or ceased between *c.* 15 and 12.5 Ma.

Woodruff & Savin (1989) proposed that the flux of warm TISW (= Indian Ocean Component of TOW) to the Southern Ocean was an important factor in meridional heat transport and inhibited the growth of the East Antarctic Ice Sheet until production ceased in the middle Miocene. Flower & Kennett (1994) have adopted this model. The TOW component in the northwest Indian Ocean was constrained vertically (Fig. 3) between overlying intermediate water (characterized by coarse-walled bolivinids) and underlying deeper water (characterized by high abundances of *N. umboniferus*) and laterally by the Chagos–Lacadive Ridge (Smart & Ramsay 1995) (Fig. 11). Nomura (1995) also concluded that TISW was absent from eastern Indian Ocean Sites 757 and 758 between *c.* 19 and 17 Ma.

Wright & Miller (1996) proposed that NCW production was a factor in controlling Neogene climate. They draw attention to a paradox in the Miocene in which a high flux of NCW (20 and *c.* 15 Ma) was associated with a warm global climate, a reduction/cessation in the production of NCW (post-15 Ma) was associated with Antarctic glaciation and renewed NCW production (after 12.5 Ma) did not signal a return to the warm conditions of the early Miocene. Investigations of early to middle Miocene ice volume changes, based on trends in benthic $\delta^{18}O$ (Miller *et al.* 1991; Wright & Miller 1992; Wright *et al.* 1992), reveal a significant increase in $\delta^{18}O$ (*c.* 1‰) between *c.* 15 Ma and *c.* 12.5 Ma. Between *c.* 20 and *c.* 14.5 Ma variations in $\delta^{18}O$ values do not exceed 0.5‰.

We propose that the solution to the Miocene paradox lies with the origin and nature of NCW. Between *c.* 20 and *c.* 13 Ma deep, outflow from the Atlantic was dominated by TOW. This interpretation is based on differences between the benthic stable isotope records and the presence/absence of smooth-walled bolivinids at Sites 555, 563 and 608. Wright *et al.* (1992) regarded the gradient in $\delta^{13}C$ values between Sites 555, 563 and 608, in a time-slice analysis for the early Miocene, as the product of variation within a single water mass. The lower benthic $\delta^{13}C$ and $\delta^{18}O$ values and the absence of smooth-walled bolivinids at Site 555 indicate that this site was not within the TOW. We speculate (Fig. 10) that this site occurred within intermediate to deep, less dense, southward flowing NCW which originated on shelf areas to the south of the Greenland–Scotland Ridge (Wold 1994). The existence of this water mass is supported by Murray (1984) who studied Paleogene and Neogene benthic foraminifera from the Rockall Plateau. Sites 563 and 608 were situated within the southward flowing deep, warm, dense saline TOW (Fig. 10) which entered the North Atlantic via the Straits of Gibraltar (Fig. 11). They have high $\delta^{13}C$ values, indicative of a young water mass, contain smooth-walled bolivinids and are characterized by more positive $\delta^{18}O$ values The latter may have originated in part from evaporation and $\delta^{18}O$ enrichment in the Tethys. The reduction/cessation in Atlantic outflow (*c.* 15–12.5 Ma) based on low intrabasinal $\delta^{13}C$ values (Wright *et al.* 1992; Wright & Miller 1996) probably relate to changes in TOW production. The convergence of the $\delta^{13}C$ and $\delta^{18}O$

time series records for Sites 555 and 563 and 608 (Figs 5 and 9) suggests that these sites occurred within a single deep water mass (NADW) from c. 13 Ma. The initiation of the southward flowing NADW probably coincided with the first significant overflow of Norwegian–Greenland Sea water across the Greenland–Scotland Ridge (Fig. 1) (Bohrmann et al. 1990). These workers suggest that this event occurred within the Iceland–Scotland region of the ridge between 13 and 11 Ma. Alternatively, Larsen et al. (1994) suggest that the overflow of Norwegian–Greenland Sea water through the Denmark Strait did not occur before c. 11.5 Ma.

The formation of NADW at c. 13 Ma was accompanied by the replacement of TOW by Mediterranean Outflow Water (MOW) (Fig. 10). The consequent reduction in meridional heat transport to high southern latitudes contributed to global cooling and to major growth of the East Antarctic Ice Sheet, at c. 13 Ma (Flower & Kennett 1994). The coeval increase in the relative abundances of N. umboniferus at Indian Ocean Sites 237 and 710, at c. 13 Ma, probably records the impact of this change on the nature of Indian Ocean deep water at depths between 2 and 4 km.

Conclusions

We have identified the Tethys Ocean as the major source of southward flowing deep water (>2 km) in the Atlantic Ocean between c. 20 and c. 13 Ma, and confirmed the occurrence of intermediate to deep Tethyan Outflow Water in the Indian Ocean between c. 20 Ma and c. 14.5 Ma. The warm, dense, saline, southward flowing Atlantic component of the TOW is, in part, equivalent to Atlantic NCW which, according to earlier workers, formed within the North Atlantic Basin. Temporal and spatial fluctuations in the high relative abundances of smooth-walled bolivinids within the Atlantic and Indian Ocean components of the TOW, between 19.5 and 16.5 Ma, coincided with phases of reduced oxygen concentration.

Between c. 20 and c. 13 Ma the North and central Atlantic Ocean was characterized by two deep, southward flowing water masses: the deep, dense , saline TOW and the overlying less dense NCW. At c. 13 Ma these water masses were replaced by the single southward flowing, cold, dense **NADW** and much less voluminous warm, saline MOW. Water mass stratification in the northwestern Indian Ocean from c. 20 and c. 14.5 Ma consisted of three water masses: an oxygenated intermediate water mass; the TOW; and a deeper, cold, dense water mass, probably an analogue of the modern Common Indian Ocean Water. The demise of the TOW at c. 14.5 Ma resulted in a two-layered water mass structure.

It has been proposed that the southward flow of Tethyan Indian Saline Water (i.e. TOW), via the Indian Ocean, was a major factor in the process of meridional heat transport to the high southern latitudes during the early to middle Miocene, and that the termination of this water mass at c. 13 Ma altered the pattern of meridional heat exchange and resulted in major growth of the East Antarctic Ice Sheet. The Indian Ocean TOW was a relatively insignificant component of the total Tethyan outflow. In our opinion, the southward flowing, high-volume TOW in the Atlantic provided the principal component in the mechanism of meridional heat transport to the high southern latitudes. The replacement of NCW and the warm, dense, saline water Atlantic TOW by a single southward flowing water mass, NADW, probably coincided with the overflow, at c. 13 Ma, of cooler and even denser Norwegian–Greenland Sea water across the Greenland–Scotland Ridge. The demise of the Atlantic TOW and, to a lesser extent, Indian Ocean TOW contributed significantly to global cooling and the expansion of the mid-Miocene Antarctic Ice Sheet.

We thank the Ocean Drilling Program for providing samples and J. Murray and E. Thomas for reviewing the original manuscript. This research was supported by an NERC/ODP Special Topic Research Grant (No. GST/02/881).

References

Adams, C. G., Gentry, A. W. & Whybrow, P. J. 1983. Dating the terminal Tethyan event. Utrecht, Micropalaeontological, Bulletin, 30, 273–278.

Belanger, P. E., Curry, W. B. & Matthews, R. K. 1981. Core-top evaluation of benthic foraminiferal isotopic ratios for palaeoceanographic interpretations. Palaeogeography, Palaeoclimatology, Palaeoecology, 33, 205–220.

Berggren, W. A., Kent, D. V. & van Couvering, J. A. 1985. Neogene geochronology and chronostratigraphy. In: Snelling, N. J. (ed.) The Chronology of the Geological Record. Geological Society, London, Memoir, 10, 211–260.

Bohrmann, G., Henrich, R. & Thide, J. 1990. Miocene to Quaternary paleoceanography in the northern North Atlantic: variability in carbonate and biogenic opal accumulation. In: Bleil, U. & Thiede, J. (eds) Geological History of the Polar Oceans: Arctic versus Antarctic. Kluwer, Netherlands, 647–675.

Boltovskoy, E. & Boltovskoy, D. 1989. Paleocene–Pleistocene benthic foraminiferal evidence of major

paleoceanographic events in the eastern South
Atlantic (DSDP Site 525, Walvis Ridge). *Marine
Micropaleontology,* **14,** 283–316.

—— & WRIGHT, R. 1976. *Recent Foraminifera.* W. Junk,
The Hague.

BREMER, M. L. & LOHMANN, G. P. 1982. Evidence for
primary control of the distribution of certain
Atlantic Ocean benthonic foraminifera by degree of
carbonate saturation. *Deep-Sea Research,* **29,**
987–998.

CORLISS, B. H. 1979. Recent deep-sea benthic foraminiferal distributions in the southeast Indian Ocean:
inferred bottom water routes and ecological
implications. *Marine Geology,* **31,** 115–138.

—— 1983. Distribution of Holocene deep-sea benthonic
foraminifera in the southwest Indian Ocean. *Deep-
Sea Research,* **30,** 95–117.

DOUGLAS, R. G. 1981. Paleoecology of continental margin
basins: a case history from the borderland of
Southern California. *Depositional Systems of Active
Continental Margin Basins.* Society of Economic
Paleontologists and Mineralogists, Pacific Section,
Short Course, 121–156.

DROOGER, C. W. 1979. Marine connections of the
Neogene Mediterranean, deduced from the
evolution and distribution of larger foraminifera.
*Annales Géologiques des Pays Helléniques Tome
Hors série,* 1979 (1), 361–369.

—— 1993. Radial foraminifera; morphometrics and
evolution. *Verhandelingen der koninklijke
Nederlandse Akademie van Wetenschappen,
Afdeeling Natuurkunde, Eerste Reeks,* **41,** 1–242.

DUPLESSY, J.-C., LALOU, C. & VINOT, A. C. 1970.
Differential isotopic fractionation in benthic
foraminifera and paleotemperatures reassessed.
Science, **168,** 250–251.

FLOWER, B. P. & Kennett, J. P. 1993. Middle Miocene
ocean-climate transition: high-resolution oxygen
and carbon isotopic records from Deep Sea Drilling
Project Site 588A, southwest Pacific. *Paleoceanography,* **8,** 811–843.

—— & —— 1994. The middle Miocene climatic
transition: east Antarctic ice sheet development,
deep ocean circulation and global carbon cycling.
*Palaeogeography, Palaeoclimatology, Palaeo-
ecology,* **108,** 537–555.

FORNACIARI, E., RAFI, D., RIO, G., VILLA, J., BACKMANN,
J. & OLAFFSON, G. 1990. Quantitative distribution
pattern of Oligocene and Miocene calcareous
nannofossils from the western equatorial Indian
Ocean. *In:* DUNCAN, R. A., BACKMAN, J., PETERSON,
L. C. *ET AL.* (eds) *Proceedings of the Ocean Drilling
Program, Scientific Results,* **115.** College Station,
TX (Ocean Drilling Program), 237–254.

GOODAY, A. J. 1994. The biology of deep-sea benthic
foraminifera: a review of some advances and their
applications in paleoceanography. *Palaios,* **9,**
14–31.

GRAHAM, D. W., CORLISS, B. H., BENDER, M. L. &
KEIGWIN, L. D. 1981. Carbon and oxygen isotopic
disequilibria of Recent benthic foraminifera.
Marine Micropaleontology, **6,** 483–497.

HAYES, D. E. 1988. Age–depth relationships and depth
anomalies in the southeast Indian Ocean and South

Atlantic Ocean. *Journal of Geophysical Research,*
93, 2937–2954.

KAIHO, K. 1994. Benthic foraminiferal dissolved-oxygen
index and dissolved-oxygen levels in the modern
ocean. *Geology,* **22,** 719–722.

KENNETT, J. P. & SHACKLETON, N. J. 1976. Oxygen
isotopic evidence for the development of the
psychrosphere 38 Myr. ago. *Nature,* **260,** 513–515.

KROOPNICK, P. M. 1985. The distribution of ^{13}C of ΣCO_2
in the world's oceans. *Deep-Sea Research,* **32,**
57–84.

LARSEN, H. C., SAUNDERS, A. D. & CLIFT, P. 1993. *Ocean
Drilling Program, Leg 152: preliminary report, east
Greenland margin.* Ocean Drilling Program,College
Station, TX.

MACKENSEN, A., GROBE, H., KUHN, G. & FÜTTERER, D. K.
1990. Benthic foraminiferal assemblages from the
eastern Weddell Sea between 68 and 73°S:
distribution, ecology and fossilization potential.
Marine Micropaleontology, **16,** 241–283.

——, FÜTTERER, D. K., GROBE, H. & SCHMIEDL, G.
1993. Benthic foraminiferal assemblages from
the eastern South Atlantic Polar Front region
between 35° and 57°S: distribution, ecology and
fossilization potential. *Marine Micropaleontology,*
22, 33–69.

——, SCHMIEDL, G., HARLOFF, J. & GIESE, M. 1995.
Deep-sea foraminifera in the South Atlantic Ocean:
ecology and assemblage generation. *Micropaleontology,* **41,** 342–358.

MANTYLLA, A. W. & REID, J. L. 1983. Abyssal
characteristics of the world's ocean waters. *Deep-
Sea Research,* **30,** 805–833.

MILLER K. G., FAIRBANKS, R. G. & MOUNTAIN, G. S. 1987.
Tertiary oxygen isotope synthesis, sea level history,
and continental margin erosion. *Paleoceanography,*
2, 1–19.

——, WRIGHT, J. D. & FAIRBANKS, R. G. 1991. Unlocking
the ice house: Oligocene–Miocene oxygen isotopes,
eustacy, and margin erosion. *Journal of Geophysical
Research,* **96,** 6829–6848.

MURRAY, J. W. 1984. Paleogene and Neogene benthic
foraminifers from Rockall Plateau. *In:* ROBERTS, D.
G., SCHNITKER, D. *ET AL.* (eds) *Initial Reports of the
Deep Sea Drilling Project,* **81.** US Government
Printing Office, Washington, DC, 503–534.

—— 1988. Neogene bottom-water masses and benthic
foraminifera in the NE Atlantic. *Journal of the
Geological Society, London,* **145,** 125–132.

—— 1991. *Ecology and Palaeoecology of Benthic
Foraminifera.* Longman, London.

——, WESTON, J .F., HADDON, C. A. & POWELL, A. D. J.
1986. Miocene to Recent bottom water masses of
the north-east Atlantic: an analysis of benthic
foraminifera. *In:* SUMMERHAYES, C. P. &
SHACKLETON, N. J. (eds) *North Atlantic Palaeoceanography.* Geological Society, London, Special
Publications, **21,** 219–230.

NOMURA, R. 1995. Paleogene to Neogene deep-sea
paleoceanography in the eastern Indian Ocean:
benthic foraminifera from ODP Sites 747, 757 and
758. *Micropaleontology,* **41,** 251–290.

PARSONS, B. & SCLATER, J. G. 1977. An analysis of the
variation of ocean floor bathymetry and heat flow

with age. *Journal of Geophysical Research,* **82,** 803–827.

PHLEGER, F .B. & SOUTAR, A. 1973. Production of benthic foraminifera in three East Pacific oxygen minima. *Micropaleontology,* **19,** 110–115.

POAG, C. W. & LOW, D. 1985. Environmental trends among Neogene benthic foraminifers at DSDP Site 548, Irish continental margin. *In:* DE GRACIANSKY, P. C., POAG, C. W. *ET AL.* (eds) *Initial Reports of the Deep Sea Drilling Project,* **80.** US Government Printing Office, Washington, DC, 489–503.

RÖGL, F. & STEININGER, F. F. 1984. Neogene Paratethys, Mediterranean and Indo-Pacific seaways. Implications for the paleobiogeography of marine and terrestrial biotas. *In:* BRENCHLY, P. (ed.) *Fossils and Climate.* Wiley, London, 171–200.

SAVIN, S. M., DOUGLAS, R. G. & STEHLI, F. G. 1975. Tertiary marine paleotemperatures. *Geological Society of America Bulletin,* **86,** 1499–1510.

——, ——, KELLER, G., KILLINGLEY, J. S., SHAUGHNESSY, L., SOMMER, M. A., VINCENT, E. & WOODRUFF, F. 1981. Miocene benthic foraminiferal isotope records: a synthesis. *Marine Micropaleontology,* **6,** 423–450.

SCHNITKER, D. 1980. Global paleoceanography and its deep water linkage to Antarctic glaciation. *Earth Sciences Reviews,* **16,** 1–20.

—— 1994. Deep-sea benthic foraminifers: food and bottom water masses. *In:* ZAHN, R. *ET AL.* (eds) *Carbon Cycling in the Glacial Ocean: constraints on the ocean's role in global change.* Springer, New York, 539–554.

SCHROEDER, C. J., SCOTT, D. B. & MEDIOLI, F. S. 1987. Can smaller benthic foraminifera be ignored in paleoenvironmental analyses? *Journal of Foraminiferal Research,* **17,** 101–105.

SEN GUPTA, B. K. & MACHAIN-CASTILLO, M. L. 1993. Benthic foraminifera in oxygen-poor habitats. *Marine Micropaleontology,* **20,** 183–201.

SETO, K. 1995. Carbon and oxygen isotopic paleoceanography of the Indian and South Atlantic Oceans – paleoclimate and paleo-ocean circulation. *Journal of Science of the Hiroshima University,* **10,** 393–485.

SHACKLETON, N. J. & KENNETT, J. P. 1975. Paleotemperature history of the Cenozoic and the initiation Antarctic glaciation: oxygen and carbon isotope analysis in DSDP Sites 277, 279, and 281. *In:* KENNETT, J. P., HOUTZ, R. E. *ET AL.* (eds) *Initial Reports of the Deep Sea Drilling Project,* **29.** US Government Printing Office, Washington, DC, 743–755.

—— & OPDYKE, N. D. 1973. Oxygen isotope and palaeomagnetic stratigraphy of equatorial Pacific Core V28-238: oxygen isotope temperatures and ice volumes on a 10^5 year and 10^6 year scale. *Quaternary Research,* **3,** 39–55.

SMART, C. W. 1992. *Ecological controls on patterns of speciation and extinction in deep-sea benthic foraminifera.* PhD thesis, University of Southampton.

—— & MURRAY, J. W. 1994. An early Miocene Atlantic-wide foraminiferal/palaeoceanographic event.

Palaeogeography, Palaeoclimatology, Palaeoecology, **108,** 139–148.

—— & RAMSAY, A. T. S. 1995. Benthic foraminiferal evidence for the existence of an early Miocene oxygen-depleted oceanic water mass? *Journal of the Geological Society, London,* **152,** 735–738.

STILLE, P., STEINMANN, M. & RIGGS, S. R. 1996. Nd isotope evidence for the evolution of the paleocurrents in the Atlantic and Tethys Oceans during the past 180 Ma. *Earth and Planetary Science Letters,* **144,** 9–19.

TCHERNIA, P. 1980. *Descriptive Regional Oceanography.* Pergamon, Oxford.

THOMAS, E. 1986a. Late Oligocene to Recent benthic foraminifers from Deep Sea Drilling Project Sites 608 and 610, northeastern North Atlantic. *In:* RUDDIMAN, W. F., KIDD, R. B., THOMAS, E. *ET AL.* (eds) *Initial Reports of the Deep Sea Drilling Project,* **94.** US Government Printing Office, Washington, DC, 997–1031.

—— 1986b. Early to middle Miocene benthic foraminiferal faunas from DSDP Sites 608 and 610, North Atlantic. *In:* SUMMERHAYES, C. P. & SHACKLETON, N. J. (eds) *North Atlantic Palaeoceanography.* Geological Society, London, Special Publications, **21,** 205–218.

—— 1992. Cenozoic deep-sea circulation: evidence from deep-sea benthic foraminifera. *In:* KENNETT, J. P. & WARNKE, D. (eds) *The Antarctic Paleoenvironment: A perspective on global change.* American Geophysical Union, Antarctic Research Series, **56,** 141–165.

WESTON, J. F. & MURRAY, J. W. 1984. Benthic foraminifera as deep-sea water-mass indicators. *In:* OERTLI, H. J. (ed.) *Benthos 1983: Second International Symposium on benthic foraminifera (Pau, 1983).* Elf-Aquitaine, ESSO REP and Total CFP, Pau, 605–610.

WISE, S.W., BREZA, J. R., HARWOOD, D. M.,WEI, W. & ZACHOS J. C. 1992. Paleogene glacial history of Antarctica in light of Leg 120 drilling results. *In:* WISE, S. W., SCHLICH, R. *ET AL.* (eds) *Proceedings of the Ocean Drilling Program, Scientific Results,* **120.** College Station, TX (Ocean Drilling Program), 1001–1030.

WOLD, C. N. 1994. Cenozoic sediment accumulation on drifts in the northern North Atlantic. *Paleoceanography,* **9,** 917–941.

WOODRUFF, F. 1985. Changes in Miocene deep-sea benthic foraminifera in the Pacific Ocean: relationship to paleoceanography. *In:* KENNETT, J. P. (ed.) *The Miocene Ocean.* Geological Society of America, Memoirs, **163,** 131–176

—— & SAVIN, S. M. 1989. Miocene deepwater oceanography. *Paleoceanography,* **4,** 87–140.

—— &—— 1991. Mid-Miocene isotope stratigraphy in the deep sea: high-resolution correlations, paleoclimatic cycles, and sediment preservation. *Paleoceanography,* **6,** 755–806.

——, —— & DOUGLAS, R. G. 1980. Biological fractionation of oxygen and carbon isotopes by Recent benthic foraminifera. *Marine Micropaleontology,* **5,** 3–11.

WRIGHT, J. D. & MILLER, K. G. 1992. Miocene stable

isotope stratigraphy, Site 747, Kerguelen Plateau. *In*: WISE, S. W. JR., SCHLICH, R. *ET AL.* (eds) *Proceedings of the Ocean Drilling Program, Scientific Results*, **120**. College Station, TX (Ocean Drilling Program), 855–866.

—— & —— 1996. Control of North Atlantic deep water circulation by the Greenland – Scotland Ridge. *Paleoceanography*, **11**, 157–170.

——, —— & FAIRBANKS, R. G. 1991. Evolution of deep-water circulation. Evidence from the late Miocene Southern Ocean. *Paleoceanography*, **6**, 275–290.

——, —— & —— 1992. Early and middle Miocene stable isotopes: implications for deepwater circulation and climate. *Paleoceanography*, **7**, 357–389.

ZACHOS, J. C., REA, D. K., SETO, K., NOMURA, R. & NIITSUMA, N. 1992. Paleogene and early Neogene deep water paleoceanography of the Indian Ocean as determined from benthic foraminifer stable carbon and oxygen isotope records. *In*: DUNCAN, R., REA, D. K., KIDD, R. B., VON RAD, U. & WEISSEL, J. K. (eds) *Synthesis of Results from Scientific Drilling in the Indian Ocean*. Geophysical Monograph, **70**, 351–385.

Mid-Cretaceous radiolarian zonation for the North Atlantic: an example of oceanographically controlled evolutionary processes in the marine biosphere?

JOCHEN ERBACHER[1] & JÜRGEN THUROW[2]

[1] Institut für Geologie und Paläontologie, Universität Tübingen, Sigwartstrasse 10, 72076 Tübingen, Germany (present address: Bundesanstalt fur Geowissenschaften und Rohstoffe, 30631 Hanover, Germany)

[2] University College of London, Department of Geological Sciences, Gower Street, London WC1E 6BT, UK

Abstract: Mid-Cretaceous (Aptian–Turonian) sediments from seven Deep Sea Drilling Project and Ocean Drilling Program sites in the North Atlantic have been investigated in order to establish a mid-Cretaceous radiolarian biostratigraphy for the North Atlantic. One of the results, a new zonation for mid-Cretaceous North Atlantic radiolaria, is presented here. From parallel investigations of calcareous sections in central Italy an accurate calibration to standard planktonic foraminiferal zonations is possible. The evolution of mid-Cretaceous radiolaria appears to be strongly related to sea-level changes, nutrient supply and the extension of the oxygen minimum zone. A model was proposed that explains extinction/radiation events of mid-Cretaceous radiolaria as being controlled by sea-level fluctuations and resulting changes of productivity. These are the parameters that control the dimensions of the oxygen minimum zone, the fluctuation of which destroys and creates habitats.

The mid-Cretaceous (Aptian–Turonian) was a period of major changes in the ocean–atmosphere system. Increases in ocean spreading documented in the Atlantic and the Pacific, coupled with an opening of new seaways and a global transgressive period modified the palaeogeography (Sheridan 1986; Larson 1991; Dercourt et al. 1993) to a broad extent. These modifications are paralleled by major faunal and floral changes in the marine biosphere (e.g. Hart 1980; Roth 1986; Wiedmann 1988; Leckie 1989; Erbacher & Thurow 1997). This paper describes the evolution of radiolaria during the mid-Cretaceous in the North Atlantic and tries to interpret radiolarian evolution patterns palaeoceanographically. Radiolaria were examined from the mid-Cretaceous sediments of North Atlantic Deep Sea Drilling Project (DSDP) and Ocean Drilling Program (ODP) Sites 137/138, 398, 417, 545, 603, 638 and 641 (Thurow 1988a; Erbacher 1994) (Fig. 1). Here we synthesize our results of these studies. The resulting new biostratigraphy of North Atlantic radiolaria is tied to other published radiolarian and foraminiferal biostratigraphies that have been established during the last two decades (Caron 1985; Sanfilippo & Riedel 1985; Schaaf 1985; Sliter 1989; O'Dogherty

1994). Until now, only the seven sections mentioned have been investigated. Nevertheless, as these sections cover large areas of the mid-Cretaceous central North Atlantic, we believe that the observed stratigraphical patterns are quite representative. This is supported by a good correlation with biostratigraphies and zonations from the western Tethys (Erbacher 1994, O'Dogherty 1994).

Only moderately to well preserved faunas from cherts, haloes, radiolarian sandstones and claystones have been studied. Radiolarian samples were treated with HCl and a detergent (REWOQUAD). Standard methods were used for washing and sieving of radiolarian samples.

Stratigraphy of the investigated DSDP/ODP sites

All the examined sections were drilled in the Aptian to Cenomanian Hatteras Formation (Jansa et al. 1979). The Hatteras-Formation comprises green to red marls, marly claystones and claystones with numerous black shale intercalations.

ERBACHER, J. & THUROW, J. 1998. Mid-Cretaceous radiolarian zonation for the North Atlantic: an example of oceanographically controlled evolutionary processes in the marine biosphere?. *In*: CRAMP, A., MACLEOD, C. J., LEE, S. V. & JONES, E. J. W. (eds) *Geological Evolution of Ocean Basins: Results from the Ocean Drilling Program*. Geological Society, London, Special Publications, **131**, 71–82.

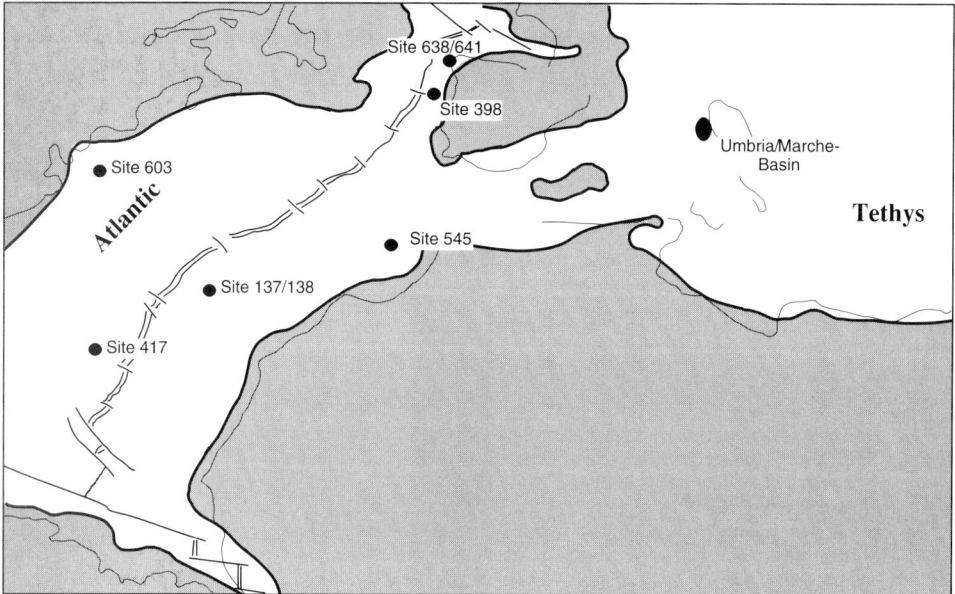

Fig. 1. Palaeogeographic reconstruction for the Cenomanian (96 Ma) showing location of investigated DSDP/ODP sites and land sections. Modified after Dercourt *et al.* (1993).

Site 398D was drilled on the slope of Vigo Seamount, 160 km west of Iberia. Detailed descriptions of the lithostratigraphy and diagenetic history were given by Shipboard Scientific Party (1979) and Thurow (1988*b*). Radiolarian faunas occur from Core 114 (*S. cabri* foraminiferal zone) to Core 56 (*R. cushmani* foraminiferal zone) (Fig. 2). The planktonic foraminiferal zonation for Site 398D has been established by Sigal (1979).

Site 545 was drilled at the base of the Mazagan Plateau, 100 km west of Casablanca (Shipboard Scientific Party 1984). We sampled Cores 56 to 28 (Late Aptian to Early Cenomanian) (Fig. 3). Radiolarian faunas occur from Cores 54 to 42 which range from the Late Aptian to the lower Late Albian (*G. algerianus* to *T. praeticinensis* foraminiferal zones) (Leckie 1984).

Site 417D was located 400 km southwest of Bermuda at the southern end of the Bermuda Rise. Detailed core descriptions can be found in Shipboard Scientific Party (1980). Cores 21 to 17 have been sampled (Fig. 4). Sufficiently well preserved radiolarians have been found in Cores 21, 20, 19 and 17. The biostratigraphy of the Cretaceous portion of Site 417D is not yet very well established. However, calcareous nanofloras from a 1 m thick black shale level at the base of Core 21 have been dated as being late Early Aptian (oceanic anoxic event (OAE) 1a) in age (Bralower 1987). This black shale horizon is followed by a cyclic

succession of red clayey marls and green radiolarian sands reaching up to Core 20, that broadly resemble time-equivalent series in the Umbria–Marche Basin of central Italy (Tornaghi *et al.* 1989). Radiolarians from these cores are of Late Aptian age which is confirmed by calcareous nanoplankton ages (*R. angustus* zone, det. Mutterlose). Planktic foraminifera are abundant in Core 17 and give a Late Albian age (*R. ticinensis* foraminiferal zone) (Miles *et al.* 1980).

Sites 137/138 are located on the upper continental rise, 1000 km west of Cape Blanc. For detailed core descriptions see Berger & von Rad (1972) and Shipboard Scientific Party (1972*a*,*b*). We processed Cores 137-16 to 8 (Fig. 4) and Core 138-5. The oldest interval yielding radiolarians is of Late Albian (*P. buxtorfi* foraminiferal zone), the youngest of Late Cenomanian age (late *R. cushmani* foraminiferal zone).

Radiolarian data for Sites 603, 638 and 641 and Site 398D, Core 56, are taken from Thurow (1988*a*).

Biostratigraphy and zonation

Many radiolarian zonations for the mid-Cretaceous have been produced during the last 20 years. Unfortunately, most of them differ from each other, and are not properly tied to calcareous plankton stratigraphies and absolute timescales. Reasons for

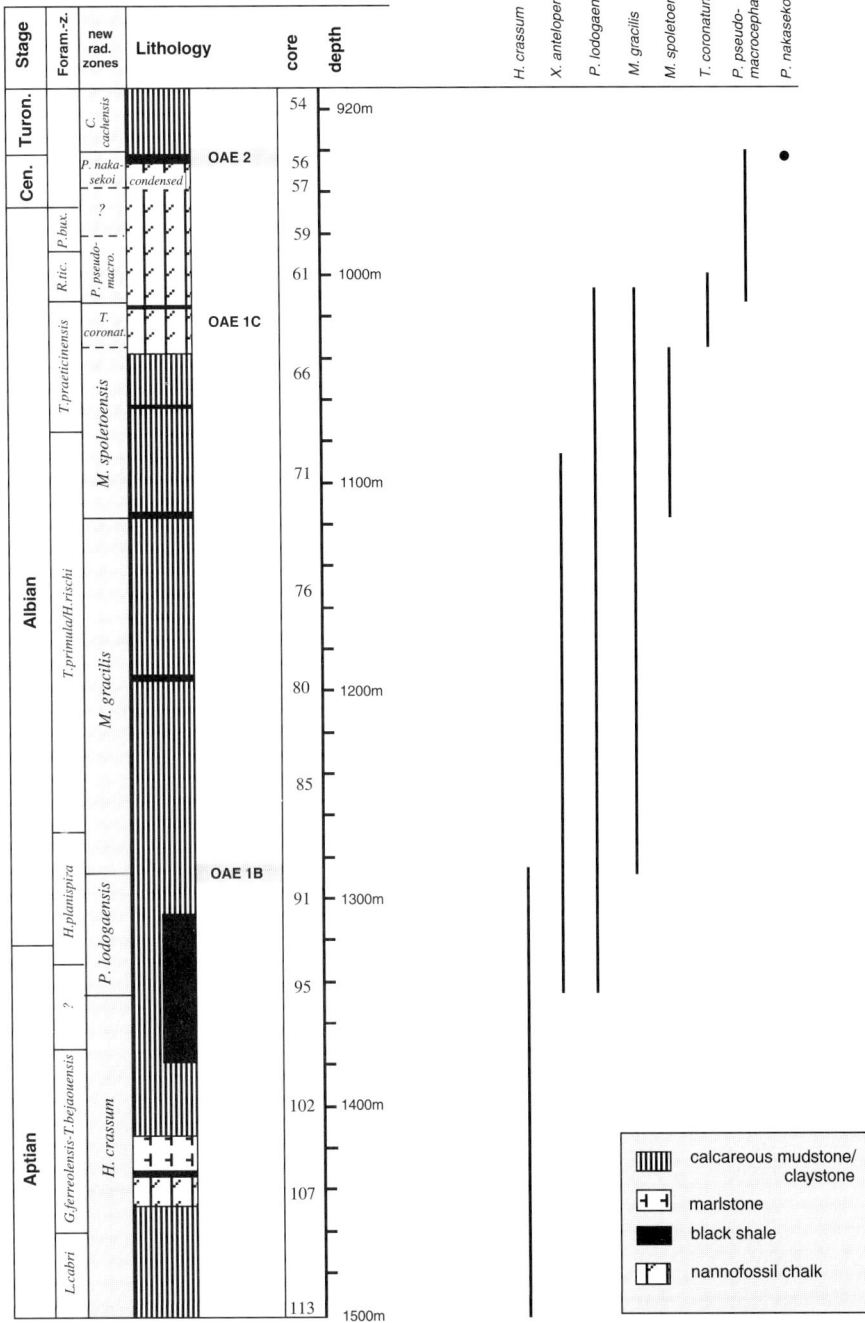

Fig. 2. Lithological section, new radiolarian zonation and ranges of index taxa for Site 398D, Vigo Seamount (Shipboard Scientific Party 1979), with the stratigraphical position of mid-Cretaceous oceanic anoxic events (OAE). Planktic foraminiferal determinations after Sigal (1979).

Leg 79, Site 545, Mazagan Plateau, NE-Atlantic

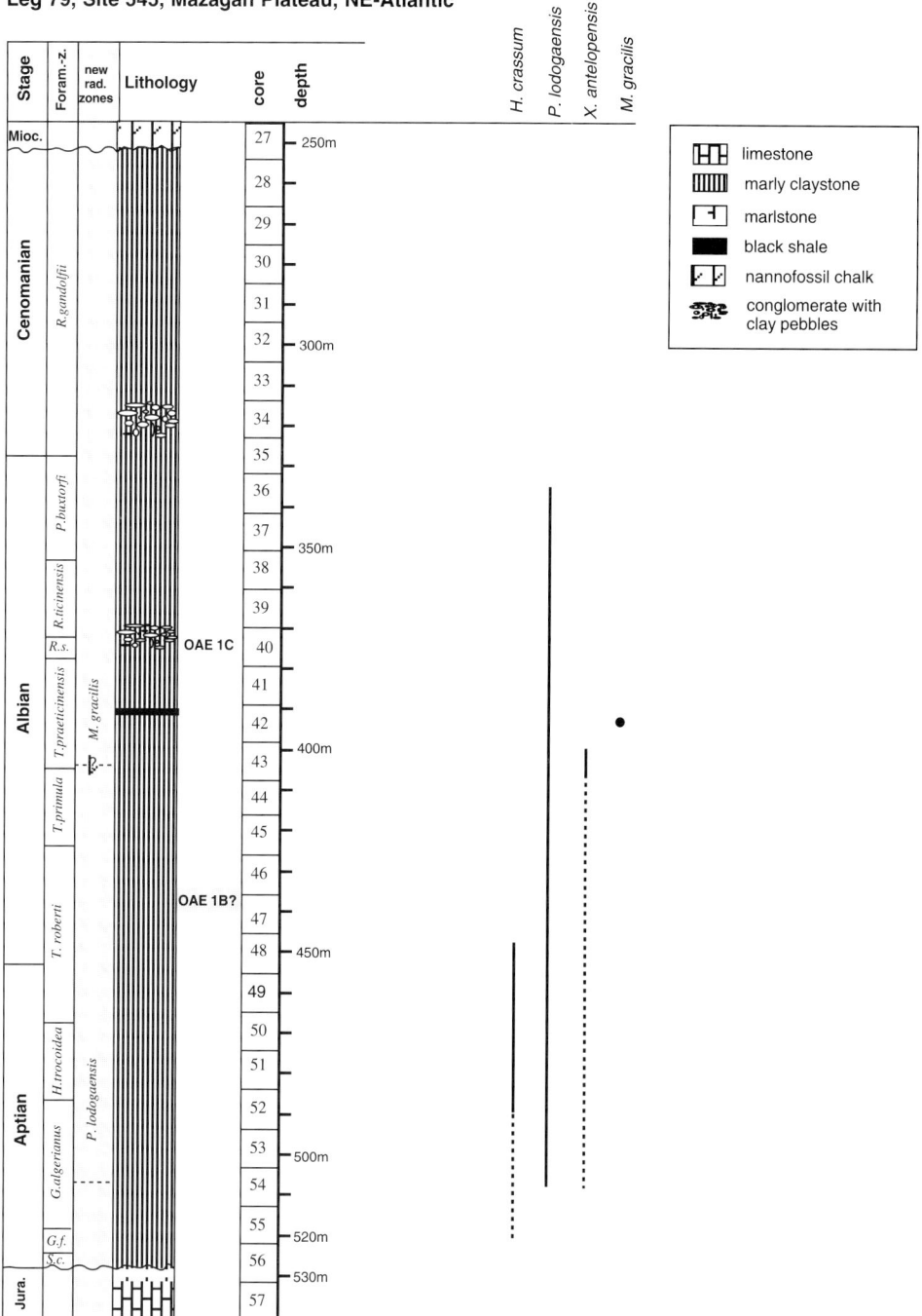

Fig. 3. Lithological section, new radiolarian zonation and ranges of index taxa for Site 545, Mazagan Plateau (Shipboard Scientific Party 1984), with the stratigraphical position of mid-Cretaceous oceanic anoxic events (OAE). Planktic foraminiferal zonation after Leckie (1984).

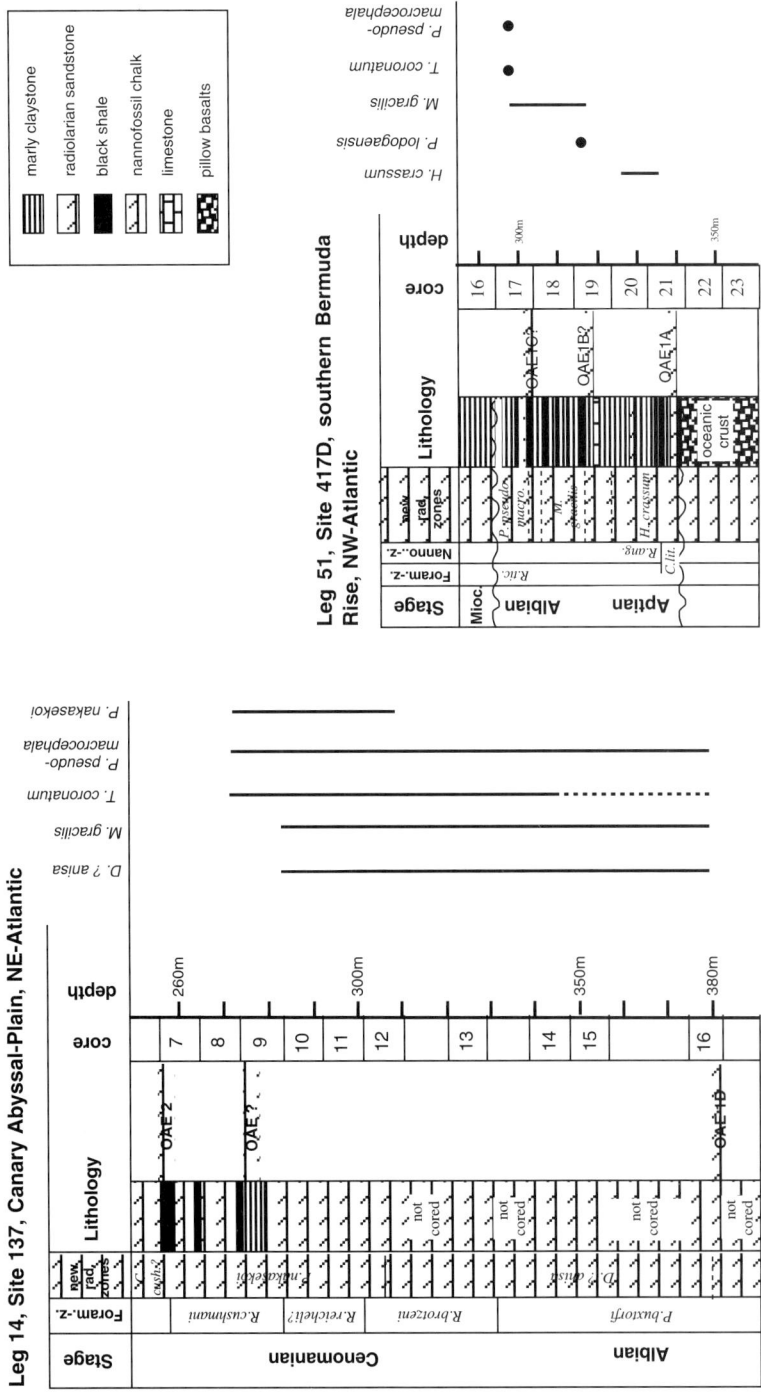

Fig. 4. Lithological sections, new radiolarian zonations and ranges of index taxa for Site 417D, southern Bermuda Rise (Shipboard Scientific Party 1980) and Site 137 (Shipboard Scientific Party 1972a,b), with the stratigraphical position of mid-Cretaceous oceanic anoxic events (OAE). Biostratigraphy for Site 417D after Miles et al. (1980), Bralower (1987) and nanofossil identifications by J. Mutterlose (Bochum).

this are: (1) a lack of calcareous plankton in many of the examined sections; (2) large variations in the state of radiolarian preservation between sites and, therefore, differing ranges of taxa; and (3) lack of a consistent taxonomy and nomenclature for Cretaceous radiolaria (compare Pessagno (1976), Pessagno (1977), Sanfilippo & Riedel (1985), Schaaf (1985), Thurow (1988a) and O'Dogherty (1994)). The intention of the proposed biostratigraphy is the use of distinct and easily identifiable, relatively solid and, therefore, quite resistant (with respect to dissolution) forms. Moreover, a manifested taxonomy is required. This is essential because of the high diversity of radiolarians, e.g. the 85 taxa described for the Aptian–Albian interval of Site 398D (see Erbacher 1994), and the continuing inconsistency of the taxonomy of mid-Cretaceous radiolaria. The result should be a biostratigraphy which can be easily applied even by non-specialists.

Accompanying studies of the biostratigraphy of sections from the Umbria–Marche Basin in central Italy (Erbacher 1994; Erbacher & Thurow 1997) enabled us to establish accurate calibrations of our biostratigraphy with the standard planktonic foraminiferal zonations (e.g. Caron 1985; Sliter 1989; Premoli Silva & Sliter 1994). The following new zonations are based on the range charts published by Erbacher (1994) and Thurow (1988a).

Recently, O'Dogherty (1994) published a comprehensive zonation for the western Mediterranean, based on the Unitary Association Method. Some of O'Dogherty's (1994) zonal markers, especially those for which the zones are named, were not observed in the North Atlantic, and have thus been shown not to be useful for zonation there. Other previously published zonations (Sanfilippo & Riedel 1985; Schaaf 1985; Thurow 1988a) also needed to be modified. For example, the zonations of Schaaf (1985) and Thurow (1988a) had to be changed completely within the Albian, because many of the first occurrences (FO) of their index taxa had changed. The zonations for other time slices, like the broad zonations of the Aptian and Cenomanian of the North Atlantic, can now be improved.

The following describes the new zonations used in this paper; in each case the base of the younger zone defines the top of the preceding (Fig. 5).

Halesium crassum zone

Definition. FO of *H. crassum* (Ozvoldova) (Fig. 6.1).
Remarks. H. crassum is described as *Angulobracchia crassa* or *A.* cf. *A. crassa* in Erbacher (1994). O'Dogherty (1994) has described *H. crassum* in the earliest Aptian of the western

Mediterranean realm. However, this form has not been observed until the *S. cabri* foraminiferal zone (early Late Aptian) in the North Atlantic. The latter radiolarian zone is often characterized by a high abundance of *Obeliscoides vinassai* (Squinabol) (Fig. 6.2).

Pseudodictyomitra lodogaensi zone

Definition. FO of *P. lodogaensis* Pessagno (Fig. 6.3).
Remarks. The base of this zone lies within the upper half of the *G. algerianus* foraminiferal zone. Another event that can be used to define the base of this zone is the FO of *Xitus antelopensis* Pessagno (Fig. 6. 5).

Mita gracilis zone (Schaaf 1985)

Definition. FO of *M. gracilis* (Squinabol) (Fig. 6.4).
Remarks. The FO of *M. gracilis* can be observed within oceanic anoxic event OAE 1b, a horizon which is characterized by an important radiolarian extinction/radiation event (Erbacher & Thurow 1997). The FO of the index taxon, or the last occurrence (LO) of *H. crassum*, is a marker for the Aptian/Albian boundary. Due to preservation biases in Early Albian sediments, *M. gracilis* first occurs within the early Late Albian (*T. praeticinensis* foraminiferal zone) in central Italy. Within the middle part of this zone, *Crolanium triangulare* (Aliev) (Fig. 6.6) makes its first appearance at Site 398D.

Mita spoletoensis zone

Definition. FO of *M. spoletoensis* (O'Dogherty) (Fig. 6.7).
Remarks. M. spoletoensis is described as *Mita* sp. B in Thurow (1988a) and Erbacher (1994). This form first occurs in the upper part of the *H. rischi/T.primula* foraminiferal zone. O'Dogherty's (1994) Spoletoensis zone ranges from the Late Albian (*R. ticinensis* foraminiferal zone) to Early Cenomanian (*R. brotzeni* foraminiferal zone) and is subdivided into three subzones (Fig. 5). However, data from Site 398D prove that the FO of *M. spoletensis* is much earlier in the Atlantic than in sections in central Italy, which are characterized by poor preservation of radiolaria in the Early to Middle Albian.

Torculum coronatum subzone

Definition. FO of *T. coronatum* (Squinabol) (Figs 6.8 and 6.9).

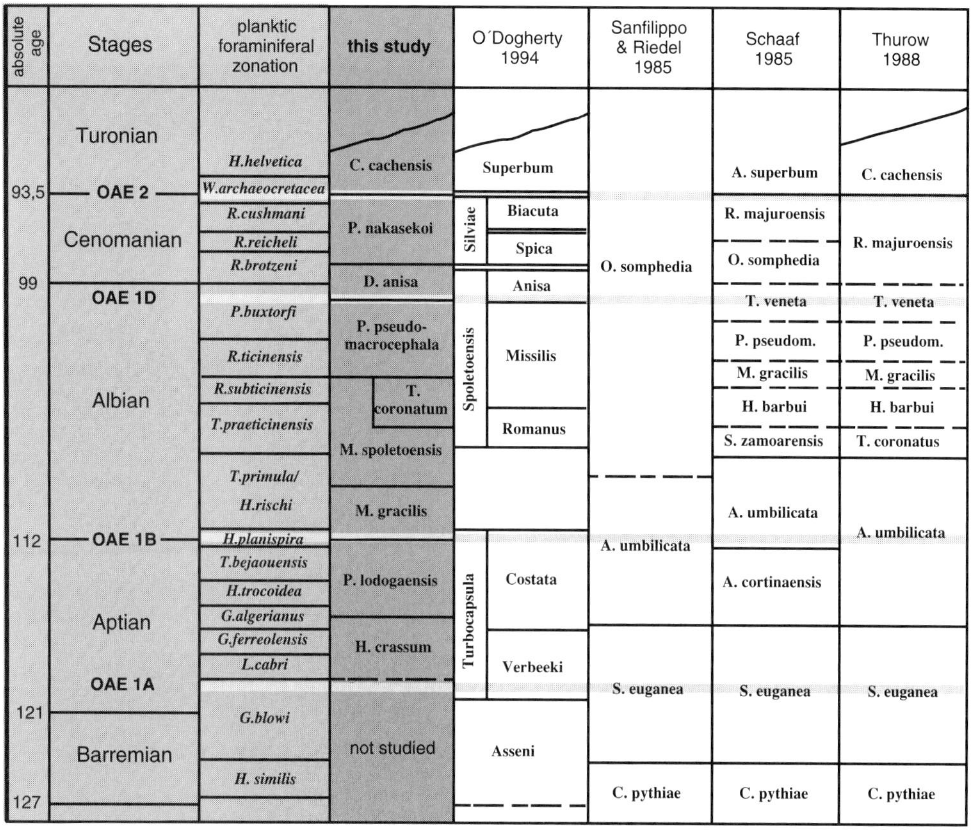

absolute age	Stages	planktic foraminiferal zonation	this study	O'Dogherty 1994	Sanfilippo & Riedel 1985	Schaaf 1985	Thurow 1988
	Turonian		C. cachensis	Superbum		A. superbum	C. cachensis
93,5	OAE 2	H.helvetica / W.archaeocretacea					
	Cenomanian	R.cushmani	P. nakasekoi	Biacuta (Silviae)		R. majuroensis	R. majuroensis
		R.reicheli		Spica (Silviae)	O. somphedia	O. somphedia	
99		R.brotzeni	D. anisa	Anisa		T. veneta	T. veneta
	OAE 1D	P.buxtorfi	P. pseudo-macrocephala	Missilis (Spoletoensis)		P. pseudom.	P. pseudom.
		R.ticinensis				M. gracilis	M. gracilis
	Albian	R.subticinensis	T. coronatum			H. barbui	H. barbui
		T.praeticinensis	M. spoletoensis	Romanus		S. zamoarensis	T. coronatus
		T.primula/ H.rischi	M. gracilis			A. umbilicata	A. umbilicata
112	OAE 1B	H.planispira			A. umbilicata		
		T.bejaouensis	P. lodogaensis	Costata (Turbocapsula)		A. cortinaensis	
	Aptian	H.trocoidea					
		G.algerianus					
		G.ferreolensis	H. crassum	Verbeeki (Turbocapsula)			
		L.cabri			S. euganea	S. euganea	S. euganea
121	OAE 1A	G.blowi	not studied	Asseni			
	Barremian	H.similis					
127						C. pythiae	C. pythiae

Fig. 5. Chart showing the correlation of the standard planktic foraminiferal zonation with composite or regional radiolarian zonations published earlier, and the new radiolarian zonation proposed here.

Remarks. *T. coronatum* is described as *Theoconus coronatus* in Thurow (1988a) and as *Theoconus* spp. group in Erbacher (1994). This form first occurs within the upper part of the *T. praeticinensis* foraminiferal zone. For reasons yet unknown, this form has not been observed in all of the sections examined in this study. However, if *T. coronatum* is present it can be used as a good zonal marker. Due to its absence in some of the investigated sites, the FO of this index taxon is only considered to define a subzone.

Pseudodictyomitra pseudomacrocephala zone (Schaaf 1985)

Definition. FO of *P. pseudomacrocephala* (Squinabol) (Fig. 6.10).
Remarks. The FO of *P. pseudomacrocephala* can be correlated with the base of the *R. ticinensis* foraminiferal zone in all examined sections.

Owing to a lack of radiolarians in the Early to Middle Cenomanian sediments of Sites 545 and 398D, the zonation of this time slice is based on samples from Site 137. However, both zones from this interval are easy to correlate with the central Italian sections.

Dorypyle ?anisa zone (O'Dogherty 1994)

Definition. FO of *D. anisa* (Foreman) (Figs. 6.11 and 6.12).
Remarks. *D. ? anisa* is described as *Podocapsa* cf. *P. guembelii* in Erbacher (1994). The FO of *D. ? anisa* can be specified within the upper part of the *P. buxtorfi* foraminiferal zone, in the top of OAE 1d (see Erbacher & Thurow 1997). This horizon is characterized by an important radiolarian extinction/radiation event (O'Dogherty 1994; Erbacher & Thurow 1997).

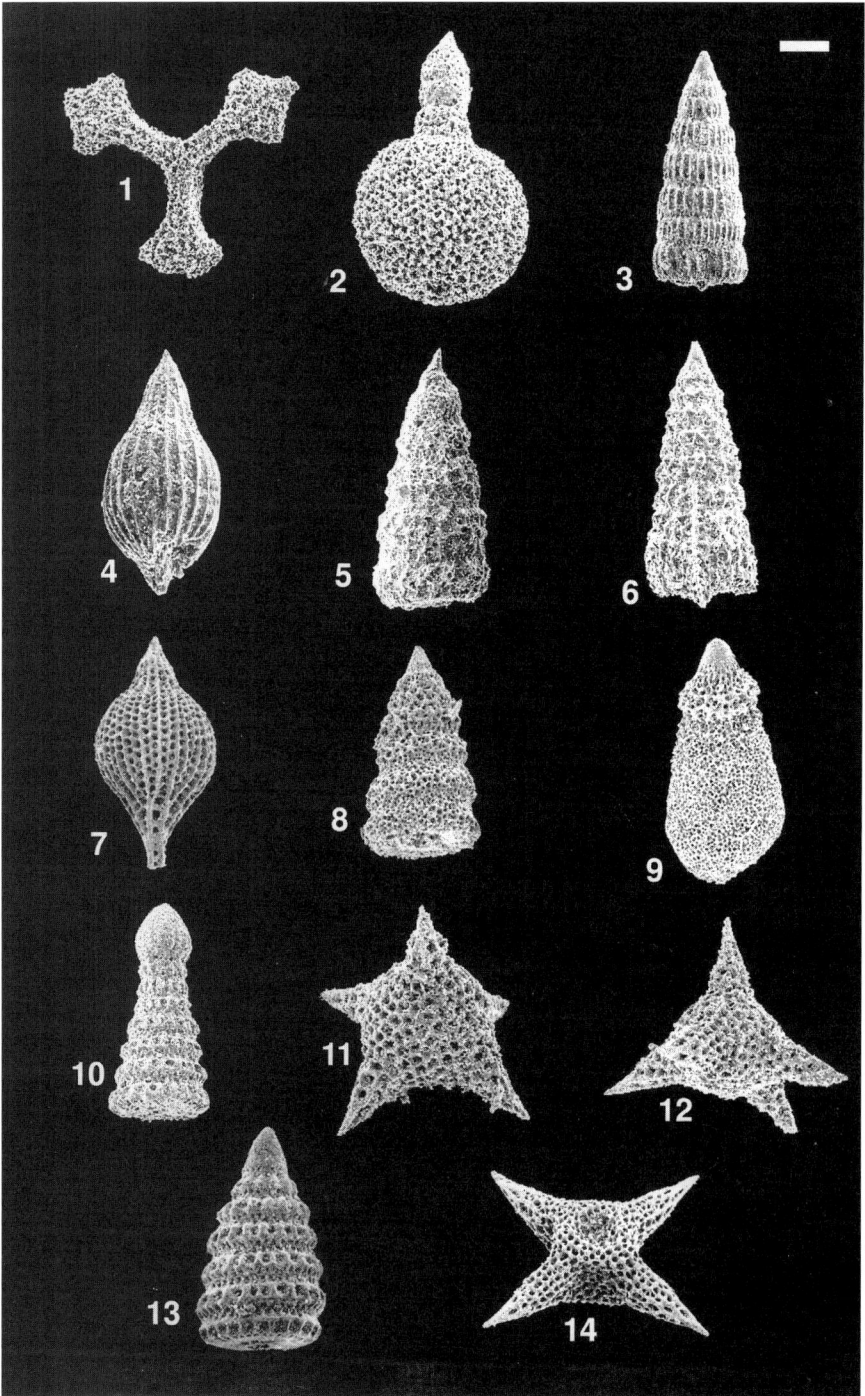

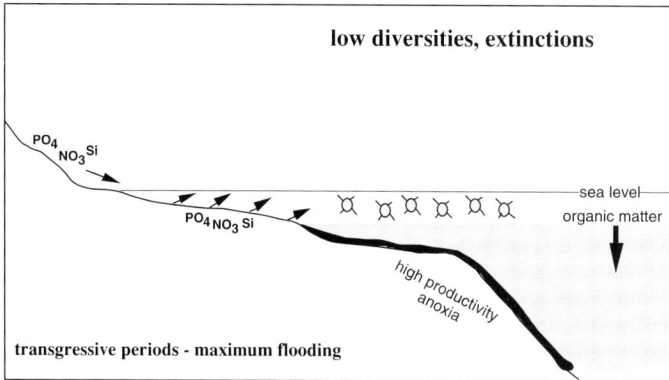

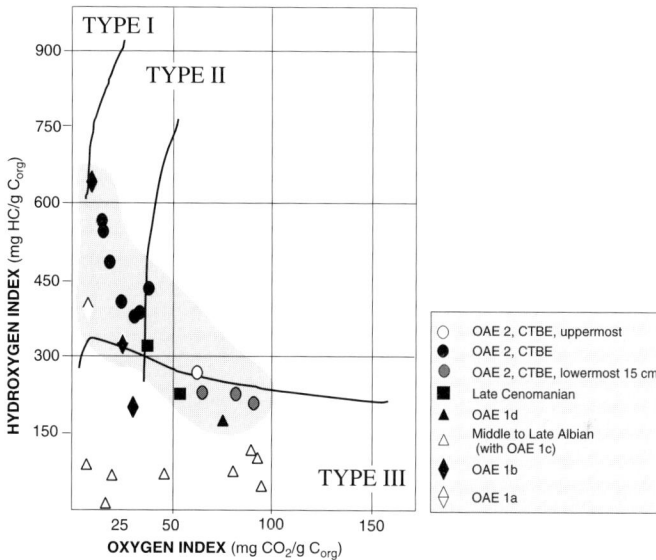

Fig. 7. Model, explaining mid-Cretaceous radiolarian extinction events. A high nutrient input leads to high productivity, an expansion of the OMZ, extinction of deeper-dwelling forms and the preservation of black shales. The diagram shows results of Rock Eval pyrolysis for sediments from the central Italian outcrops, plotted in the HI/OI diagram. Classification of kerogen after Espitalié, *et al.* (1985). Data for the regional OAE 1c are from Pratt & King (1986). Shaded area indicates where black shales are paralleled by extinction/radiation events and positive $\delta^{13}C$-values. Modified after Erbacher *et al.* (1996).

Fig. 6. Index taxons of radiolarian zones and some other forms of high stratigraphic value. (1) *Halesium crassum* (Ozvoldova), Late Aptian, Site 417D, 20/1, 119–121, Scale: 100 µm. (2) *Obeliscoides vinassai* (Squinabol), Late Aptian, Site 417D, 21/1, 141–143, Scale: 100 µm. (3) *Pseudodictyomitra lodogaensis* Pessagno, Late Aptian, Site 398D, 95/2, 138–140, Scale: 60 µm. (4) *Mita gracilis* (Squinabol), Middle Albian, Site 398D, 80/1, 50-53, Scale: 70 µm. (5) *Xitus antelopensis* Pessagno, Late Aptian, Site 398D, 95/2, 138–140, Scale: 50 µm. (6) *Crolanium triangulare* (Aliev), Middle Albian, Site 398D, 80/1, 50–53, Scale: 80 µm. (7) *Mita spoletoensis* (O'Dogherty), Late Albian, Gorgo Cerbara, central Italy, Scale: 80 µm. (8) *Torculum coronatum* (Squinabol), Late Albian, Gorgo Cerbara, central Italy, Scale: 100 µm. (9) *Torculum coronatum* (Squinabol), Late Albian, Gorgo Cerbara, central Italy, Scale: 150 µm. (10) *Pseudodictyomitra pseudomacrocephala* (Squinabol), Late Albian, Site 137, 16/2, 18–20, Scale: 90 µm. (11) *Dorypyle ? anisa* (Foreman), Middle Cenomanian, Gorgo Cerbara, central Italy, Scale: 80 µm. (12) *Dorypyle ?anisa* (Foreman), Middle Cenomanian, Gorgo Cerbara, central Italy, Scale: 80 µm. (13) *Pseudodictyomitra nakasekoi* Taketani, Late Cenomanian, Lozzo, northern Italy, Scale: 70 µm. (14) *Crucella cachensis* Pessagno, Early Turonian, Monte Petrano, central Italy, Scale: 70 µm.

Pseudodictyomitra nakasekoi zone

Definition. FO of *P. nakasekoi* Taketani (Fig. 6.13).
Remarks. The FO of *P. nakasekoi* lies in the middle of the *R. brotzeni* foraminiferal zone. The *P. nakasekoi* zone seems to be equivalent to O'Dogherty's (1994) Silvia zone. However, *Dactyliosphaera silviae* Squinabol has been described from the North Atlantic Site 398D as *Pseudoaulophacus*(?) sp. D by Thurow (1988*a*).

Crucella cachensis zone (Thurow 1988*a*)

Definition. FO of *Crucella cachensis* Pessagno (Fig. 6.14).
Remarks. The FO of *C. cachensis* marks the Cenomanian–Turonian boundary. The base of this zone coincides with the onset of the *W. archaeocretacea* foraminiferal zone. *C. cachensis* is one of the first new taxa occurring after the significant extinction event during OAE 2 (Thurow 1988*a*; Marcucci Passerini *et al.* 1991; O'Dogherty 1994).

Palaeoceanographic interpretations

Each zonation is controlled by the biostratigraphic pattern of the investigated group. Regarding the zonation proposed here, it is evident that every OAE (Jenkyns 1980; Arthur *et al.* 1990) correlates with a zonal boundary representing the coexistence of basin-wide OAEs and major radiolarian extinction/radiation events (Erbacher & Thurow 1997).

Organic matter and stable isotope investigations on mid-Cretaceous sections in central Italy have shown that major radiolarian extinction/radiation events appear during times of elevated $\delta^{13}C$ values. Furthermore, they co-occur with black shales that are dominated by marine organic matter (type II kerogen, see Fig. 7). Reviewing literature data on North Atlantic and western Tethyan carbonate platforms, these horizons seem to be time-equivalent to major drowning events and, therefore, to rises of sea level (Arnaud-Vanneau 1986; Jenkyns 1991; Scott 1993; Föllmi *et al.* 1994). These observations point to an obvious link between increases of sea-level and productivity during radiolarian extinction/radiation events (Fig. 7). We believe that it is the spatial dimension of the oxygen minimum zone (OMZ), that is responsible for radiolarian extinction and radiation. The leaching of nutrients from soils and enhanced nutrient cycling during a relative sea-level rise leads to an increase in marine productivity and thus to an extension of the OMZ. Consequently, deeper-dwelling taxa loose their habitats and disappear. The onset of contraction of the OMZ during highstand and relative sea-level fall creates new habitats paralleled by radiolarian radiation. This radiation is presumably forced by a decrease in marine productivity. Therefore, we can call it radiation by competition. A more comprehensive discussion of this model has been published in Erbacher (1994) and Erbacher *et al.* (1996).

This scenario would explain the extraordinary faunal changes around OAE 1b (base of the *M. gracilis* radiolarian zone), OAE 1d (base of the *D. anisa* radiolarian zone) and OAE 2 (base of the *C. cachensis* radiolarian zone). OAE 1a, which was not sampled in North Atlantic cores due to a lack of radiolarian faunas, shows similar patterns in the central Italian sections.

Moreover, this good correlation between the evolution of marine planktic protozoa and sea-level changes allows sequence stratigraphy to be tied to high-resolution planktonic biostratigraphy.

We thank Ch. Hemleben, G. Schmiedl and A. von Gyldenfeldt (all Tübingen) for discussions and helpful comments. The reviews by B. Funnel and S. Haslett are gratefully acknowledged. The project has been supported by the Deutsche Forschungsgemeinschaft.

References

ARNAUD-VANNEAU, A. 1986. Transgressive events and renewal of benthic foraminifera on the Lower Cretaceous platforms of southern France. *Bulletin des Centres de Recherches Exploration Production Elf-Aquitaine*, **10**, 405–420.

ARTHUR, M. A., BRUMSACK, H.-J., JENKYNS, H. C. & SCHLANGER, S. O. 1990. Stratigraphy, Geochemistry, and paleoceanography of organic carbon-rich Cretaceous sequences. *In:* GINSBURG, R. N. & BEAUDOIN, B. (eds) *Cretaceous Resources, Events and Rythms.* Kluwer, Dordrecht, 75–119.

BERGER, W. H. & VON RAD, U. 1972. Cretaceous and Cenozoic sediments from the Atlantic Ocean. *In:* HAYES, D. E., PIMM, A. C. ET AL. (eds) *Initial Reports of the Deep Sea Drilling Project*, **14**. US Govt Printing Office, Washington, 787–954.

BRALOWER, T. J. 1987. Valanginian to Aptian calcareous nannofosil stratigraphy and correlation with the M-sequence magnetic anomalies. *Marine Micropaleontology*, **11**, 293–310.

CARON, M. 1985. Cretaceous planktic foraminifera. *In:* BOLLI, H. M., SAUNDERS, J. B. & PERCH-NIELSEN, K. (eds) *Plankton Stratigraphy.* Cambridge University Press, Cambridge, 17–86.

DERCOURT, J., RICOU, L. E. & VRIELYNCK, B. (eds) 1993. *Atlas Tethys Palaeoenvironmental Maps.* Gauthier-Villars, Paris.

ERBACHER, J. 1994. Entwicklung und Paläoozeanographie mittelkretazischer Radiolarien der westlichen Tethys (Italien) und des Nordatlantiks. *Tübinger Mikropaläontologische Mitteilungen*, **12**.

—— & THUROW, J. 1997. Influence of oceanic anoxic

events on the evolution of mid-Cretaceous radiolaria in the North Atlantic and western Tethys. *Marine Micropalaeontology*, **30**, 139–158.

——, —— & LITTKE, R. 1996. Evolution patterns of radiolaria and organic matter variations – a new tool to identify sea-level changes in mid-Cretaceous pelagic environments. *Geology*, **24**, 499–502.

ESPITALIÉ, J., DEROO, G. & MARQUIS, F. 1985. La pyrolyse Rock Eval et ses applications. *Révue Institute Francais Petrole*, **10**, 755–784.

FÖLLMI, K. B., WEISSERT, H., BISPING, M. & FUNK, H. 1994. Phosphogenesis, carbon-isotope stratigraphy, and carbonate-platform evolution along the Lower Cretaceous northern Tethyan margin. *Geological Society of America Bulletin.* **106**, 729–746.

HART, M. B. 1980. A water depth model for the evolution of the planktonic Foraminiferida. *Nature*, **286**, 252–254.

JANSA, L. F., ENOS, P., TUCHOLKE, B. E., GRADSTEIN, F. M. & SHERIDAN, R. E. 1979. Mesozoic–Cenozoic sedimentary formations of the North American Basin; western North Atlantic. *In:* TALWANI, M., HAY, W. & RYAN, W. B. F. (eds) *Deep Drilling Results in the Atlantic Ocean: Continental Margins and Paleoenvironment.* Maurice Ewing Series, **3**, 1–57.

JENKYNS, H. C. 1980. Cretaceous anoxic events: from continents to oceans. *Journal of the Geological Society, London*, **137**, 171–188.

—— 1991. Impact of Cretaceous sea level rise and anoxic events on the Mesozoic carbonate platform of Yugoslavia. *AAPG Bulletin*, **75**, 1007–1017.

LARSON, R. L. 1991. Geological consequences of superplumes. *Geology*, **19**, 963–966.

LECKIE, R. M. 1984. Mid-Cretaceous planktonic foraminiferal biostratigraphhy off central Morocco, Deep Sea Drilling Project Leg 79, Sites 545 and 547. *In:* HINZ, K., WINTERER, E. L. *ET AL.* (eds) *Initial Reports of the Deep Sea Drilling Project*, **79**. US Govt Printing Office, Washington, 579–620.

—— 1989. A paleoceanographic model for the early evolutionary history of planktonic foraminifera. *Palaeogeography Palaeoclimatology Palaeoecology*, **73**, 107–138.

MARCUCCI PASSERINI, M., BETTINI, P., DAINELLI, J. & SIRUGO, A. 1991. The "Bonarelli Horizon" in the central Apennines (Italy): radiolarian biostratigraphy. *Cretaceous Research*, **12**, 321–331.

MILES, G. A., NORR, W. N. *ET AL.* 1980. Planktonic foraminifers from the Bermuda Rise, Deep Sea Drilling Project, Legs 51, 52 and 53. *In:* DONNELLY, T. W., FRANCHETEAU, J. *ET AL.* (eds) *Initial Reports of the Deep Sea Drilling Project*, **50, 51, 52**. US Govt Printing Office, Washington, 791–813.

O'DOGHERTY, L. 1994. Biochronology and paleontology of middle Cretaceous radiolarians from Umbria-Marche Appennines (Italy) and Betic Cordillera (Spain). *Mémoires de Géologie (Lausanne)*, **21**, 351.

PESSAGNO, E. A., JR. 1976. Radiolarian zonation and stratigraphy of the Upper Cretaceous portion of the Great Valley Sequence, California Coast Range. *Micropaleontology*, **2**, 95 S.

PESSAGNO, E. A. J. 1977. *Lower Cretaceous Radiolarian Biostratigraphy of the Great Valley Sequence and*

Franciscan Complex, California Coast Ranges. Cushman Foundation for Foraminiferal Research Special Publication, **15**, 5–87.

PRATT, L. M. & KING, J. D. 1986. Variable marine productivity and high eolian input recorded by rhythmic black shales in mid-Cretaceous pelagic deposits from Central Italy. *Paleoceanography*, **4**, 507–522.

PREMOLI SILVA, I. & SLITER, W. V. 1994. Cretaceous planktonic foraminiferal biostratigraphy and evolutionary trends from the Bottacione section, Gubbio, Italy. *Paleontographica Italia*, **81**, 2–90.

ROTH, P. H. 1986. Mesozoic palaeoceanography of the North Atlantic and Tethys Oceans. *In:* SUMMERHAYES, C. P. & SHACKLETON, N. J. (eds) *North Atlantic Palaeoceanography.* Geological Society, London, Special Publication, **21**, 299–320.

SANFILIPPO, A. & RIEDEL, W. R. 1985. Cretaceous Radiolaria. *In:* BOLLI, H. M., SAUNDERS, J. B. & PERCH-NIELSEN, K. (eds) *Plankton stratigraphy.* Cambridge University Press, Cambridge, 573–630.

SCHAAF, A. 1985. Un nouveau canevas biochronologique du Cretace inferieur et moyen: les biozones a radiolaires. *Sciences Geologiques Bulletin*, **38**, 227–269.

SCOTT, R. W. 1993. Cretaceous carbonate platform, U.S. Gulf Coast. *AAPG Memoir*, **56**, 97–109.

SHERIDAN, R. E. 1986. Pulsation tectonics as the control of North Atlantic palaeoceanography. *In:* SUMMERHAYES, C. P. & SHACKLETON, N. J. (eds) *North Atlantic Palaeoceanography.* Geological Society, London, Special Publication, **21**, 255–275.

SHIPBOARD SCIENTIFIC PARTY. 1972a. Site 137. *In:* HAYES, D. E., PIMM, A. C. *ET AL.* (eds) *Initial Reports of the Deep Sea Drilling Project*, **14**. US Govt Printing Office, Washington, 85–134.

—— 1972b. Site 138. *In:* HAYES, D. E., PIMM, A. C. *ET AL.* (eds) *Initial Reports of the Deep Sea Drilling Project*, **14**. US Govt Printing Office, Washington, 135–155.

—— 1979. Site 398. *In:* SIBUET, J.-C., RYAN, W. B. F. *ET AL.* (eds) *Initial Reports of the Deep Sea Drilling Project*, **47**. US Govt Printing Office, Washington, 23–233.

—— 1980. Site 417. *In:* DONNELLY, T. W., FRANCHETEAU, J. *ET AL.* (eds) *Initial Reports of the Deep Sea Drilling Project*, **50, 51, 52**. US Govt Printing Office, Washington, 23–350.

—— 1984. Site 545. *In:* HINZ, K., WINTERER, E .L. *ET AL.* (eds) *Initial Reports of the Deep Sea Drilling Project*, **79**. US Govt Printing Office, Washington, 81–177.

SIGAL, J. 1979. Chronostratigraphy and ecostratigraphy of Cretaceous formations recovered on DSDP Leg 47B, Site 398. *In:* SIBUET, J.-C., RYAN, W. B. F. *ET AL.* (eds) *Initial Reports of the Deep Sea Drilling Project*, **47**. US Govt Printing Office, Washington, 287–326.

SLITER, W. V. 1989. Biostratigraphic zonation for Cretaceous planktonic foraminifers examined in thin section. *Journal of Foraminiferal Research.* **19**, 1–19.

THUROW, J. 1988a. Cretaceous radiolarians of the North Atlantic Ocean: ODP Leg 103 (Sites 638, 640 and 641) and DSDP Legs 93 (Site 603) and 47B (Site

398). *In:* BOILLOT, G., WINTERER, E. L. *ET AL.* (eds) *Proceedings of the Ocean Drilling Program Scientific Results*, **103**. College Station, TX, Ocean Drilling Program, 379–418.

—— 1988*b*. Diagenetic history of Cretaceous radiolarians, North Atlantic Ocean (ODP Leg 103 and DSDP Holes 398D and 603B). *In:* BOILLOT, G., WINTERER, E. L. *ET AL.* (eds) *Proceedings of the Ocean Drilling Program Scientific Results*, **103**. College Station, TX, Ocean Drilling Program, 531–555.

TORNAGHI, M. E., PREMOLI SILV, I. & RIPEPE, M. 1989. Lithostratigraphy and planktonic foraminiferal biostratigraphy of the Aptian–Albian "Scisti a Fucoidi" in the Piobbico core, Marche, Italy: background for cyclostratigraphy. *Rivista Italiana di Paleontologia e Stratigrafia*, **95**, 223–264.

WIEDMANN, J. 1988. Plate tectonics, sea level changes, climate – and the relationship to Ammonite evolution, provincialism, and mode of life. *In:* WIEDMANN, J. & KULLMANN, J. (eds) *Cephalopods – Present and Past*. Schweizerbart, Stuttgart, 737–765.

Low-latitude Plio-Pleistocene temporal abundance variations in the radiolarian *Cycladophora davisiana* Ehrenberg: stratigraphic and palaeoceanographic significance

SIMON K. HASLETT[1] & BRIAN M. FUNNELL[2]

[1] *Quaternary Research Unit, Faculty of Applied Sciences, Bath Spa University College, Newton Park, Newton St Loe, Bath BA2 9BN, UK*

[2] *School of Environmental Sciences, University of East Anglia, Norwich NR4 7TJ, UK*

Abstract: The abundance variation of the radiolarian *Cycladophora davisiana* was recorded throughout the Plio-Pleistocene Olduvai Time-slab of Ocean Drilling Program (ODP) Sites 677, 709, 847, 850 and 851, in the equatorial Pacific and Indian Oceans. Data were plotted on independently erected timescales. Twelve peak-abundance events were recognized, although not all events were recognized at all the sites. Most of the events, where they are recorded at different sites, were synchronous. However, events recorded at Site 709 in the Indian Ocean were found to be approximately 15 ka older than those events recorded at Pacific sites. This age difference may be due to either diachronous *C. davisiana* events or a slight inaccuracy in the erected timescale for ODP Site 709 compared with those of the eastern Pacific. The previously demonstrated synchroneity of *C. davisiana* events in the Late Quaternary leads us to believe that the ODP Site 709 timescale may be inaccurate. Thus, recognition of correlatable *C. davisiana* events in the equatorial Pacific and Indian Oceans permits the further evaluation and fine-tuning of high-resolution timescales, enabling accurate comparisons to be made of sediment sequences on a regional, and possibly oceanic and interoceanic scale. Of the 12 *C. davisiana* events recognized, only one event occurred at all the investigated sites. This event occurred at *c.*1.8 Ma and immediately above the internationally defined Pliocene–Pleistocene boundary, and may thus prove to be a new useful biostratigraphic marker for the Tertiary/Quaternary boundary. *C. davisiana* is primarily characteristic of high-latitude water masses, therefore the cyclical abundance peaks of this species noted at the low-latitude sites investigated here, represent the injection of deep cold water of high-latitude origin.

High-latitude ocean sediments are often lacking in calcareous microfossils, therefore stratigraphical analysis based on $CaCO_3$ variation, $\delta^{18}O$ isotopes, or evolutionary or faunal events of calcareous microfossil groups, such as foraminifera and calcareous nanofossils, cannot be performed. To compensate for this, stratigraphies based on siliceous microfossils, such as radiolaria, diatoms and silicoflagellates, have been extensively developed (see Sanfilippo *et al.* (1985), Barron (1985) and Perch-Nielsen (1985) for reviews). This is done primarily through the use of zonal schemes, based largely on first and last occurrences of distinct morphotypes. However, in addition to this, it has been shown that temporal abundance variations in the radiolarian species *Cycladophora davisiana* Ehrenberg throughout the Late Pliocene and Quaternary are remarkably consistent, and can be correlated on interhemispheric and interoceanic scales (Morley 1980, 1987; Morley & Hays 1979, 1983).

Despite the usefulness of *C. davisiana* in high-latitude stratigraphy, it is not widely known that the same method can also be successfully used in low-latitude sediments, supplementing information obtained from other stratigraphical sources. In this paper we report on the application of *C. davisiana* stratigraphy as a means of independently evaluating high-resolution timescales developed by other methods for Ocean Drilling Program (ODP) sites in the equatorial eastern Pacific and Indian Oceans.

Material and methods

Five ODP sites from three legs were examined in the present study and are described below. Legs 111 (Site 677) and 138 (Sites 847, 850, and 851) drilled in the eastern equatorial Pacific (see Fig. 1 for location), whilst Leg 115 (Site 709) drilled in the equatorial Indian Ocean.

HASLETT, S. K. & FUNNELL, B. M. 1998. Low-latitude Plio-Pleistocene temporal abundance variations in the radiolarian *Cycladophora davisiana* Ehrenberg: stratigraphic and palaeoceanographic significance. *In*: CRAMP, A., MACLEOD, C. J., LEE, S. V. & JONES, E. J. W. (eds) *Geological Evolution of Ocean Basins: Results from the Ocean Drilling Program*. Geological Society, London, Special Publications, **131**, 83–89.

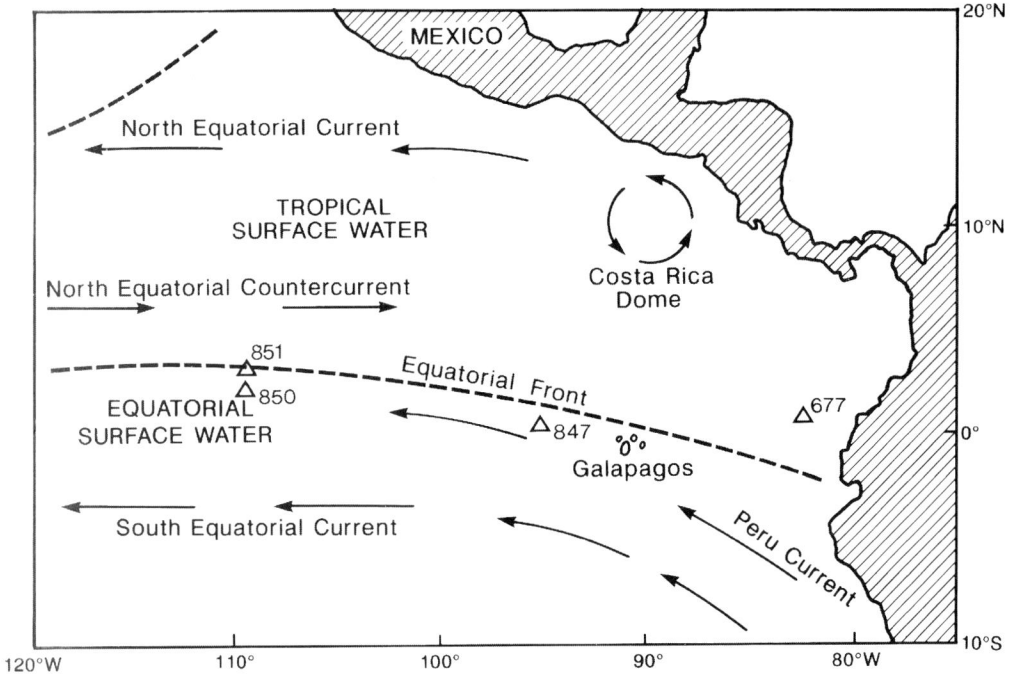

Fig. 1. Location of the ODP Sites 677, 847, 850 and 851 in the eastern equatorial Pacific. The major oceanographic currents of the region are marked.

Site 677

This site is located in the Costa Rica Rift, 1°12.138′N, 83°44.220′W (Panama Basin) and was drilled at a water depth of 3461.2 m (Shipboard Scientific Party 1988). The recovered sequence represents Late Miocene to Recent deposition and comprises oozes with a high siliceous content. Ninety-two samples were examined from Hole 677A, at 10 cm intervals, spanning 67.10 to 76.80 mbsf (metres below sea floor).

Site 709

This site is located in the equatorial Indian Ocean, in a small basin near the summit of Madingley Rise, a regional topographic high between the Carlsberg Ridge and the western Mascarene Plateau. Its position is 03°54.9′S, 60°33.1′E. Hole 709C was investigated here where the water depth is 3040.8 m Thirty samples were examined at 5 cm intervals, spanning a depth from 16.4 to 18.35 mbsf. The basic lithology consists of foraminifera-bearing nanofossil ooze, largely homogenous but with some bioturbation.

Site 847

This site is located 0°11.59′N, 95°19.20′W within the present-day equatorial divergence zone in the eastern Pacific (Shipboard Scientific Party 1992). Holes 847B and C were investigated here. They were drilled at a water depth of 3334.3 m below sea level and recovered diatom nanofossil ooze spanning the late Miocene to Holocene. Hole 847B recovered 242.19 m, and 847C recovered 225.71 m, of core. Fifty-two samples were examined for radiolaria from this site, at 15 cm intervals, spanning 56.68–64.25 mcd (metres composite depth).

Site 850

This site is located 1°17.83′N, 110°31.29′W within the present-day equatorial divergence zone. Holes 850A and B were investigated here. They were drilled at a water depth of 3798 m below sea level and recovered nanofossil ooze with varying proportions of other microfossil constituents, spanning late Miocene to Holocene (Shipboard Scientific Party 1992). Thirty-three samples were examined for radiolaria at this site, at 10 cm intervals, spanning 35.3–38.5 mcd.

Site 851

This site is located 2°46.22′N, 110°34.31′W at the northern edge of the present-day westward-flowing south equatorial current. Hole 851B was investigated here. It was drilled at a water depth of 3772 m below sea level, and recovered primarily nanofossil ooze, spanning the late Miocene to Holocene (Shipboard Scientific Party 1992). Thirty-six samples were examined for radiolaria at this site, at 10 cm intervals, spanning 32.9–36.4 mcd.

Sample preparation and methodology

The samples were processed for radiolaria by dissolving the carbonate fraction with dilute hydrochloric acid, washing over a 63 μm sieve, and mounting the residue in Canada Balsam to be viewed on a transmitted light stereomicroscope. The abundance of *C. davisiana* was measured quantitatively, and is described as a percentage of the counted radiolarian fauna (which in every sample consists of more than 300 identified specimens). The raw radiolarian data for Sites 677 and 709 are given by Haslett (1994*a*) and Haslett *et al.* (1994*b*) respectively; data for the Leg 138 sites are included in Haslett (1996).

The chrono-section chosen for this high-resolution timescale analysis is referred to as the 'Olduvai Time-slab' (*c.* 1.75–2 Ma), as it can be readily identified by reference to the Olduvai submagnetochronozone (subchron), although it is not precisely coextensive with it (there is some overlap with the reversed Matuyama chron at either end). The general concept is to look at continuous variation in siliceous microfossil abundance over a restricted time frame, which is not as restricted as that involved in time slices and is less subject to sample selection error or possible unrepresentativeness of time-slice studies, thus ultimately giving a better indication of palaeoceanographical evolution over time.

Timescales

Site 677

The timescale attributed to the samples from ODP Hole 677A has been derived from Shackleton *et al.* (1990) as follows.

1. Sample intervals and their corresponding depths below sea floor (mbsf) have been related to adjusted depth by reference to their table 1, thus:

   Sample 677A/8/1/010-012 from 63.3 mbsf = 67.3 mcd;

   Sample 677A/8/7/030-032 from 72.5 mbsf = 76.5 mcd;
   Sample 677A/9/1/000-002 from 72.7 mbsf = 76.7 mcd;
   Sample 677A/9/6/140-142 from 81.6 mbsf = 85.6 mcd;

   with intervening values intercalated.

2. Age control points have been obtained by reference to their table 3, thus:

   70.69 mcd = 1.736 Ma; 71.4 mcd = 1.757 Ma; 72.89 mcd = 1.788 Ma; 73.89 mcd = 1.809 Ma; 74.6 mcd = 1.829 Ma; 75.89 mcd = 1.851 Ma; 76.29 mcd = 1.880 Ma; 76.69 mcd = 1.910 Ma; 77.69 mcd = 1.944 Ma; 79.55 mcd = 1.983 Ma; 81.99 mcd = 2.021 Ma;

   with intervening values interpolated.

3. $\delta^{18}O$ stage 63 has previously been identified at approximately 72 mcd, and stage 78 at approximately 87.5 mcd (Shackleton *et al.* 1990, figure 2), indicating that our sequence should run from approximately $\delta^{18}O$ stage 62 at 71.1 mcd (= 1.748 Ma) to approximately stage 75 at 80.81 mcd (= 2.003 Ma). These time values compare with ages of 1.616 Ma for stage 62, and 1.883 Ma for stage 75 derived by Ruddiman *et al.* (1989) from Atlantic ODP Site 607. The difference of approximately 150–200 ka is attributable to the direct tuning to the astronomical timescale achieved by Shackleton *et al.* (1990), compared with the control provided by the radiometrically determined ages of magnetic reversals by Ruddiman *et al.* (1989).

There is one core break in the sequence, between 76.5 and 76.7 mcd. No adjustment in the mcd for a gap or replication in the coring record at this juncture was made by Shackleton *et al.* (1990), but the age control points given in their table 3 imply a hiatus or slowing of sedimentation between 75.89 and 76.69 mcd (see Haslett *et al.* 1994*a*). Oxygen isotope data for this site are given in Shackleton & Hall (1989).

Site 709

In their study of the Pliocene oxygen isotope stratigraphy of ODP Hole 709C, Shackleton & Hall (1990) commented on the surprisingly featureless planktonic isotope record which they had obtained, and they plotted their data on an approximate timescale obtained by linear interpolation between biostratigraphic datum levels. Subsequent identification of selected $\delta^{18}O$ stages in the planktonic

Table 1. *Age control points used to construct ODP Hole 709C timescale*

Depth (mbsf)	Age (Ma)
16.05	1.734
16.85	1.787
17.25	1.847
18.05	1.938
19.15	2.110

Data from N. J. Shackleton & S. Crowhurst (pers. comm.)

isotope record of Hole 709C enabled the allocation of age control points to the 709C record (N. J. Shackleton & S. Crowhurst 1991, pers. comm.), and it is these control points which have been used in plotting data against time for the present paper (Table 1).

Sites 847, 850 and 851

Our timescales for Sites 847, 850 and 851 were obtained from an age–depth calibration between mcd (Shipboard Scientific Party 1992) and the astronomically tuned GRAPE record (Shackleton *et al.* 1995). Oxygen isotope data fro Site 851 are given in Ravelo & Shackleton (1995); however, there are no published isotope records for Sites 847 and 850 covering the timespan under investigation here.

Results

The raw *C. davisiana* abundance percentage data for Sites 677, 709, 847, 850 and 851 are displayed in Fig. 2. The curves were carefully examined and an attempt has been made to identify distinctive peak-abundance events which are numbered 1–12. The age of the numbered events in each site has been tabulated in Table 2 for cross-reference and correlation between sites.

Of the five sites, Sites 677 and 847 contain the majority of identified events, with eight (Nos. 1–5

Table 2. *Age* Cycladophora davisiana *abundance events throughout the Olduvai Time-slab of ODP Sites 677, 709, 847, 850 and 851.*

Event no.	677	709	847	850	851
1	1.755	1.77	.	.	—
2	1.801	1.815	1.8	1.8	1.8
3	1.815	–	1.815	–	–
4	1.84	1.855	1.84	1.84	–
5	1.86	–	1.86	1.865	–
6	–	–	1.875	1.875	–
7	–	–	1.89	1.89	–
8	–	1.93	1.915	1.915	1.91
9	–	.	1.93	–	1.93
10	1.95	.	1.95	.	1.95
11	1.97	.	.	.	.
12	1.995	.	.	.	.

Age (Ma); . = not sampled; – = event not encountered. The events were read off Fig. 2.

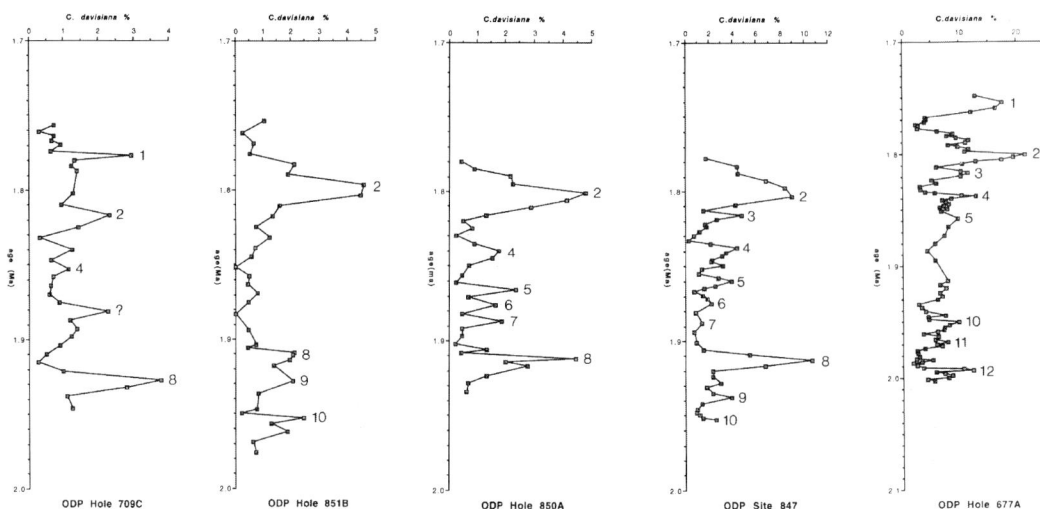

Fig. 2. Abundance variation of *Cycladophora davisiana* throughout the Olduvai Time-slab of ODP Sites 677, 847, 850, 851 and 709. Abundance events are numbered 1–12.

and 10–12) and nine (Nos. 2–10) events being recognized respectively. The three events not reported from Site 847 (Nos. 1, 11 and 12) are expected to occur above and below the section examined for that site; however, the four events missing from Site 677 (Nos. 6–9) should occur in the middle of the Olduvai section, but are not preserved in that interval due to the low sediment accumulation rate or hiatus mentioned earlier.

Site 850 yielded six events (Nos. 2 and 4–8) with four events (Nos. 1 and 10–12) expected to occur above and below the studied section at this site, and with two events (Nos. 3 and 9) not recognized within the section. Sites 851 and 709 both record four events (Nos. 1, 2, 4 and 8, and 2 and 8–10 respectively).

Only one event (No. 2) is recognized at all five sites and two events (Nos. 4 and 8) are recognized at four sites. Another two events (Nos. 5 and 10) are present at three sites, five events (Nos. 1, 3, 6, 7 and 9) are recognized at two sites, and two events (Nos. 11 and 12) are recognized only at Site 677, probably because a longer stratigraphical section was examined at this site.

Discussion

The *C. davisiana* abundance curves recorded at the five ODP sites enable an evaluation to be made between the timescales devised for Site 677 (Skackleton *et al.* 1990), Site 709 (N. J. Shackleton & S. Crowhurst 1991, pers. comm.), and Sites 847, 850 and 851 (Shackleton *et al.* 1995).

There appears to be excellent agreement within the eastern equatorial Pacific sites, with the majority of events occurring at the individual sites being synchronous. Two minor exceptions to this concern event No. 5 at Site 850, which appears to lead the same event at Sites 677 and 847 by *c.* 0.005 Ma, and event No. 8 at Site 851, which appears to lag the same event at Sites 847 and 850, again by *c.* 0.005 Ma (i.e. = one sample interval).

At Site 709, however, the ages assigned to *C. davisiana* events recognized there appear to lead event ages at the eastern Pacific sites by *c.* 0.015 Ma. In detail, event No. 1 at Site 709 occurs at 1.77 Ma, whereas it occurs at 1.755 Ma at Site 677; event No. 2 at Site 709 occurs at 1.815 Ma, whilst in all the eastern Pacific sites this event occurs at 1.8–1.805 Ma; event No. 4 at Site 709 occurs at 1.855 Ma, but the equivalent event at Sites 677, 847 and 850 is dated at 1.84 Ma; finally, event No. 8 at Site 709 occurs at 1.93 Ma, whereas at Sites 847 and 850 (but not Site 851 as mentioned above) event No. 8 occurs at 1.915 Ma.

Results from the *C. davisiana* analysis at Site 709 suggest one of the following alternatives.

(a) *C. davisiana* events are diachronous between the Pacific and Indian Oceans. Indeed the first appearance of *C. davisiana* in the Pacific, Indian and Atlantic Oceans is diachronous (see Haslett 1994*b*).

(b) The timescale developed for Site 709, by identifying selected $\delta^{18}O$ stages in a rather featureless planktonic isotope record, has produced slight inaccuracies which require identification and adjustment. This idea is supported by studies of late Pleistocene *C. davisiana* curves (see above) which maintain that abundance variations are synchronous, and the fact that events identified at Site 709 consistently lag the eastern equatorial Pacific sites by approximately 15 ka; arguably, a more random, rather than systematic, relationship between Site 709 and the eastern equatorial Pacific sites would be expected if the age difference between events were due to *C. davisana* diachroneity.

The presence of *C. davisiana* abundance events at the studied sites is variable, with only one event occurring at all five sites (see Results). This one event, which is synchronous at 1.8 to 1.805 Ma at both the Pacific and (with adjustment) at the Indian Ocean sites, corresponds closely with the position of the Plio-Pleistocene boundary, as identified in the Vrica type-section, dated at 1.81 Ma by Hilgen (1991). Thus, this widespread and distinctive *C. davisiana* event may constitute a new biostratigraphical marker for the Plio-Pleistocene boundary.

The varying number of events recognized at each of the sites is attributable to palaeoecological factors affecting the occurrence of *C. davisiana*. *C. davisiana* is a deep cold-water, high-latitude species (Lombari & Boden 1985). However, its range also extends into low latitudes, where it inhabits intruded cold water masses. Thus, *C. davsiana* is an indicator of deep cold-water, with its abundance variation being a crude palaeotemperature proxy; it is also the dominant member of an early Pleistocene glacial radiolarian assemblage in the eastern equatorial Pacific (Haslett 1992). Therefore, *C. davisiana* peaks (events) in the present study reflect intrusions of deep cold water into low-latitude areas. Therefore, more, and higher magnitude events would be expected at sites closer to the source of cold-water input.

In the eastern equatorial Pacific, the Peru Current acts as the main conveyor of high-latitude deep cold water, where winds blowing offshore Peru and Ecuador drive coastal upwelling, propagating a tongue of advected cold water westwards. Sites 677 and 847, which lie within the area directly affected by the Peru Current, record the highest number of events. However, at Sites 850 and 851, and 709, equatorial upwelling is divergence-driven, sourced by local deep water, rather than water of higher-latitude provenance. Thus, at these sites, only *C. davisiana* events generated by intense cooling events are recorded. The recognition of a *C. davisiana* event at all sites at 1.8–1.805 Ma, in both the Pacific and Indian Oceans, indicates that a major global cooling event occurred in association with the Plio-Pleistocene boundary conditions.

Conclusions

1. This study has shown that low-latitude peak-abundance events of the radiolarian *Cycladophora davisiana* are generally synchronous, and may be used to evaluate and fine-tune timescales for high-resolution regional, and possibly oceanic and interoceanic, correlation.
2. From the present-day distribution and known

ecology of *C. davisiana*, it can be inferred that *C. davisiana* events reflect a response to global cooling. Equatorial sites directly located close to deep cold-water sources (e.g. Peru Current), are more 'sensitive', recording more cooling events than sites located away from these areas.

3. Only one *C. davisiana* event was recorded at all of the five sites investigated. This event occurred at 1.8–1.805 Ma and is associated with the Plio-Pleistocene boundary. It may be that, at least in tropical sediments, this event can be used as a new biostratigraphic marker for the Tertiary/Quaternary boundary. The *C. davisiana* event at the Plio-Pleistocene boundary probably reflects a major global cooling event.

This work was funded by NERC ODP Special Topic Grant: GST/02/583. N. J. Shackleton made an essential contribution to our construction of the timescales used in this paper. We are also grateful to C. Nigrini and J.-P. Caulet for providing constructive reviews of an earlier draft of this paper.

References

BARRON, J. 1985. Miocene to Holocene planktic diatoms. *In*: BOLLI, H. M., PERCH-NIELSEN, K. & SAUNDERS, J. B. (eds) *Plankton Stratigraphy*. Cambridge University Press, Cambridge, 763–809.

HASLETT, S. K. 1992. Early Pleistocene glacial–interglacial radiolarian assemblages from the eastern equatorial Pacific. *Journal of Plankton Research*, **14**, 1553–1563.

—— 1994a. High-resolution radiolarian abundance data through the Late Pliocene Olduvai subchron (1.75–2 Ma) of ODP Hole 677A (Panama Basin, eastern equatorial Pacific). *Revista Española de Micropaleontologia*, **26**, 127–162.

—— 1994b. Plio-Pleistocene radiolarian biostratigraphy and palaeoceanography of the mid-latitude North Atlantic (DSDP Site 609). *Geological Magazine*, **131**, 57–66.

—— 1996. Radiolarian faunal data through the Plio-Pleistocene Olduvai subchron of ODP Leg 138 Sites 847, 850, and 851 (eastern equatorial Pacific). *Revista Española de Micropaleontologia*, **28**, 225–256.

——, FUNNELL, B. M. & DUNN, C. L. 1994a. Calcite preservation, palaeoproductivity and the radiolarian *Lamprocyrtis neoheteroporos* Kling in Plio-Pleistocene sediments from the eastern equatorial Pacific. *Neues Jahrbuch für Geologie und Paläontologie, Monatshefte*, **1994**, 82–94.

——, FUNNELL, B. M., BLOXHAM, K. S. & DUNN, C. L. 1994b. Plio-Pleistocene palaeoceanography of the tropical Indian Ocean: radiolarian and $CaCO_3$ evidence. *Journal of Quaternary Science*, **9**, 199–208.

HILGEN, F. J. 1991. Astronomical calibration of Gauss to Matuyama Sapropels in the Mediteranean and implication for geomagnetic polarity time scale. *Earth and Planetary Science Letters*, **104**, 226–244.

LOMBARI, G. & BODEN, G. 1985. *Modern Radiolarian Global distributions*. Cushman Foundation for Foraminiferal Research, Special Publication, No. **16A**.

MORLEY, J. J. 1980. Analysis of the abundance variations of the subspecies *Cycladophora davisiana*. *Marine Micropaleontology*, **5**, 205–214.

—— 1987. Comparison of the Pleistocene records of the radiolarian *Cycladophora davisiana* at high-latitude sites of the Deep Sea Drilling Project. *Initial Reports of the Deep Sea Drilling Project*, **94**, 889–894.

—— & HAYS, J. D. 1979. *Cycladophora davisiana*: a stratigraphic tool for Pleistocene North Atlantic and interhemispheric correlation. *Earth & Planetary Science Letters*, **44**, 383–389.

—— & —— 1983. Oceanographic conditions associated with high abundances of the radiolarian *Cycladophora davisiana*. *Earth & Planetary Science Letters*, **66**, 63–72.

PERCH-NIELSEN, K. 1985. Silicoflagellates. *In*: BOLLI, H. M., PERCH-NIELSEN, K. & SAUNDERS, J. B. (eds) *Plankton Stratigraphy*. Cambridge University Press, Cambridge, 811–846.

RAVELO, A. C. & SHACKLETON, N. J. 1995. Evidence for surface-water circulation changes at Site 851 in the eastern tropical Pacific Ocean. *Proceedings of the Ocean Drilling Program, Scientific Results*, **138**, 503–514.

RUDDIMAN, W. F., RAYMO, M. E., MARTINSON, D. G., CLEMENT, B. M. & BACKMAN, J. 1989. Pleistocene evolution: Northern hemisphere ice sheets and North Atlantic ocean. *Paleoceanography*, **4**, 353–412.

SANFILIPPO, A., WESTBERG-SMITH, M. J. & RIEDEL, W. R. 1985. Cenozoic radiolaria. *In*: BOLLI, H. M., PERCH-NIELSEN, K. & SAUNDERS, J. B. (eds) *Plankton*

Stratigraphy. Cambridge University Press, Cambridge, 631–712.

SHACKLETON, N. J. & HALL, M. A. 1989. Stable isotope history of the Pleistocene at ODP Site 677. *Proceedings of the Ocean Drilling Program, Scientific Results*, **111**, 295–316.

—— & —— 1990. Pliocene oxygen isotope stratigraphy of Hole 709C. *Proceedings of the Ocean Drilling Program, Scientific Results*, **115**, 529–538.

——, BERGER, A. & PELTIER, W. R. 1990. An alternative astronomical calibration of the lower Pleistocene timescale based on ODP Site 677. *Transactions of the Royal Society of Edinburgh, Earth Sciences*, **81**, 251-261.

——, CROWHURST, S., HAGELBERG, T., PISIAS, N. & SCHNEIDER, D. A. 1995. A new Neogene time scale: application to Leg 138 Sites. *Proceedings of the Ocean Drilling Program, Scientific Results*, **138**, 73-101.

SHIPBOARD SCIENTIFIC PARTY. 1988. Sites 677 and 678. *Proceedings of the Ocean Drilling Program, Initial Results*, **111**, 253–346.

—— 1992. Leg 138. *Proceedings of the Ocean Drilling Program, Initial Results*, **138**.

Intra-interglacial cold events: an Eemian–Holocene comparison

M. MASLIN[1,2], M. SARNTHEIN[2], J.-J. KNAACK[2], P. GROOTES[3]
& C. TZEDAKIS[4]

[1] *Environmental Change Research Centre, Department of Geography, University College London, 26 Bedford Way, London WC1H 0AP, UK*
[2] *Geologisch-Paläontologisches Institut, Universität Kiel, Olshausenstr. 40, 24098 Kiel, Germany*
[3] *AMS C14 Laboratory, Universität Kiel, Olshausenstr. 40, 24098 Kiel, Germany*
[4] *Godwin Institute of Quaternary Research, Department of Geography, University of Cambridge, Downing Place, Cambridge CB2 9EA, UK*

Abstract: Rapid oscillations between warm and cold climates have been found in the oxygen isotope record of the Greenland Ice-core Project (GRIP) ice core during the Eemian/Marine oxygen Isotope Stage (MIS) 5e. In contrast, the variability in Greenland Ice Sheet Project 2 (GISP2) ice core is significantly different and some Atlantic deep-sea records suggest no such climate variations. We present here a high-resolution (50–300 years) set of marine proxies from the low-latitude east Atlantic margin (ODP Site 658), which suggest that in general the Eemian was climatically very similar to the Holocene. We, however, observe that the upwelling intensity off the West African coast was greatly reduced during the early Eemian, corresponding to the very mild climate observed in the European lake records. We observe that MIS 5e contains one significant short cold spell (<400 years), which is marked by a reduction of upper North Atlantic deep water ventilation. We suggest this cold event may correlate with the cold interval found in the European terrestrial records. The cause of the intra-Eemian event was likely to be the freshening and cooling observed in the Norwegian Sea. This brief cold spell, however, did not affect the overall stability of MIS 5e, and moreover it has an analogue event in the Holocene 'Sub-Boreal' period. Marine and terrestrial records thus seem to be incompatible with those of the GRIP ice core record, supporting the suggestion that the GRIP record has been altered by ice tectonics.

The climate records of the Greenland Ice-core Project (GRIP) (Dansgaard *et al.* 1993) and Greenland Ice Sheet Project 2 (GISP2) (Grootes *et al.* 1993) ice cores recovered from Summit, Greenland, agree in detail for the last 110 ka (Peel 1995). Before 110 ka they disagree, posing the question of whether the rapid oxygen isotope variations observed in the Marine oxygen Isotope Stage (MIS) 5e or Eemian section of the GRIP core (Dansgaard *et al.* 1993) represents a true global or local climate signal, or an effect of ice tectonics (Grootes *et al.* 1993). Determining whether these variations are real is essential as the Eemian, with a documented higher sea level than at present, may provide the closest analogue to the present 'Holocene' interglacial and future climate change. Lake records from continental Europe, the Massif Central, France (Thouveny *et al.* 1994) and Bispingen, Germany (Field *et al.* 1994), suggest that the Eemian climate was possibly more variable

than the Holocene, and have been used to support the validity of the GRIP oxygen isotope variations. Some oceanic records from the northeast Atlantic (McManus *et al.* 1994) and the Bahamas Outer Ridge (Keigwin *et al.* 1994) indicate very little or no variability during MIS 5e. In contrast, records from both the northeast Norwegian Sea and west of Ireland show a cooling and freshening of the North Atlantic in the middle of MIS 5e (Cortijo *et al.* 1994).

We present here high-resolution results (50–300 years sample interval) from Ocean Drilling Program (ODP) Site 658 off West Africa, a location chosen because of its sensitivity to changes in upwelling and deep-water ventilation. The aim of this study is: (1) to test whether the Eemian climate variability indicated by the GRIP oxygen isotope and the European pollen records can be found in marine records; (2) to compare the marine record of the Holocene and Eemian to test whether there are

MASLIN, M., SARNTHEIN, M., KNAACK, J.-J., GROOTES, P. & TZEDAKIS, C. 1998. Intra-interglacial cold events: an Eemian–Holocene comparison. *In*: CRAMP, A., MACLEOD, C. J., LEE, S. V. & JONES, E. J. W. (eds) *Geological Evolution of Ocean Basins: Results from the Ocean Drilling Program.* Geological Society, London, Special Publications, **131**, 91–99.

91

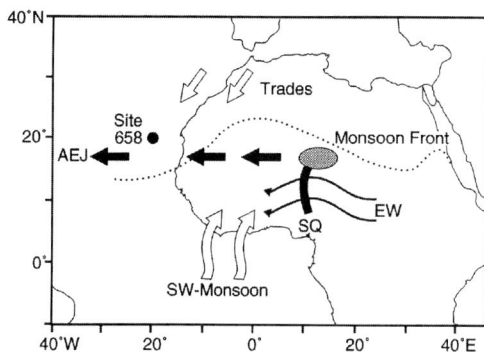

Fig. 1. Schematic diagram of the present meteorological patterns over northwest Africa during the summer months (Tiedemann *et al.* 1994), and the position of ODP Site 658. EW = easterly waves, SQ = squall lines, and AEJ = African jet stream. The elliptical shape indicates dust clouds associated with the squall line.

significant differences between the two inter-glacials; and (3) to explain features of the Eemian record.

Study area

ODP Site 658 is located 165 km offshore of northwest Africa (20°45′N, 18°35′W, 2263 m water depth, see Fig. 1) and has an average sedimentation rate of 15 cm ka^{-1} for the last 700 ka. The sedimentation rate at Site 658 is at least twice as high between 115 and 135 ka as previous marine MIS 5e records (McManus *et al.* 1994; Keigwin *et al.* 1994; Cortijo *et al.* 1994). This site is situated below a perennial cell of coastal upwelling driven by the northeast trade winds, which produces cold and nutrient-rich surface waters with a modern temperature as low as 16°C (Sarnthein *et al.* 1982). Site 658 is thus an ideal location to monitor changes in trade wind-influenced upwelling and upper deep-water ventilation.

Methods

The first coarse resolution stable isotope records for Site 658 were published by Tiedemann (1991) and Sarnthein & Tiedemann (1989, 1990). We have resampled the Eemian section identified by Sarnthein & Tiedemann (1989) in hole 658A and hole 658B (at core breaks) to provide a resolution of approximately 200 years. The stable isotope measurements of ODP Site 658 holes A and B were made on either 30 planktonic foraminifera *G. inflata* tests (250–315 µm fraction) or ten specimens of the epibenthic species *C. wuellerstorfi* (315–400 µm) using a Finnigan MAT 251 at the C-14 Laboratory of the Institut für Kernphysik,

University Kiel. The detailed account of the composite depth calculations linking holes A and B is given in Tiedemann (1991) and Sarnthein & Tiedemann (1989). Molecular analysis of the alkenones was made at 1–2 cm intervals down core, representing 50–100 years, and provides the bases of the U^k_{37} index (Eglinton *et al.* 1992), which was calibrated to summer sea-surface temperature (Rosell 1994). The age model for the Eemian was in general constructed to fit the *C. wuellerstorfi* d^{18}O record to the SPECMAP curve (Martinson *et al.* 1987). The MIS 5/6 boundary was shifted to an older age than the SPECMAP curve, as suggested by Sarnthein & Tiedemann (1990), based on sedimentation rate constraints. Without this shift the sedimentation rate during the Termination II would be four times higher than either the glacial or the Eemian and there is no support from any of the productivity or sedimentary proxies for such a spike (Tiedemann 1991). This shift has since been confirmed by coral ages (Szabo *et al.* 1994; Stirling *et al.* 1995) suggesting that the initial sea-level rise for glacial Termination II may have been as early as 132 ka.

Ice core data

The variations in the Greenland oxygen isotope and the Vostok deuterium ice core records (see Fig. 1) indicate mainly air temperatures (e.g. Dansgaard 1964; Jouzel *et al.* 1987; Jouzel 1994) though there could be additional factors such as changes in evaporative sources, e.g. Atlantic versus Pacific Ocean, and/or in seasonality of precipitation (Charles *et al.* 1994). Global circulation models suggest, though, that the temperature shifts inferred from the ice core data may differ from the true variation by only about 30% (Waelbroeck *et al.* in press). The Eemian controversy is due to the rapid fluctuations in δ^{18}O recorded in the GRIP ice core (Dansgaard *et al.* 1993). Claims have not been made about the Eemian climate from the GISP2 ice core because inclined ice layers have been found below 110 ka (Grootes *et al.* 1993). Inclined layers indicate that the ice has been disturbed and there is no way to distinguish simple tilting from folding or slippage. It has been suggested that the deeper parts of the GRIP ice core record, in particular the Eemian, may also have been affected by ice tectonics (e.g. Grootes *et al.* 1993; Taylor *et al.* 1993; Boulton 1993; Peel 1995). Johnsen *et al.* (1995) have reported layers tilted up to 20° within the MIS 5c section of the GRIP ice cores, which is precisely where the correlation between GRIP and GISP2 breaks down. Moreover, the recent GISP2–GRIP Joint Workshop in Wolfeboro, New Hampshire, USA in September 1995, concluded from bipolar data on ice-trapped atmospheric CH$_4$

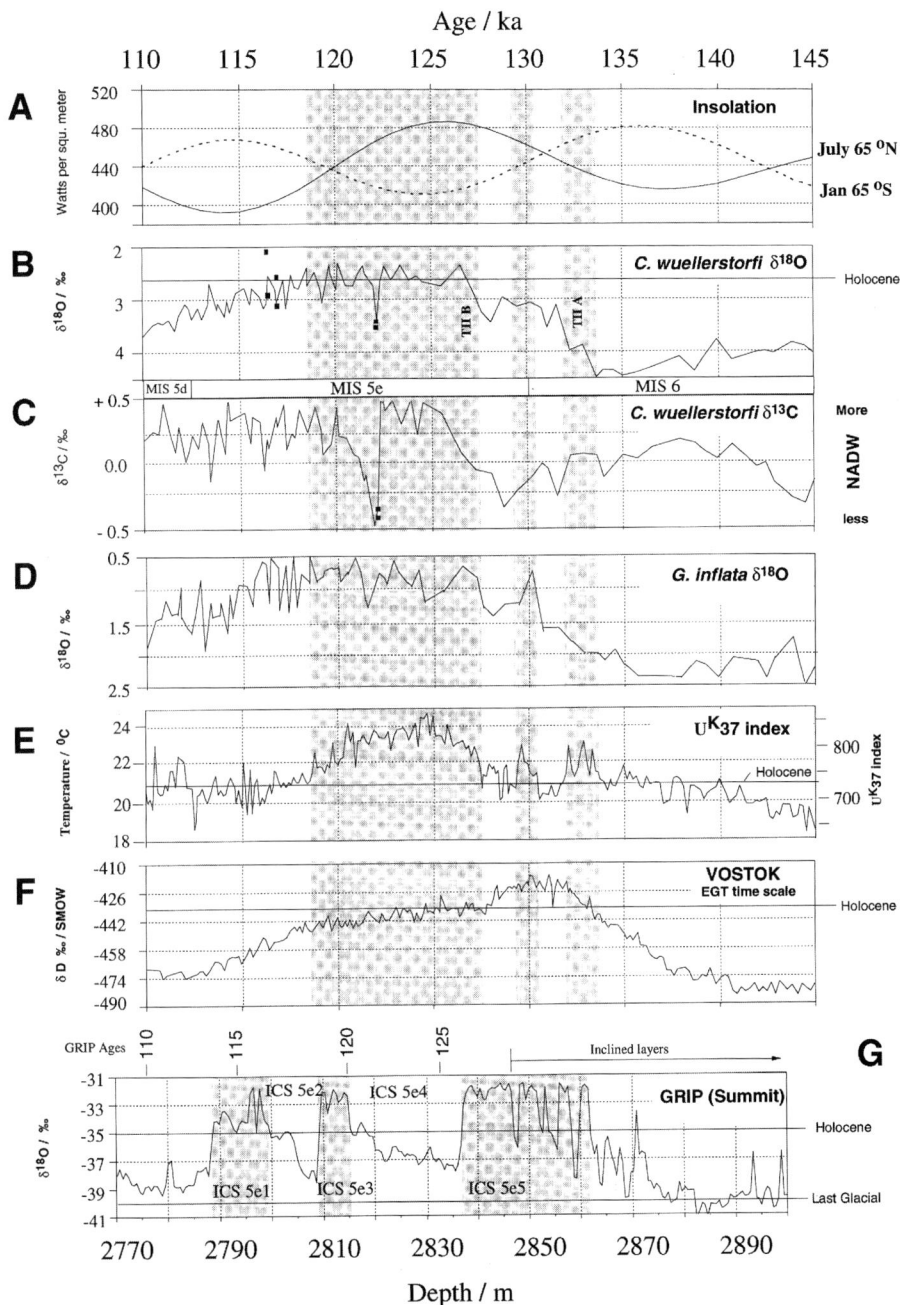

Fig. 2. Comparison of the Eemian or MIS 5e records of: (A) the insolation at 65°N and 65°S calculated by Berger & Loutre (1991); (B) Site 658 oxygen isotope record of the benthic foraminifera *C. wuellerstorfi* (TIIA = Termination IIA, TIIB = Termination IIB); (C) Site 658 carbon isotope record of the benthic foraminifera *C. wuellerstorfi*; (D) Site 658 oxygen isotope record of the planktonic foraminifera *G. inflata*; (E) Site 658 U$^k_{37}$ index (Eglinton *et al.* 1992) and summer sea-surface temperature calibration; (F) Vostok ice core deuterium record on the EGT timescale (Jouzel *et al.* 1987); (G) the GRIP ice core oxygen isotope record against depth (m) with age scale above (Dansgaard *et al.* 1993).

and $\delta^{18}O$ and more detailed work on the structural properties of the cores, that both the GRIP and GISP2 ice cores had suffered stratigraphic disturbances in ice older than 110 ka (Peel 1995). It is still, however, far from being completely accepted that the GRIP ice core Eemian record is invalid.

It is noted that the Vostok 'extended glaciological timescale' (EGT) age model has an error range of +/– 10%. There are also still considerable doubts over the deep parts of the GRIP timescale (Grootes *et al.* 1993) as it relies on uniform thinning even in the deformed sections of the core. Comparison between the ice core records with the marine and terrestrial records is, therefore, difficult. However, the $\delta^{18}O$ records of the atmospheric oxygen trapped in the ice cores (Bender *et al.* 1994; Sowers & Bender 1995) do demonstrate that undisturbed ice core records can be convincingly correlated to each other and to other records. If we assume, therefore, that the age models are compatible, then the variability of the GRIP Ice Core Substage (ICS) 5e5 and the peak temperatures in the Vostok record seem to occur during marine Termination II. It has, however, been suggested that the early warming observed in the Vostok ice core could be due to the influence of southern hemisphere variations in insolation, see Fig. 2 (Genthon *et al.* 1987; J. Jouzel pers. comm.). This is similar to the early southern hemisphere warming found for the last deglaciation by Sowers & Bender (1995). The GRIP 'cold events' ICS 5e4 and ICS 5e2 seems to occur within the mid- to late Eemian. ICS 5e4 occurs during a plateau in the Vostok temperatures (Fig. 2F) and during the highest summer sea-surface temperature (SST) at Site 658 (Fig. 2E), while ICS 5e2 occurs during a drop in both the Vostok and Site 658 temperature records.

Subtropical East Atlantic upwelling

During MIS 5e the SST off northwest Africa, as recorded by the U^k_{37} index at Site 658 (Eglinton *et al.* 1992), culminated at 24°C compared to approximately 21°C (Zhao *et al.* 1993) during the Holocene (Fig. 3). This Eemian high in SST could have been the result of a curtailment of coastal upwelling which persisted for over 6000 years, from 126–120 ka. This suggests that either the trade winds' strength was reduced or their direction was different during the Eemian compared with the Holocene. It is interesting to note that this 6000 year SST plateau is coeval with a period of relatively little change in the benthic $\delta^{18}O$ and is very similar to the temperature plateau seen in the Vostok ice core record (Fig. 2). Moreover this alteration in the trade wind intensity or direction coincides with the mild early Eemian climate over

Europe (Thouveny *et al.* 1994; Field *et al.* 1994; Tzedakis *et al.* 1994), see Fig. 4. Reduced upwelling also suggests that surface water productivity was diminished as few nutrients were being brought up from intermediate water depths. Short-term variations in the U^k_{37} record amounting to almost 3°C within a few hundred years characterize Termination II, and are absent from Termination I. Comparison of records from two nearby cores (20 m apart) showed that these variations are reproducible (Zhao *et al.* 1993). However, no such rapid fluctuations in SST are observed between about 126 and 120 ka.

Site 658 oxygen isotope records

The benthic foraminifera (*C. wuellerstorfi*) $\delta^{18}O$ structure of Termination II closely resembles Termination I (see Fig. 3), being composed of two deglacial steps and a Younger Dryas-style event in between. The exception is the larger amplitude in Termination II and a peak MIS 5e (Eemian) value higher than the Holocene, indicating that the Eemian was milder, with a higher sea level (at least 4–7 m; Chappel & Shackleton 1986). The benthic $\delta^{18}O$ record of MIS 5e / Eemian is generally smooth, but we have found one significant 'cool' event at 122 ka. This $\delta^{18}O$ deviation has been confirmed by duplicate measurements and it is not at a core break. At first it was thought that this event may have been due to reworking, but the core photographs and the core description show that Site 658 Core A3, section 3-4 is continuous, homogeneous siliceous nano-ooze with no signs of disturbance. Moreover the magnetic susceptibility records of Site 658 Core 658 A3 are reproduced perfectly in the neighbouring hole in Site 658 Core B3 (Tiedemann 1991). In addition it would be difficult to rework benthic foraminifera downslope because of the location of the site, and it has been shown that *C. wuellerstorfi* does not occur much further upslope (Ganssen & Lutze 1982).

The short (<400 years) cold spike near 122 ka has an amplitude of approximately 1‰ in $\delta^{18}O$, twice as large as the intra-Holocene variation at this site. There are, however, minor late Holocene $\delta^{18}O/\delta^{13}C$ events in the Site 658 records (Fig. 2), for example between 4500 and 4800 years BP, which is at the beginning of a significant period of climatic deterioration (Pachur & Kröpelin 1987; Frenzel *et al.* 1992) in the northern hemisphere defined in Europe as the Sub-Boreal (see Fig. 3). Moreover, variations of a magnitude similar to the Eemian cold spike have been found in cores in the surrounding area during Heinrich event 1 and near the end of the Younger Dryas (Sarnthein *et al.* 1994). Since the Eemian excursion is unlikely to be

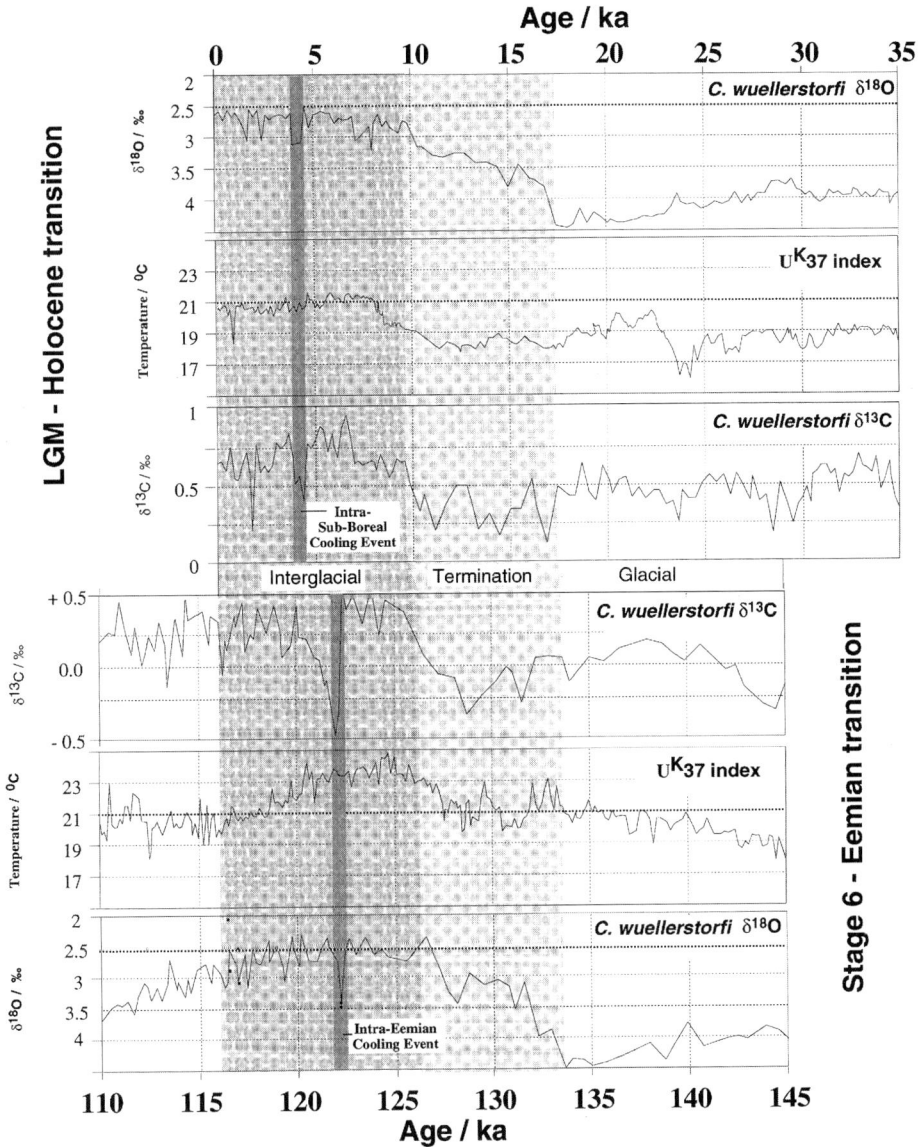

Fig. 3. Comparison of the oxygen and carbon isotope records of the benthic foraminifera *C. wuellerstorfi* and the U^k_{37} index from the Stage 6 – Eemian transition (ODP Site 658 holes A and B) and the Last Glacial Maximum (LGM) - Holocene transition (ODP Site 658 hole C). Both age scales are calendar years. Note that the carbon isotope records of the Stage 6 – Eemian and the LGM – Holocene are offset by approximately 0.4‰ due to the long-term 400 ka cyclicity observed at this site (Sarnthein *et al.* 1982; Tiedemann 1991; Sarnthein & Tiedemann 1989, 1990).

a short-term advance of continental ice sheets, we infer a cooling (by 3°C) and/or circulation changes of the upper deep waters at about 2300 m. Indeed, coeval with the $\delta^{18}O$ excursion we observe a minor excursion in the planktonic *G. inflata* $\delta^{18}O$ record and a major reduction in the epibenthic *C. wuellerstori* $\delta^{13}C$ values (Figs 2 and 3).

Site 658 benthic carbon isotope record

Keigwin *et al.* (1994), as many colleagues before them (e.g. Duplessy *et al.* 1988; Sarnthein *et al.* 1994), have used epibenthic $\delta^{13}C$, which reflects bottom-water $\delta^{13}C$, to deduce past changes in the ventilation of Atlantic deep waters. The $\delta^{13}C$ fluctuations of *C. wuellerstorfi* at Site 658 near

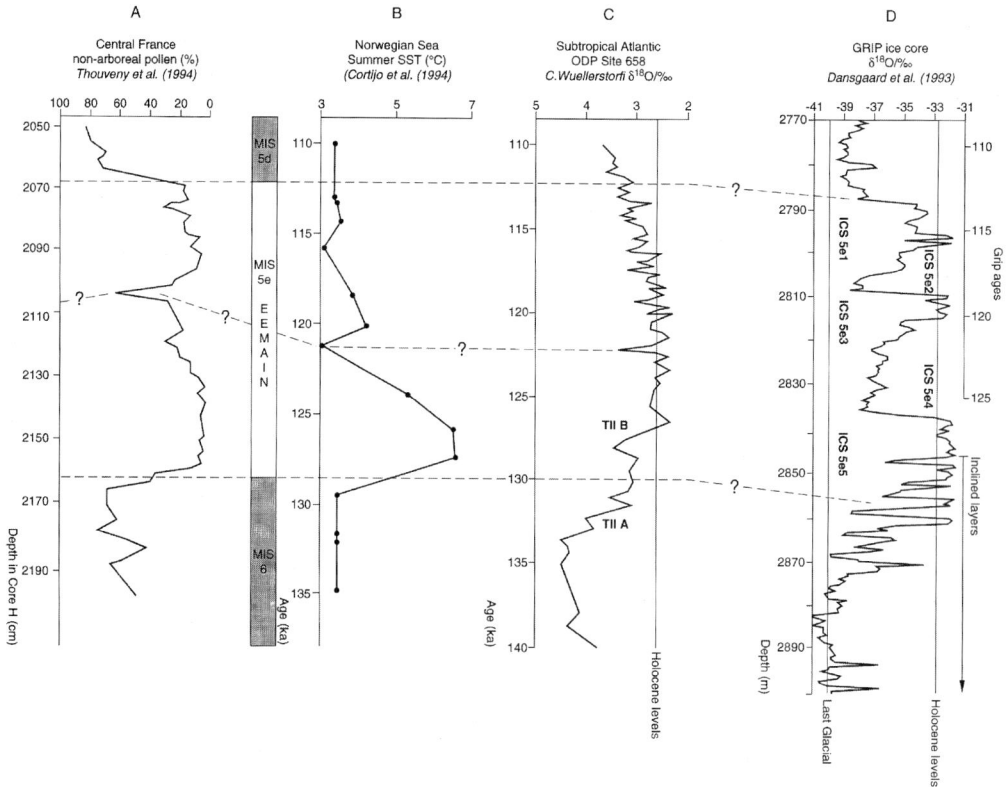

Fig. 4. Comparison of the pollen record from (A) the Lac du Bouchet, France (Thouveny et al. 1994), (B) the marine records from the Norwegian Sea (Cortijo et al. 1994), (C) the subtropical Atlantic (this study) and (D) the ice core records from GRIP (Dansgaard et al. 1993). A tentative correlation of the intra-Eemian cooling events has been made between the records.

122 ka amount to about 1‰, which is more than the usual glacial-to-interglacial variation (Figs 2 and 3). In summary, these variations of the $\delta^{13}C$ signal are caused by the contributions of: (1) admixture of Antarctic source deep water to the North Atlantic deep water (NADW) from below; (2) the shift of isotopically light carbon between the terrestrial biosphere (e.g. Duplessy et al. 1988; Maslin et al. 1995), the atmospheric CO_2 (Marino et al. 1992; Lynch-Stieglitz & Fairbanks 1994) and the ocean; and (3) local surface water productivity changes that can produce, via organic fluff layers, $\delta^{13}C$ excursions of up to 0.45‰ (Curry et al. 1988; Sarnthein et al. 1994). Preliminary calcium carbonate, organic carbon and opal (not shown) records from Site 658 may indicate a minor increase in surface water productivity, despite the fact that U^k_{37} SST estimates suggest little change in the generally low Eemian upwelling intensity. We suggest that part of the pronounced intra-Eemian $\delta^{13}C$ event may have been caused by the 'fluff

layer' effect due to enhanced surface water productivity. However, this can only account for a maximum shift of 0.45‰ in the C. wuellerstorfi $\delta^{13}C$ record (Curry et al. 1988; Sarnthein et al. 1994), therefore the remaining 0.55‰ must represent a significant reduction of the upper NADW ventilation. The exact duration of this $\delta^{13}C/\delta^{18}O$ event cannot be assessed because bioturbation mixing has smoothed and broadened the signal (Trauth 1995), but it seems to be on the order of hundreds of years.

Terrestrial records

Continental records, despite the uncertainty over the climatic mechanisms involved, have shown oscillations within a number of Eemian pollen sequences (e.g. Tzedakis et al. 1994). Records derived from Lac du Bouchet, France, show fluctuations in tree pollen percentages associated with changes in sediment magnetic susceptibility and organic carbon content and provide further evidence for climate variability during the Eemian (Thouveny *et al.* 1994). Although these records have been used to lend support to the notion of Eemian instability, detailed correlation with the GRIP record has proved difficult, because of the lack of a sufficient chronology for the pollen sequences and the current belief that the Eemian GRIP core record has been disturbed. However, correlations between the terrestrial and marine pollen sequences have provided some chronological control over this period and suggest a significant divergence in the implied timing and length of cold events recorded in the GRIP and pollen records. In the GRIP record the first major cold episode begins at about 126 ka and persists for a few thousand years. No cooling event is detected so early in the pollen records. Interglacial maximum conditions occurred during the 'yew phase' north of the Alps which was coeval with an 'olive phase' and maximum expansion of Mediterranean vegetation in southern Europe (van Andel & Tzedakis in press). The olive event is also evident in marine pollen records from the eastern Mediterranean and occurred contemporaneously with the deposition of sapropel S5, itself associated with increased monsoonal activity during maximum northern hemisphere insolation (125–126 ka). Following this period, temperate conditions persisted during a phase of hornbeam expansion across Europe, and only after this do the pollen sequences show the first evidence of climate instability. The implication, therefore, is that the cold events recorded in various pollen sequences as a decrease of thermophilous trees or the expansion of open herbaceous vegetation, occurred much later than the disputed GRIP events and appear to have persisted for only hundreds of years rather than thousands. The cold oscillation recorded after the hornbeam expansion at Luc du Bouchet appears to correlate much more convincingly in terms of timing (about 122 ka) and duration with the intra-Eemian cooling event seen in the marine records.

The Holocene $\delta^{13}C/\delta^{18}O$ event around 4.5–4.8 ka is similar to the intra-Eemian cooling event (Fig. 3) in that they both occur after the interglacial peak and signal the beginning of a trend of climatic deterioration. Although no such abrupt climatic oscillations are reported in Holocene pollen records, this is the period when there was an elm decline in Europe (about 5000 ^{14}C years or 5700 calendar years) and the hemlock decline in North America (about 4700 [4]C years or 5300 calendar years). Both of these have been attributed to specific pathogen attacks (e.g. Rackham 1980; Davis 1981; Peglar & Birks 1993), but with the evidence from the Eemian it may now be worth considering these declines in terms of climate deterioration, or at least its effect on the spread of the epidemics.

Discussion

As discussed previously, precise age correlation between the marine, terrestrial and ice core climatic records is not possible. In general, though, the Vostok record (Fig. 1) does not depict any intra-Eemian cold events, while the two potential cold events of the GRIP core (ICS 5e2 and ICS 5e4, Fig. 2) have a duration of thousands of years, hence are incompatible with the single, short-term event observed in our Site 658 records. However, the European lake pollen records (e.g. Tzedakis *et al.* 1994) do indicate a mid-Eemian cooling. In particular the reduced arboreal pollen signal found in the central French lake records (Thouveny *et al.* 1994) matches precisely the short duration (about 300 years) and possibly the timing (5 ka after the onset of the peak Eemian) of the cold event observed in our Site 658 records (see Fig. 3).

Further support for this observed intra-Eemian cooling event comes indirectly from coral reef records. The new data of Stirling *et al.* (1995) from Western Australia and their review of other published high-precision U-series coral dates indicate that the last interglacial period lasted at least between 130 and 117 ka. Moreover their work suggests that the main episode of coral reef building, globally, was confined to a narrow range of dates between 127 and 122 ka. What is of great interest is that the period of reef building seems to end at the same time as the intra-Eemian cold event begins, i.e. at about 122 ka.

The mid-Eemian isotope event near 122 ka is also coeval, with the intense cooling and freshening event observed by Cortijo *et al.* (1994) west of Ireland and in the northern Norwegian Sea (Fig. 3). Today a freshening of the North Atlantic Drift and the Norwegian Current of this order of magnitude would result in a reduction of deep-water formation in the Nordic Seas (e.g. Bryan 1986; Dickson *et al.* 1990; Seidov & Maslin in press), a reduced NADW flow and an enhanced penetration of the low ventilated Antarctic bottom water. We suggest that a similar chain of events may have occurred in the mid-Eemian. Based on the absence of ice rafted debris and other evidence of melting icebergs (Keigwin *et al.* 1994; McManus *et al.* 1994) during the peak Eemian, the observed freshening of the surface waters and the corresponding reduction of deep-water formation cannot be ascribed to a mystical ice surge event but rather to the overlapping processes cited by Cortijo *et al.* (1994): (1) the increased incursion of 'fresh' North Pacific water via the Bering Strait during times of raised sea level (Shaffer & Bendtsen 1994); and (2) an enhanced precipitation over the North Atlantic and Nordic Seas following the northern hemisphere insolation maximum.

What we cannot at present explain is why after the mid-Eemian cooling event the surface waters of the subtropical Atlantic and the climate of central France (Thouveny *et al.* 1994) returned to their pre-Eemian event conditions, while the Nordic Seas (Cortijo *et al.* 1994) and more northerly and central pollen records indicate a cooler late Eemian (Field *et al.* 1994; van Andel & Tzedakis in press).

Conclusion

Site 658 proxy records and other Eemian marine and terrestrial records indicate that there is no support for the climate changes implied from the GRIP ice core records. This supports the suggestion that the GRIP record has been altered by ice tectonics (Peel 1995). It is not suggested, however, that the Eemian was climatically stable, as evidence is presented here for a significant intra-Eemian cold event. This cold event has no direct analogue in

magnitude in the Holocene record (Fig. 2), but there are many similarities with the cool event in Sub-Boreal. The Eemian cold event lasted no longer than 400 years and may have had a profound short-term effect, but we find no evidence of a major long-term influence on the northern hemisphere climate, i.e. it did not lead to the termination of the Eemian, though some records indicate a harsher late Eemian. We therefore conclude that the Eemian / MIS 5e was, at least in the subtropics and probably globally, much more stable than implied by the GRIP ice core records. Therefore we suggest that the Eemian / MIS 5e still provides the best analogue for present and future climate change both for studying long-term interglacial climate stability as well as intra-interglacial climate events.

Great thanks to H. Erlenkeuser, J. Hennings, J. Jouzel, J-C. Duplessy, G. Eglinton, M. Trauth, L-J. Wang and D. Seidov for discussion and data. We would like to acknowledge the support of the DFG, EU project 'Environment', German National Program of Climate Research and the Department of Geography at UCL. We would also like to thank N. McCave and H. Schulz for their constructive reviews.

References

BENDER, M. *ET AL.* 1994. Climate correlations between Greenland and Antarctica during the past 100,000 years. *Nature,* **372**, 663–665.

BERGER, A. & LOUTRE, M. F. 1991. Insolation values for the climate of the last 10 million years. *Quaternary Science Review,* **10**, 297–315.

BOULTON, G. S. 1993. Greenland ice sheets: Two cores are better than one. *Nature,* **366**, 507–508.

BRYAN, F. 1986. High latitude salinity effects and interhemispheric thermohaline circulations. *Nature,* **323**, 301–304.

CHAPPEL, J. & SHACKLETON, N. J. 1986. Oxygen isotopes and sea level. *Nature,* **324**, 137–140.

CHARLES, C. D., RIND, D., JOUZEL, J., KOSTER, R. D. & FAIRBANKS, R. G. 1994. Glacial-Interglacial changes in moisture sources for Greenland: influences on the ice core record of climate. *Science,* **263**, 508–511.

CORTIJO, E. *ET AL.* 1994. Eemian cooling in the Norwegian Sea and North Atlantic ocean preceding continental ice-sheet growth. *Nature,* **372**, 446–449.

CURRY, W. B., DUPLESSY, J-C., LABEYRIE, L. D. & SHACKLETON, N. J. 1988. Changes in the distribution of $\delta^{13}C$ of deep water ΣCO_2 between the last glaciation and the Holocene. *Paleoceanography,* **3**, 327–337.

DANSGAARD, W. 1964. Stable isotopes in precipitation. *Tellus,* **16**, 436–468.

—— *ET AL.* 1993. Evidence for general instability in past climate from a 250-kyr ice-core record. *Nature,* **364**, 218–220.

DAVIS, M. B. 1981. Outbreaks of forest pathogens in Quaternary history. *In*: BARADWAJ, D., VISHNU-

MITTRE, C. & MAHESHWARI, H. (eds) *Fourth International Palynological Conference Proceedings Volume III.* Birbal Sahni Institute of Palaeobotany, Lucknow, 216–228.

DICKSON, R .R., GMITROWICZ, E. M. & WATSON, A. J. 1990. Deep-water renewal in the northern North Atlantic. *Nature,* **344**, 848–850.

DUPLESSY, J-C., SHACKLETON, N. J., FAIRBANKS, R. J., LABEYRIE, L. D., OPPO, D. & KALLEL, N. 1988. Deepwater source variation during the last climatic cycle and their impact on the global deepwater circulation. *Paleoceanography,* **3**, 343–360.

EGLINTON, G., BRADSHAW, S., ROSELL, A., SARNTHEIN, M., PFLAUMANN, U. & TIEDEMANN R. 1992. Molecular record of secular sea surface temperature changes on 100-year timescales for glacial terminations I, II, and IV. *Nature,* **356**, 423–425

FIELD, M., HUNTLEY, B. & MÜLLER, H. 1994. Eemian climate fluctuations observed in a European pollen record. *Nature,* **371**, 779–782.

FRENZEL, B., PECSI, M. & VELICHKO, A. A. (eds) 1992. *Atlas of Paleoclimates and Paleoenvironments of the Northern Hemisphere.* Gustav Fischer Verlag and Geograph. Res. Inst. Hungar. Acad. Sci., Stuttgart and Budapest.

GANSSEN G. & LUTZE, G. 1982. The aragonite compensation depth at the Northeastern Atlantic continental margin. *'Meteor' Forsch.-Ergebnisse,* **36**, 57–59.

GENTHON, C. *ET AL.* 1987. Vostok ice core: climatic response to CO_2 and orbital forcing changes over the last climatic cycle. *Nature,* **329**, 414–418.

GROOTES, P. M., STULVER, M., WHITE, J. W. C., JOHNSEN,

S. & JOUZEL, J. 1993. Comparison of oxygen isotope records from GISP2 and GRIP Greenland ice cores. *Nature*, **366**, 552–554.

JOHNSEN, S. *ET AL.* 1995. The Eem stable isotope record along the GRIP ice core and its interpretation. *Quaternary Research*, **43**, 117–124.

JOUZEL, J. 1994. Ice cores north and south. *Nature*, **372**, 612–613.

—— *ET AL.* 1987. Vostok ice core: a continuous isotope temperature record over the last climatic cycle (160,000 years). *Nature*, **329**, 403–408.

KEIGWIN, L., CURRY, W., LEHMAN, S. & JOHNSEN, S. 1994. The role of the deep ocean in North Atlantic climate change between 70 and 130 kyr ago. *Nature*, **371**, 323–325.

LYNCH-STIEGLITZ, J. & FAIRBANKS, R. G. 1994. A conservative tracer for glacial ocean circulation from carbon isotope and palaeo-nutrient measurements in benthic foraminifera. *Nature*, **369**, 308–310.

MCMANUS, J. *ET AL.* 1994. High-resolution climate records from the North Atlantic during the last interglacial. *Nature*, **371**, 326–329.

MARINO, B., MCELROY, M., SALAWITCH, R. J. & SPAULING, W. G. 1992. Glacial-to-interglacial variations in the carbon isotope composition of atmospheric CO_2. *Nature*, **357**, 461–466..

MARTINSON, D. G. *ET AL.* 1987. Age dating and the orbital theory of the ice ages: development of a high resolution 0-300,000-year chronostratigraphy. *Quarternary Research*, **27**, 1–29.

MASLIN, M. A., ADAMS, J., THOMAS, E., FAURE H. & HAINES-YOUNG R. 1995. Estimating the carbon transfer between the oceans, atmosphere and the terrestrial biosphere since the last glacial maximum. *Terra Nova*, **7**, 358–366.

PACHUR, H-J. & KRÖPELIN, S. 1987. Wadi Howar: Paleoclimatic evidence from extinct river systems in southeastern Sahara. *Science*, **237**, 298–300.

PEEL, D. 1995. Ice cores: Profiles of the past. *Nature*, **378**, 234.

PEGLAR, S. M. & BIRKS, H. J. B. 1993. The mid-Holocene *Ulmus* fall at Diss Mere, South-East England – disease and human impact? *Vegetation History and Archaeobotany*, **2**, 61–65.

RACKHAM, O. 1980. *Ancient Woodland: its history, vegetation and uses in England.* Edward Arnold, London.

ROSELL I MELE, A. 1994. *Long-chain alkenones, alkyl alkenoates and total pigment abundances as climatic proxy-indicators in the Northeastern Atlantic.* PhD thesis, University of Bristol.

SARNTHEIN, M. & TIEDEMANN, R. 1989. Toward a high resolution stable isotope stratigraphy of the last 3.4 million years: Sites 658 and 659 off northwest Africa. *In*: RUDDIMAN, W. F., SARNTHEIN, M., BALDAUF, J. ET AL. (eds) *Proceedings of ODP, Scientific Results,* **108**. College Station, TX (Ocean Drilling Program), 167–187.

—— & —— 1990. Younger Dryas-style cooling events at glacial terminations I-IV at ODP Site 658: Associated benthic $\delta^{13}C$ anomalies constrain meltwater hypothesis. *Paleoceanography*, **5**, 1041–1055

—— *ET AL.* 1982. Atmospheric and oceanic circulation patterns off Northwest Africa during the past 25 million years. *In*: RAD, U. *ET AL.* (eds) *Geology of the Northwest African Continental Margin.* Springer Verlag, New York, 545–604.

——, WINN, K., JUNG, S. J. A., DUPLESSY, J.-C., LABEYRIE, L., ERLENKEUSER, H. & GANSEN, G. 1994. Changes in east Atlantic deepwater circulation over the last 30,000 years: Eight time slice reconstructions. *Paleoceanography*, **9**, 209–268

SEIDOV, D. & MASLIN, M. A. 1996. Seasonally ice free glacial Nordic Seas without deep water ventilation. *Terra Nova*, **8**, 245–354.

SHAFFER, G. & BENDTSEN J. 1994. Role of the Bering Strait in controlling North Atlantic ocean circulation and climate. *Nature*, **367**, 354–357.

SOWERS, T. & BENDER, M. 1995. Climate records covering the last deglacial. *Science*, **269**, 210–214.

STIRLING, C. H., ESAT, T. M., MCCULLOCH, M. T & LAMBECK, K. 1995. High-precision U-series dating of corals from Western Australia and implications for the timing and duration of the last interglacial. *Earth and Planetary Science Letters*, **135**, 115–130.

SZABO, B., LUDWIG, K., MUHS, D. & SIMMONS, K. 1994. Thorium-230 ages of corals and duration of the last interglacial sea level high stand on Oahu, Hawaii. *Science*, **266**, 93–96.

TAYLOR, K. C. *ET AL.* 1993. Electrical conductivity measurements from GISP2 and GRIP Greenland ice cores. *Nature*, **366**, 549–552.

THOUVENY, N. *ET AL.* 1994. A high resolution of the last climate cycle in Western Europe from magnetic susceptibility in Maar lake sequences. *Nature*, **371**, 503–507.

TIEDEMANN, R. 1991. *Acht Millionen Jahre Klimageschichte von Nordwest Afrika und Palaeozeanographie des angrenzenden Atlantiks: Hochaufosende Zeitreihen von ODP Sites 658–661.* PhD thesis, Berichte Report No. **46**, Geol.-Paleaont. Inst., Univ. Kiel, Kiel, Germany.

——, SARNTHEIN, M. & SHACKLETON, N. J. 1994. Astronomic timescale for the Pliocene Atlantic $\delta^{18}O$ and dust flux records of ODP Site 659. *Paleoceanography*, **9**, 619–638.

TRAUTH, M. 1995. *Bioturbate Signalverzerrung hockauflosender paläoozeanographischer Zeitreihen* (Bioturbational signal distortion of high-resolution paleoceanographic time-series). PhD thesis, Berichte Report No. **74**, Geol.-Paleaont. Inst., Univ. Kiel.

TZEDAKIS, P. C., BENNETT, K. D. & MAGRI, D. 1994. Climate and the pollen record. *Nature*, **370**, 513.

VAN ANDEL, T. H. & TZEDAKIS, P. C. 1996. European Palaeolithic landscapes 140,000–30,000 years ago. *Quaternary Science Reviews*, **15**, 481–500.

WAELBROECK, C. *ET AL.* in press. A comparison of the Vostok ice deuterium record and series from Southern Ocean core MD 88-770 over the last two interglacial cycles. *Climate Dynamics*.

ZHAO, M., ROSELL, A. & EGLINTON, G. 1993. Biomarkers in Hole 658C. *Palaeogeography, Palaeoclimalology and Palaeoecology*, **103**, 57–65.

Two 30 000 year high-resolution greyvalue time series from the Santa Barbara Basin and the Guaymas Basin

MICHAEL SCHAAF[1] & JÜRGEN THUROW[2]

[1] *Geologisches Institut, Ruhr Universität, 44801 Bochum, Germany*
Present address: Shell International Exploration and Production BV, EPT-HM,
Volmerlaan 8, PO Box, 602280 AB Rijswijk, The Netherlands
[2] *Department of Geological Sciences, UCL, London WC1E 6BT, UK*

Abstract: Two 30000 year records of sedimentary greyvalues from the Santa Barbara Basin and the Guaymas Basin are presented. The records exhibit substantial variability on all timescales from millennial to annual. Strong shifts towards darker greyvalues parallel the postglacial sea-level rise. Approximately 1500 year cycles occur persistently in both records. Spectral analyses proved the existence of Gleissberg (86 year), sunspot (11 year, 22 year), and El Niño (3–9 year). cycles in almost all cores from both sites. In particular El Niño Southern Oscillation forcing could be traced back until the end of the last glacial cold period. Both time series show a high degree of similarity both in the time and the frequency domain, indicating equivalent system responses to external forcing events.

Both the Santa Barbara Basin (SBB) and the Guaymas Basin (GuB) are located in the northern East Pacific Ocean. The SBB lies between the California coast to the north and the Northern Channel Islands to the south (Fig. 1a). It forms the northernmost in a row of semi-enclosed basins in the California Bight area. The basin is separated from the open Pacific Ocean by two sills at 450 and 250 metres below sea level (mbsl), respectively (Thornton 1984). The SBB was drilled by the Ocean Drilling Program (ODP) in a water depth of 576.5 mbsl, during Leg 146, in 1992 (Kennett *et al.* 1994). The GuB is located approximately 400 km north of the mouth of the Gulf of California (Fig. 1b). The Gulf of California is formed into a series of deep basins whose sill depths shoal from 3000 mbsl at the entrance to 1600 mbsl in the north (Alvarez-Borrego & Lara-Lara 1991). The GuB is the northernmost of these deep basins. The slope of the GuB was drilled by the Deep Sea Drilling Project (DSDP) at a depth of 655 mbsl, during Leg 64, in 1979/1980 (Curray *et al.* 1982).

The waters of the California Bight and the Gulf of California are characterized by persistent high surface water productivity sustained through seasonally strong coastal upwelling of nutrient-rich cold water (Reimers *et al.* 1990; Santamaria-del-Angel *et al.* 1994). High bioproductivity, together with the proximity of the basins and slopes to terrigenous sediment sources, yields sedimentation

rates between 0.05 and 2.5 mma^{-1}. (Soutar & Crill 1977; Soutar *et al.* 1982), substantially higher than in other coastal areas. Oxygenation of the large excess production of organic matter, raining from the surface waters down to the sediment/water interface, induces a distinct oxygen minimum zone (<0.1 ml O_2/l) (Anderson *et al.* 1987) between 300 and 1000 mbsl. The absence of burrowing benthic organisms, caused by the lack of oxygen, allows the preservation of primary depositional structures in the basin sediments. Where the sedimentation rates are sufficiently high, and variations of seasonal forcing are pronounced, individual seasonal laminae are preserved (Soutar & Crill 1977; Baumgartner *et al.* 1991). Two to three (depending on the location) of such laminae form an annual varve. Lange *et al.* (1987) and Schimmelmann *et al.* (1990) found lamina formation and preservation in the SBB to change with interannual climatic variability, such as the El Niño Southern Oscillation (ENSO). These authors reported enhanced preservation of organic matter in laminae during anti-El Niño years and vice versa during El Niño years. Similar variations of organic carbon concentration and lamina formation/preservation were reported for the SBB by Stein & Rack (1995) and Schaaf & Thurow (1995), also for lower frequency climatic variability, from decades to millennia. Finally Stein & Rack (1995) for the SBB and Keigwin & Jones (1990) and Calvert *et al.*

SCHAAF, M. & THUROW, J. 1998. Two 30 000 year high-resolution greyvalue time series from the Santa Barbara Basin and the Guaymas Basin. *In*: CRAMP, A., MACLEOD, C. J., LEE, S. V. & JONES, E. J. W. (eds) *Geological Evolution of Ocean Basins: Results from the Ocean Drilling Program*. Geological Society, London, Special Publications, **131**, 101–110.

101

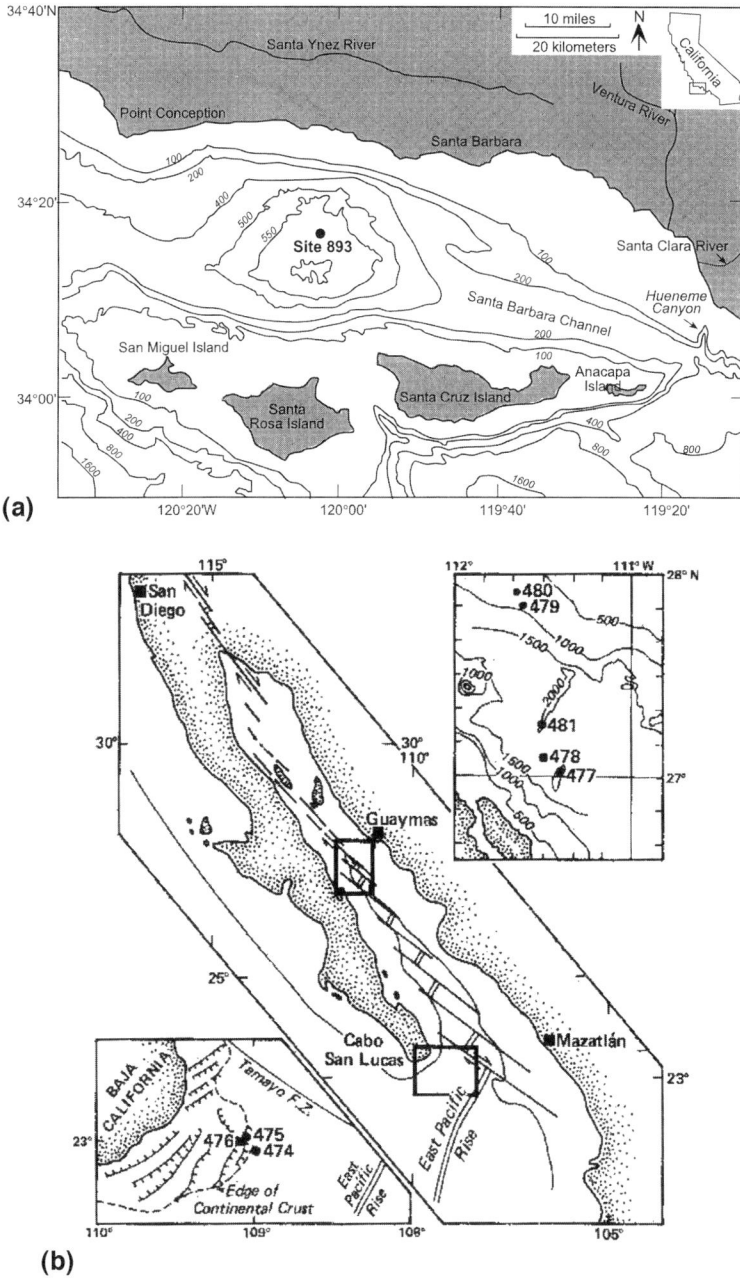

Fig. 1. (**a**) Location map of the Santa Barbara Channel. ODP Site 893 is indicated in the centre of the Santa Barbara Basin (after Kennett *et al.* 1994). (**b**) Location map of the Gulf of California and Site 480 (modified after Curray *et al.* 1982).

(1992) for the GuB reported strong variations of organic matter preservation/accumulation between glacial and interglacial periods.

During the past years numerous studies have investigated the basins and slopes of the East Pacific Ocean. The scope of these studies grasps almost all aspects of physical, chemical, climatological and oceanographic processes, on timescales from centuries (e.g. Lange *et al.* 1987; Schimmelmann *et al.* 1990) to several hundred thousand years (e.g. Kennett 1995). The sample resolutions vary from annual to millennial. However, no such study has yet presented a long high-resolution record (subannual resolution) of either the SBB or the GuB. Either the resolution was high and the investigated records short or vice versa. We will present here two sedimentary greyvalue records from the SBB and the GuB, respectively. The records were derived by applying the new technique of Digital Sediment Colour Analysis (DSCA) (Schaaf & Thurow 1994). Both records span the last 30 000 years and have a resolution of approximately 11 measurements per millimetre of sediment.

Analysis methods

For a detailed description of the digitizing procedures used to obtain the two records from the SBB and the GuB, the reader is referred to Schaaf & Thurow (1994, 1995, 1997). In this section we will only give a brief overview of the techniques developed and used by us.

The cores were digitized in 30-cm sections using an OptoTech 8-bit, single CCD, colour scanner. The scanner was run with the Cirrus 1.5 software package for the Macintosh. All images were stored as Tag Image File Format (TIFF) files which can be displayed and manipulated by many commercial software packages (e.g. Adobe Photoshop, NIH Image, Digital Darkroom). The effective resolution chosen (i.e. the *real* resolution on the core surface) for all of the images is 290 dots per inch (dpi), equivalent to one measurement every 0.09 mm. Despite its colour ability, the digitizer was run in 8-bit greyscale mode, i.e. colour is expressed as 256 shades of grey. Owing to an inevitable change in the system set-up, the direction of colour conversion changed between the SBB and the GuB work. Whilst for the SBB data greyvalue 0 = black and 255 = white, for the GuB data this relationship is reversed. Unequal distribution of the light intensity on the core surface (e.g. owing to linear changes of light emission by the lamps used for illumination) was corrected for mathematically. The constancy of the light intensity throughout an entire digitizing session was monitored using a light meter and frequent standard scans.

To obtain the highest resolution time-series data sets, the digital images were loaded into NIH Image and greyvalue readings were taken along a 1-pixel-wide line downcore. The resolution of the time series is determined by the image resolution, in this case 290 dpi (= 1 pixel/0.09 mm). Once the time series were derived, all further data handling was done using our self-developed software 'Time-Series-Assistant', including the removal of artificial trends, cracks, gasvoids and band-pass filtering. Two high band-pass filters were used, simple moving average and a newly developed vector-space filter (Schaaf & Thurow 1994). Application of the vector-space filter results in a shift of greyvalues to higher values; the direction of the data, however, remains unchanged. The data range of 256 values originally is expanded according to the dimension of the vector-space filter used. The amount of the shift depends on the dimension used for filtering. A detailed discussion of the vector-space filter, filter algorithm, possible applications and results, is beyond the scope of this paper and is presented in Schaaf & Thurow (1994).

The greyvalue time series

The greyvalue records for the SBB and the GuB are presented in Fig. 2. The isotopic stage boundaries (Martinson *et al.* 1987) are indicated next to the plots. A shift of greyvalues between the last glacial cold period (LGCP) and present day is apparent. For the SBB, greyvalues during the LGCP exhibit reduced variance; in general greyvalues are lighter during the glacial. The GuB greyvalue data also have reduced variance during the LGCP, but the difference is less pronounced than in the SBB. The GuB greyvalues are also lighter during the LGCP. Neither record shows variability during the Younger Dryas cold event. Between the glacial Termination and approximately 6000 a BP the greyvalues in both records shift towards darker values, typical for the Holocene. The values stabilize at this dark level thereafter, until the core tops. Cyclic variability is visible in both records with millennial to annual timescales. The longest cycles visible have wavelengths of about 1500 ± 200 years. Five such cycles are clearly visible in the Holocene portion of the records. Millennial scale, sub-Milankovitch cycles (1500 to 5000 years) have recently been described in the SBB by Behl & Kennett (1995) from sediment fabric data. These authors connected them to the Dansgaard–Oeschger events common in ice cores and North Atlantic sediments. The 1500 year mid-cycle length is also in good agreement with the Bond cycles, reported in North Atlantic sedimentary records (Bond *et al.* 1993). Superimposed on these low-frequency variations are higher-order changes with much reduced amplitude. Figure 3a and b shows the high-resolution end of the variability scale. The seasonal cycle is clearly visible. Subpeaks within individual seasons are the expression of individual bloom events of single diatom species (Kemp 1990; Kemp pers. comm. 1995). Two seasonal cycles form an annual varve, indicated in the graphs by the shading. Four to seven such annual varves show

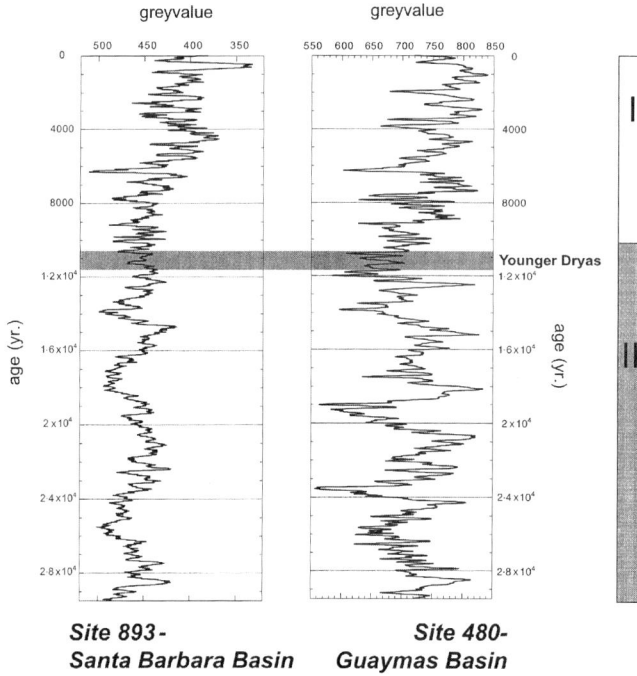

Fig. 2. Greyvalue records for the Santa Barbara Basin and the Guaymas Basin. The isotopic stage boundaries (Martinson *et al.* 1987) are indicated next to the plots.

bundling into larger cycles. Most probably this is the sedimentary greyvalue expression of the influence of the ENSO cycle on sedimentation (Lange *et al.* 1987). The characteristic recurrence periods of ENSO-type forcing are between three and nine years (Diaz & Markgraf 1992, and references therein). Both greyvalue time series, for the SBB and the GuB, exhibit a continuous superposition of cyclic variability from glacial/interglacial to seasonal. The appearance of both curves indicates substantial similarity between both records, in terms of the greyvalue expression to climatic and oceanographic forcing.

Spectral analysis

We performed spectral analysis on selected, well laminated portions of both records. The data were band-pass filtered to remove very long and very short trends. The resulting power spectra are presented in Fig. 4a to f and Fig. 5a to d. In particular, for Fourier transformations the very high resolution of the greyvalue time series proved to be a powerful tool for resolving even the highest frequency oscillations in the sedimentation records. The age assignments to the peaks are based on ^{14}C dates (Ingram & Kennett 1995) and lamina

counting performed directly on the greyvalue linescans.

A prominent peak present in all power spectra with suitable length has a frequency between 83 and 96 years, in very good agreement with the 86 year secular cycle of solar activity (Landscheidt 1987). Sunspot activity cycles (11 year and 30 year; Schove 1987; Landscheidt 1987) are weakly present only in single sections of the SBB core suite, yet they form distinct spectral peaks in the GuB cores. ENSO events have been described from the sedimentary records of many East Pacific basins by many authors (e.g. Lange *et al.* 1987; Schimmelmann *et al.* 1990; Baumgartner *et al.* 1991). All descriptions of such events, from various sites throughout the eastern tropical Pacific, however, have been restricted to the last millennium. We have run spectral analyses on greyvalue time series from the upper Pleistocene to recent. Spectral peaks related to today's ENSO (3–9 years) phenomena are abundant throughout most of the Fourier analyses, back to 13000 a BP, close to the last glacial maximum (LGM). The large variety of peaks for the very restricted El Niño frequency range underlines the descriptions by Philander (1983) of varying frequencies for ENSO also in the last millennium. The question of whether

greyvalue

(a) (resolution = 11 pixel/mm)

= dark laminae

Leg 146-893A-1H-2, 22 to 25cm

greyvalue

(b) (resolution = 13 pixel/mm)

= dark laminae

Leg 64-480-1-3, 0 to 2.3 cm

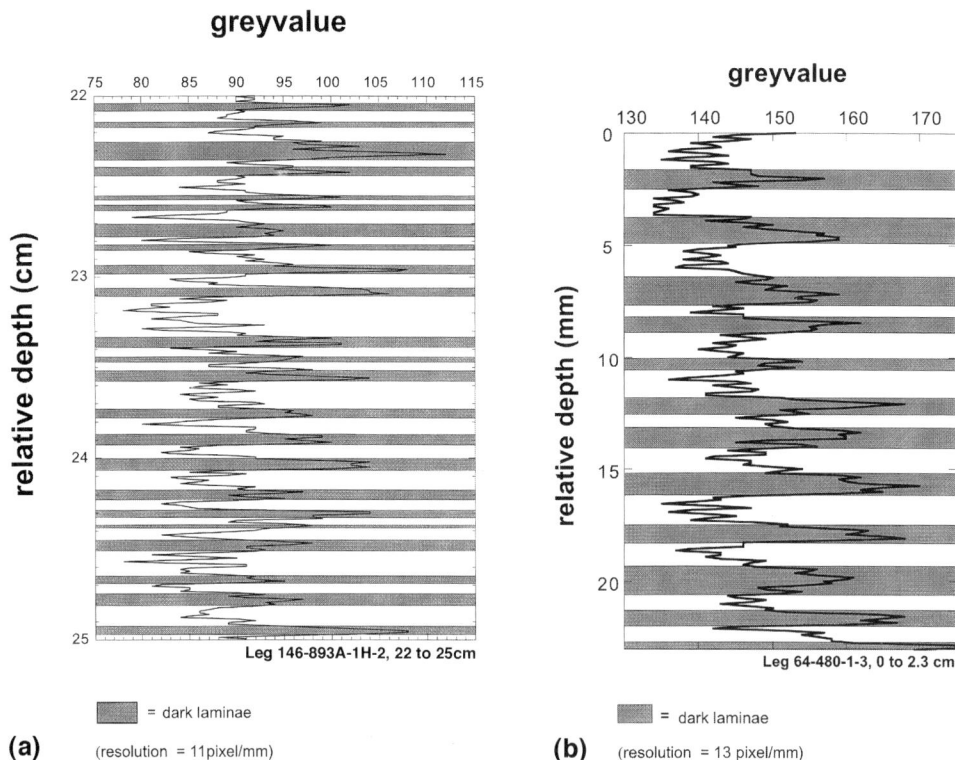

Fig. 3. (**a**) Greyvalue record of 3 cm well-laminated sediments from Site 893A-1-2, plotted together with the occurrence of lamination. (**b**) Greyvalue record of 2.3 cm well-laminated sediments from Site 480-1-3, plotted together with the occurrence of lamination. In both plots, note the bundling of several laminae into sets.

a pattern (probably a non-linear pattern) underlies the occurrence of ENSO and its duration with time cannot be answered here. The influence of ENSO events on today's SBB and GuB sedimentation has been described in detail by Lange *et al.* (1987), Reimers *et al.* (1990) and Baumgartner *et al.* (1991). It is most likely that these were similar for the other ENSO events during the last 13 000 years. The response of the greyvalue parameter to ENSO-type forcing, i.e. a shift of the base greyvalue, on which the seasonal signal is superimposed, towards lighter values for El Niño years and vice versa for anti-El Niño years, remains unchanged from the most modern until the glacial portions of the records. Proof of this similarity in response, however, requires detailed microscopic, SEM and high-resolution chemical studies of the related sediments, which are beyond the scope of this preliminary study.

Lamination is found in the power spectra wherever distinctly present in the cores. If the lamination is clearly proven to be a response to

seasonal forcing, i.e. two to three laminae form one annual varve, the lamination peaks may be used to tune the time series and correct the age assignments made previously based on the ^{14}C ages. The occurrence of three spectral peaks related to lamination indicates slight high-frequency variations of the sedimentation rates.

Concluding discussion

The behaviour of sedimentary greyvalues from two basins in the East Pacific Ocean, the SBB and the GuB, was examined. The greyvalue records generated for this study have a resolution never achieved in previous studies spanning an equivalently long time period. Our aim was to show how sedimentary colour, expressed as greylevels, changes with the changing climatic and oceanographic boundary conditions. In particular we attempted to prove the utilization of the extremely high resolution of our time series as a recorder of

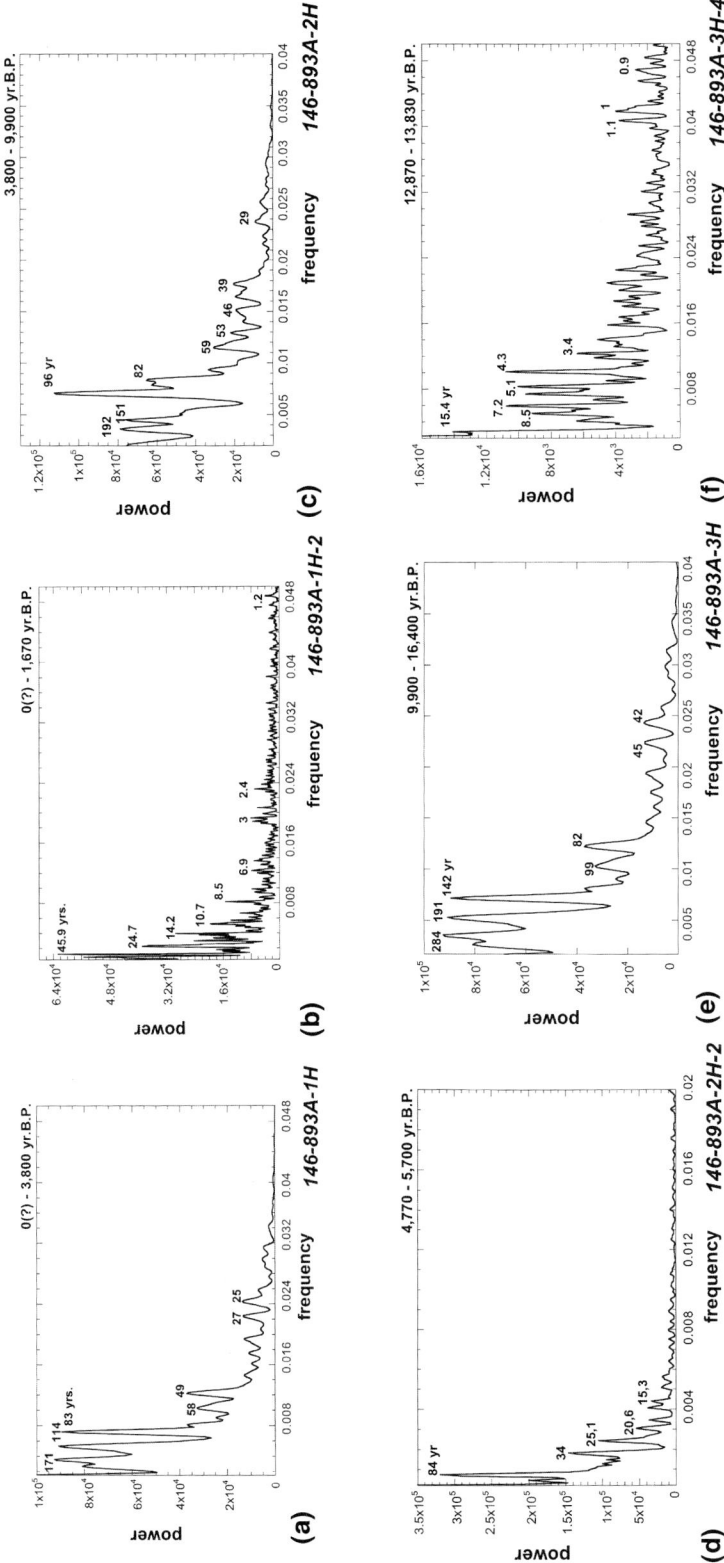

Fig. 4. Power spectra of six sections from Hole 893A. The age span of the power spectrum is given above each graph. Major spectral peaks are marked with the corresponding periods next to the peak. Plots a, c, and e are spectra of entire cores, plots b, d, and f are spectra of single segments.

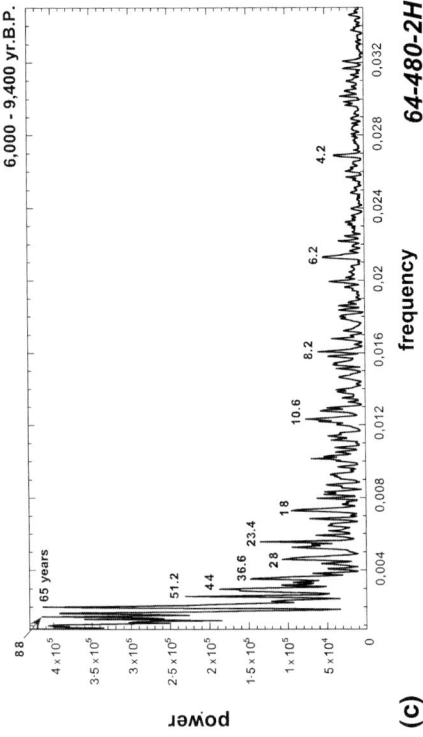

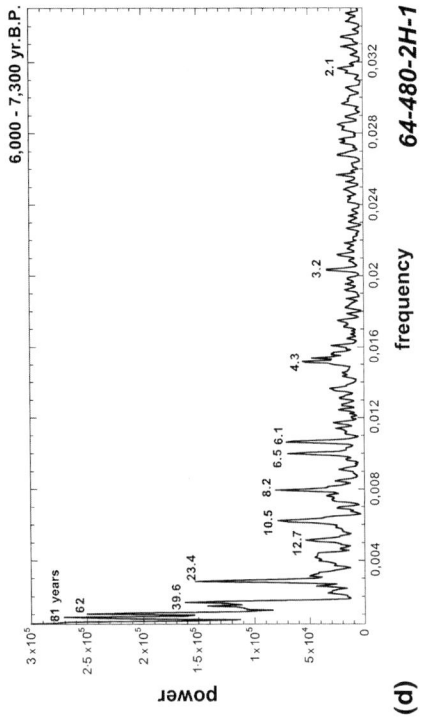

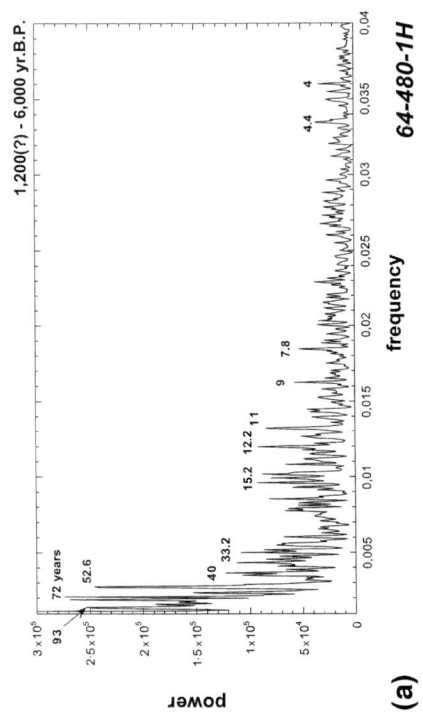

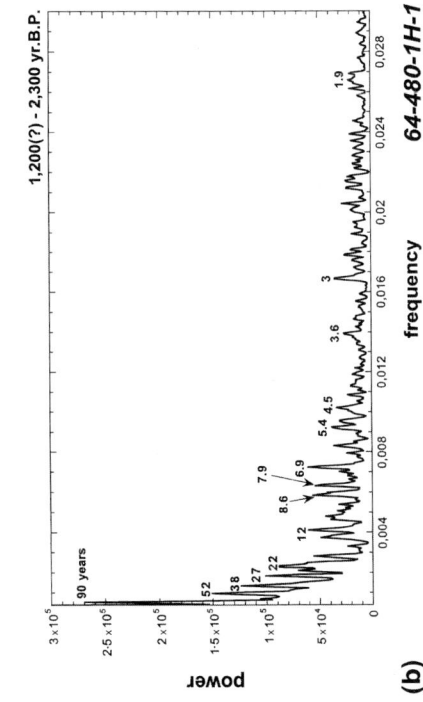

Fig. 5. Power spectra of four sections from Site 480. The age span of the power spectrum is given above each graph. Major spectral peaks are marked with the corresponding periods next to the peak. Plots a and c are spectra of entire cores, plots b and d are spectra of single segments.

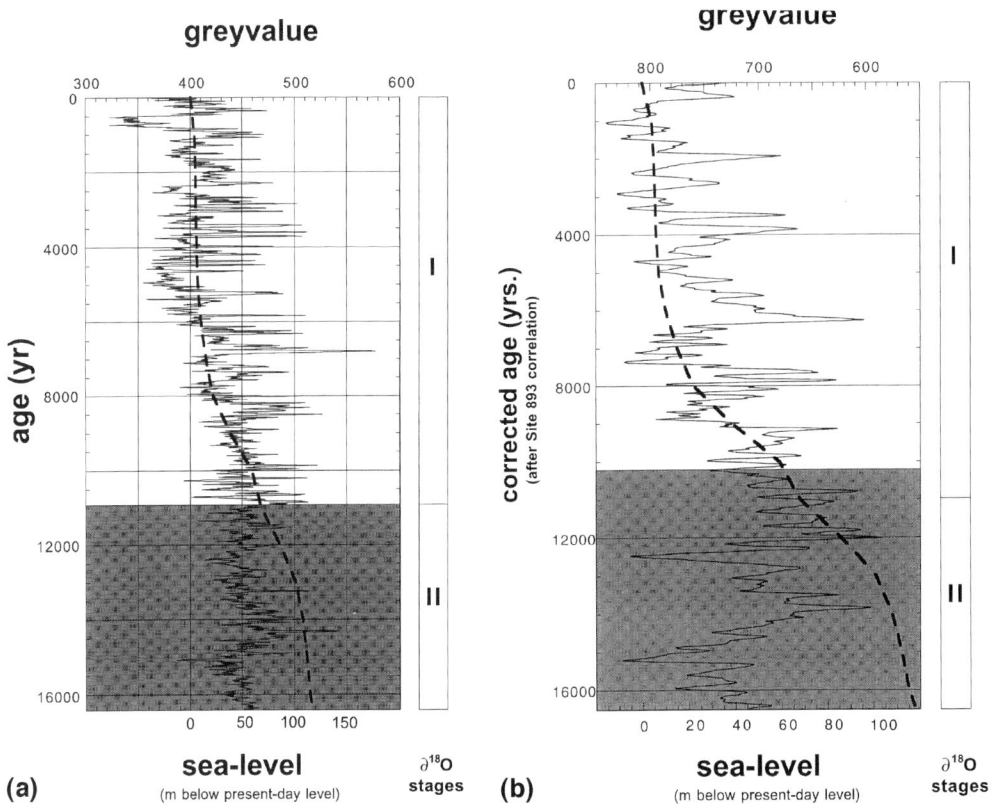

Fig. 6. (a) The greyvalue time series of ODP Site 893 until 16500 a BP plotted together with the eustatic sea-level curve after Fairbanks (1989). (b) The greyvalue time series of DSDP Site 480 until 16500 a BP plotted together with the eustatic sea-level curve after Fairbanks (1989). The glacial cold stage II is indicated by the shaded area in the graphs.

high-frequency oceanographic and/or climatic forcing.

Broad scale trends in the greyvalue time series were found to comply with known large-scale forcing variabilities. For the Holocene, both records exhibit shifts from light, glacial greyvalue levels, towards darker levels during most Holocene portions of core. These shifts commence around 6000 a BP. The trend is in good agreement with the rise in sea level after the LGM (Fairbanks 1989). Also the sea-level trend changes from a rapid to a more gradual rise around 6000 a BP (Fig. 6a and b). Stein & Rack (1995) have demonstrated similar trends in total organic carbon data for the SBB, indicating a gradual shift towards higher organic carbon accumulation/preservation in the basin sediments with increasing sea-level height. Thus the trends towards darker greyvalues seems to be related to this gradual shift of carbon flux. The similarity of this colour modulation between the

SBB and the GuB hints at broad-scale equivalence in system response to the sea-level change. A number of approximately 1500 year cycles were observed in both greyvalue time series. Cyclic alteration of sediment structure with similar periods have recently been reported for the SBB by Behl & Kennett (1995). A connection between the Dansgaard–Oeschger cycles, operating with similar periods on North Atlantic sediments, and the bioturbation cycles of Behl & Kennett (1995) was speculated by these authors. Our data provide independent proof for the existence of such 1500 year cycles, back to 30 000 a BP. In addition we could show that such oscillations force sedimentation not only in the SBB, but also in the GuB, thus arguing against a local origin, but rather for a regional or possibly even global origin of forcing.

Both long- and short-term cyclicity is recorded in the greyvalue time series. Whilst the long-term oscillations, such as the (potential) Dansgaard–

Oeschger variability, can be recorded by many direct and proxy measurement, high-frequency, short-term shifts and cycles are only well documented in long high-resolution time series. With the resolution of the data presented here, it was possible to track back high-frequency climatic variability through the Holocene, into the upper Pleistocene. The most prominent cycles found in the SBB and the GuB records were the approximately 86 year Gleissberg oscillations of solar activity. Such cycles are found at each site back into the top of the LGCP. The cycles are recorded in greyvalue time series as rhythmic shifts between lighter and darker values, potentially reflecting changes in organic matter export to the sediments, i.e. anoxia and productivity. The recently much-discussed El Niño cycles were the second most prominent cyclic colour modulation at the high-frequency end of the spectrum. Again, both basins are similar in terms of variability in the ENSO frequency band. Greyvalues in both cores are shifted between light and dark during anti-ENSO and ENSO years, respectively. Thus the greyvalues record the lowered export production of biogenic matter during ENSO years as lighter values. Our long records allowed us to establish the existence of ENSO-type cyclicity in both basins back through the Holocene and (in the case of the SBB) into the LGCP, until 13 000 a BP. The El Niño must hence have remained stable and unaffected by either the build-up and decay of land- and sea-ice, the strongly altered wind speeds, or the change in sea level between glacial and interglacial times. This is a reasonable assumption, since free air mass and water exchange over the ENSO core region (the tropical Pacific), which are the key factors of any ENSO event (Enfield 1992), remained possible despite the sea-level and wind alteration induced by the glaciation (COHMAP Members 1988). Also, since the greyvalue response within the ENSO frequency band has not changed with time, it appears that the El Niño may have operated in a similar fashion over the past several thousand years, as it has during this century. The equivalence between the greyvalue records from the GuB and the SBB indicate a strong link in system response between the two basins. Our study has shown that, despite their geographical separation (approximately 1000 km), they react in much the same fashion to external forcing on timescales between years and millennia. Through band-pass filtering of the greyvalue time series it should be possible to document the development of ENSO and potentially generate a time series of El Niño events throughout the Holocene.

We wish to thank R. Merrill and J. Beck from ODP-TAMU for their help during the digitizing process. The ODP-TAMU computer department, especially J. Olsen and J. Cochrane, are thanked for their fast and unbureaucratic help with data handling and storage. We thank O. Podlaha, and N. Chalimourda for their suggestions for programming. This work was financially supported by the DFG-ODP grant Ve112/9 to J. Veizer and travel grants of the DFG, and the Ruth and Gerd Massenberg Stiftung to M. Schaaf.

References

ALVAREZ-BORREGO, S. & LARA-LARA J. R. 1991. The physical environment and primary productivity of the Gulf of California. *In*: DAUPHIN, P. & SIMONEIT, B. (eds) *The Gulf and Peninsula Province of the Californias*. AAPG Memoir, **47**, 555–567.

ANDERSON, R. Y., HEMPHILL-HALLEY, E. & GARDNER, J. V. 1987. Persistent late Pleistocene–Holocene upwelling and varves off the coast of California. *Quaternary Research*, **28**, 307–313.

BAUMGARTNER, T. R., FERREIRA-BARTRINA, V., COWEN, J. & SOUTAR, A. 1991. Reconstruction of twentieth century varve chronology from the central Gulf of California. *In*: DAUPHIN, P. & SIMONEIT, B. (eds) *The Gulf and Peninsula Province of the Californias*. AAPG Memoir, **47**, 603–616.

BEHL, R. J. & KENNETT, J. P. 1995. Oceanographic and ecologic manifestations of brief interstadials (Dansgaard-Oeschger events) in Santa Barbara Basin, Northeast Pacific. *ICP V conference, Halifax, Nova Scotia, Abstracts*, 182.

BOND, G., BROECKER, W., JOHNSEN, S., MCMANUS, J., LABEYRIE, L., JOUZEL, J. & BONANI, G. 1993. Correlations between climate records from North Atlantic sediments and Greenland ice. *Nature*, **365**, 143–146.

CALVERT, S. E., BUSTIN, R. M. & PEDERSEN, T. F. 1992. Lack of evidence for enhanced preservation of sedimentary organic matter in the oxygen minimum of the Gulf of California. *Geology*, **20**, 757–760.

COHMAP MEMBERS. 1988. Climatic changes of the last 18,000 years: observations and model simulations. *Science*, **241**, 1043–1052.

CURRAY, J. R., MOORE, D. G. *ET AL.* 1982. *Initial, Reports DSDP*, **64**. US Govt Printing Office, Washington.

DIAZ, H. F. & MARKGRAF, V. (eds) 1992. *El Niño, Historical and Paleoclimatic Aspects of the Southern Oscillation*. Cambridge University Press, Cambridge.

ENFIELD, D. B. 1992. Historical and prehistorical overview of El Niño/Southern Oscillation. *In*: DIAZ, H. F. & MARKGRAF, V. (eds) *El Niño, Historical and Paleoclimatic Aspects of the Southern Oscillation*. Cambridge University Press, Cambridge, 95–117.

FAIRBANKS, R. G. 1989. A 17,000-year glacio-eustatic sea level record: influence of glacial melting rates on the Younger Dryas event and deep-ocean circulation. *Nature*, **342**, 637–642.

INGRAM, B. L. & KENNETT, J. P. 1995. Radiocarbon chronology and planktonic – benthic foraminiferal

[14]C age differences in Santa Barbara Basin sediments (Hole 893A). *In*: KENNETT, J. P., BALDAUF, J. *ET AL. Proceedings ODP, Scientific Results*, **146**. College Station, TX (Ocean Drilling Program), 19–27.

KEIGWIN, L. D. & JONES, G. A. 1990. Deglacial climatic oscillations in the Gulf of California. *Paleoceanography*, **5**, 1009–1023.

KEMP, A. E. S. 1990. Sediment fabrics and variation in lamination style in Peru continental margin upwelling sediments. *In*: SUESS, E., VON HUENE R., *ET AL. Proceedings ODP, Initial Reports*, **112**. College Station, TX (Ocean Drilling Program), 43–58.

KENNETT, J. P. 1995. Latest Quaternary benthic oxygen and carbon isotope stratigraphy: Hole 893A, Santa Barbara Basin, California. *In*: KENNETT, J. P., BALDAUF, J. *ET AL. Proceedings ODP, Scientific Results*, **146**. College Station, TX (Ocean Drilling Program), 3–18.

——, BALDAUF, J. G. *ET AL.* 1994. *Proceedings ODP, Initial Reports*, **146** (Pt.2). College Station, TX (Ocean Drilling Program).

LANDSCHEIDT, T. 1987. Long-range forecast of solar cycles and climate change. *In*: RAMPINO, M. R., SANDERS, J. E., NEWMAN, W. S., KÖNIGSSON, L. K. (eds) *Climate*. Van Nostrand Reinhold, New York, 421–446.

LANGE, C. B., BERGER, W. H., BURKE, S. K., CASEY, R. E., SCHIMMELMANN, A., SOUTAR, A. & WEINHEIMER, A. L. 1987. El Niño in Santa Barbara Basin: Diatom, radiolarian and foraminiferan responses to the "1983 El Niño" event. *Marine Geology*, **78**. 153–160.

PHILANDER, S. G. H. 1983. El Niño Southern Oscillation phenomena. *Nature*, **302**, 295–301.

REIMERS, C. E., LANGE, C. B., TABAK, M. & BERNHARD, J. M. 1990. Seasonal spillover and varve formation in the Santa Barbara Basin, California. *Limnology and Oceanography*, **35**, 1577–1585.

SANTAMARIA-DEL-ANGEL, E., ALVAREZ-BORREGO, S. & MÜLLER-KARGER, F. E. 1994. Gulf of California biogeographic regions based on coastal zone colour scanner imagery. *Journal of Geophysical Research*, **99**(C4), 7411–7421.

SCHAAF, M. & THUROW, J. 1994. A fast and easy method to derive highest-resolution time series data sets from drillcores and rock samples. *Sedimentary Geology*, **94**, 1–10.

—— & —— 1995. Late Pleistocene–Holocene climatic cycles recorded in Santa Barbara Basin Sediments: Interpretation of colour density logs from Site 893. *In*: KENNETT, J. P., BALDAUF, J. *ET AL. Proceedings ODP, Scientific Results*, **146**. College Station, TX (Ocean Drilling Program), 31–44.

—— & —— 1997. Tracing short cycles in long records: the study of inter-annual to inter-centennial climate change from long sediment records, examples from the Santa Barbara Basin. *Journal of the Geological Society, London*, **154**, 613–622.

SCHIMMELMANN, A., LANGE, C. B. & BERGER, W. H. 1990. Climatically controlled marker layers in Santa Barbara Basin sediments and fine-scale core-to-core correlation. *Limnology and Oceanography*, **35**, 165–173.

SCHOVE, D. J. 1987. Sunspot cycles and weather history. *In*: RAMPINO, M. R., SANDERS, J. E., NEWMAN, W. S. & KÖNIGSSON, L. K. (eds) *Climate*. Van Nostrand Reinhold, New York, 355–378.

SOUTAR, A. & CRILL, P. A. 1977. Sedimentation and climatic patterns in the Santa Barbara Basin during the 19th and 20th centuries. *Geological Society of America Bulletin*, **88**, 1161–1172.

——, JOHNSON, S. R., TAYLOR, E. & BAUMGARTNER, T. R. 1982. X-Radiography of Hole 480: procedures and results. *In*: CURRAY, J. R., MOORE, D. G. *ET AL. Initial. Reports. DSDP*, **64**. US Govt Printing Office, Washington, 1183–1190.

STEIN, R. & RACK, F. R. 1995. A 160,000-year high resolution record of quantity and composition of organic carbon in the Santa Barbara Basin (ODP-Site 893). *In*: KENNETT, J. P., BALDAUF, J. *ET AL. Proceedings ODP, Scientific Results*, **146**. College Station, TX (Ocean Drilling Program), 125–138.

THORNTON, S. E. 1984. Basin model for hemipelagic sedimentation in a tectonically active continental margin: Santa Barbara Basin, California Continental Borderland. *In*: STOW, D. A. V. & PIPER, D. J. W. (eds) *Fine-grained Sediments: Deep-water Processes and Facies*. Geological Society, London, Special Publications, **15**, 377–394.

Equatorial western Atlantic Ocean circulation changes linked to the Heinrich events: deep-sea sediment evidence from the Amazon Fan

MARK MASLIN

Environmental Change Research Centre, Department of Geography, University College London, 26 Bedford Way, London WC1H 0AP, UK,

Abstract: The Amazon Fan is an excellent area to obtain climatic records of the last glacial–interglacial cycle. Glacial sedimentation rates ranging from 1 to over 50 m ka^{-1} provide an opportunity to obtain marine records approaching the resolution of the Greenland ice cores. Ocean Drilling Program Leg 155 Sites 932 and 933 from the Eastern Middle Amazon Fan complex have been studied at a resolution of 50 cm (100–400 years). The identification of palaeomagnetic excursions, distinctive patterns in the palaeomagnetic remanence intensity records and AMS ^{14}C dates have provided an age framework for both Sites 932 and 933. However, due to extensive reworking in the Site 933 records, it was only possible to construct a detailed age model for Site 932. Planktonic foraminiferal stable oxygen and carbon isotope records of six species were obtained for both sites.

The planktonic foraminiferal carbon isotope records of both Sites 932 and 933 show a distinct negative deviation during Termination I (13–15 calendar ka). We suggest this could have been caused by an increase in the sediment discharge of the Amazon River because of the deglaciation of the Andes and/or the release of significant quantities of gas hydrates as large parts of the Amazon Fan sediment column failed because of increased sea level. Positive δ^{18}O deviations are also observed in the Site 932 records which seem to be coeval with the Heinrich events. It is suggested that the enhanced ice-rafting in the North Atlantic during the Heinrich events increased the latitudinal thermal gradient and thus the zonal component of the wind system. This reduced the northward penetration of the Inter-Tropical Convergence Zone, curtailing the cross-equatorial export of the North Brazilian Coastal Current (NBCC) and resulting in a permanent NBCC-retroflection. This increased the surface water salinity of the Amazon Fan and thus led to the observed positive δ^{18}O deviations.

The Amazon Fan has developed since the early Miocene, because of the massive increase in sediment transport of the Amazon River following the Andean uplift (Castro *et al.* 1978; Hoorn *et al.* 1995; Curry *et al.* 1995). During Pleistocene interglacial periods, the Amazon River terrigenous sediment is deposited predominantly on the inner continental shelf (Flood *et al.* 1995; see Fig. 1), so that only pelagic sediments accumulate on the fan. During glacial periods the lower sea level causes the outflow of the Amazon River to shift to the northeast, allowing terrigenous sediment to feed into the Amazon Canyon and hence into the fan complex (Damuth & Fairbridge 1970; Damuth & Kumar 1975). Changing sea level controls the sedimentation regime of the Amazon Fan. It has been suggested that rapid changes in sea level may have destabilized the shelf and the Amazon Fan sediments, causing the formation of extensive mass-transport deposits (Maslin & Mikkelsen in press; Mikkelsen & Maslin this volume).

This switching on and off of terrigenous sediment supply to the Amazon Fan has a major impact on sedimentation rates. On average the Holocene sedimentation rates are similar to open ocean rates, about 2 to 5 cm ka^{-1}, whereas the glacial sedimentation rates range from 1 to >50 m ka^{-1} (Mikkelsen *et al.* in press). The glacial sediments of the Amazon Fan thus provide us with unique opportunities to study past climate. First, the sedimentation rates are so high that it becomes possible to obtain ocean records at an age resolution similar to that of the Greenland ice cores (Dansgaard *et al.* 1993). Second, it is possible to monitor changes of climate, hydrology and vegetation of the Amazon Basin during the last glacial, as well as climate and circulation changes of the western equatorial Atlantic.

MASLIN, M. 1998. Equatorial western Atlantic Ocean circulation changes linked to the Heinrich events: deep-sea sediment evidence from the Amazon Fan. *In:* CRAMP, A., MACLEOD, C. J., LEE, S. V. & JONES, E. J. W. (eds) *Geological Evolution of Ocean Basins: Results from the Ocean Drilling Program.* Geological Society, London, Special Publications, **131**, 111–127.

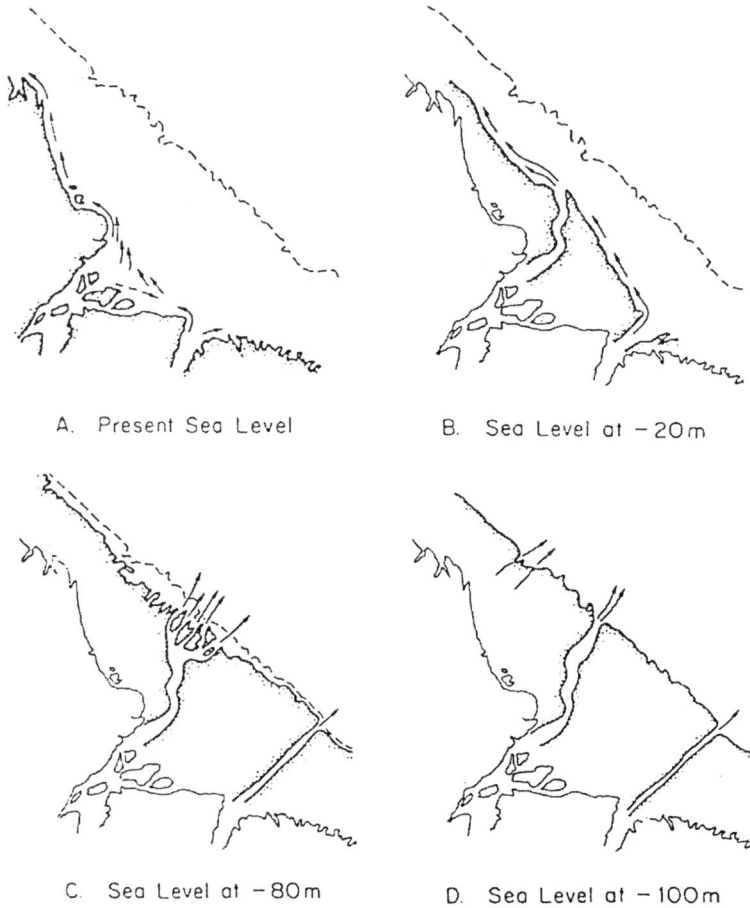

A. Present Sea Level   B. Sea Level at −20m

C. Sea Level at −80m   D. Sea Level at −100m

Fig. 1. Model of Quaternary sedimentation off the Amazon Basin during high and low sea-level stands. Arrows indicate the main routes of sediment dispersal. Dashed line refers to the present 100 m isobath, which approximates the shelf edge (after Milliman *et al.* 1975).

The high sedimentation rates, however, also make the study of Amazon Fan sediments a challenge, because the very dynamic nature of deep-sea fans makes them notorious for being reworked. The major concerns are local slumping, flows, turbidity currents and erosion as well as deposition of older material from higher up the Amazon Fan complex and from the continental shelf and slope. Great care is therefore required when choosing site locations and when interpreting the results, although previous work has demonstrated that good climatic records can be obtained from the Amazon Fan, despite these drawbacks (e.g. Damuth 1975, 1977; Showers & Bevis 1988).

One of the key palaeoceanographic objectives of the Ocean Drilling Program (ODP) Leg 155 was to try to utilize the exceptional sedimentation rates to obtain high-resolution climate records in order to investigate interglacial–glacial changes in the circulation of the western equatorial Atlantic Ocean (Flood *et al.* 1995). A key component of the North Atlantic heat and salinity budget is the North Brazil Coastal Current (NBCC) which is the only surface current that crosses the equator (Metcalf & Stalcup 1967; Richardson & Walsh 1986; see Fig. 2). From December to June, wind stress variation causes an increase in the NBCC transport; the NBCC may extend into the Guyana Current which links in with the Caribbean Current (Picaut *et al.* 1985; Philander & Pacanowski 1986). The NBCC therefore influences the temperature and salinity of the sources of the Florida Current and the Gulf Stream (Levitus 1982) and thus the characteristics of the surface waters reaching the Nordic Seas, and the deep water formed there. Between July and November, the NBCC turns eastward (retroflects) into the eastward flowing North Equatorial Counter-Current (NECC), switching off this cross-equatorial transport (Fig. 2).

During low sea-level stands, the Amazon River would have discharged directly into relatively deep water, and mixing of the river plume into the coastal water might have occurred more slowly than at present, allowing the build-up of more extensive freshwater lenses. Planktonic foraminiferal oxygen isotope records from piston cores recovered on the eastern Amazon Fan, have indicated a number of well developed negative deviations during the late glacial and early Holocene. These have been interpreted as evidence of Amazon River palaeodischarge events (Showers & Bevis 1988). Such deviations seem to be less common on the western part of the fan, suggesting that they may instead represent reduced activity of the NBCC. These isotopic events potentially represent periods of reduced cross-equatorial ocean heat and salt transport, both important components of global oceanic circulation.

The aim of this study is to produce high-resolution stable isotopic records for the last glacial from the Amazon Fan to investigate. (1) the occurrence of planktonic foraminiferal oxygen and carbon isotope deviations, and (2) the Amazon River discharge and the circulation of the equatorial western Atlantic Ocean.

Site selection

Sites 932 (5°12.7′N, 47°1.8′W, water depth 3334 m) and 933 (5°5.8′N, 46°48.7′W, water depth 3366 m), see Fig. 3, were selected for this stable isotope study for the following reasons: (1) both sites in the shallower sections contain levee mud sequences making them suitable for palaeoceanographic studies; (2) the sites have glacial sedimentation rates high enough to provide high-resolution records with a time resolution of between 100 and 400 years at the initial 50 cm sampling intervals; 3) the sedimentation rates are also low enough that planktonic foraminifera can be found in sufficient quantities for stable isotope analysis; and (4) the sites are approximately 25 km apart at a similar water depth, therefore it will be possible to verify isotope excursions by comparing the records of the two sites. This is important as one of the aims of this study is to investigate short intense climatic changes and a key problem with Amazon Fan sediments is that climatic records can

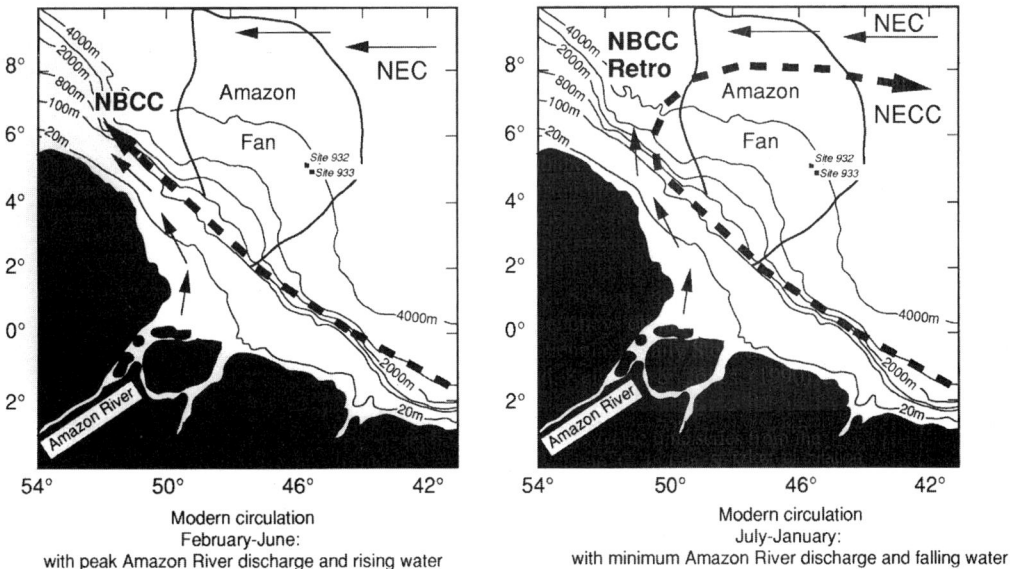

Fig. 2. Sketch map showing the present-day seasonal variation in the surface circulation of the western equatorial Atlantic. It is suggested in this study that the retroflection of the NBCC may have had a longer seasonal occurrence compared with the modern circulation during the last glacial, and during some periods it may have been permanent. Note that the glacial coastline of Brazil has been moved to the 100 m contour line due to the lower sea level. NBCC = North Brazil Coastal Current, NEC = North Equatorial Current, NECC = North Equatorial Counter-Current, NBCC Retro = retroflection of the NBCC.

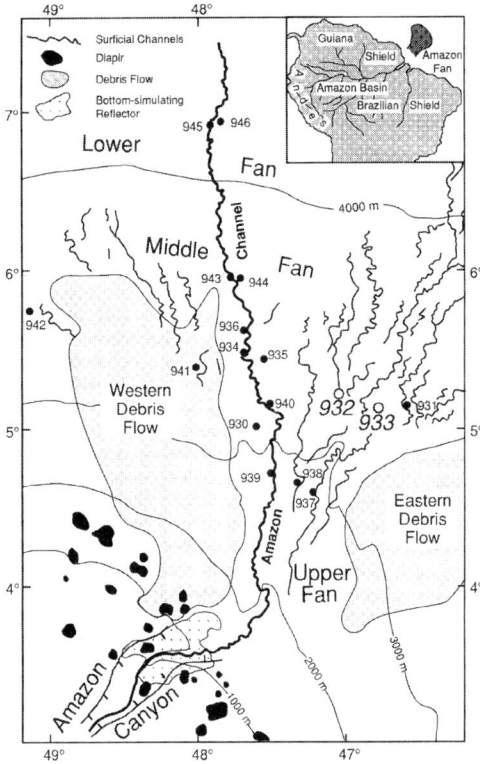

Fig. 3. Location map of the ODP Leg 155 Amazon Fan Sites.

analytical reproducibility based on replicate measurements of an internal laboratory standard is ±0.04‰ for $\delta^{13}C$ and ±0.07 for $\delta^{18}O$. All results were calibrated to peedee belemnite (PDB) scale using the National Bureau of Standards (NBS) 19 standard. In many cases the lack of calcium carbonate, due to the dilution effect of the extremely high sedimentation rates (1–50 m ka^{-1}), resulted in the measurement of very few specimens (no samples of less than six specimens were run); the gas pressure from the mass spectrometer was checked to make sure there was sufficient material for a reliable measurement. The six planktonic foraminiferal species which were analysed, and the discrete size fraction from which the specimens were picked were: *Globigerinoides ruber* (250–300 µm), *G. trilobus* (300–350 µm), *G. sacculifer* (300–350 µm), *Neogloboquadrina dutertrei* (300–350 µm), *Pulleniatina obliquiloculata* (350–400 µm), and *Globorotalia truncatulinoides* (300–350 µm). These six species were chosen primarily for their relatively high abundance in the Amazon Fan sediments (Flood *et al.* 1995) and also because they represented a range of water depths from the near-surface dwellers down to thermocline dwellers. Moreover, the numbers of foraminiferal species and specimens varied greatly throughout the site 932 and 933 sediments because of the large fluctuations in sedimentation rate and thus dilution by terrestrial sediment; therefore by analysing all the key planktonic species it may be possible to obtain a composite isotopic record for the last glacial. Variations in both the number of planktonic foraminifera per sample and low numbers of specimens per sample can be sources of considerable error (Trauth 1995). The extreme sedimentation rates of the Amazon Fan, however, negate there being any bioturbation effects. However, the greatest problem with interpreting any isotopic record recovered from the Amazon Fan, is the effect of turbidity currents and general reworking.

contain features that appear to be significant 'events' but are due to reworking, especially turbidites.

Methods

Samples of 20 cm^3 were taken onboard ship approximately every 50 cm; care was taken not to take samples from disturbed sections or areas identified as turbidites (Flood *et al.* 1995). The samples were freeze-dried and then wet-sieved through a 63 µm mesh sieve. The samples were then dry-sieved at discrete intervals of between 250 and 400 µm depending on which planktonic foraminiferal species was picked; each species was picked from a consistent size fraction. The oxygen and carbon isotopes of the planktonic foraminifera were measured using a Finnigan MAT 251 mass spectrometer at Kiel University, Germany. The Kiel laboratory uses a Carbo-Kiel preparation line which is an automated system in which acid is added to individual samples and reacted at 90°C. The

Age models

One of the major challenges of trying to reconstruct past climate from fan sediments is building a reliable age model. Unfortunately the sediments at Sites 932 and 933 have so few foraminifera that detailed AMS ^{14}C dating was not possible. The age models of Sites 932 and 933 were, thus, primarily based on palaeomagnetics. Conventional magnetic stratigraphy of the Leg 155 Amazon Fan sediments was not possible because of the relatively young ages of the sediments recovered (<460 ka; Maslin & Mikkelsen in press; Mikkelsen & Maslin this volume), so other less-utilized magnetostratigraphic techniques were used (Cisowski & Hall in press). These include identification of short periods of anomalous geomagnetic polarity and patterns in the magnetic remanence intensity.

The Lake Mungo excursion (32 calendar ka; Merrill & McElhinny 1983) and the Laschamp excursion (43 calendar ka) have been identified by a peak in relative intensity and inclination in Holes 932A and 933A (Cisowski & Hall in press). Twelve intensity features in the Amazon Fan sediments (see Table 1) have also been identified for Holes 932A and 933A (see Flood et al. (1995), Site 933, figure 15). Tric et al. (1992) were able to date, using oxygen isotope stratigraphy and radio-carbon dating, changes in the 'stacked' palaeoremanence intensity records of four Mediterranean cores. By careful comparisons of the Amazon Fan remanence intensity records and those of Tric et al. (1992), Cisowski (1995) has estimated the date in calendar years of each of these events.

It was possible to obtain enough planktonic foraminifera to AMS ^{14}C date three samples from Hole 932A. These samples were dated at the new Leibniz Labor fur Altersbestimmung und Isotopenforschung, Kiel University. The ages are given as 'conventional' ^{14}C ages calculated with a 5568 year, 'Libby half-life' and the results were corrected for sample preparation contamination by subtracting the activity measured in Eemian foraminifera processed in the same way. A general ocean reservoir correction of −400 years has been made, though it is noted that this may differ both because of location and changes in the rate of deep-water ventilation in the past. The first sample (932A-1H-1, 19–24 cm, 0.19 metres below sea floor (mbsf)) was chosen to date the top of Hole 932A, and consisted of 2000 G. trilobus tests in the > 350 µm size range and gave an age of 3970 ± 90 years. This sample confirms that a significant part of the Holocene was recovered at Sites 932. The second sample (933A-1H-1, 70–74 cm, 0.70 mbsf) was chosen to date the top of Hole 933A, and consisted of 1500 G. trilobus tests in the >250 µm size range and gave an age of 11910 ± 90 years. This suggests that some of the Holocene was recovered at Site 933. The third sample (932A-3H-6, 20–24 cm, 23.20 mbsf) was chosen to confirm the dating and identification of the Lake Mungo excursion in Hole 932A and consisted of 1350 tests which were a mixture of N. dutertrei and G. truncatulinoides in the >250 µm size fraction and gave an age of 32730 +320/–300 years. Despite the low number of tests and the necessity of using more than one species, the AMS ^{14}C date does confirm that the Lake Mungo event (24.9 mbsf) has been correctly identified.

One biostratigraphic datum was identified in Hole 932A, the disappearance of the planktonic foraminifera P. obliquiloculata (Figs 4 and 5) at

Table 1. *Age control points, based on palaeomagnetic remanence intensity features and palaeomagnetic excursions*

	Age (calendar ka)	Hole depth (mbsf)	
		932A	933A
Palaeomagnetic remanence intensity features*			
Peak A	9	5‡	4
Peak B	17	8‡	12 (?)
Peak C	26	15‡	25
Peak D	27	18‡	35 (?)
Peak E	30	22‡	45 (?)
Low L1	39	26–28 (?)	?
Peak F	42.5	30 (?)	?
Peak G	49	35‡	?
Peak H	54	40‡	?
Low L2	66	44‡	?
Peak I	68	45‡	?
Palaeomagnetic excursions†			
Lake Mungo	32	26‡	82
Laschamp	43	27.5‡	?

*Dates based on Tric et al. (1992)
†Dates based on McHargue et al. (1995)
‡Control points used in the Site 932 age model

Estimated age (calendar ka)

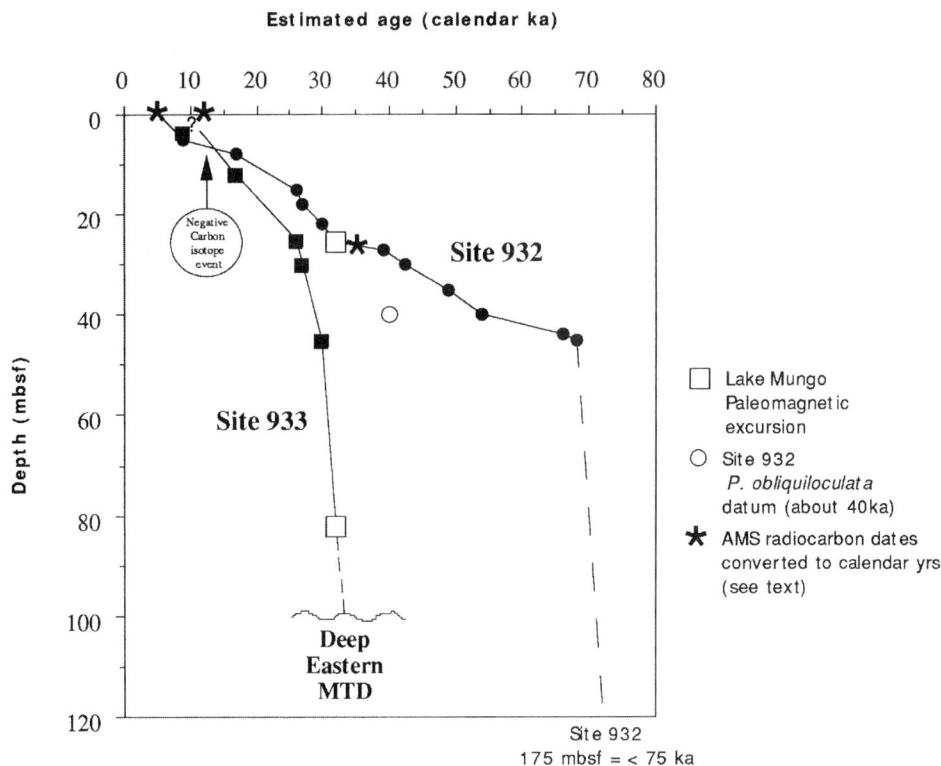

Fig. 4. Age (calendar ka) vs depth (mbsf) plot for Sites 932 and 933, illustrating the palaeomagnetic-based age model for each site (see Table 1), the AMS ^{14}C dates converted to calendar years (see text) and the *P. obliquiloculata* datum identified in Site 932.

about 40 ka (Flood *et al.* 1995). The identified *P. obliquiloculata* datum in Hole 932A does not agree with the dates provided by the palaeomagnetic remanence intensity features; in fact the last *P. obliquiloculata* test is found 10 m below where it would be predicted to occur. This could be the result of the following. (1) The *P. obliquiloculata* datum is diachronous across the equatorial Atlantic and may vary between 35 and 50 ka (Prell & Damuth 1978; Flood *et al.* 1995; R. Schneider 1995, pers. comm.). (2) The disappearance of *P. obliquiloculata* may be an artefact of very low numbers of planktonic foraminifera being present in the glacial sections of Site 932, as *P. obliquiloculata* usually makes up less than 5% of the glacial equatorial Atlantic planktonic foraminiferal assemblage. This may be further compounded by the rapid changes of sedimentation rates, so that the disappearance of *P. obliquiloculata* in Hole 932A may be the result of dilution rather than climatic control. (3) The remanence intensity features may be incorrectly identified or dated.

The age models of Holes 932A and 933A were constructed by linearly interpolating between the age points identified from the remanence intensity features and the Lake Mungo and Laschamp excursions (see Fig. 4). The dating of Hole 933A is extremely weak, because of: (1) the difficulty in identifying the remanence intensity features; (2) the conflict between the magnetic remanence intensity feature A and the AMS ^{14}C date for the top of the core; and (3) curtailment of the continuous sediment record of Site 933 by the Deep Eastern Mass-transport Deposit (MTD). By extrapolating the sedimentation rate above the Lake Mungo event identified in Hole 933A it seems the top of the Deep Eastern MTD is no older than 33 ka. Because of the difficulty of dating Hole 933A, no attempt at present has been made to construct an age model for this site.

The age model for Hole 932A is better constrained, as nine remanence intensity features and two excursions have been identified. The Holocene AMS ^{14}C date was converted to calendar years using the method outlined in Sarnthein *et al.* (1994)

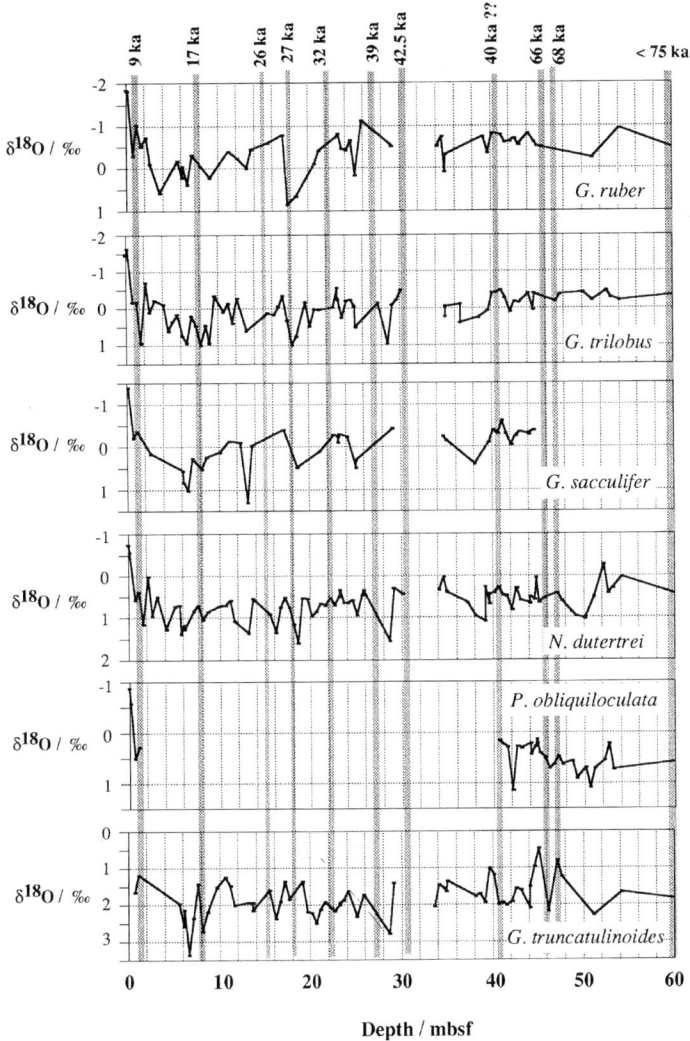

Fig. 5. Planktonic foraminifera oxygen isotope records for Hole 932A for the top 60 mbsf; age points are from the palaeomagnetic-based age model (see Table 1 and Fig. 4).

and used to constrain the top of the age model. The deepest intensity feature identified in Site 932 is at 45 m (68 ka); however, Hole 932A was drilled down to 170 mbsf. Biostratigraphic evidence, i.e. the absence of the *G. menardii* complex (Mikkelsen *et al.* in press), suggests the base of Hole 932A is younger than 85 ka which would place the base of Site 932 at the end of the last interglacial or beginning of the last glacial Sedimentological and seismic work suggests that the sediment between 47 and 170 mbsf, Unit IIb, is of a very similar character, mainly silty clay and part of a levee crest of the Channel levee system 6B. These results are confirmed by the organic chemistry of the

sediments, which, from the carbon isotopes of total organic carbon and the lignin concentrations, indicate that all the organic carbon in Unit IIb is terrestrial, which is characteristic of last glacial Amazon Fan sediments. Therefore, the base of Site 932 can be no older than 75 ka, which would suggest a sedimentation rate of over 18 m ka^{-1}. This is consistent firstly with the lack of planktonic foraminifera within the sediments, which within the Amazon Fan is usually an indicator of dilution of massive amounts of terrestrial material, and secondly the sedimentation rate is similar to that found in the adjacent Site 933 below a depth of 25 mbsf (see Fig. 4).

Results

Site 932

Fairly complete planktonic foraminiferal oxygen and carbon isotope records for the top 60 mbsf have been obtained for *G. ruber, G. trilobus, N. dutertrei* and *G. truncatulinoides*; only between 29.5 and 31.0 mbsf were the abundances of all planktonic foraminifera so low that no isotope analyses could be obtained (Figs 5 and 6). The top two samples, at 0.19 and 0.74 mbsf, have interglacial-type oxygen isotope values and represent the low sedimentation rate Holocene sediments, whereas samples from all

other depths have heavy glacial-type oxygen isotope values (Fig. 5). The *P. obliquiloculata* isotope record is curtailed by its disappearance from the equatorial Atlantic at about 40 ka, although it reappears at about 11 ka. The isotope records of *G. sacculifer* and *G. truncatulinoides* are patchy, because of the low overall abundance of planktonic foraminifera and their low percentage occurrence within the glacial planktonic foraminiferal assemblage. The most significant features of the planktonic carbon isotopic records (see Fig. 6) are the negative excursions. One occurs at about 2 mbsf during the deglacial, seen most clearly in

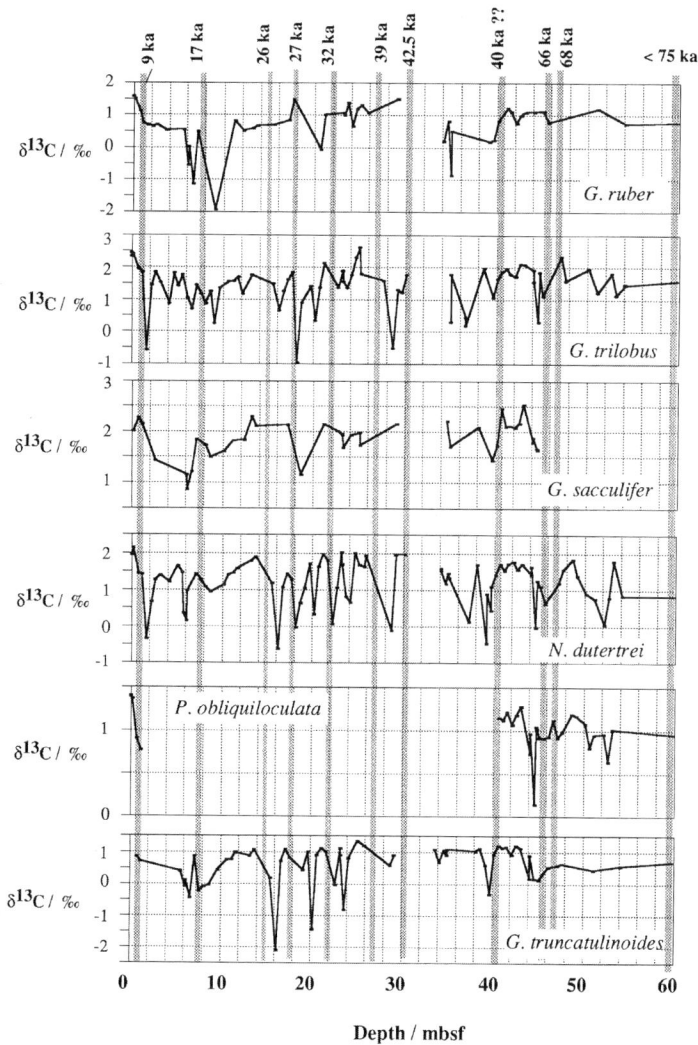

Fig. 6. Planktonic foraminifera carbon isotope records for Hole 932A for the top 60 mbsf; age points are from the palaeomagnetic-based age model (see Table 1 and Fig. 4).

the *G. trilobus* and *N. dutertrei* records. The other is a group of excursions between 16 and 24 mbsf, and is seen most clearly in the G. *truncatulinoides* and *N. dutertrei* records.

Site 933

Fairly complete planktonic foraminiferal oxygen and carbon records have been obtained for five of the six species for the top 35 mbsf (Figs 7 and 8). Below 35 mbsf the sedimentation rate is so high that there are very few planktonic foraminifera, and results below 35 mbsf were obtained from the very

large core-catcher samples (70 to >100 cm³). The top sample, at 0.21 mbsf, has interglacial-type oxygen isotope values and represents the low-sedimentation rate Holocene sediment, whereas all the other samples above the Deep Eastern MTD have heavy glacial-type oxygen isotope values (Fig. 7). The *Pulleniatina obliquiloculata* isotope record is curtailed by its disappearance from the equatorial Atlantic at about 40 ka, although it reappears at about 11 ka. The glacial planktonic foraminiferal oxygen isotope records do contain significant positive and negative excursions (Figs 7 and 8). All five planktonic oxygen isotope records

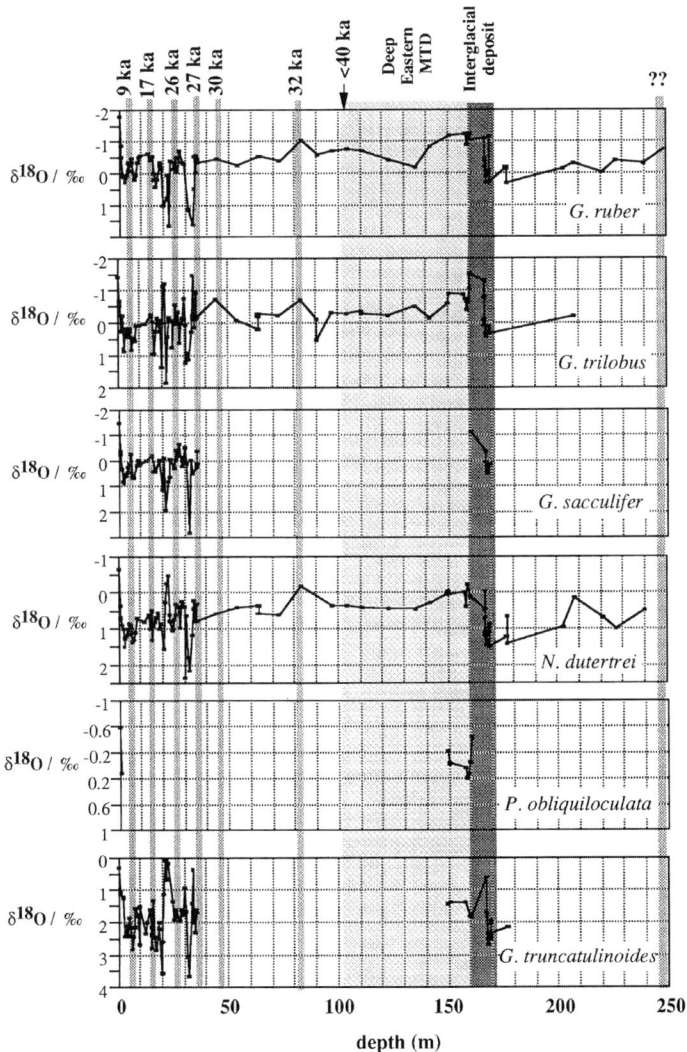

Fig. 7. Planktonic foraminifera oxygen isotope records for Hole 933A; age points are from the palaeomagnetic-based age model (see Fig. 3).

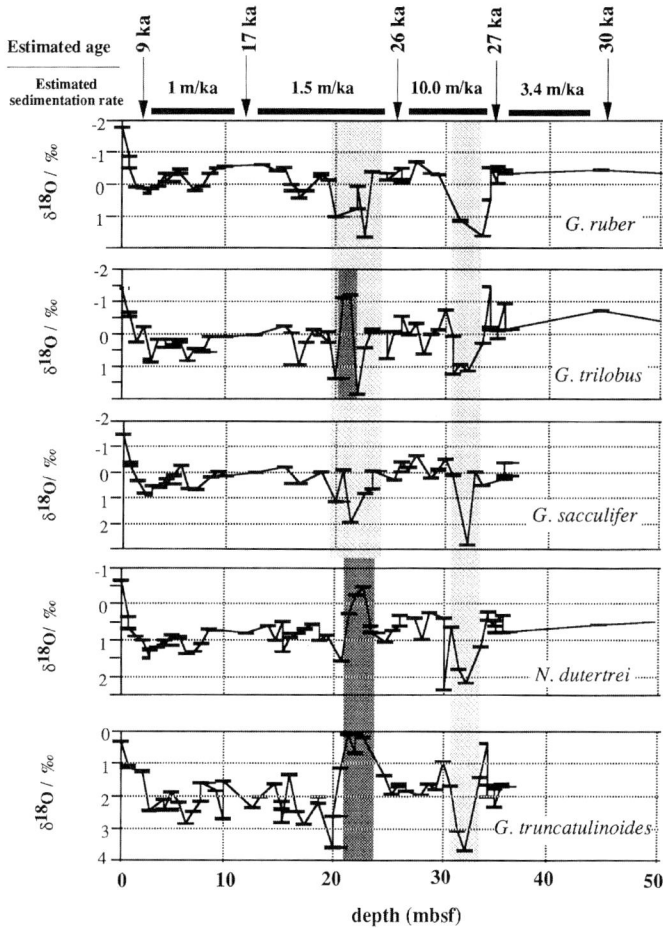

Fig. 8. Planktonic foraminifera oxygen isotope records for Hole 933A for the top 50 mbsf; age points are from the palaeomagnetic-based age model (see Fig. 3). Lightly shaded regions indicate positive isotope deviations, while the dark shading indicates negative isotope deviations.

have a positive excursion (up to 1.5‰) at about 32 mbsf (Fig. 8). In comparison, *G. ruber* and *G. sacculifer* have positive excursions at 22 mbsf, whereas *G. trilobus, N. dutertrei* and *G. truncatulinoides* have negative excursions. The negative excursion values are equal to or lower than the Holocene values, which would represent a massive increase in surface water temperature (6.5°C) or a freshening (1.5‰, Duplessy *et al.* 1992), both of which are unlikely in the western tropical Atlantic. Events of such a magnitude are not seen in the isotope records of Hole 932A, but they may have been missed because of the much lower sedimentation rates; the excursion at 32 mbsf in Hole 933A lasted less than 500 years in the age model. A detailed examination of the cores revealed no evidence of turbidites or other forms of

reworking (Flood *et al.* 1995; Hiscott 1995, pers. comm.). However, this does not exclude the possibility that the samples were affected by very fine turbidites which are extremely hard to identify (Flood *et al.* 1995). So at present it is not known whether these excursions are real or are caused by secondary calcification and/or reworking. The opposite deviations in the records at 22 mbsf and the values above those of the Holocene, however, strongly suggest reworking or secondary calcification.

The major feature of the Hole 933A carbon isotopic records (Fig. 9) is the massive negative deviation of up to 2‰, at about 7 mbsf. According to the age model this is between 9 and 17 ka. This is very similar to the deviation recorded in Hole 932A and there are no accompanying deviations in the oxygen isotope records and no sedimentological evidence to suggest that this is not a real feature.

The oxygen isotope records of planktonic foraminifera in the Deep Eastern MTD and the underlying interglacial deposit at Site 933 are shown in Fig. 7. The planktonic foraminiferal oxygen isotope results indicate that the interglacial deposit has 'lighter' values than the underlying glacial levee muds. However, these values do not, in general, reach the same level as the Holocene. This may indicate either that the deposit represents the very beginning or end of an interglacial period or that there was reworking of heavier glacial material into

the deposit. Maslin & Mikkelsen (in press; Mikkelsen & Maslin this volume) have suggested that both are likely because: (1) the nanofossil evidence suggests the interglacial deposit is a curtailed sequence; and (2) the reworking hypothesis is supported by the benthic foraminiferal oxygen isotopes results (Vilela & Maslin in press), as only glacial values were recorded in the interglacial deposit.

Because of the sparsity of planktonic foraminifera in the Deep Eastern MTD, only the isotope records of *G. ruber*, *G. trilobus* and *N. dutertrei* were possible (Fig. 7). All three indicate

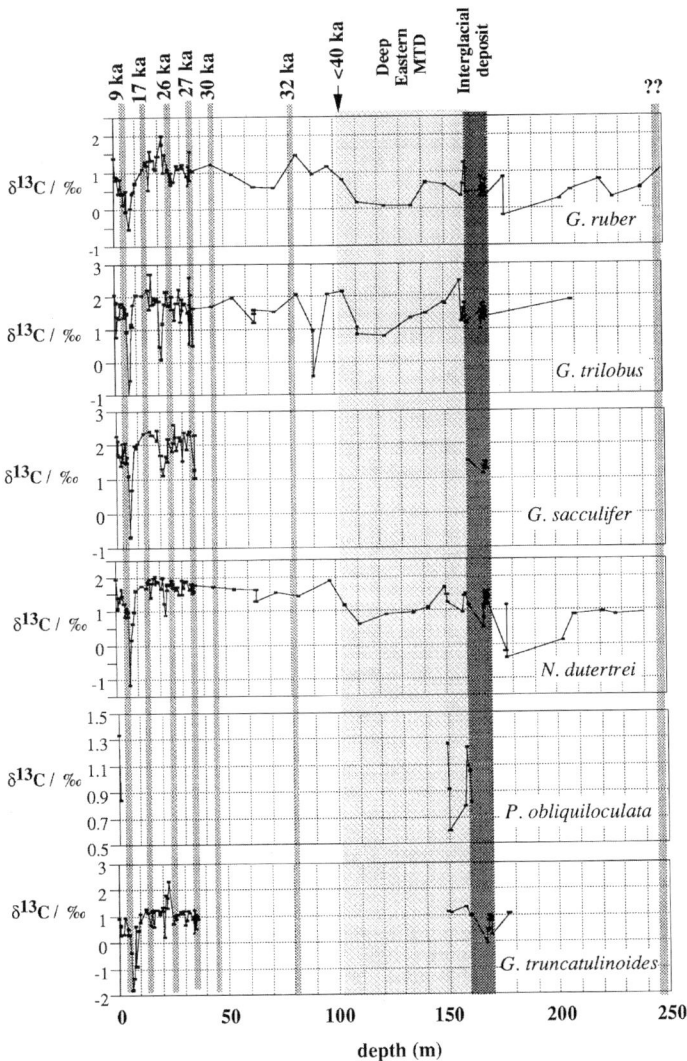

Fig. 9. Planktonic foraminifera carbon isotope records for Hole 933A; age points are from the palaeomagnetic-based age model (see Fig. 3).

that the foraminifera in the MTD have oxygen isotope values halfway between the Holocene and Last Glacial Maximum (LGM) values. This indicates either that the sediment from the Deep Eastern MTD was mainly formed during early oxygen stage 3 (or any other early glacial period) when sea level was still relatively high, or that there is a lot of mixing between interglacial and glacial material within the MTD. The abyssal benthic foraminifera are absent from the Deep Eastern MTD, except for the very top, where they give values heavier than for the LGM. Again this could reflect reworking of older material or secondary calcification of the tests.

Discussion

Deglacial negative carbon isotopic event

The planktonic foraminiferal carbon isotopic records of Sites 932 and 933 contain a significant negative deviation (0.5–2‰) at about 12–15 ka (Figs 4, 6 and 9). The average glacial–interglacial change in oceanic carbon isotope composition is about 0.4‰ (Maslin *et al.* 1995a), so the negative deviation observed in the Amazon Fan sediment must have a more local cause. There could be three possible causes of the carbon isotope event: (1) a change in surface water productivity; (2) an increase in the terrestrial organic carbon discharged by the Amazon River; or (3) significant release of gas hydrates into the water column.

The only way surface water productivity could have caused the negative carbon isotope event is if it crashed, releasing all the ^{12}C-enriched organic carbon stored in the photosynthate. The diatom, nanofossil and planktonic foraminiferal species-abundance (Maslin & Mikkelsen in press; Mikkelsen & Maslin this volume) evidence from the Amazon Fan sediments suggests that glacial surface water productivity was extremely low, near 'blue' ocean levels and there is no evidence for radical changes. It is therefore difficult to conceive a mechanism which could have reduced productivity any lower thus causing the isotopic event.

Another explanation for this negative deviation is an increase in the amount of 'light' terrestrial organic carbon being brought into the Atlantic Ocean by the Amazon River. Bird *et al.* (1992) have estimated the average isotopic content of the organic matter entering the Amazon River from the Amazon Basin to be between −27 and −30‰, compared with the present surface ocean average of about 2‰ (Maslin *et al.* 1995a). If it is assumed that the carbon isotope shift was between 1 and 2‰, it would mean an average relative increase of 5% terrestrial organic matter in the water column.

To increase the relative amount of terrestrial organic matter in the water column either surface water productivity must have dropped, which has already been discounted, or the sediment discharge of the Amazon River must have increased. The timing of the carbon isotopic event, 12–15 calendar ka, is similar to the first major temperature increase observed in the Huascaran (Peru) ice core δ^{18}O record, i.e. 13 to 15 calendar ka (Thompson *et al.* 1995). The increase in the Huascaran (Peru) ice core δ^{18}O record also coincides with first major rise in sea level documented by Fairbanks (1989), i.e. meltwater event I, suggesting a connection between the deglaciation of the Andes and the ice sheets of the northern hemisphere. The first stage in the deglaciation of the Andes may have, therefore, caused an increase in the Amazon River discharge and thus increased the organic matter sediment load to the western equatorial Atlantic. With a temporary increase in the discharge of the Amazon River, a freshening and therefore lighter δ^{18}O signal would be expected in the Amazon Fan sediments. However the carbon negative event seems to have been coeval with a positive deviation of the oxygen isotope records. This could be explained by the following: (1) only the sediment load of the Amazon River increased because of meltwater flushing; and/or (2) the Amazon River discharge did increase but Sites 932 and 933 are too far away to register the effect; and/or (3) other circulation changes involving the NBCC mask the discharge signal; or (4) there was no increase in terrestrial organic matter discharge by the Amazon River and the negative carbon isotope event was caused by increased gas hydrates in the water column.

The third possible cause of the negative carbon isotope event is the release of significant amounts of gas hydrates into the water column. It has been suggested by Maslin & Mikkelsen (in press; Mikkelsen & Maslin this volume) that sea-level changes could have destablized gas hydrate reservoirs in the Amazon Fan. Maslin & Mikkelsen (in press; Mikkelsen & Maslin this volume) have placed the last occurrence of the near-surface Eastern and Western Debris Flows (Flood *et al.* 1995) during Termination I (13–15 calendar ka), coincidental with the carbon isotope event. The current suggestion is that the sea-level rise associated with the initial stages of the last deglaciation destabilized the gas hydrate reservoirs and overburdened the already overconsolidated Amazon Fan sediments causing massive failure. To give some scale of the slope failure that occurred, each one of these debris flows covers an area of up to 25 000 km^2 and reaches a maximum thickness of 200 m, and consists of approximately 50 000 gigatonnes of sediment. At the deglaciation at approximately the same time as the carbon isotope event, two of these massive failures occurred. If gas hydrates are the primary cause then it must be considered that there are many such releases as the planktonic $\delta^{13}C$ records of both Sites 932 and 933 show many such negative deviations. The most significant occur between about 26 and 32 ka (see Figs 6 and 9).

Glacial positive oxygen isotope events and their link with the Heinrich events

To better assess the oxygen isotope records of Site 932 they were combined into a composite record. This was done by removing the smaller data gaps and the inconsistencies in the sampling by applying a Gaussian interpolation using weighted duplicates with a sample spacing of 50 cm (same resolution as the sampling) and a window size of 150 cm to the

isotopic records. Only the most complete oxygen isotope records were used; in the case of Site 932 that included *G. ruber, G. trilobus, N. dutertrei* and *G. truncatulinoides*. These records were interpolated, normalized (maximum range –1 to +1) and were then combined by averaging the four different records. The combined and normalized planktonic oxygen isotope record was placed on the Site 932 age model and is shown in Fig. 10. However, it must be remembered that this record has many drawbacks: (1) it has combined isotopic records of planktonic species from very different depth habitats, thus the inclusion of species that lie just above or within the thermocline means that it will not be possible to monitor salinity changes at the surface; and (2) there are numerous data gaps in each of the individual records, which could cause events or features to be missed or artificially enhanced.

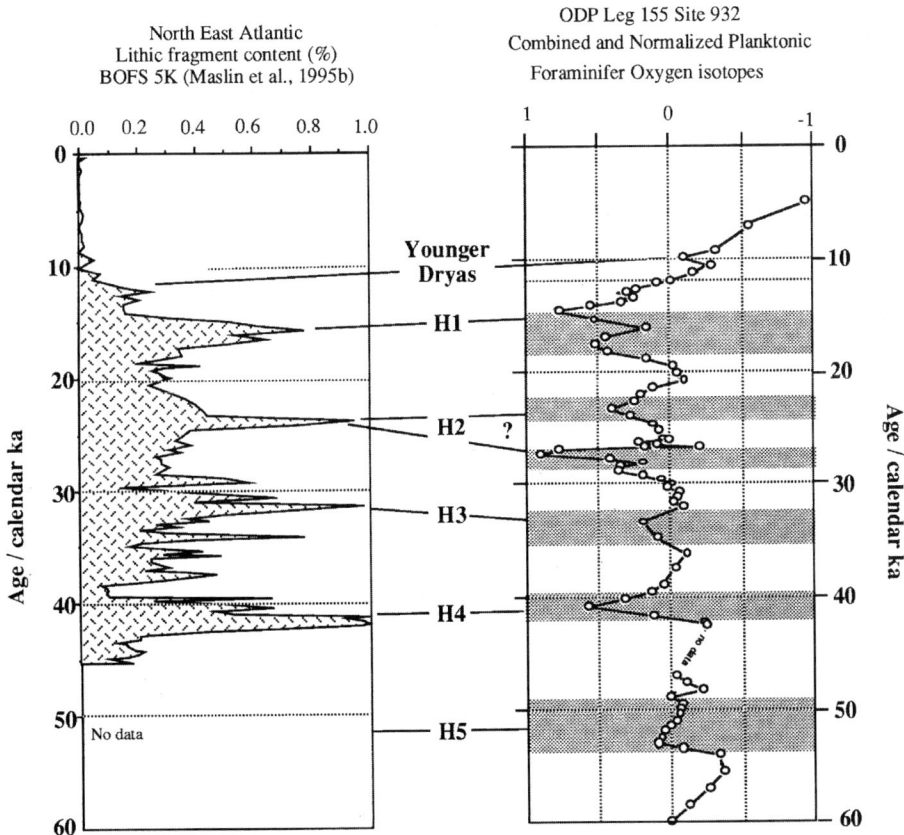

Fig. 10. Comparison of the combined and normalized planktonic foraminifera oxygen isotope records from Hole 932A, with the Heinrich events (H1–H5). The Heinrich events are illustrated by the percentage lithic content of the coarse fraction (> 63 µm) of BOFS 5K (50°41.3'N, 21°51.9'W, water depth 3547 m; McCave 1989), which indicates when there was a significant increase in the amount of ice-rafted debris being deposited in the North Atlantic (Maslin *et al.* 1995*b*). Unfortunately this record only covers the last 45 ka. The timing of Heinrich event 5 is taken from Bond *et al.* (1992, 1993).

It is suggested that if the Site 932 positive $\delta^{18}O$ deviations are real features, they may have been caused by an increase in surface water salinity because of an increase in the intensity and the prolongation of the retroflection of the NBCC. The main controlling force on the direction and intensity of the NBCC is the prevailing winds, i.e. the South East Trades (Picaut et al. 1985; Philander & Pacanowski 1986). The global wind system during the last glacial was more zonal because of the greater thermal gradient between the high and low latitudes (e.g. COHMAP Members 1988; Lautenschlager & Herterich 1991; Lautenschlager et al. 1992; Seidov & Maslin 1996; Seidov et al. 1996). More zonal glacial South East Trades would results in reduced penetration of the Inter-Tropical Convergence Zone (ITCZ) and thus the NBCC into the northern hemisphere (see Fig. 11). If the NBCC was periodically prevented from entering the northern hemisphere this would reduce the export of salinity and may have led to heavier surface water oxygen isotopes, i.e. the positive $\delta^{18}O$ deviations.

The timing of the positive oxygen isotope excursion compares well with that of the Heinrich events (Heinrich 1988; Bond et al. 1992, 1993; Maslin et al. 1995b; see Fig. 10). If there is a causal

link between the two records I suggest that increased ice-rafting (Robinson et al. 1995; Maslin et al. 1995b; see Fig. 12) in the North Atlantic would have further increased the enhanced glacial–latitudinal thermal gradient, further increasing the zonal component of the wind system (Lautenschlager & Herterich 1991; Lautenschlager et al. 1992; Seidov & Maslin 1996; Seidov et al. 1996). This would have curtailed the cross-equatorial export of the NBCC and may have resulted in continual NBCC-retroflection (Fig. 11). This would have produced a feedback mechanism, as the reduced salt and heat input from the NBCC into the Caribbean and thus the Gulf Stream would have reduced the sinking potential of the North Atlantic surface waters. Any reduction in the deep formation of the North Atlantic would have increased the latitudinal thermal gradient (Seidov & Maslin 1996; Seidov et al. 1996) further reducing the NBCC penetration into the North Atlantic.

Further work on high-resolution records from the western Amazon Fan and the gateway between the western equatorial Atlantic and the Caribbean are required to test this theory linking Heinrich events via the tropical wind system, to the circulation patterns of the equatorial western Atlantic. Until then the possible link between the equatorial

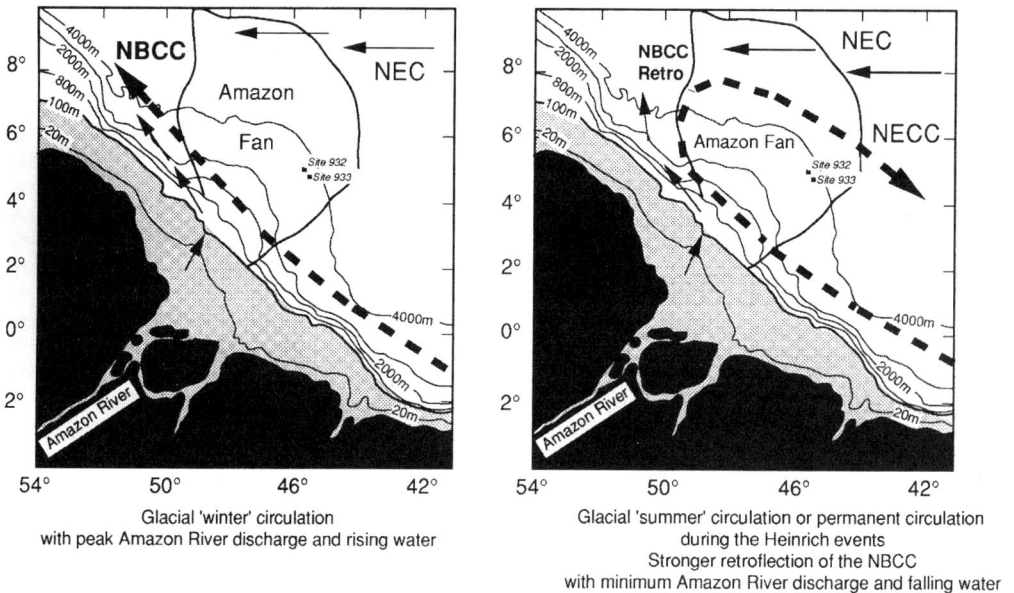

Glacial 'winter' circulation
with peak Amazon River discharge and rising water

Glacial 'summer' circulation or permanent circulation
during the Heinrich events
Stronger retroflection of the NBCC
with minimum Amazon River discharge and falling water

Fig. 11. Sketch map showing the possible glacial period seasonal variation in the surface circulation of the western equatorial Atlantic. It is suggested in this study that the retroflection of the NBCC may have had a longer seasonal occurrence compared with the modern circulation during the last glacial period and that during the Heinrich events it may have been permanent. Note that the glacial coastline of Brazil has been moved to the 100 m contour line due to the lower sea level. NBCC = North Brazil Coastal Current, NEC = North Equatorial Current, NECC = North Equatorial Counter-Current, NBCC Retro = retroflection of the NBCC.

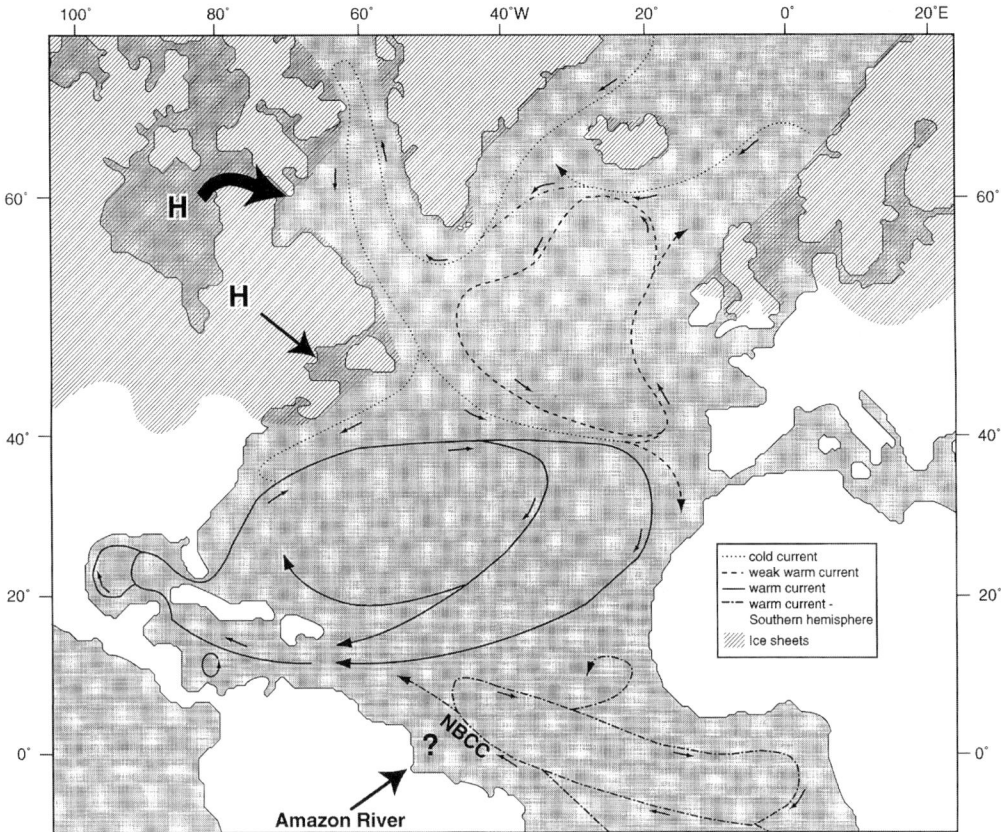

Fig. 12. Reconstruction of the surface water circulation of the Atlantic Ocean during the last glacial maximum (Robinson *et al.* 1995; Maslin *et al.* 1995; Seidov *et al.* 1996; Seidov & Maslin, 1996), illustrating the unknown exchange between the NBCC and the North Atlantic. Note also the two arrows on the North American continent suggesting two possible routes for the collapse of the Laurentide ice sheet during the Heinrich events.

western Atlantic circulation and the Heinrich events remains essentially speculative (Fig. 11).

Conclusions

The Amazon Fan provides the opportunity to obtain a climatic record of the last glacial–interglacial cycle at a resolution approaching that of the Greenland ice cores. This preliminary study indicates that there are major drawbacks to reconstructing past climate from fan sediments, i.e. reworking and the difficulty of age modelling. It is stressed that precautions such as isotope analysis of multiple planktonic foraminiferal species and detailed sedimentological work (Flood *et al.* 1995) are required if complete records are to be obtained and possible sections of reworking are to be identified. The planktonic foraminiferal carbon

isotope records of Sites 932 and 933 have provided evidence of a distinct negative deviation during Termination I (13–15 calendar ka). I suggest this could have been caused by an increase in the sediment discharge of the Amazon River because of the deglaciation of the Andes and/or the release of significant quantities of gas hydrates as large parts of the Amazon Fan sediment column failed because of increased sea level. Positive deviations in the composite Site 932 oxygen isotope record coincide with the timing of the Heinrich events. I speculatively suggest that increased ice-rafting during the Heinrich events would have increased the latitudinal thermal gradient, increasing the zonal component of the wind system. This would have reduced the penetration of the ITCZ into the northern hemisphere and thus curtailed the cross-equatorial export of the NBCC, and may have resulted in continual NBCC-retroflection.

The authors are grateful for the help and support of M. Sarnthein, N. Mikkelsen and N. Shackleton. We gratefully acknowledge the co-operation of H. Cordt and H. Erlenkeuser, who supervised the operation of the mass spectrometer in Kiel. The Deutsche Forschungs-gemeinschaft supported this study. We also wish to thank both E. Thomas and A. Cramp for detailed and con-structive reviews.

References

BIRD, M. I., FYFE, W. PINHEIRO-DICK, D. & CHIVAS, A. R. 1992. Carbon isotope indicators of catchment vegetation in the Brazilian Amazon. *Global Biogeochemical Cycles*, **6**, 293–306.

BOND, G., HEINRICH, H., BROECKER, W. *ET AL*. 1992. Evidence for massive discharges of icebergs into the North Atlantic ocean during the last glacial. *Nature*, **360**, 245–249.

——, BROECKER, W., JOHNSEN, S. *ET AL*. 1993. Correlation between climate records from North Atlantic sediments and Greenland ice. *Nature*, **365**, 143–147.

CASTRO, J. C., MIURA, K. & BRAGA, J. A. E. 1978. Stratigraphic and structural framework of the Foz do Amazonas Basin. *Proceedings of Annual Offshore Technology Conference*, **3**, 1843–1847

CISOWSKI, S. M. 1995. Synthesis of magnetic remanence correlation, Leg 155. *Proceedings of ODP, Initial Reports*, **155**. College Station, TX (Ocean Drilling Project).

—— & HALL, F. in press. An examination of the paleointensity record and geomagnetic excursions recorded in Leg 155 cores. *Proceedings of ODP, Scientific Reports*, **155**. College Station, TX (Ocean Drilling Program).

COHMAP MEMBERS. 1988. Climatic change of the last 18,000 years: observations and model simulations. *Science*, **241**, 1043–1052.

CURRY, W., SHACKLETON, N., RICHTER, C. & SHIPBOARD SCIENTIFIC PARTY, 1995. *Proceedings of ODP, Initial Reports*, **154**. College Station, TX (Ocean Drilling Project).

DAMUTH, J. E. 1975. Quaternary climate change as revealed by calcium carbonate fluctuations in western Equatorial Atlantic sediments. *Deep-Sea Research*, **22**, 725–743.

—— 1977. Late Quaternary sedimentation of the western Equatorial Atlantic. *Geological Society of America Bulletin*, **88**, 695–710.

——, & FAIRBRIDGE, R. W. 1970. Equatorial Atlantic deep-sea arkosic sands and ice age aridity in tropical South America. *Geo-Marine Letters*, **3**, 109–117.

—— & KUMAR, N. 1975. Amazon Cone, morphology, sediments, age, and growth pattern. *Geological Society of America Bulletin*, **86**, 863–878.

DANSGAARD, W. *ET AL*. 1993. Evidence for general instability in past climate from a 250-kyr ice-core record. *Nature*, **364**, 218–222.

DUPLESSY, J.-C., LABEYRIE, L., ARNOLD, M., PATERNE, M., DUPRAT, J. & VAN WEERING, T. 1992. Changes in surface water salinity of the North Atlantic Ocean during the last deglaciation. *Nature*, **358**, 485–488.

FAIRBANKS, R. G. 1989. A 17,000 year glacio-eustatic sea level record: influence of glacial melting rates on the Younger Dryas event and deep-ocean circulation. *Nature*, **342**, 637–642.

FLOOD, R., PIPER, D., KLAUS, A. & SHIPBOARD SCIENTIFIC

PARTY. 1995. *Proceedings of ODP, Initial Reports*, **155**. College Station, TX (Ocean Drilling Project).

HEINRICH, H. 1988. Origin and consequences of cyclic ice rafting in the northeast Atlantic Ocean during the past 130,000 years. *Quaternary Research*, **29**, 142–152.

HOORN, C., GUERRERO, J., SARMIENTO, G. & LORENTE, M. 1995. Andean tectonic as a cause for changing drainage paterns in Miocene northern South America. *Geology*, **23**, 237–240.

LAUTENSCHLAGER, M. & HERTERICH K. 1991. Atmospheric response to Ice Age conditions: climatology near the earth's surface. *Journal of Geophysical Research*, **95**, 22547–22557.

——, MIKOLAJEWICZ, U., MAIER-REIMER, E. & HEINZE, C. 1992. Application of ocean models for the interpretation of atmospheric general circulation model experiments on the climate of the last glacial maximum. *Paleoceanography*, **7**, 769–782.

LEVITUS, S. 1982. *Climatological Atlas of the World Ocean*. NOAA Professional Paper **13**.

MCCAVE, I. N. 1989. *RRS Discovery 184 BOFS 1989 Leg 3 Cruise Report*. University of Cambridge.

MCHARGUE, L. R., DAMON, P. E. & DONAHUE, D. J. 1995. Enhanced cosmic-ray production of 10Be coincident with the Mono Lake and Laschamp geomagnetic excursions. *Geophysical Research Letters*, **22**, 659–662.

MASLIN, M .A. & MIKKELSEN, N. in press. The mass-transport deposits and interglacial sediments of the Amazon Fan: Age estimates and Fan dynamics. *Proceedings of ODP, Scientific Reports*, **155**. College Station, TX (Ocean Drilling Program).

——, ADAMS, J., THOMAS, E., FAURE, H. & HAINES-YOUNG, R. 1995a. Estimating the carbon transfer between the oceans, atmosphere and the terrestrial biosphere since the last glacial maximum. *Terra Nova*, **7**, 358–366.

——, SHACKLETON, N. & PFLAUMANN, U. 1995b. Surface water temperature, salinity and density changes in the N.E. Atlantic during the last 45,000 years: Heinrich events, deep water formation and climatic rebounds. *Paleoceanography*, **10**, 527–544.

MERRILL, R. T & MCELHINNY, M. W. 1983. *The Earth's Magnetic Field*: its history, origin, and planetary perspective. Academic Press, London.

METCALF, W. G. & STALCUP, M. C. 1967. Origin of the Atlantic Equatorial Undercurrent. *Deep-Sea Research*, **72**, 4959–4975.

MIKKELSEN, N. & MASLIN, M. 1998. Timing of the late Quaternary Amazon Fan Complex mass-transport deposits. *This volume*.

——, ——, GIRAUDEAU, J. & SHOWERS, W. in press. Biostratigraphy and sedimentation rates of the Amazon Fan. *Proceedings of ODP, Scientific Reports*, **155**. College Station, TX (Ocean Drilling Program).

MILLIMAN, J. D., SUMMERHAYS, C. P. & BARRETTO, H. T. 1975. Quaternary sedimentation on the Amazon continental margin: a model. *Geological Society of America Bulletin*, **86**, 610–614.

PHILANDER, S. & PACANOWSKI, R. 1986. A model of the seasonal cycle in the tropical Atlantic. *Journal of Geophysical Research*, **91**, 192–206.

PICAUT, J., SERVAIN, J., LECONTE, P., SEVA, M., LUKAS, S. & ROUGIER, G. 1985. *Climatic Atlas of the Tropical Atlantic Wind Stress and Sea Surface Temperature 1964–1979*. University of Hawaii Press, Honolulu.

PRELL, W. L. & DAMUTH, J. E. 1978. The climate related diachronous disappearance of Pulleniatina obliquiloculata in late Quaternary sediments of the Atlantic and Caribbean. *Marine Micropaleontology*, **3**, 267–277.

RICHARDSON, P. & WALSH, D. 1986. Mapping climatological seasonal variations of surface currents in the tropical Atlantic using ship drifts. *Journal of Geophysical Research*, **91**, 537–550.

ROBINSON, S., MASLIN, M. A. & McCAVE, I. N. 1995. Magnetic susceptibility variations in Late Pleistocene deep-sea sediments of the North East Atlantic: Implications for ice rafting and palaeo-circulation at the Last Glacial Maximum. *Paleoceanography*, **10**, 221–250.

SARNTHEIN, M., WINN, K., JUNG, S. J. A., DUPLESSY, J.-C., LABEYRIE, L., ERLENKEUSER, H. & GANSEN, G. 1994. Changes in east Atlantic deepwater circulation over the last 30,000 years: Eight time slice reconstructions. *Paleoceanography*, **9**(2), 209–268.

SEIDOV, D. & MASLIN, M. 1996. Seasonally ice free glacial nordic seas without deep water ventilation. *Terra Nova*, **8**, 245–254.

——, SARNTHEIN, M., STATTEGGER, K., PRIEN, R. & WEINELT, M. 1996. North Atlantic ocean circulation during the Last Glacial Maximum and subsequent meltwater event: A numerical model. *Journal of Geophysical Research*, **101**, 16305–16332.

SHOWERS, W. J. & BEVIS, M. 1988. Amazon Cone isotope stratigraphy: evidence for the source of the tropical meltwater spike. *Paleogeography, Paleoclimatology, Paleoecology*, **64**, 189–199.

THOMPSON, L. G., MOSLEY-THOMPSON, E., DAVIS, M. E. ET AL. 1995. Late glacial stage and Holocene tropical ice core records from Huascaran, Peru. *Science*, **269**, 46–50.

TRAUTH, M. 1995. *Bioturbate Signalverzerrung hockauflosender paläoozeanographischer Zeitreihen* (Bioturbational signal distortion of high-resolution paleoceanographic time-series). PhD thesis, Ber. No. **74** (Geol.-Paleaont. Inst. Univ. Kiel).

TRIC, E. ET AL. 1992. Paleointensity of the geomagnetic field during the last 80,000 years. *Journal of Geophysical Research*, **97**, 9337–9351.

VILELA, C. & MASLIN, M. A. in press. Benthic and planktonic foraminifera assemblage and stable isotope results from the mass-flow sediments in the Amazon Fan. *Proceedings of ODP, Scientific Research*, **155**. College Station, TX (Ocean Drilling Program).

Timing of the late Quaternary Amazon Fan Complex mass-transport deposits

MARK MASLIN[1] & NAJA MIKKELSEN[2]

[1] *Environmental Change Research Centre, Department of Geography, University College London, 26 Bedford Way, London WC1H 0AP, UK*

[2] *Geological Survey of Denmark and Greenland, Thoravej 8, DK-2400 Copenhagen NV, Denmark*

Abstract: The Amazon Fan at the mouth of the Amazon River is a large, highly structured muddy deep-sea fan. Sediment recovered by Ocean Drilling Program Leg 155 has allowed us to reconstruct the structure of the late Quaternary Amazon Fan Complex. A significant sedimentary component of the complex are the mass-transport deposits (MTDs) formed by catastrophic failure of the continental slope. There are two sets of MTDs: the near-surface MTDs and the deep MTDs. The near-surface MTDs are divided into the Eastern and Western Debris Flows which are capped by Holocene sediments. The deep debris flows are divided into the Unit R MTD in the western Amazon Fan complex, and the Deep Eastern MTD in the eastern Amazon Fan Complex. The MTDs each cover an area of up to 25 000 km^2, reach a maximum thickness of 200 m, and consist of approximately 50 000 gigatonnes of sediment. A characteristic of the MTDs is the inclusion of reworked glacial and interglacial material from both the continental slope and the surrounding fan. The relatively thin interglacial deposits below the MTDs are from the nanofossil zones CN14b and CN15a and thus younger than 460 ka. Bio-, seismic- and magnetostratigraphy and sedimentation rate constraints have been used to date the top of both the near-surface and the deep MTDs. It seems that the near-surface Western and Eastern Debris Flows were last active during Termination I. The Deep Eastern MTD was last active at about 33 ka, while the Unit R MTD was last active at about 45 ka. It is inferred that the interglacial deposits beneath these MTDs were formed during Oxygen Isotope Stages 7, 9 and 11. We speculate that the MTDs were triggered by climatically induced changes in sea level, and that the interglacial deposits may have acted as slip planes for the MTDs.

The Amazon Fan has developed since the early Miocene, due to the massive increase in the sediment transport of the Amazon River following the Andean uplift (Castro *et al.* 1978; Curry *et al.* 1995). At present, and during other interglacial periods, Amazon River terrigenous sediments are deposited predominantly on the inner continental shelf (Flood *et al.* 1995). Only pelagic sediments, therefore, accumulated on the Amazon Fan during interglacial periods. During glacial periods the lower sea level caused the outflow of the Amazon River to shift to the northeast and to feed sediment directly into the Amazon Canyon and hence into the Amazon Fan Complex (Damuth & Fairbridge 1970; Damuth & Kumar 1975). Changing sea levels may also have destabilized the shelf and fan sediments, causing many slumps and flows. The most important of these, in terms of volume of sediment, are the huge debris flows or mass-transport deposits (MTDs). These deposits have

been recognized on seismic lines both near the surface of the Amazon Fan (Fig. 1) and also deep within the fan complex (Figs 2 and 3). Associated features found beneath the deep MTDs are the interglacial deposits, in this paper termed collectively the 'deep carbonate units'. These deposits are volumetrically not as prominent as the MTDs, but they apparently play an active role in the dynamics of the MTDs and thus the evolution of the Amazon Fan Complex. Moreover, the deep carbonate units can be dated using biostratigraphy, so that the timing and the amount of the stratigraphical column removed by the MTDs can be estimated.

Mass-flow deposits and MTDs can cover areas of hundreds of square kilometres. From detailed side-scan sonar mapping it has become apparent that the MTDs are a significant component of all continental margins. The surficial parts of some of these deposits, e.g. the Amazon Fan, have been investigated by acoustic and coring methods (e.g.

MASLIN, M. & MIKKELSEN, N. 1998. Timing of the late Quaternary Amazon Fan Complex mass-transport deposits. *In*: CRAMP, A., MacLEOD, C. J., LEE, S. V. & JONES, E. J. W. (eds) *Geological Evolution of Ocean Basins: Results from the Ocean Drilling Program.* Geological Society, London, Special Publications, **131**, 129–150.

129

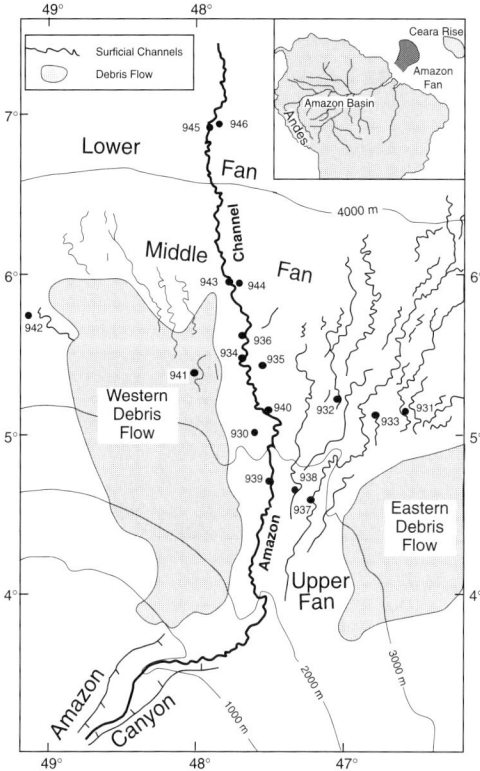

Fig. 1. Map of the Amazon Fan and the sites drilled during Leg 155 (adapted from Flood *et al.* 1995).

near-surface MTDs, termed the Eastern and Western 'Debris Flows' by Damuth *et al.* (1988), each extends to approximately 200 km downslope and 100 km in width and about 100–200 m in thickness (Figs 1 and 2). Two major deep MTDs were cored during Leg 155 (Fig. 3). The first, the 'Unit R Debris Flow' (Manley & Flood 1988), underlies the Upper Levee Complex which was formed during the last glacial, and was recovered at Sites 935, 936 and 944 (Fig. 2). Unit R was found to have levee muds directly below the contact surface. In addition, a sand unit of approximately 40 m was recovered below the levee muds at Sites 936 and 944. Below the sandy unit and directly below the levee muds in Site 935, interglacial carbonate-rich units were recovered. The second MTD on the eastern fan, here termed the Deep Eastern MTD, overlies the crest of the Bottom Levee Complex and was cored at Sites 931 and 933. This ancient MTD was found to have interglacial sediments directly below the contact surface (Fig. 3). The interglacial sediment units are present in seven of the sites and are characterized by a fairly high content of carbonate (up to almost 40%) and by the presence of calcareous nanofossils and foraminifera. A full description of all the units recovered by ODP Leg 155 can be found in Flood *et al.* (1995).

Methods

Stratigraphy/age assignments

Age assignments are based on biostatigraphic analyses of calcareous nanofossils, planktonic foraminifera and palaeomagnetic dates. Two short geomagnetic features were used for regional correlations: the Lake Mungo excursion at about 32 ka and the Blake event at about 105 ka (Cisowski & Hall in press). Correlation attempts between the various sites have to a lesser degree been supplemented by seismic interpretations (Flood *et al.* 1995).

Calcareous nanofossils

The nanofossil assemblages were described directly from smear slides prepared from unprocessed sediment. The slides were prepared using standard methods. A small amount of sediment was smeared onto a glass slide with a drop of water and dried on a hot-plate and subsequently mounted with a drop of Norland Optical Adhesive on a cover glass. The slides were examined by means of standard light microscope techniques under crossed nicols and transmitted light at 1000×. Critical identifications were confirmed by scanning electron microscopy.

Damuth *et al.* 1988). Until Leg 155, however, the deeper parts of the MTDs and, as important, the underlying sediments have not been accessible. This has meant that interpretation of the dynamics and the ultimate cause of these mass-transport deposits has been at best speculative. The aim of this study is to investigate the biostratigraphical and stable isotopic evidence for (1) the age of the sediment directly above the MTDs, to provide an estimate of the youngest occurrence of the deposits, (2) the age and characteristics of the deep carbonates recovered below the two deep MTDs, (3) the dynamics of the Amazon Fan Complex in the late Quaternary, and (4) possible causes of the initiation of the MTDs.

Sedimentary components recovered from the Amazon Fan

Ocean Drilling Program (ODP) Leg 155 penetrated MTDs at eight sites and recovered sediment from the entire unit of the MTDs and underlying sediment at seven of these sites (Fig. 2). The two

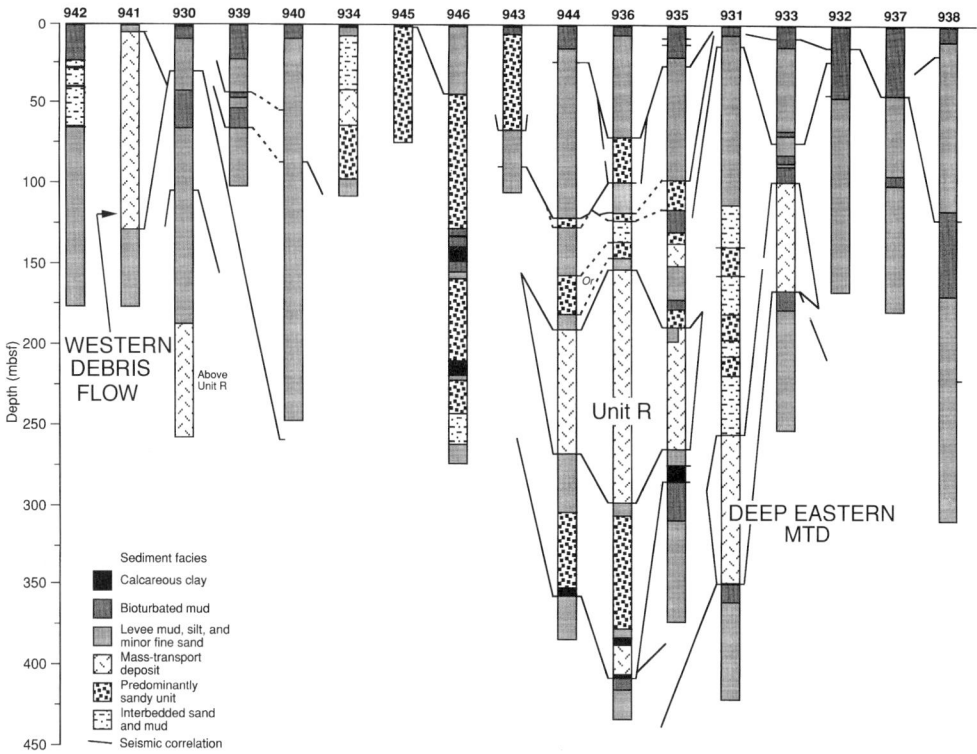

Fig. 2. Bar chart indicating the sediment types recovered at each of the Leg 155 sites. Complete mass-transport deposits and associated calcareous clays were recovered at Sites 931, 933, 935, 936 and 944 (adapted from Flood *et al.* 1995).

The state of preservation of the nanofossil assemblages was characterized by the code system of Curry *et al.* (1995):

G = Good (little or no evidence of dissolution and/or secondary overgrowth of calcite; diagnostic characters fully preserved);

M = Moderate (dissolution and/or secondary overgrowth; partially altered primary morphological characteristics; however, nearly all specimens can be identified at the species level);

P = Poor (severe dissolution, fragmentation, and/or secondary overgrowth with primary features largely destroyed; many specimens cannot be identified at the species level and/or generic level).

Pleistocene calcareous nanofossil datums are sparse. The zonal scheme of Bukry (1973, 1978) and Okada & Bukry (1980) includes only two datums for the time period covered by the sequences drilled, namely the first occurrence of *Emiliania huxleyi* at 260 ka and the onset of the *E.*

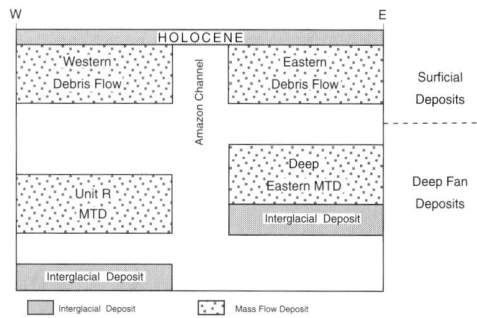

Fig. 3. Schematic diagram showing the relationship between the mass-transport deposits (MTD) and the interglacial hemipelagic deposits (the 'deep carbonates') of the Amazon Fan.

huxleyi acme at approximately 85 ka (Fig. 4). The sediments recovered during Leg 155 are all younger than 460 ka as *Pseudoemiliania lacunosa* was not observed. Modifications to improve the

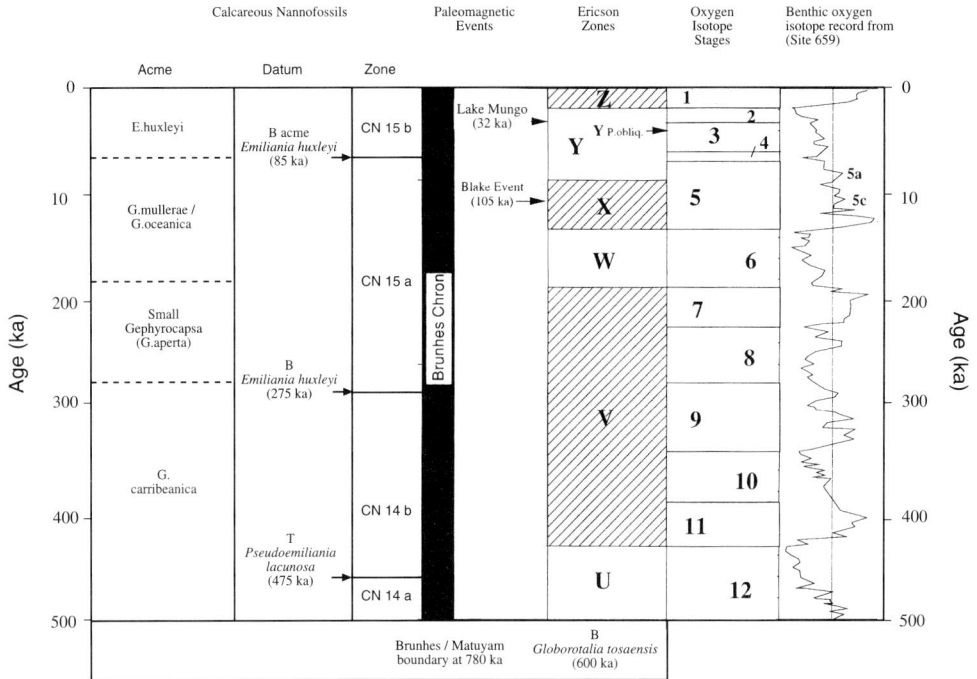

Fig. 4. Summary of the nanofossil, palaeomagnetic and planktonic foraminifera (Ericson zones) stratigraphy compared to age and the classical oxygen isotope stages illustrated by the the isotope curve of Tiedemann *et al.* (1994).

standard nanofossil resolution of the Pleistocene were recently made by Takayama & Sato (1987) and Raffi *et al.* (1993). Pujos & Giradeau (1993), Weaver (1993) have subsequently worked with assemblage changes especially related to the morphotypes of *Gephyrocapsa* to refine the subdivision of the Pleistocene.

Planktonic foraminiferal assemblages

Onboard ship samples were taken routinely from the core-catcher (*c.* 200 cm³) as well as discrete samples at important lithological boundaries (*c.* 20 cm³). The samples were prepared depending on the degree of lithification. Unlithified ooze was washed with tap water over a 63 μm sieve. Samples with cohesive lumps were first soaked in a weak solution of calgon, then gently stirred in a beaker with a magnetic stirrer, and then washed over a 63 μm sieve. Samples were air-dried at 65°C in an oven. Preservation in the greater than 150 μm fraction was estimated from a visual examination of the dried sample. All planktonic foraminiferal species found in the samples were quantified according to the following categories:

A = Abundant (greater than 30%);
C = Common (15–30%);
F = Few (3–15%);
R = Rare (greater than 0% but less than 3%);
B = Barren (0%).

The sequential stratigraphy of the Ericson zones (Ericson & Wollin 1956; Ericson *et al.* 1961) was applied to the youngest sediment sequences where these were continuous and undisturbed (Fig. 4). The Ericson zones are based on the appearance of the *menardii* complex in interglacial deposits and the absence of it in the glacial deposits. Otherwise, owing to the very young age of the Amazon Fan sediments recovered on Leg 155, the primary use of the planktonic foraminiferal assemblages was to differentiate glacial and interglacial sediments, using the *menardii* complex as a signal of interglacial sediments. The youngest Atlantic Ocean planktonic foraminifera true datum is the last appearance of *Globorotalia tosaensis* at 0.6 Ma, which was not found in any of the sediment recovered during Leg 155. One abundance shift datum that has been shown to be reliable, apart from the Ericson zones, is the disappearance of *Pulleniatina obliquiloculata* ($Y_{p.\ obliq.}$) at approximately 40 ka.

Planktonic and benthic foraminiferal stable isotopes

The oxygen and carbon isotopes of benthic and planktonic foraminifera were measured. The samples were freeze-dried, wet-sieved through a 63 µm mesh sieve, and then dry-sieved at convenient intervals between 250 and 400 µm from which the foraminifera were picked. Two benthic species were selected for analysis: *Cibicides wuellerstorfi* and *Uvigerina* sp. When the rare abundance of foraminifera, caused by the extremely high sedimentation rates (1–50 m ka^{-1}), resulted in the measurement of only one specimen, the gas pressure from the mass spectrometer was checked to make sure that there was sufficient material for a reliable measurement. Six planktonic foraminiferal species were analysed: *Globigerinoides ruber, G. trilobus, G. sacculifer, Neogloboquadrina dutertrei, Pulleniatina obliquiloculata* and *Globorotalia truncatulinoides*. The numbers of species and specimens varied greatly throughout the sediment recovered. In the case of low numbers of specimens (no sample of less than six specimens was measured), the mass spectrometer gas pressure was checked especially carefully to ensure reliable measurements. It should be noted that variations in both the number of planktonic foraminifera per sample and low numbers of specimens per sample can be sources of considerable error (Trauth 1995). As a consequence of the high sedimentation rates of the Amazon Fan, however, the bioturbation component of the stable isotope errors is greatly reduced. All the errors listed above were considered when interpreting the isotopic record.

Results

The overall correlation of the sites studied in this paper has been facilitated by the combined use of bio- and magnetostratigraphy as well as sedimentary constraints (Flood *et al.* 1995). The general correlation of the eastern fan Sites 933, 931 and for comparison Site 932 is shown in Fig. 5, while the general correlation between Sites 935, 936 and 944 is shown in Fig. 6. The MTDs have been identified primarily by their sedimentary and seismic characteristics (Flood *et al.* 1995). The interglacial sediments were identified by their high calcium carbonate content, the high abundance of calcareous nanofossils (Fig. 7), the presence of the *menardii* complex (Figs 8 and 9), and by the light,

Holocene-like, planktonic and benthic isotope values (Figs 10 to 13).

Both the nanofossil assemblages and the planktonic foraminiferal assemblages indicate that below both the Unit R and the Deep Eastern MTD there are interglacial sediments (Flood *et al.* 1995). The debris-flow material sits directly on the interglacial sediments in the case of the Deep Eastern MTD (Sites 931 and 933; see Fig. 5). In contrast there are levee muds and in some cases sandy units between the Unit R MTD and the interglacial sediments (Sites 935, 936, and 944; see Fig. 6).

Nanofossils

Well preserved calcareous nanofossil assemblages are found only in the interglacial deposits of the Amazon Fan Complex and in clasts of interglacial origin within the MTDs. In contrast, the matrix of the MTDs as well as the glacial sediments is almost barren of nanofossils. According to the nanofossil evidence, all the interglacial deep carbonates recovered during Leg 155 are younger than 460 ka, as there is a lack of *Pseudoemiliania lacunosa*. Small clasts with reworked Miocene and Pliocene nanofossils were found within the MTDs (Flood *et al.* 1995).

The MTDs interrupt the sedimentary sequence and thus sequential stratigraphy based on changes in planktonic foraminiferal assemblages or stable isotopes could not be used for dating. The dating of the sediments below the MTDs was therefore primarily done using the nanofossils, whereas foraminifera were mainly used to date the top of the MTDs.

The overall abundance of nanofossils distinctively points out the interglacial 'deep carbonate' sequences (Fig. 5) in the Amazon Fan. Rich nanofossil assemblages are thus found below the Unit R in the deep carbonate units recovered in Sites 935, 936 and 944 and below the Deep Eastern MTD in Sites 931 and 933. The interglacial sequences range in thickness from almost 3 m at Site 944 to approximately 8 m at Site 933. An exception is the curtailed interval of 0.2 m in Site 931.

The nanofossil assemblages of the deep carbonate units represent the zones CN14b and CN15a (Fig. 7). The oldest assemblage from CN14b is referred to the *G. carribeanica* acme of Weaver (1993) and Pujos & Giradeau (1993). The abundance of small *Gephyrocapsa* specimens in the low-diversity assemblages increases relative to *G. carribeanica* at the CN14b–15a boundary. In the upper part of CN15a, which is referred to Isotope Stage 5 based on isotope studies of the continuous section in Site 942 off the fan (Showers *et al.* 1995),

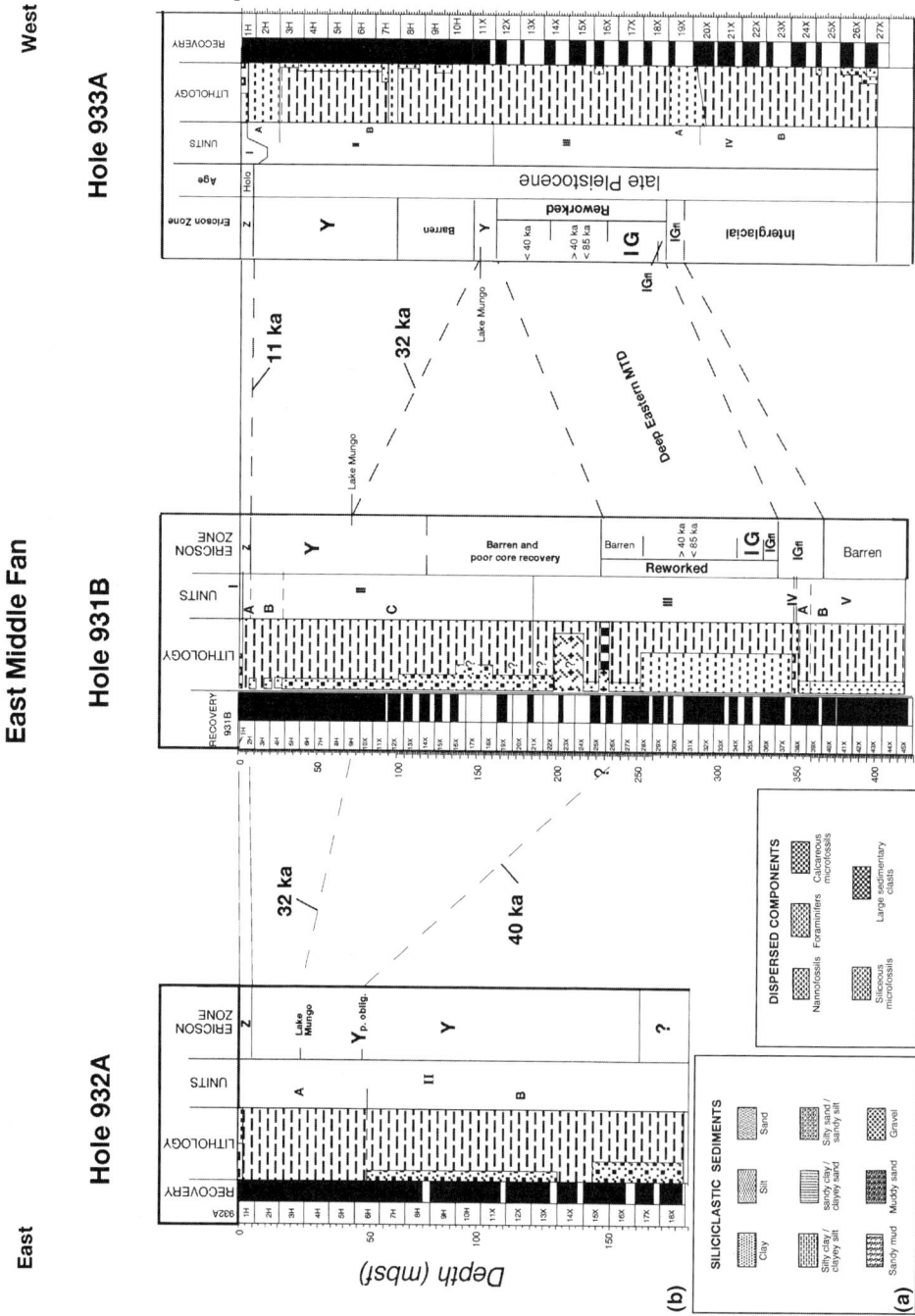

Fig. 5. (a) Lithological key for (b) and Fig 6. (b) Correlation between Holes 931B, 932A and 933A based on sedimentology, biostratigraphy and magnetostratigraphy. IG = interglacial deposit, IGfl = interglacial deposit containing *Globorotalia tumida flexuosa* (modified and updated from Flood *et al.* 1995).

Middle Fan

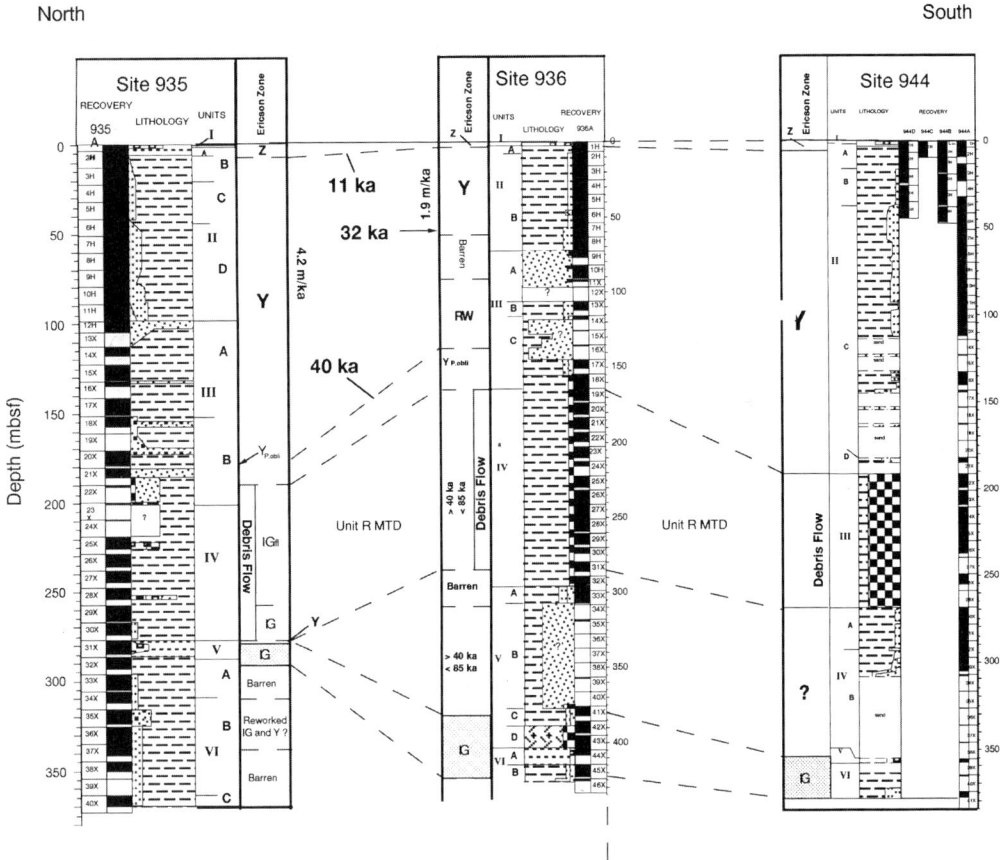

Fig. 6. Correlation between Sites 935, 936 and 944 based on sedimentology, biostratigraphy and magnetostratigraphy. IG = interglacial deposit, IGfl = interglacial deposit containing *Globorotalia tumida flexuosa* . For lithology key see Fig. 5a. (modified and updated from Flood *et al.* 1995).

the diversity of the nanofossil assemblages increases significantly. Some of the interglacial sections contain a number of calcium-carbonate-poor intervals with a high abundance of terrigenous material, and we suggest that these intervals are small turbidites (see Discussion). Delineation of the CN14b–CN15a boundary (first occurrence of *E. huxleyi* at 260 ka) in the interglacial deposits is further hampered by a substantial reworking in at least the top and bottom of the sequences caused by the movements of the surrounding/imbedding mass-flow deposits.

The deep carbonate unit at Site 931 comprises a 0.2 m section of interglacial material spanning the CN14b–15a boundary. The low abundance of nanofossils, the poorly preserved specimens and high abundance of terrigenous material signal the

transition from the bottom of an interglacial sequence to the underlying glacial sediments, where the main part of the interglacial sequence apparently has been removed by the overlying MTD. Site 933 provides a less carbonate-rich but also less disturbed interglacial sequence and the nanofossil evidence suggests that the majority of the sediment is from CN15a, with evidence for some CN14b material at the base of the deposit.

Sites 935, 936 and 944 all have deep carbonate units below the Unit R MTD; however, unlike Sites 931 and 933 there is not a direct contact between the MTD and the underlying interglacial deposits. At Site 935 there is a short sequence of levee muds between the Unit R MTD and the interglacial unit which furnishes a rich *G. carribeanica* acme of nanozone CN14b. At Site 936 there is evidence for

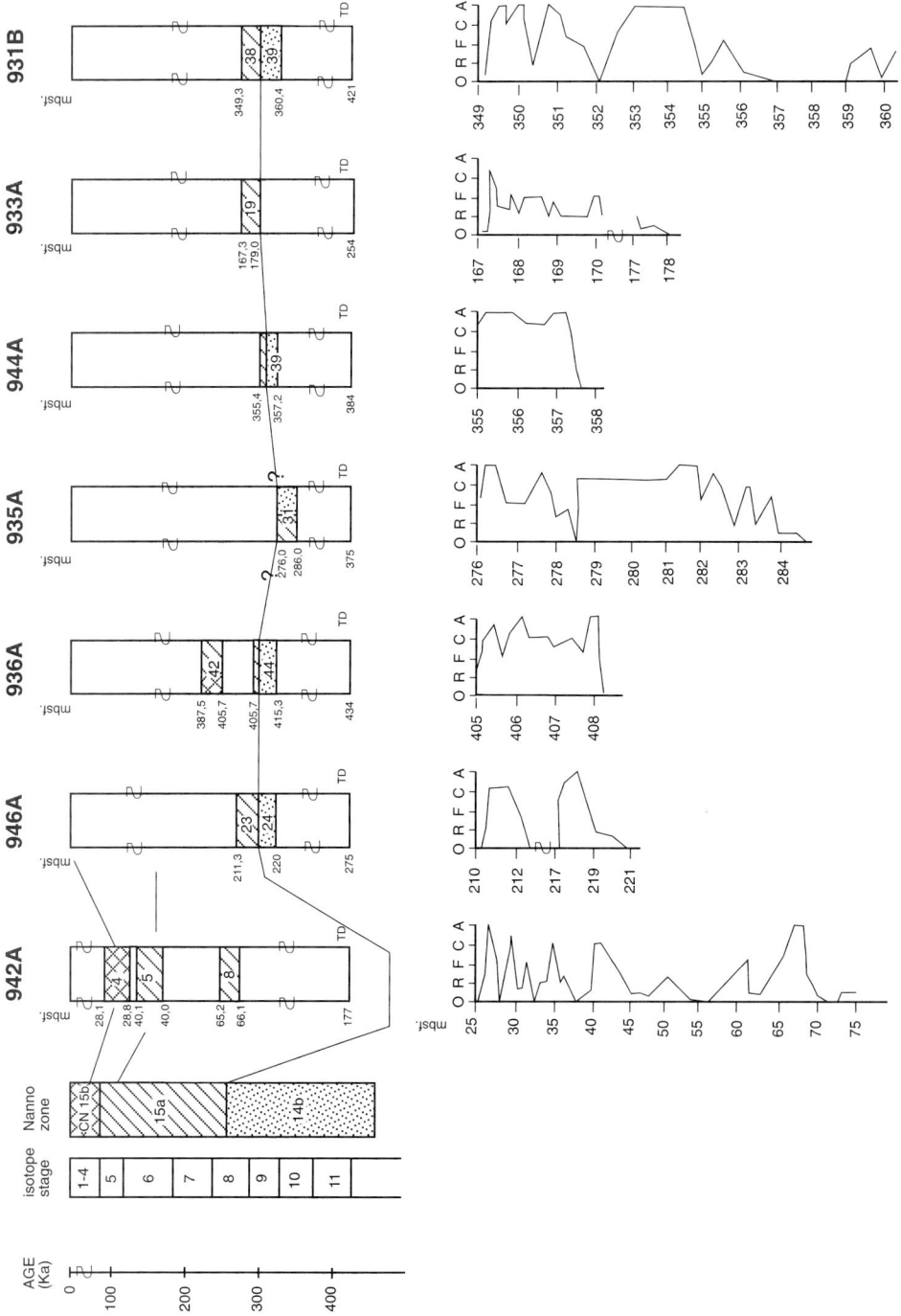

Fig. 7. Tentative correlation of the interglacial deep carbonate units of the Amazon Fan sections based on calcareous nanofossils. The abundance of calcareous nanofossils is listed as A = Abundant, C = Common, F = Few, R = Rare, O = barren. TD = total depth of drill hole. Numbers in the schematic holes refer to core numbers and the depth intervals (mbsf) to the carbonate units described in Flood *et al.* (1995).

a sandy unit and a small MTD between the Unit R MTD and the interglacial deposit. The top 50 cm of the interglacial deposit (Core 936A-44X), just below the debris flow, furnishes a rather poorly preserved assemblage which represents zone CN15a, while the remaining 3.7 m are well preserved and represent CN14b. The nanofossil assemblages in Core 936A-42X (Subunit VD) are of interglacial origin but heavy reworking is indicated by the occurrence of older assemblages (CN14b) above younger (CN15a). This clearly suggests that this interglacial section is not *in situ* and this raises the question of how much of the interglacial material in the Amazon Fan is actually found *in situ*.. At Site 944 there is a sequence of levee muds between the Unit R MTD and the underlying interglacial deposit, where the top 8 cm represents zone CN15a and the remaining 1.92 m CN14b (Fig. 7).

Clasts and blocks within the Unit R MTD at Sites 936 and 944 contain nanofossil assemblages from both CN15a and CN14b. Because of the lack of a direct contact between the interglacial deposits and the Unit R MTD it is difficult to date the base of the MTD. The best estimate is that it is between the age of the youngest material incorporated into the Unit R MTD and the underlying interglacial unit. The dating of the tops of both deep MTDs by foraminifera is supported by the nanofossil stratigraphy, which indicates that in general the sediment above the MTDs is from nanofossil zone CN15b, i.e. younger than 85 ka.

Foraminifera

The 'deep carbonate' units recovered below both the Unit R and the Deep Eastern MTDs had clearly interglacial-type planktonic foraminiferal assemblages. Many of the samples examined contained the *menardii* complex as well as *Globorotalia hexagonus* and *Globorotalia tumida flexuosa* (Figs 8 and 9), both of which occur usually only during

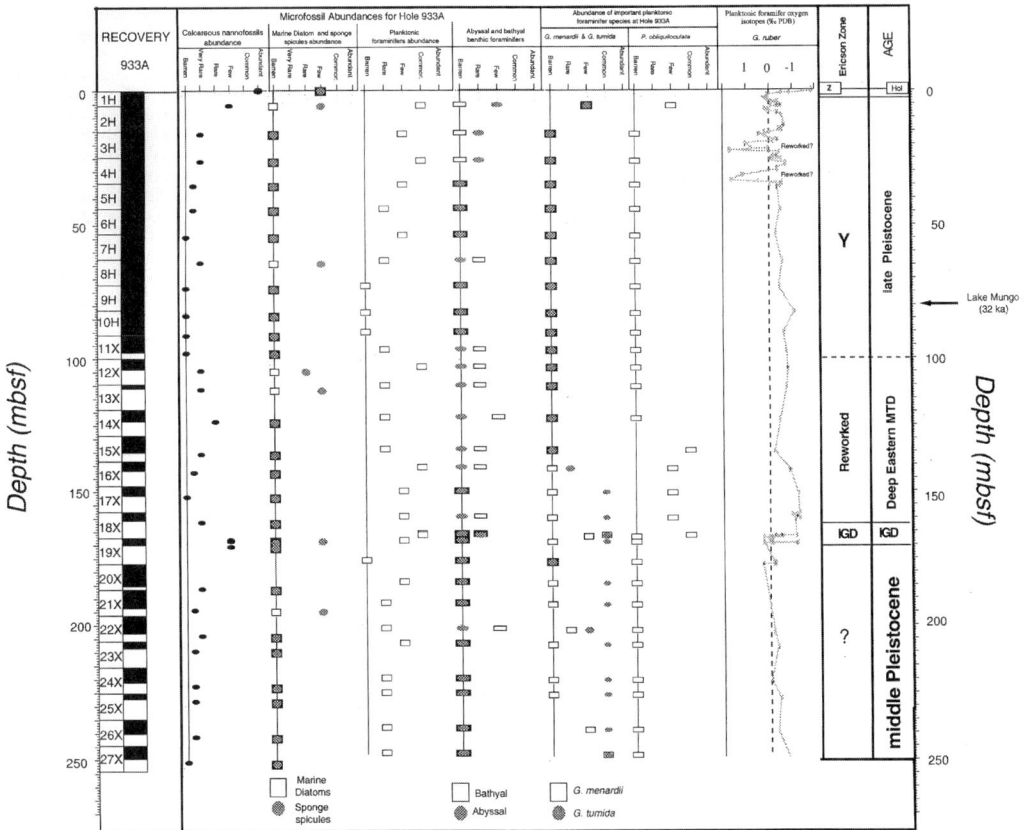

Fig. 8. Hole 933A: biostratigraphy and oxygen isotope record of *G. ruber.* IDG = interglacial deposit (modified and updated from Flood *et al.* 1995).

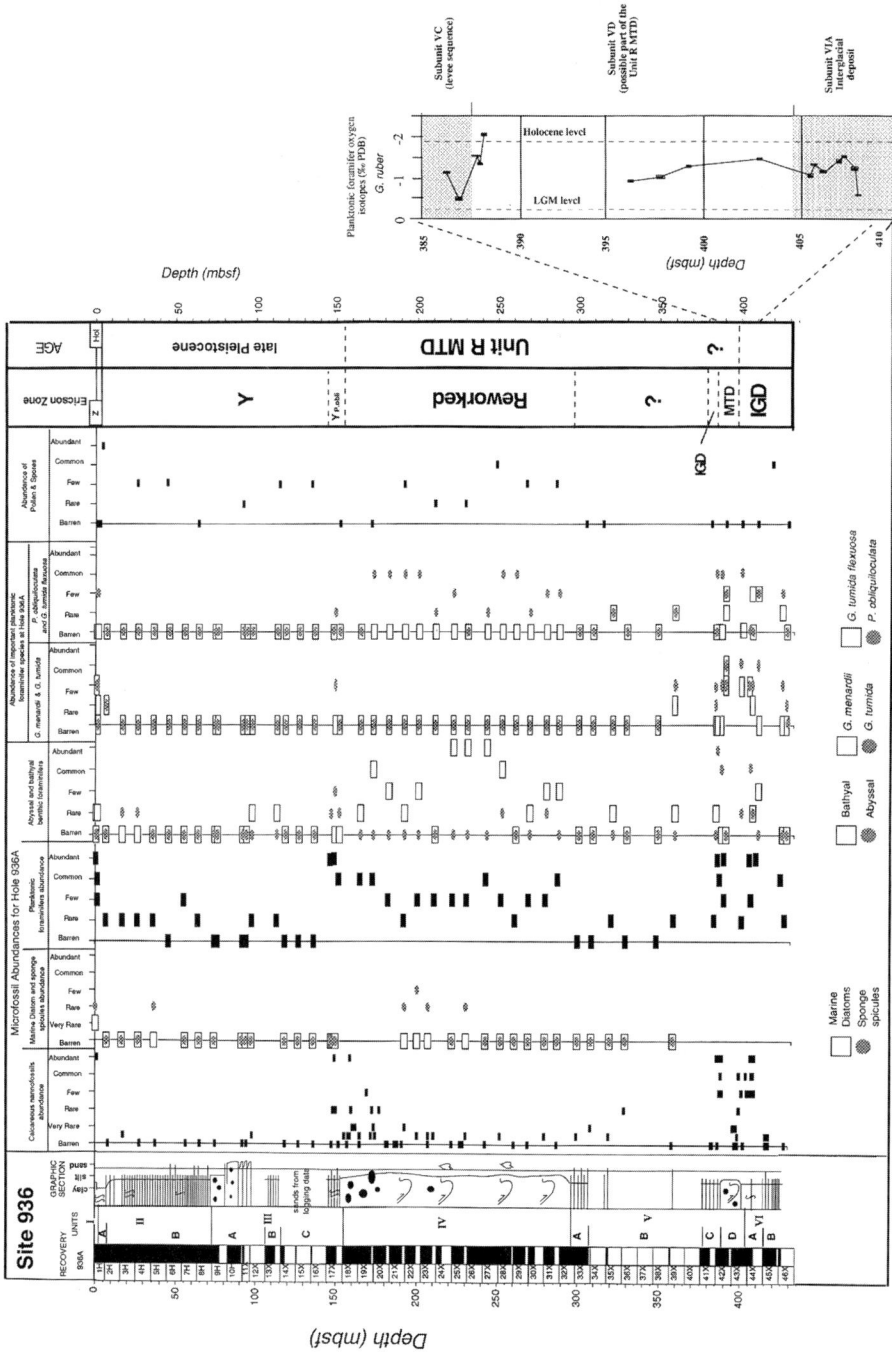

Fig. 9. Site 936: biostratigraphy and oxygen isotope record of *G. ruber*. IDG = interglacial deposit, MTD = mass-transport deposit (modified and updated from Flood *et al.* 1995).

peak interglacial conditions. Despite *G. tumida flexuosa* abundance supposedly peaking in Oxygen Isotope Stage 5e (see review in Flood *et al.* 1995), it was found that the disturbed nature of the 'deep carbonates' prevented any true quantitative estimate of the amount of *G. tumida flexuosa* within any one deposit. The occurrence of *G. tumida flexuosa* and *G. ruber* (pink) dates the deep carbonates to younger than 430 ka (Flood *et al.* 1995), supporting the nanofossil evidence.

The Deep Eastern MTD and the planktonic foraminiferal assemblages of Sites 931 and 933 have similar patterns down the sedimentary section (Fig. 5). The planktonic foraminiferal assemblages also indicate that the top of the debris flow contains sediment that was originally formed during a peak glacial period, as both *P. obliquiloculata* and the *menardii* complex are absent (Fig. 8). In the lower half of the Deep Eastern MTD *P. obliquiloculata* is abundant, while the *menardii* complex is still absent, suggesting a shift to at least a sediment contribution from material of early glacial age. This is because *P. obliquiloculata* is indicative of interglacial and early glacial sediments, while the *menardii* complex is indicative of full interglacial conditions (Ericson & Wollin 1956, 1968; Ericson *et al.* 1961). Near the base of the Deep Eastern MTD, *G. tumida* appears (Fig. 8), suggesting either a separate distal interglacial sediment source or reworking of the underlying interglacial material up into the MTD. This biostratigraphic evidence supports the sedimentary evidence that the MTD could indeed be multiple events or block slides (Piper *et al.* in press).

In contrast to the Deep Eastern MTD, the Unit R MTD has different foraminiferal assemblages at Site 936 compared with Sites 935 and 944 (Fig. 6). Another difference between Unit R and the Deep Eastern MTD is that at each site the planktonic foraminiferal assemblage in Unit R is consistent throughout the sequence (compare Figs 5 and 8 with 6 and 9). At Site 935 and 944 Unit R contains the *menardii* complex and *P. oliquiloculata*, suggesting a significant sediment input from interglacial and early glacial material. However, at Site 936 the *menardii* complex is absent, though *P. oliquiloculata* is present, suggesting an early glacial sediment source (Fig. 9). It must thus be postulated that a single MTD can contain very differently aged sediments across a distance of less than 15 km (the distance between Sites 935 and 936).

Stable isotope results

The oxygen isotope records of planktonic foraminifera in the Deep Eastern MTD and the underlying interglacial deposit at Site 933 are shown in Figs 8 and 10. The planktonic foraminiferal oxygen isotope results indicate that the interglacial deposit has 'lighter' values than the underlying glacial levee muds. However, these values do not in general reach the same level as the Holocene. This may indicate that either the deposit represents the very beginning or end of an interglacial period, or, as previously suggested, there was reworking of heavier glacial material into the deposit. We suggest both are likely: firstly, because the nanofossil evidence suggests the interglacial deposit is a curtailed sequence; secondly, the reworking hypothesis is supported by the benthic foraminiferal oxygen isotopes results, as only glacial values were recorded in the interglacial deposit (Fig. 11). It is also important to note that the oxygen isotope values of the reworked benthic foraminifera are much heavier (>0.5‰) than found during the Last Glacial Maximum (LGM). This indicates that either they originated from a more

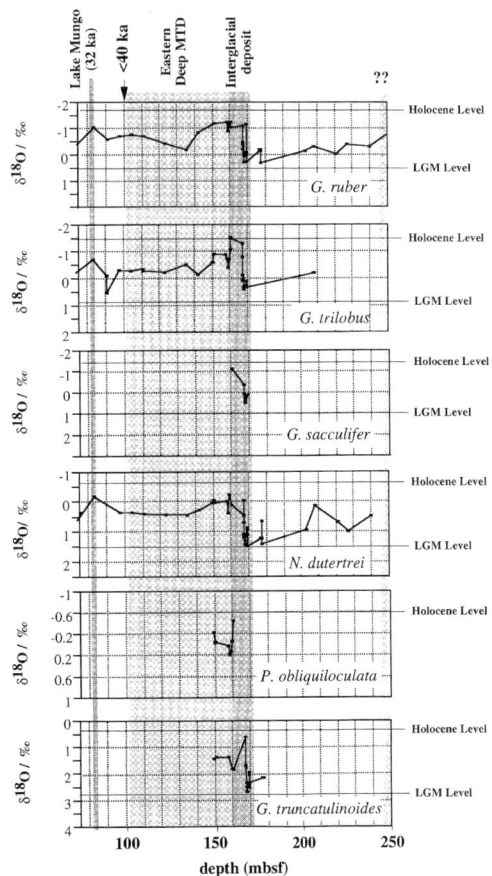

Fig. 10. Planktonic foraminiferal oxygen isotope records of Site 933.

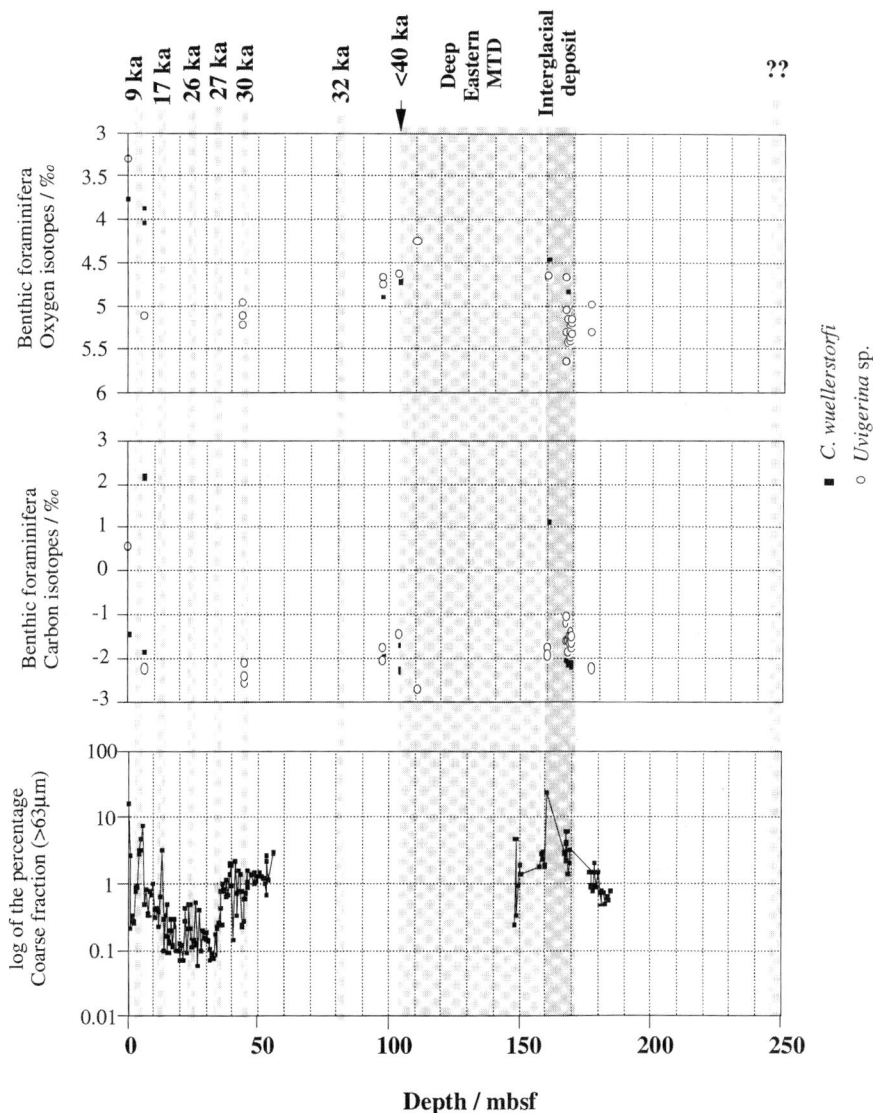

Fig. 11. Benthic foraminiferal oxygen isotope records of Site 933.

intense glacial period (e.g. Oxygen Isotope Stage 6) or they have been affected by secondary calcification. No evidence of secondary calcification was observed when the specimens were picked, but many were abraded and of poor quality.

The sparsity of planktonic foraminifera in the Deep Eastern MTD means that it was only possible to obtain isotope records of *G. ruber, G. trilobus* and *N. dutertrei*. All three records indicate that the foraminifera in the MTD have oxygen isotope values halfway between the Holocene and LGM values (Maslin 1998). This means either that the sediment from the Deep Eastern MTD was mainly

formed during early Oxygen Isotope Stage 3 (or any other early glacial period) when sea level was still relatively high, or that there was a lot of mixing between interglacial and glacial material within the MTD. The abyssal benthic foraminifera are absent from the Deep Eastern MTD, except for the very top, where their values were heavier than for the LGM. These data may again be interpreted as reworking of older material or secondary calcification of the tests.

The sedimentation at Site 936 is more complicated than at Site 933, because of poor core recovery and the apparent levee mud sequences between the Unit R MTD and the underlying interglacial deposits. In the section covered by Cores 936A-41X to 45X an interval of levee muds was recovered overlying a short MTD which in turn overlies an interglacial deposit on top of more levee muds (Flood *et al.* 1995). What is uncertain is whether the small MTD is part of the Unit R MTD, despite there being a sequence of levee muds in between. The planktonic foraminiferal oxygen isotopes suggest that only at the top of Subunit VD (388 metres below sea floor (mbsf)) are there foraminifera which were formed during interglacial conditions (Fig. 12). This suggests that the foraminifera in the interglacial deposit are either reworked or that the interglacial deposit was formed during a period when sea level was not at its maximum height. Reworking is supported by nanofossil evidence. Moreover the planktonic for-aminifera within the Subunit VD MTD have values very similar to that of the underlying interglacial deposit (Fig. 12), indicating reworking of that material into the MTD. The benthic foraminiferal oxygen isotope values in all three deposits (Subunits VC, VD and VIA) are very similar, indicating values similar to that found during the LGM (Fig. 13). This indicates a significant amount of reworking in all three subunits. The only

exception is one *C. wuellerstorfi* result within the Subunit VD MTD which has an interglacial value.

The oxygen isotope record of *G. sacculifer* in the main Unit R MTD (Showers *et al.* 1995) again indicates that either the sediment was formed mainly during early Oxygen Isotope Stage 3 (or any other early glacial period) when sea level was still relatively high or that there is a substantial mixing between interglacial and glacial material within the Unit R MTD.

Discussion

Problems of dating the interglacial deposits

The age estimation of the MTDs is based on the fossil-bearing hemipelagic sediments found above and below them. The Deep Eastern MTD rests directly on interglacial deposits which are here considered autochthonous (*in situ*), whereas the Unit R MTD is separated from the interglacial deposits by up to 75 m of sediment (Flood *et al.* 1995). Thin carbonate sequences and clasts are, however, also found within sections considered to be MTDs (e.g. within Unit VD in Hole 936A). This therefore suggests that some or all of the other deeper interglacial deposits drilled in the Amazon Fan Complex could be allochthonous, i.e. that the up to 8 m thick interglacial deposits in the deeper parts of the sites have been slumped as one undisturbed unit and therefore do not show any internal distortion. No sites penetrate more than 470 mbsf and the deeper part of the Amazon Fan Complex thus remains unexplored at present. Even though some nanofossil datums occur randomly at the top and bottom of the interglacials, and the benthic isotope data show a mixture of interglacial and glacial values within the interglacial deposits, all the sedimentological evidence suggests that the interglacial deposits are relatively undisturbed and in place (Flood *et al.* 1995; Piper *et al.* in press). This is supported by the fact that the boundary between the interglacial deposits and the under-lying glacial levee muds is very similar in characteristics to the Holocene–last glacial boundary found in the uppermost part of each of the sites. Therefore, until it can be proved otherwise, we consider the deep carbonates as being *in situ*.

Another problem is whether all the sediments within the 'deep carbonate' are pelagic. Assuming that an average interglacial period in the late Quaternary lasts about 40 ka and that open-ocean sedimentation rates in the less productive tropical western Atlantic range from 1 to 5 cm ka^{-1}, then the interglacial deposits found in the Amazon Fan Complex should be at maximum 2 m thick. At Sites 933, 935 and 936 the deep interglacial sediments recovered below the MTD are between 3 and 8 m

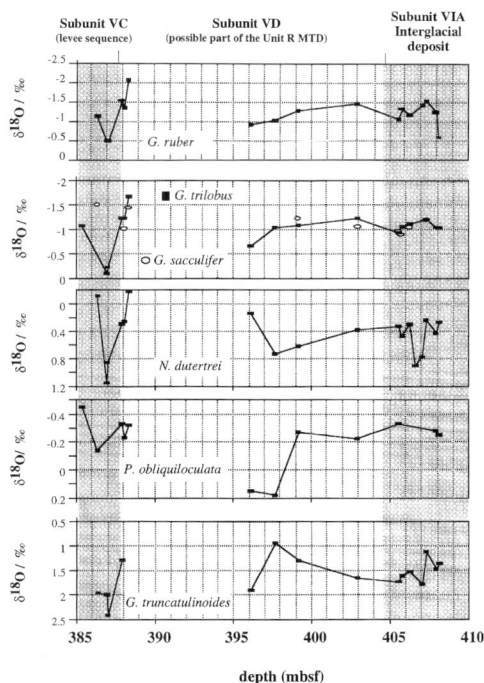

Fig. 12. Planktonic foraminiferal oxygen isotope records of Site 936.

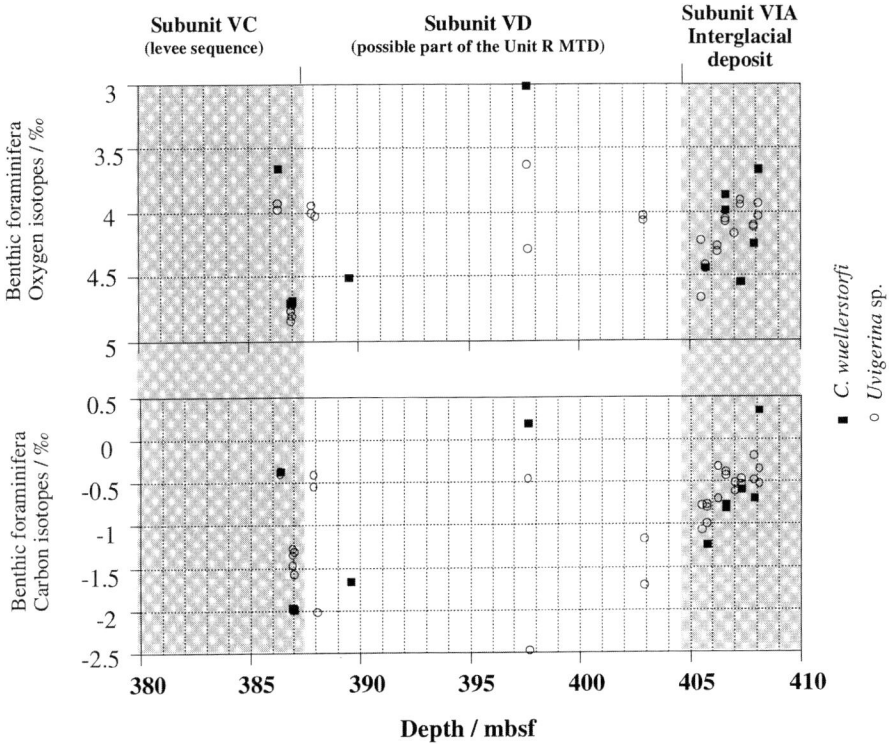

Fig. 13. Benthic foraminiferal oxygen isotope records of Site 936.

(Flood *et al.* 1995). This suggests that a considerable amount of the 'deep carbonates' is reworked material. We suggest that the reworked material in the deep carbonates is probably in the form of fine turbidites, analogous with what has been documented for the last interglacial section (Oxygen Isotope Stage 5) recovered at Site 942. In Site 942 well over half the interglacial sediments were shown to be fine turbidites (Showers *et al.* 1995). The possible presence of turbidites in the deep interglacial deposits of the Amazon Fan may explain the highly variable abundances of nanofossils found in each of the deposits (Fig. 7). This also compounds the problems of dating these deposits as not only is there reworking from the MTDs above and the glacial sediments below, but there is also reworked material in the form of turbidites. This significant reworked component is confirmed by the benthic foraminiferal oxygen isotopes at Sites 933 and 936 (Figs 11 and 13). In both cases there are benthic foraminifera with peak glacial oxygen isotope values in the heart of the interglacial deposit. The isotopic evidence corroborates the nanofossil data that there are sections with low abundances, poorly preserved

specimens, and a high abundance of terrigenous material within each of the deep carbonates, consistent with the presence of turbidites. The very short deep carbonate sequence (0.2 m) recovered at Site 931 (Fig. 7) apparently represents the very bottom of an interglacial deposit and the boundary with the underlying glacial sequence. This situation may illustrate the mechanism by which the MTDs are sliding on a weaker and less consolidated substratum of the carbonate-rich interglacial deposits. This mechanism would incorporate interglacial material in the lower part of the MTDs, as seen in Sites 933, 935, 936 and 944 (Flood *et al.* 1995).

The Calcium Carbonate Compensation Depth (CCD) in the Atlantic Ocean is today at a depth of approximately 4.5 km (Berger 1978) and will therefore not have any profound impact on even the deeper sites. The shoaling of the CCD to approximately 3.5 km during glacial periods may have influenced the deeper sites on the fan and thus partly accounts for the low carbonate content in the glacial sections. However, we suggest that the overriding factor is dilution by the huge sedimentation rates caused by the influx of terrestrial material, i.e. up to three orders of magnitude compared to interglacial conditions (5 cm ka^{-1} compared to 50 m ka^{-1}).

Inferred Oxygen Isotope Stages of the interglacial deposits

The nature of the Amazon deep-sea fan complex does not allow for a traditional subdivision of the sediments based on either biostratigraphy, sequence stratigraphy or isotope stratigraphy due to the very disrupted nature of the sedimentary record. The combination of biostratigraphic data, nanofossil assemblage data, isotope records, sedimentological data and seismic data is used in this study to establish the geochronology of the fan, which we tentatively propose within an isotope stage framework.

The predominance of the CN15a assemblage in Sites 931 and 933 with CN14b material at the base suggests that this interglacial deposit was formed during either Oxygen Isotope Stage 5 or 7. As the CN15a–14b boundary is within Oxygen Isotope Stage 8, we suggest, because of the CN14b material in the base of both deposits, that the interglacial deposit is from Stage 7 (Fig. 14). A comparison of the nanofossil assemblages from the Sites 931–933 deep carbonate and last interglacial material at Site 942 indicates distinct differences in assemblage composition, supporting the inference that the Sites 931–933 deep carbonate was not formed during Stage 5.

Sites 935, 936 and 944 all have deep carbonate units below the Unit R MTD; however, unlike Sites 931 and 933 there is not a direct contact between the MTD and the underlying interglacial deposits. In the case of Site 936 the top 50 cm of the interglacial deposit represents CN15a, while the remaining 3.7 m represents CN14b. In contrast in Site 944 the top 8 cm represents CN15a and the remaining 1.92 m CN14b (Fig. 7). We infer that the predominance of the CN14b nanofossil assemblage dates the interglacial deposit as either Oxygen Isotope Stage 9 or 11. At Site 935 within the interglacial deposit there is only evidence for CN14b (Fig. 7), therefore it must have been formed during either Oxygen Isotope Stage 9 or 11.

The seismic evidence suggests that the interglacial deposit at Site 935 is below that in Sites 936 and 944, and we postulate that the interglacial deposit at Sites 936 and 944 may have been formed during Oxygen Isotope Stage 9, while that at Site 935 may have been formed during Stage 11. This is somewhat supported by the fact that significant amounts of the younger CN15a material are found in interglacial deposits at Sites 936 and 944, while none is found in the proposed older interglacial deposit at Site 935.

Clasts and blocks within the Unit R MTD at Sites 936 and 944 contain nanofossil assemblages from both CN15a (distinctive Stage 7/8 assemblage) and CN14b. Because of the lack of a direct contact between the interglacial deposits and the Unit R MTD it is difficult to date the base of the MTD. The best estimate we have is that it must lie between the youngest material incorporated into the Unit R MTD and the underlying interglacial unit. Therefore we suggest that at Sites 936 and 944 the base of the Unit R MTD is dated between Oxygen Isotope Stage 7 and 9 (see Fig. 14).

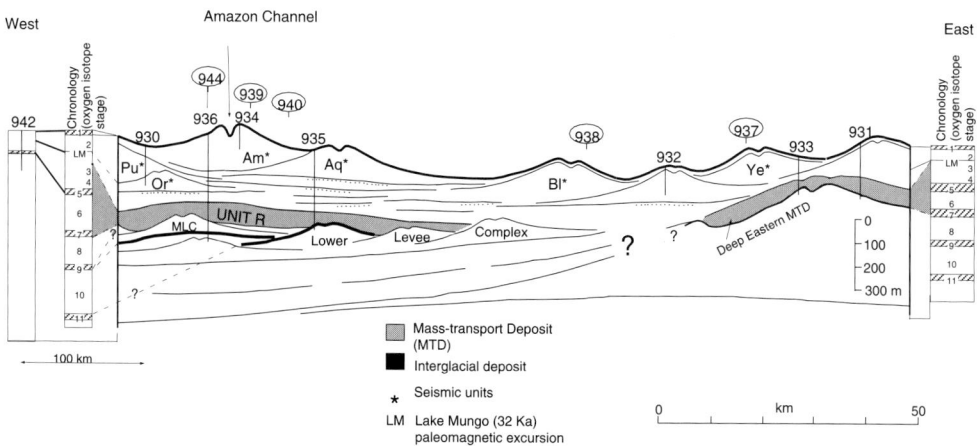

Fig. 14. Theoretical model of the overall stratigraphy of the mass-transport units (MTDs) and interglacial deep carbonates drilled on the Amazon Fan during ODP Leg 155. Initial unit correlations are based on seismic stratigraphy of Flood *et al.* (1995).

Timing of the MTDs

Knowledge of the age of the sediment directly above the MTDs is essential to the understanding of causation, as this indicates the latest time that the MTDs could have been formed. The Lake Mungo palaeomagnetic event is detected above the Deep Eastern MTD in both Sites 931 and 933 (Cisowski & Hall in press), thus the MTD is older than 32 ka. The absence of both the *menardii* complex and *P. obliquiloculata* in the foraminiferal assemblages of the intervening sediment between the Lake Mungo palaeomagnetic event and the top of the MTD, suggests that the MTD may be younger than 40 ka. Assuming, therefore, that the foraminiferal assemblages are complete, despite the low abundances, then the Deep Eastern MTD was last active between 30 and 40 ka. The sedimentation rate between the palaeomagnetic intensity variation datum of 30 ka and the Lake Mungo excursion at 32 ka (Cisowski & Hall in press; Maslin 1998) in Site 933 is approximately 15 m ka^{-1} (Fig. 5). If it is assumed that the sedimentation rate below the Lake Mungo event is similar to that above, then the top of the Deep Eastern MTD is no older than 33 ka.

The Unit R MTD is harder to date as there are significant high-amplitude reflection-parallel (HARP) units above the MTD (Flood *et al.* 1995), consequently the Lake Mungo event was not detected at any of the three sites. However, at Sites 935 and 936 pockets of hemipelagic sediment were found within the HARP unit. These hemipelagic sequences contained *P. oliquiloculata* but not the *menardii* complex, thus dating these sections between 40 and 85 ka. Only in the case of Site 936 is there hemipelagic mud in contact with the MTD, inferring that the Unit R MTD was last active between 40 and 85 ka. At Site 935 there is a direct contact between the levee muds and the MTD. If it is assumed that this is not an erosional contact, and the average late glacial sedimentation rate for Site 935 is 4 m ka^{-1}, then the top of the Unit R MTD is about 45 ka.

The dating of the tops of both deep MTDs is confirmed by the nanofossil stratigraphy, which indicates that in general the sediment above the MTDs is from nanofossil zone CN15b, i.e. younger than 85 ka. The key fact, however, is that the two deep MTDs occurred at different times, i.e. pre- and post- 40 ka.

It has been shown that the Eastern and Western near-surface 'Debris Flows' are capped by Holocene sediments (Damuth *et al.* 1988; Showers & Bevis 1988). Detailed sampling of the Holocene sediments above the Western Debris Flow at Site 941 indicates the presence of *P. obliquiloculata, G. tumida* and *G. menardii* (Flood *et al.* 1995; Fig. 15).

As there is no interglacial material found in the top 25 m of the underlying MTD then it seems that the Western Debris Flow was last active during the early Holocene. The presence of *P. obliquiloculata* dates the assemblage as Holocene, while *G. tumida* dates the sediment younger than 7.5 ^{14}C ka, while *G. menardii* dates the top of the MTD as younger than 6.3 ^{14}C ka. It has been shown at Site 934 (Flood *et al.* 1995) that mid-Holocene material can be reworked by coring disturbances down the whole length of the first core. Moreover, the sedimentological evidence at Site 941 suggests that the Holocene section contains a number of turbidites (Flood *et al.* 1995). Therefore, a conservative estimate of the age on top of the Western Debris Flow is between 13 and 8 calendar ka. Seismic stratigraphy suggests that the near-surface Western and Eastern Debris Flows are coeval (Damuth *et al.* 1988; Flood *et al.* 1995), and thus it appears that they were last active during Termination I.

Essential to the discussion of cause, is whether the MTDs represent single or multiple events. Moreover, if they represent multiple events are these co-occurring or widely separated in time? If the MTDs do represent multiple events occurring over a long period of time then there could be various causes for each of the successive events. Studies of the sedimentological characteristics (Flood *et al.* 1995), the consolidation properties and the biostratigraphy (Figs 8 and 9) of the MTDs indicate that there are distinct blocks or units within each of the debris flows. There is, however, still an ongoing debate as to whether these blocks are within the same event or represent separate mass-transport events. For the purposes of discussion we assume that the MTDs could be caused by either single or multiple events, but that if they are multiple events they occur within a short enough time period to be assigned a single cause.

Sources of the material within the MTDs

Two hypotheses have been suggested for the sources of the MTDs (Piper *et al.* in press). Some seismic data suggest that the MTDs represent channel-levee sediment that has been deformed essentially *in situ*. In contrast, based on benthic foraminifera, Vilela & Maslin (in press) suggest that a significant amount of the MTD originated in the middle to upper bathyal depth with some shelf contribution.

Supporting evidence for the interpretation of in-place formation of the MTD comes from the incorporation of carbonate-rich interglacial sediments in the base of some of the MTDs (Piper *et al.* in press), and the lithological similarities between some of the MTD blocks and the fine-

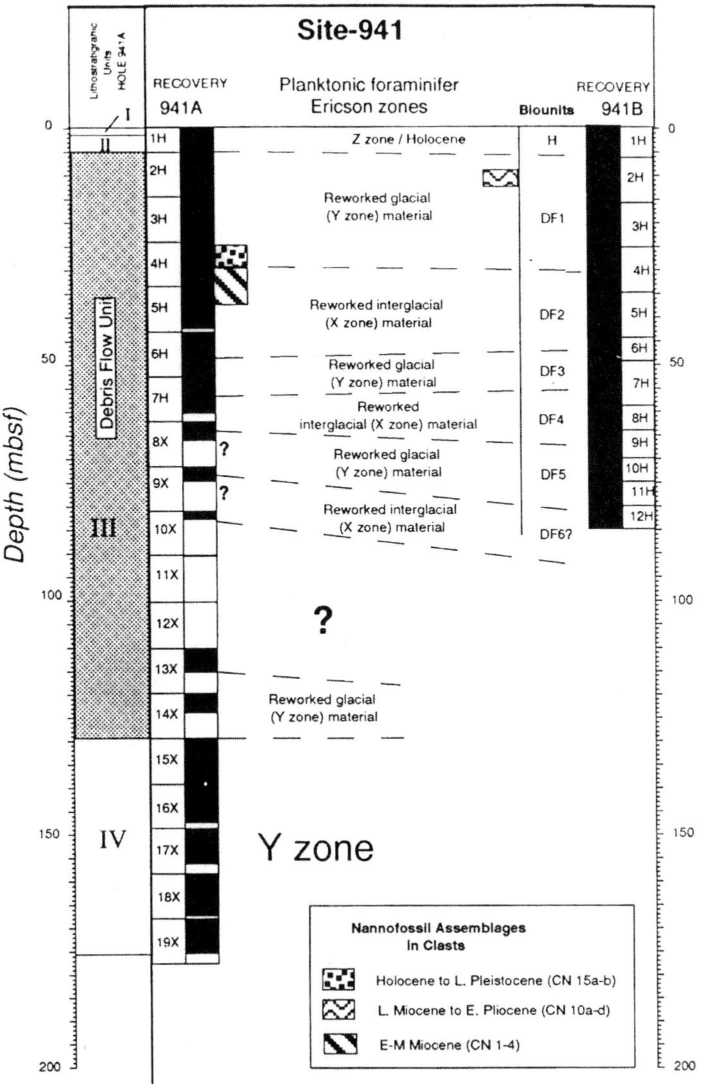

Fig. 15. Biostratigraphy of Site 941, indicating the alternating sediment type within the near-surface Western Debris Flow.

grained levee deposits, including a distinctive clay mineral assemblage reported at Site 931 (Piper *et al.* in press). However, these observations are not inconsistent with a significant distal source of the MTDs. We have thus suggested that the carbonate-rich interglacial deposits have acted as the main slip plane for the movement of the MTDs on the fan, and that incorporation of interglacial sediment within the base of the MTDs should therefore be expected. The transport of such large amounts of

sediment from a distal area would also be expected to include blocks torn off the adjacent levees. A more radical view that we have also proposed is that the interglacial deposits themselves may not be *in situ* and that they may also have been transported to the deep-sea fan sites either coevally with the MTDs or earlier. However, until this has been further investigated we adhere to the general idea that the main bodies of interglacial deposits are found *in situ*.

Speculation as to the causes of the MTDs

One of the most intriguing questions to come out of the work on the Amazon Fan is what caused the MTDs. There have been extensive investigations into large debris flows around the world, such as the Saharan and Canary Debris Flows and the vast turbiditic infill areas in the Madeira Abyssal Plain on the NW African margin (Roberts & Cramp 1996). These have produced a large range of suggested causes including earthquakes (e.g. Kastens 1984), high sedimentation rates due to proximity to fluvial inputs (e.g. Prior & Suhayda 1979), sea-level fluctuations (e.g. Weaver &

Kuijpers 1983), progressive failure (e.g. Moore 1961), creep (e.g. Moore et al. 1994), and gas hydrate destabilization (e.g. Kayen & Lee 1991). We have used these theories to speculate from the currently available evidence which of them is most applicable to the Amazon MTDs.

The first line of evidence is the good correlation between the timing of the Amazon Fan MTDs and changes in sea level (see Fig. 16). The deep MTDs seem to correlate with periods of rapidly decreasing sea level. We speculate that rapidly lowering sea level could have destabilized the sediments of the continental slope by: (1) funnelling more Amazon River sediment directly into the fan complex

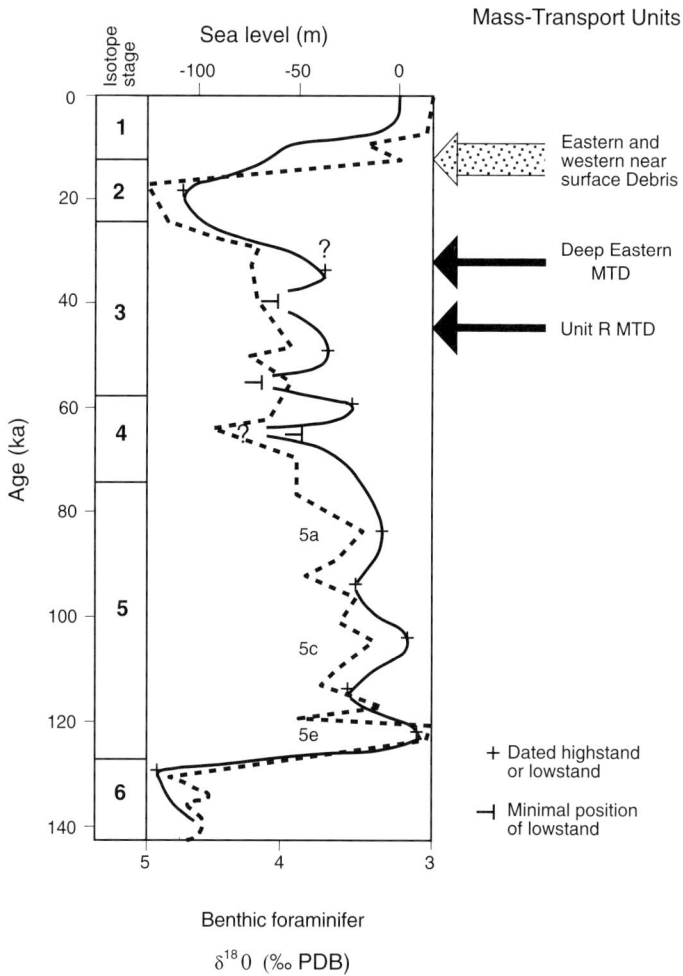

Fig. 16. Comparison of the terrestrially derived 'eustatic' variations in sea level (Flood et al. 1995) with the timing of the mass-transport deposits drilled on the Amazon Fan. Note that the Deep Eastern MTD (and Unit R MTD) occur during periods of falling sea level, while the Eastern and Western near-surface Debris Flows occur during a period of rapid rise in sea level.

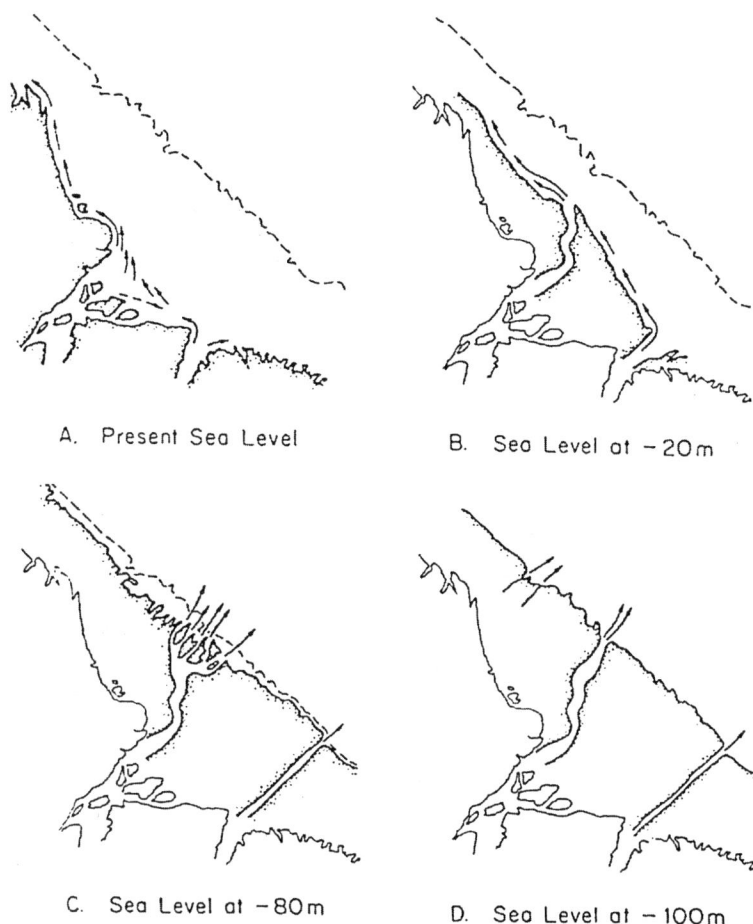

A. Present Sea Level

B. Sea Level at −20m

C. Sea Level at −80m

D. Sea Level at −100m

Fig. 17. Model of Quaternary sedimentation off the Amazon Basin during high and low sea-level stands. Arrows indicate the main routes of sediment dispersal. Dashed line refers to the present 100 m isobath, which approximates the shelf edge (after Milliman *et al.* 1975).

instead of onto the continental shelf; (2) exposing and undercutting sediment piles on the continental shelf and slope, built up during the previous interglacial and early glacial period; and (3) causing the decompression and sublimation of gas hydrates reported from the shelf area (Kvenvolden & Barnard 1993; Kayen & Lee 1991; Kvenvolden 1988, 1996). Evidence was found for the presence of gas hydrates in the Amazon Fan sediments (Flood *et al.* 1995).

The sea-level lowering hypothesis is supported by the fact that both of the deep MTDs contain continental shelf and slope benthic foraminifera (Vilela & Maslin in press). This suggests that the initial failure of the sediment column must have begun at a relatively shallow water depth. Moreover, the timing of the two MTDs at 33 ka and

45 ka corresponds to periods of a rapidly lowering sea level from approximately −30 m to at least −70 m (see Fig. 16). When compared to the Milliman *et al.* (1975) reconstruction of the drainage pattern of the Amazon River at different sea levels, the magnitude of the drops in sea level at 45 and 33 ka could have switched the drainage pattern from Fig. 17B to Fig. 17C, thus (1) rapidly eroding the continental shelf sediments, and (2) depositing sediment directly into the fan complex from the river and the erosion, so destabilizing the sediments already there.

The sea level/climate-driven hypothesis is further supported by the formation of the near-surface Eastern and Western Debris Flows between 12 and 14 ka, i.e. during the beginning of deglaciation. We speculate that there could be three main causes which may have acted independently or concurrently. (1) The rising sea level could have undercut the exposed sediment on the continental shelf causing it to fail. (2) Sediment failure may have been exacerbated by overburdened shelf sediments resulting from an increase in the Amazon River discharge caused by the deglaciation of the Andes. Evidence for this Amazon River discharge event may come from bulk organic matter and planktonic foraminiferal calcite carbon isotopes from Leg 155 which show a significant increase in terrestrial carbon in the surface waters between 12 and 14 ka (a shift of greater than 2‰; Maslin *et al.* in press). Also the first major warming of the Andes ice sheet occurred at about 14 ka according to the Huascaran ice cores (Thompson *et al.* 1995). (3) Release of gas hydrates could have been caused by their destabilization resulting from the lowering of sea level throughout the glacial, then their release due to the rapid changes in pressure because of changes in sea level and sediment distribution on the continental slope (Nisbet 1990; Nisbet pers. comm. 1996). Alternatively the documented increase in bottom and intermediate water temperatures of over 4°C during the last deglaciation may have caused the release of gas hydrates. It is interesting to note that coeval with the MTD formation there is a major rise in CH_4 recorded in the ice cores (Chappellaz *et al.* 1990). We suggest that if gas hydrates were a cause of the Amazon failure events then this indicates that degassing of the continental margins may have made a significant contribution to this rise in atmospheric CH_4 and thus the whole feedback process of deglaciation.

The ultimate cause of the Amazon Fan MTDs is still open for debate. Roberts & Cramp (1996) concluded that the Canary Debris Flow off the coast of NW Africa was caused by climatically driven changes in sea level and creep. In contrast, though we believe that the Amazon Fan MTDs were similarly caused by sea-level changes, we suggest that gas hydrate release and/or fluvial outwashing of the Amazon Basin were the

secondary triggering mechanisms of the MTDs. Further research on the Amazon Fan sediments and other deep-sea fans will hopefully be able to prove or disprove the hypothesis that climatically driven sea-level changes control the formation of MTDs on coastal margins.

Summary

1. All sediments recovered during Leg 155, that were not included as clasts in a mass-transport deposit, are younger than 460 ka.
2. Interglacial deposits recovered from beneath the major mass-transport deposits cover the interval of nanofossil zones CN14b to CN15a. Inferences from seismic and reworking characteristics of these deposits suggest that the interglacial deposits may represent Oxygen Isotope Stages 7, 9 and 11.
3. Use of biostratigraphy, magnetostratigraphy and sedimentation rate constraint to date the top of the mass-transport deposits suggests that (a) the Deep Eastern MTD was last active at 33 ka, and (b) the Unit R MTD was last active at 45 ka. Biostratigraphy clearly demonstrates that the two deep MTDs were active at different times, pre- and post- 40 ka.
4. Using biostratigraphy and seismic correlations it is suggested that the Eastern and Western near-surficial Debris Flows were last active during the transitional period between the last glacial and the Holocene.
5. It is suggested that climate-induced changes in sea level caused Amazon Fan mass-transport deposits. Secondary triggering mechanisms could be gas hydrate release, increase in fluvial sediment discharge and sediment undercutting.

We would like to especially thank J. Giradeau for his support and advice, and also the Shipboard Party. We would also like to acknowledge the funding provided by the Danish National Science Research Council, the DFG (Deutsche Forschungsgemeinschaft) and the Department of Geography, UCL. We would like to thank the reviewers A. Cramp and R. Kidd.

References

BERGER, W. H. 1978. Sedimentation of deep-sea carbonate; maps and models of variations and fluctuations. *Journal of Foraminiferal Research*, **8**(4), 286–302.

BUKRY, D. 1973. Low-latitude coccolith biostratigraphic zonation. *In*: EDGAR, N. T., SAUNDERS, J. B. *ET AL.* (eds) *Initial Reports of DSDP*, **15**. US Govt Printing Office, Washington, 685–703.

—— 1978. Biostratigraphy of Cenozoic marine sediment by calcareous nanofossils. *Micropaleontology*, **24**, 44–60.

CASTRO, J. C., MIURA, K. & BRAGA, J. A. E. 1978.

Stratigraphic and structural framework of the Foz do Amazonas Basin. *10th Annual Offshore Technology Conference*, Houston, **3**, 1843–1847.

CHAPPELLAZ, J., BAMOLA, J., RAYNAUD, J., KOROTBEVICH, D. & LORIUS, C. 1990. Ice core record of atmospheric methane over the past 160,000 years. *Nature*, **345**, 127–131.

CISOWSKI, S. & HALL, F. in press. An examination of the paleointensity record and geomagnetic excursions recorded in Leg 155 cores. *Proceedings of ODP, Scientific Reports*, **155**. College Station, TX (Ocean Drilling Program).

CURRY, W. B., SHACKLETON, N. J., RICHTER, C. *ET AL.* 1995. *Proceedings of ODP, Scientific Reports*, **154**. College Station, TX (Ocean Drilling Program).

DAMUTH, J. E. & FAIRBRIDGE, R. W. 1970. Equatorial Atlantic deep-sea arkosic sands and ice-age aridity in tropical South America. *Geological Society of America Bulletin*, **81**, 189–206.

—— & KUMAR, N. 1975. Amazon cone: morphology, sediments, age and growth pattern. *Geological Society of America Bulletin*, **86**, 863–878.

——, FLOOD, R. D., KOWSMANN, R. O., BELDERSON, R. H. & GORINI, M. A. 1988. Anatomy and growth pattern of Amazon deep-sea fan as revealed by long- range side-scan sonar (GLORIA) and high-resolution seismic studies. *AAPG Bulletin*, **72**, 885–911.

ERICSON, D. B. & WOLLIN, G. 1956. Correlation of six cores from the equatorial Atlantic and the Caribbean. *Deep-Sea Research*, **3**, 104–125.

—— & —— 1968. Pleistocene climates and chronology in deep-sea sediments. *Science*, **162**, 1227–1234.

——, EWING, M., WOLLIN, G. & HEEZEN, B. C. 1961. Atlantic deep-sea sediment cores. *Geological Society of America Bulletin*, 72:193-286.

FLOOD, R. D., PIPER, D. J. W., KLAUS, A. *ET AL.* 1995. *Proceedings of ODP, Initial reports,* 155. College Station, TX (Ocean Drilling Program).

KASTENS, K. A. 1984. Earthquakes as a triggering mechanism for debris flows and turbidites on the Calabrian Ridge. *Marine Geology*, **55**, 13–33.

KAYEN, R. E. & LEE, H. J. 1991. Pleistocene slope instability of gas hydrate-laden sediments on the Beaufort Sea margin. *Marine Geotechnology*, **10**(1–2), 125–142.

KUEHL, S. A., NITTROUER, C .A. & DeMASTER, D. J. 1982. Modern sediment accumulation and strata formation on the Amazon continental shelf. *Marine Geology*, **49**, 279–300.

KVENVOLDEN, K. A. 1988. Methane hydrate – a major reservoir of carbon in the shallow geosphere? *Chemical Geology*, **71**, 41–51.

—— 1996. Review paper: A review of the geochemistry of methane in natural gas hydrates. *Organic Geochemistry*, **23**, 997–1008.

—— & BARNARD, L. A. 1993. Hydrates of natural gas in continental margins. *In*: WATKINS, J. S. & DRAKE, C. L. (eds) *Studies in Continental Margin Geology.* AAPG Memoir, **34**, 631–640.

MANLEY, P. L. & FLOOD, R. D. 1988. Cyclic sediment deposition within Amazon Deep-Sea Fan. *AAPG Bulletin*, **72**, 912–925.

MASLIN, M. 1998. Equatorial western Atlantic Ocean circulation changes linked to the Heinrich events: deep-sea sediment evidence from the Amazon Fan. *This volume.*

MASLIN, M. A., ADAMS, J., THOMAS, E., FAURE, H. & HAINES-YOUNG, R. 1995. Estimating the carbon transfer between the oceans, atmosphere and the terrestrial biosphere since the last glacial maximum. *Terra Nova*, **7**, 358–366.

——, BURNS, S., ERLENKEUSER, H. & HOHNEMANN, C. in press. Stable isotope records from ODP Sites 932 and 933. *Proceedings of ODP, Scientific Reports*, **155**. College Station, TX (Ocean Drilling Program).

MILLIMANN, J., SUMMERHAYES, C. & BARRETTO, H. 1975. Quaternary sedimentation on the Amazon continental margin: A model. *Geological Society of America Bulletin*, **86**, 610–614.

MOORE, D. G. 1961. Submarine slumps. *Journal of Sedimentology Petrology*, **31**, 342–357.

MOORE, J. G., NORMARK, W. R. & HOLCOMB, R. T. 1994. Giant Hawaiian underwater landslides. *Science*, **264**, 46–47.

NISBET, E. G. 1990. The end of the ice age. *Canadian Journal of Earth Science*, **27**, 148–157.

OKADA, H. & BUKRY, D. 1980. Supplementary modification and introduction of code numbers to the low-latitude coccolith biostratigraphic zonation (Bukry, 1973; 1975). *Marine Micropaleontology*, **5**, 321–325.

PIPER, D. *ET AL.* in press. Mass transport deposits of the Amazon Fan. *Proceedings of ODP, Scientific Reports*, **155**. College Station, TX (Ocean Drilling Program).

PRIOR, D. B. & SUHAYDA, J. N. 1979. Applications of infinite slope analysis to subaqueous sediment instability, Mississippi Delta. *Engineering Geology*, **14**, 1–10.

PUJOS, A. & GIRADEAU, J. 1993. R partition des Noelaerhabdaceas (nanofossiles calcaires) dans le Quaternaire moyen et suprieur des ocans Atlantique et Pacifique. *Oceanologica Acta*, **16**, 349–362.

RAFFI, I., BACKMAN, J., RIO, D. & SHACKLETON, N. J. 1993. Plio-Pleistocene nanofossil biostratigraphy and calibration to oxygen isotope stratigraphies from Deep Sea Drilling Project Site 607 and Ocean Drilling Program Site 677. *Paleoceanography*, **8**, 387–408.

ROBERTS, J. A. & CRAMP, A. 1996. Sediment stability on the western flanks of the Canary Islands. *Marine Geology*, **134**, 13–30.

SHOWERS, W. J., GENNA, B., PRICE, P., FLOOD, R. & PIPER, D. 1995. Amazon continental margin high resolution records of western tropical Atlantic circulation over the past 130 ky. *International Conference on Paleoceanography V, Halifax, Canada, Program and Abstract Volume*, 139.

SHOWERS, W. J. & BEVIS, M. 1988. Amazon cone isotopic stratigraphy: Evidence for the source of the tropical freshwater spike. *Palaeogeography, Palaeoclimatology, Palaeoecology*, **64**, 189–199.

SOWERS, T. & BENDER, M. 1995. Climate records covering the last deglacial. *Science*, **269**, 210–214.

TAKAYAMA, T. & SATO, T. 1987. Coccolith biostratigraphy of the North Atlantic Ocean, Deep Sea Drilling Project Leg 94. *In*: RUDDIMAN, W. F., KIDD, R. B., THOMAS, E. *ET AL., Initial Reports of DSDP*, **94**(Pt. 2). US Govt Printing Office, Washington, 651–702.

THOMPSON, L. G., MOSLEY-THOMPSON, E., DAVIS, M. E. *ET AL.* 1995. Late glacial stage and Holocene tropical ice core records from Huascaran, Peru. *Science*, **269**, 46–50.

TIEDEMANN, R., SARNTHEIN, M. & SHACKLETON, N. J. 1994. Astronomic timescale for the Pliocene Atlantic ~18O and dust flux records of ODP Site 659. *Paleoceanography*, **9**, 619–638.

TRAUTH, M. 1995. *Bioturbate Signalverzerrung hockauflosender paloozeanographischer Zeitreihen*

(Bioturbational signal distortion of high-resolution paleoceanographic time-series). PhD thesis, Ber. No. **74**, Geol.-Paleaont. Inst., Univ. Kiel.

VILELA, C. & MASLIN, M. in press. Benthic foraminifera and stable isotope results from the mass-transport deposits in the Amazon Fan. *Proceedings of ODP, Scientific Reports*, **155**. College Station, TX (Ocean Drilling Program).

WEAVER, P. P. E. 1993. High resolution stratigraphy of marine Quaternary sequences. *In*: HAILWOOD, E. W. A. & KIDD, R. B. (eds) *High Resolution Stratigraphy*. Geological Society, London, Special Publications, **70**, 137–153.

—— & KUIJPERS, A. 1983. Climate control of turbidite deposition on the Maderia Abyssal Plain. *Nature*, **306**, 360–366.

Mineralogy and geochemistry of Bay of Bengal deep-sea fan sediments, ODP Leg 116: evidence for an Indian subcontinent contribution to distal fan sedimentation

STEPHEN F. CROWLEY[1], DORRIK A. V. STOW[2] & IAN W. CROUDACE[2]

[1] *Department of Earth Sciences, University of Liverpool, PO Box 147, Liverpool L69 3BX, UK*

[2] *Department of Geology, University of Southampton, Southampton SO9 5ND, UK*

Abstract: Sediments recovered during Ocean Drilling Program Leg 116 (Bay of Bengal deep-sea fan) fall into three mineralogically and geochemically distinct groups. The first of these groups (Group I), characterized by quartz–mica-rich turbidites, constitutes the largest proportion of distal fan sediments. Trace element patterns are similar to modern River Ganges suspended sediment and are consistent with a Himalayan (meta)sedimentary/granitic source. The second group of sediments (Group II) is represented by organic carbon-rich, smectite–kaolinite turbidites. Trace element data reveal significant enrichment in compatible and ferromagnesian elements consistent with a significant contribution from a basaltic crustal source. Although mixing of a basaltic source with granitic crust can account for specific geochemical relationships, mixing of these components cannot account for observed rare earth element (REE) patterns. REE data are best explained by mixing of basaltic detritus (e.g. Deccan Trap basalts of central India) with Precambrian tonalitic crust of the Indian subcontinent. A third group of carbonate-rich sediments (Group III), containing a low-latitude marine fauna, is characterized by a clastic component similar in composition to smectite–kaolinite turbidites. Although carbonate-rich sediments are superficially similar to Group II turbidites, geochemical data indicate a reduced contribution from basaltic crustal sources compared with smectite–kaolinite turbidites. A possible southern India/Sri Lankan provenance has been assigned to Group III turbidites on the basis of faunal content and geochemical composition, although insufficient information exists to substantiate this using geochemical data alone.

The distribution of lithofacies and provenance-sensitive geochemical signatures suggests that the relative contribution of Himalayan and Indian subcontinent sources to distal fan sedimentation varied with time. Controls on sediment supply include variations in uplift, weathering and erosion rates, eustatic sea-level changes, and switching of major distributary fan channels. Although a degree of correlation is observed between the occurrence of coarser-grained Himalayan-derived turbidites and relative low-stands, the relationship is patchy. Assuming that the rate of sediment supply from the Himalayas to the Bay of Bengal remained relatively constant, the most likely controls on distal fan sedimentation are thought to be related to an interplay between sea-level change and channel switching. As a consequence attempts to reconstruct major Himalayan tectonic and climatic events based on data obtained from a record of distal fan sedimentation may be unreliable due to the discontinuous nature of Himalayan sediment supply to distal fan sites.

The relationship between uplift and erosion of the Himalayan mountain chain and the growth of the Bay of Bengal deep-sea fan has been well documented on the basis of geophysical and sedimentological observations (Curray & Moore 1971; Emmel & Curray 1984; Curray 1994; Einsele *et al.* 1996). Mineralogical studies of Deep Sea Drilling Project (DSDP) Site 218 cores, shallow piston cores and dredge samples recovered from locations across the fan confirm the obvious importance of Himalayan erosion to fan sedimentation (Kolla & Biscaye 1973; Ingersoll & Suczek 1979; Bouquillon *et al.* 1989). Exceptions to a purely Himalayan provenance for Holocene and Pleistocene distal fan deposits have, however, been invoked by Goldberg & Griffin (1970), Kolla & Biscaye (1973) and Bouquillon *et al.* (1989) to account for the occurrence of smectite–kaolinite-rich sediments which have mineralogical affinities with basaltic and (or) high-grade metamorphic

CROWLEY, S. F., STOW, D. A. V. & CROUDACE, I. W. 1998. Mineralogy and geochemistry of Bay of Bengal deep-sea fan sediments, ODP Leg 116: evidence for an Indian subcontinent contribution to distal fan sedimentation. *In*: CRAMP, A., MACLEOD, C. J., LEE, S. V. & JONES, E. J. W. (eds) *Geological Evolution of Ocean Basins: Results from the Ocean Drilling Program.* Geological Society, London, Special Publications, **131**, 151–176.

crustal sources. Proposed sources for these sediments include the Deccan Trap basalts of central India and Precambrian high-grade metamorphic and igneous terrains exposed in central/southern India and Sri Lanka (Goldberg & Griffin 1970; Kolla & Biscaye 1973).

Investigation of Neogene sedimentation on the distal portion of the Bengal Fan by the Ocean Drilling Program (ODP Leg 116) resulted in the acquisition of a wide range of mineralogical and geochemical data (Cochran *et al.* 1989, 1990). Most of these data support the conclusion that

quartz–mica–sodic plagioclase–chlorite–amphibole turbidites, which constitute much of the recovered core material, were derived from a source terrain dominated by rapid mechanical weathering and erosion of differentiated upper crustal rocks ([meta]sediments and granites) typical of the Himalayas (Bouquillon *et al.* 1990; Brass & Raman 1990; Yokoyama *et al.* 1990). In contrast, mineralogical and palaeomagnetic studies (Brass & Raman 1990; Sager & Hall 1990; Yokoyama *et al.* 1990; Aoki *et al.* 1991; Amano & Taira 1992) indicate that organic-rich (>1.0% total organic

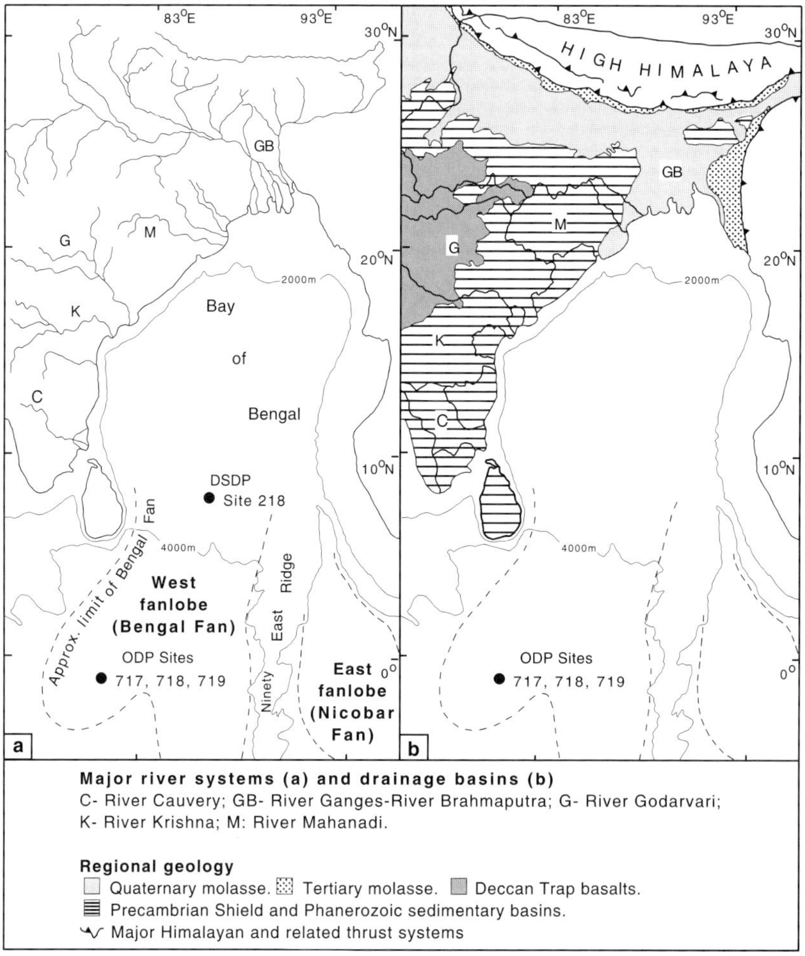

Fig. 1. Location map showing: (**a**) the geographical position of Leg 116 and major fluvial systems of the Himalayas and the Indian subcontinent; (**b**) regional geology and major drainage basin boundaries. The predominantly easterly directed drainage pattern of Indian subcontinent rivers is thought to have evolved in response to thermal uplift, underplating and rifting of the Indian shield during eruption of end-Cretaceous Deccan Trap flood basalts (Cox 1989). Continued uplift and erosion (Radhakrishna 1993) has resulted in considerable reduction in the geographical distribution of Deccan Trap basalts which originally covered much of the Indian subcontinent (Mahoney 1988).

carbon) sediments containing high concentrations of smectite and kaolinite were derived from the Indian subcontinent.

Bouquillon *et al.* (1990) and more recent publications (Derry & France-Lanord 1991; Derry *et al.* 1993; France-Lanord *et al.* 1993) have, however, drawn into question the importance of an Indian subcontinent contribution to distal fan sedimentation. They suggest, on the basis of similarities between the isotopic composition (^{143}Nd/^{144}Nd, ^{87}Sr/^{86}Sr) of Himalayan crust and Neogene fan sediments, that virtually all distal fan turbidites were derived from the same Himalayan sources (High Himalaya Crystalline Series, with minor contributions from the Lesser Himalaya and the Tibetan Sedimentary Series). Differences in mineralogy (i.e. quartz–mica-rich *versus* smectite–kaolinite-rich sediments) are attributed to variations in the extent of chemical weathering (through either prolonged sediment storage of alluvium in floodplains or climate change) of Himalayan detritus prior to turbidite sedimentation. Importantly, France-Lanord *et al.* (1993) attribute differences in distal fan sediment mineralogy (and by implication the degree of chemical weathering) to the effects of major tectonic and climatic changes related to Himalayan uplift events. As a consequence, correct interpretation of the provenance of distal fan sediments and, in particular, the origin of smectite–kaolinite turbidites has important implications for understanding factors controlling:

(1) tectonic and climatic change in sediment source terrains (specifically the Himalayas); (2) the evolution of sediment supply to the Bengal Fan; and (3) depositional sedimentary processes contributing to sediment accumulation in distal fan environments.

In this paper we address the question of sediment provenance on the basis of the mineralogy and trace element geochemistry of distal fan turbidites. These data are used in conjunction with published mineralogical and isotopic information to investigate the extent of chemical weathering of sediment source terrains, and identify the bulk crustal composition of source terrains supplying sediment to the distal fan. In addition, we utilize provenance-sensitive geochemical indicators to examine the temporal evolution of sediment supply to the distal fan with respect to Himalayan uplift and erosion, glacio-eustatic sea-level change, and behaviour of fan distributary channel systems.

Lithostratigraphy and sedimentology of distal Bengal Fan sediments

During ODP Leg 116 (Cochran *et al.* 1989) three closely spaced sites (717–719) were cored on the distal portion of the Bay of Bengal deep-sea fan (Fig. 1) resulting in the recovery of sediments ranging in age from Lower Miocene (*c.* 17.5 Ma) to Present (Fig. 2). All three sites penetrated similar

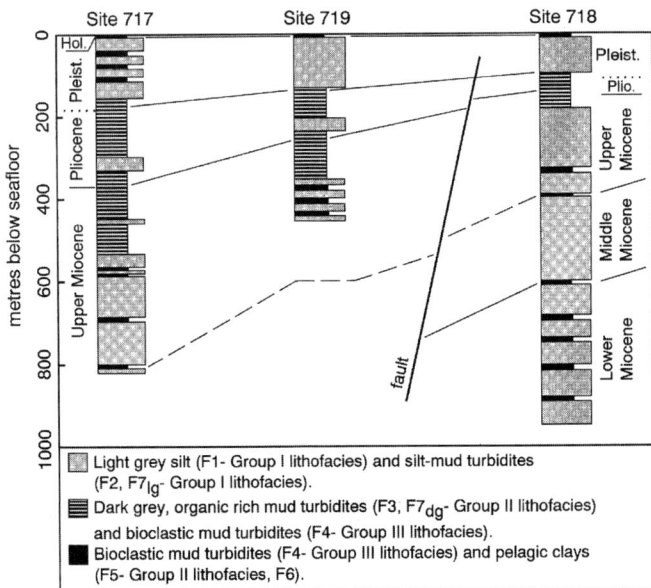

Fig. 2. Simplified lithostratigraphic logs for Sites 717, 718 and 719 (modified from Cochran *et al.* 1989; Cochran 1990).

Table 1. *Outline description of distal Bengal Fan lithofacies*

Facies	Lithology*	Range of mean grain sizes† (ϕ)	Mineralogy‡	Description
Silt, silt-mud turbidites (F1)	Litharenite–wacke–shale	4.32–7.51	Quartz, sodic plagioclase, K-feldspar, muscovite, biotite, chlorite, amphibole, calcite, dolomite	Sharp-based, normally graded, organic-poor (<0.5% TOC), light grey silt-to-mud turbidites. Rare bioturbation
Organic-poor mud turbidites (F2)	Shale–calcareous shale	6.31–7.84	Quartz, sodic plagioclase, K-feldspar, muscovite, biotite, chlorite, amphibole, calcite, dolomite	Sharp-based, normally graded, organic-poor (<0.5% TOC), light grey mud turbidites. Sparse bioturbation
Organic-rich mud turbidites (F3)	Fe-shale–calcareous Fe shale	5.99–7.98	Quartz, plagioclase, smectite, kaolinite, illite, calcite, pyrite	Sharp-based, normally graded, organic-rich (>0.5% TOC), dark grey mud turbidites. Abundant authigenic pyrite. Localized bioturbation
Biogenic mud turbidites (F4)	Calcareous Fe-shale ($F4_{green}$)–limestone ($F4_{green}$, $F4_{white}$)	5.52–7.95	Calcite, aragonite, quartz, plagioclase, smectite, kaolinite, illite, pyrite	$F4_{green}$: sharp-based, normally graded, greenish-grey mud turbidites containing 10–50% biogenic carbonate. Localized bioturbation. $F4_{white}$: sharp-based, normally graded, white turbidites containing >60% biogenic carbonate
Pelagic clays (F5)	Fe-shale ($F5_{red}$)–shale ($F5_{grey}$).	7.51–7.90	Quartz, plagioclase, smectite, kaolinite	Intensely bioturbated, organic-poor (<0.5% TOC), red and light grey mud
Pelagic calcareous clays (F6)	Calcareous Fe-shale	7.25–7.43	Calcite, aragonite, quartz, plagioclase, smectite, kaolinite, illite, pyrite	Intensely bioturbated, white muds containing abundant (>30%) biogenic carbonate
Structureless muds (F7)	Fe-shale ($F7_{dark\ grey}$)–shale ($F7_{light\ grey}$)	6.96–7.84	$F7_{dark\ grey}$: quartz, plagioclase, smectite, kaolinite, illite, calcite, pyrite. $F7_{light\ grey}$: quartz, sodic plagioclase, K-feldspar, muscovite, biotite, chlorite, amphibole, calcite, dolomite	$F7_{dark\ grey}$: sharp-based (organic-rich, pyritic) 'hemiturbidites'. Localized intense bioturbation. $F7_{light\ grey}$: sharp-based (organic-poor) mud 'hemiturbidites'. Localized intense bioturbation.

*Lithological classification scheme of Herron (1988)
†Data from Balson & Stow (1990)
‡Qualitative whole-rock X-ray diffraction. Minerals listed in order of estimated abundance

sedimentary sequences, but revealed significant differences in sediment thickness due to faulting and syndepositional intraplate deformation (Cochran 1990). The sediments were resolved into ten lithofacies (the following abbreviations are used: F1, F2, F3, F4$_{gr}$, F4$_{w}$, F5$_{r}$, F5$_{g}$, F6, F7$_{lg}$, F7$_{dg}$) during shipboard examination of core material. Detailed descriptions of these facies are reported in Stow *et al.* (1990). Grain-size characteristics, mineralogy and other features are summarized in Table 1.

In addition to primary depositional features, a variety of minor diagenetic processes were also recognized (Cochran *et al.* 1989; Aoki *et al.* 1991). These include: (1) the formation of authigenic pyrite (commonly associated with organic carbon-rich facies F3 and F4$_{gr}$) and a single calcite cemented horizon; (2) localized development of 'chemical fronts' formed by the relocation of redox-sensitive elements (Fe, Mn) in organic carbon-rich turbidites (F3); and (3) dissolution of siliceous and carbonate microfossils.

Sampling and analytical techniques

Forty-six samples were selected from cores recovered at ODP Sites 717 (0–740 metres below sea floor (mbsf); 0–8.9 Ma) and 718 (437–954 mbsf; 9.5–17.0 Ma) in order to provide data

coverage through the available stratigraphy. All samples were dried at 105°C and crushed (<63 mm) in an agate mill. Qualitative mineralogical composition of each sample was determined by whole-rock X-ray diffraction (XRD). Analysis was performed on randomly orientated powders packed into aluminium mounts and scanned from 3 to 63°2θ.

Major element concentrations were determined by X-ray fluorescence (XRF) on fused glass discs (Harvey *et al.* 1973) at the University of Nottingham. The instrument was calibrated by analysing 12 international standard rocks (NIST and NIM standards) together with samples. Reference values for these standards were taken from Govindaraju (1989). Analytical precisions (see Harvey *et al.* 1973) are <1% for all major elements except Mg. Typical major element concentrations obtained from XRF analysis of international standards (G1, W1) run as 'unknowns' using this procedure are reported in Table 2. No attempt was made to remove water-soluble contaminants (notably salts precipitated from residual pore fluids) prior to analysis so that XRF-determined Na$_2$O data include Na present in both NaCl and sodic aluminosilicate phases. In order to correct Na$_2$O for salt contamination, soluble (non-silicate fraction) Na was measured by atomic absorption spectrophotometry following

Table 2. *Analyses of international standards by X-ray fluorescence (major elements) and instrumental neutron activation (trace elements)*

	G1*		W1*			JB1a*			
	XRFS analysis[†]	Certified value	XRFS analysis[†]	Certified value		INAA analysis	σ_{n-1} (n = 6)	Detection limit	Certified value
SiO$_2$	72.55	72.46	52.57	52.55	Co	39.4	0.2	0.3	39.5
Al$_2$O$_3$	14.40	14.23	14.83	14.99	Cr	418	4	3	415
TiO$_2$	0.27	0.25	1.06	1.07	Cs	1.1	0.1	0.2	1.2
Fe$_2$O$_3$^{T}	1.93	1.96	11.09	11.18	Hf	3.43	0.05	0.1	3.4
MgO	0.34	0.39	6.63	6.62	Rb	42	2	8	41
CaO	1.32	1.38	10.94	10.94	Sc	29.14	0.10	0.06	29
Na$_2$O	3.30	3.33	2.12	2.13	Ta	1.99	0.03	0.10	2.0
K$_2$O	5.49	5.48	0.66	0.64	Th	8.82	0.09	0.20	8.8
MnO	0.03	0.03	0.17	0.17	U	1.60	0.11	0.50	1.6
P$_2$O$_5$	0.08	0.09	0.14	0.14	La	37.8	0.9	0.3	38
					Ce	66.7	0.8	1.0	67
					Nd	25	1	3	24
					Sm	5.18	0.10	0.05	5.2
					Eu	1.49	0.03	0.05	1.5
					Tb	0.71	0.02	0.05	0.70
					Tm	0.34	0.01	0.15	0.34
					Yb	2.05	0.07	0.10	2.0
					Lu	0.33	0.01	0.10	0.33

*Certified values from Potts *et al.* (1992)
[†]Data from Harvey *et al.* (1973)

dissolution in deionized water (McLennan *et al.* 1990) and the data reported as Na_2O^* (XRF Na_2O – water-soluble Na_2O). Estimates of relative precision and accuracy based on analysis of artificial mixtures of silica sand and reagent grade NaCl were <5% for Na contents ranging between 0.5 and 2 wt%.

Total organic carbon (TOC), inorganic carbon (TIC) and total sulphur (TS) contents were measured using a Carlo Erba C-H-N-S element analyser at the University of Southampton (TOC, TIC) and the University of Leeds (TS). Detection limits are 0.01 wt% and 0.005 wt% for carbon and sulphur respectively, with relative precisions and accuracies of <5% for both techniques at concentrations above 1 wt%.

Trace element concentrations were determined by instrumental neutron activation analysis (INAA). Samples were irradiated for 24 h under a neutron flux of 1×10^{12} cm^{-2} s^{-1}. Counting was carried out at the University of Southampton by collecting spectra simultaneously using a Canberra Industries 16% P-type HPGe and a Canberra Industries planar Ge spectrometer at 7 and 30 days after irradiation. Data were processed using Canberra APOGEE software. Element concentrations were obtained following corrections for flux variation and calibrated by using international reference samples NIM-G and BHVO. Detection limits, estimates of analytical reproducibility (σ_{n-1}) based on replicate analysis of the JB1a standard, and certified values for JB1a (Potts *et al.* 1992) are reported in Table 2.

Mineralogical and geochemical results

In order to simplify the presentation of analytical results (a tabulation of all data reported in this paper is provided in the Appendix), the range of individual lithofacies has been reduced to the following three groups based on general facies associations (see Stow *et al.* 1990), similarities in mineralogy, major element geochemistry and principal component analysis of major element data (Fig. 3): Group I. – F1, F2, F7$_{lg}$; Group II. – F3, F5$_r$, F7$_{dg}$; and Group III. – F4$_{gr}$. Variations in the composition of individual facies are, however, discussed where appropriate.

Mineralogy

Qualitative whole-rock XRD analysis of distal fan turbidites largely confirms the results of more detailed studies conducted by Bouquillon *et al.* (1990) and Brass & Raman (1990), although data obtained during this investigation provide complementary information on the mineralogy of these sediments.

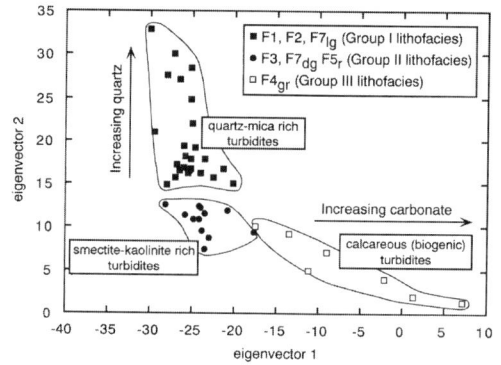

Fig. 3. Principal component plot for distal Bengal Fan sediments. Eigenvectors are calculated from anhydrous (LOI-free), normalized major element data (MnO and P_2O_5 were not included due to the low concentration of these oxides). Eigenvectors, eigenvalues and variance proportions for the first three principal components are listed below. Large eigenvalues for the first two principal components (V1 and V2) account for a significant proportion of the total variability of the data. V1 effectively discriminates calcareous from non-calcareous sediments, while V2 discriminates between $Fe_2O_3^T$–Al_2O_3-rich (smectite–kaolinite) and SiO_2–K_2O–Na_2O^*-rich (quartz–mica–feldspathic) sediments. Fields outlining the three lithofacies groupings are hand-drawn. Na_2O^* is Na_2O present in silicate minerals only (see text for details).

Eigenvectors	V1	V2	V3
SiO_2	−0.313	0.461	0.269
Al_2O_3	−0.442	−0.222	−0.148
TiO_2	−0.291	−0.476	0.272
$Fe_2O_3^T$	−0.255	−0.543	0.135
MgO	−0.336	−0.019	−0.613
CaO	0.474	−0.047	−0.196
K_2O	−0.354	0.301	−0.444
Na_2O^*	−0.308	0.351	0.449
Eigenvalue	4.088	2.188	1.287
Variance proportion (%)	51.1	27.3	16.1

Group I. Samples of Group I turbidites are characterized by the occurrence of mica (muscovite, biotite), chlorite, amphibole, sodic plagioclase and potassium feldspar. Quartz and variable (but generally minor) quantities of calcite and dolomite occur in all samples. Variations in mineralogical abundances are most probably related to the grain size of sediment reaching the distal portion of the fan and the unmixing of density current components during deposition. As a consequence the proportion of quartz and feldspar (as indicated by relative peak XRD heights) increases relative to sheet silicate minerals, calcite and dolomite with increasing sediment grain size (estimated from visual inspection of hand specimens).

Group II. The mineralogy of Group II turbidites differs considerably from samples of Group I. Diffraction patterns reveal the occurrence of variable amounts of smectite and (or) mixed-layer clays, kaolinite, illite and pyrite. Quartz and feldspar occur in all samples, but are present in reduced concentrations compared to Group I due to the higher clay mineral content of Group II sediments. Calcite and dolomite are present in the majority of samples, although carbonates rarely contribute significantly to bulk mineralogy.

Group III. Group III turbidites contain large concentrations of biogenic carbonate and are characterized by the occurrence of calcite and aragonite. Associated silicate phases include quartz, plagioclase, smectite, kaolinite and illite/mica. Minor quantities of dolomite and pyrite are present in virtually all cases. Importantly, those $F4_{gr}$ samples containing relatively low carbonate abundances reveal whole-rock XRD patterns similar to those recorded from Group II turbidites.

Major element geochemistry

Samples recovered from Leg 116 cores are primarily composed of silt and clay grade sediment (Balson & Stow 1990) and, as an alternative to standard petrographic schemes, have been classified using SandClass (Herron 1988, see Fig. 4). The scheme effectively separates samples on the basis of lithic to feldspar content ($Fe_2O_3^T/K_2O$), the proportion of quartz to sheet silicate minerals (SiO_2/Al_2O_3) and carbonate abundance (CaO). Data plotted using SandClass (Fig. 4) reveal three dominant sediment populations corresponding to the facies grouping recognized on the basis of lithological characteristics, XRD mineralogy and principal component analysis of major element data (see Fig. 3).

Group I. Group I turbidites are characterized by a linear array of compositions ranging from litharenites to shales (three calcareous Group I samples also plot on this trend). This array is attributed to the progressive hydraulic separation (unmixing) of quartz and feldspar from sheet silicate minerals during deposition of turbidites from sediment density currents. Compositional relationships expressed in terms of a correlation matrix (Table 3a) show that TiO_2, $Fe_2O_3^T$, MgO and K_2O are positively correlated with Al_2O_3 and negatively correlated with SiO_2. Interelement relationships are consistent with the occurrence of essentially two 'independently differentiable components' (Argast & Donnelly 1987). This results in the concentration of TiO_2, $Fe_2O_3^T$, MgO and K_2O in sheet silicates (chlorite, biotite, muscovite) and the separation of SiO_2 (quartz) from other elements through depositional processes. Na_2O^* is negatively correlated with all major elements apart from SiO_2 and indicates that sodium probably occurs largely in sodic plagioclase. In contrast, CaO is only weakly correlated with other elements suggesting that Ca is distributed between a variety of phases (e.g. calcite, dolomite, apatite, plagioclase). TOC contents are

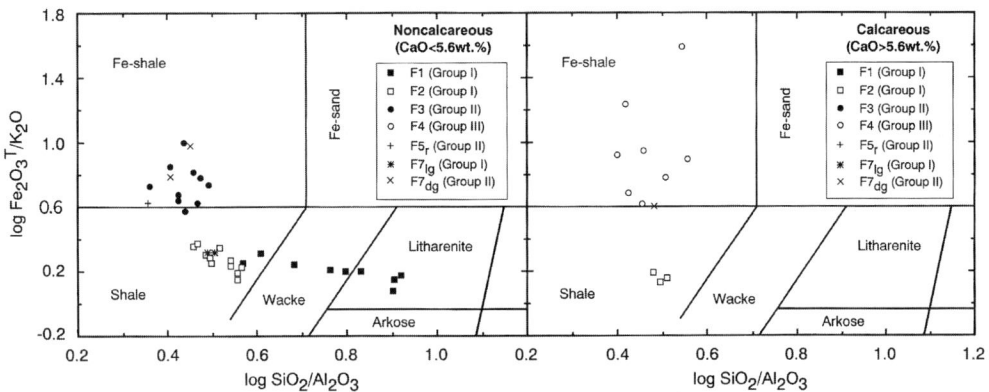

Fig. 4. Geochemical classification of distal Bengal Fan sediments using SandClass (Herron 1988). Brief lithofacies descriptions are provided in Table 1. Lithofacies F4 are characterized by high CaO contents (>5.6 wt% CaO) compared to most other fan deposits and define a third distinct geochemical grouping comprising calcareous sediments rich in biogenic carbonate.

Table 3. *Pearson product–moment correlation matrix*

	SiO$_2$	Al$_2$O$_3$	TiO$_2$	Fe$_2$O$_3^T$	MgO	CaO	K$_2$O	Na$_2$O*	MnO	P$_2$O$_5$	S
(a) Group I lithofacies (n = 25)											
SiO$_2$	–										
Al$_2$O$_3$	–0.892	–									
TiO$_2$	–0.791	0.841	–								
Fe$_2$O$_3^T$	–0.777	0.931	0.854	–							
MgO	–0.962	0.801	0.717	0.662	–						
CaO	–0.206	–0.244	–0.197	–0.418	0.323	–					
K$_2$O	–0.887	0.908	0.666	0.807	0.832	–0.042	–				
Na$_2$O*	0.720	–0.532	–0.432	–0.432	–0.710	–0.398	–0.667	–			
MnO	–0.596	0.649	0.604	0.604	0.474	–0.129	0.480	–0.240	–		
P$_2$O$_5$	–0.566	0.480	0.329	0.329	0.617	0.214	0.368	–0.282	0.245	–	
S	–0.243	0.367	0.373	0.599	0.141	–0.401	0.342	–0.105	0.192	–0.084	–
(b) Group II lithofacies (n = 14)											
SiO$_2$	–										
Al$_2$O$_3$	0.187	–									
TiO$_2$	0.219	0.441	–								
Fe$_2$O$_3^T$	–0.135	–0.256	0.347	–							
MgO	0.671	–0.173	–0.271	–0.152	–						
CaO	–0.566	–0.486	–0.689	–0.484	–0.240	–					
K$_2$O	0.347	–0.144	–0.589	–0.562	0.631	0.274	–				
Na$_2$O*	0.161	0.166	0.618	0.477	–0.388	–0.576	–0.595	–			
MnO	–0.119	0.088	0.234	0.079	0.169	–0.138	0.170	–0.055	–		
P$_2$O$_5$	–0.545	–0.338	–0.454	–0.398	–0.191	0.718	0.159	–0.465	–0.231	–	
[†]S	–0.449	–0.395	0.205	0.843	–0.249	–0.200	–0.414	0.375	0.290	–0.55	–
(c) Group III lithofacies (n = 7)											
SiO$_2$	–										
Al$_2$O$_3$	0.965	–									
TiO$_2$	0.986	0.941	–								
Fe$_2$O$_3^T$	0.861	0.932	0.860	–							
MgO	0.843	0.944	0.835	0.922	–						
CaO	–0.966	–0.996	–0.949	–0.937	–0.945	–					
K$_2$O	0.969	0.960	0.933	0.811	0.863	–0.946	–				
Na$_2$O*	0.964	0.920	0.927	0.746	0.798	–0.913	0.984	–			
MnO	–0.844	–0.871	–0.863	–0.783	–0.866	0.898	–0.798	–0.778	–		
P$_2$O$_5$	–0.802	–0.701	–0.774	–0.530	–0.575	0.727	–0.779	–0.873	0.663	–	
S	–0.283	–0.095	–0.292	0.000	0.089	0.071	–0.300	–0.370	–0.182	0.379	–

[†]S: Inter-element correlations calculated using F3 and F7$_{dg}$ data only

uniformily low and rarely exceed 0.5 wt%. Highest TOC contents are associated with fine-grained lithofacies (F2, F7$_{lg}$) reflecting the hydraulic separation of organic matter from coarser-grained components during deposition. Sulphur concentrations are similarly low (generally <0.2 wt%) and indicate the presence of only minor quantities of detrital sulphide and sulphate minerals (Fig. 5).

Group II. Smectite–kaolinite turbidites form a broad cluster of data which plot away from the trend defined by Group I samples and fall predominantly within the Fe-shale field of SandClass (Fig. 4). Fewer significant interelement relationships are defined by linear correlations (Table 3b) due to the absence of differentiable components

and wide variations in the geochemistry of detrital phases present in these sediments (Aoki *et al.* 1991). Weak positive correlations between SiO$_2$, MgO and K$_2$O may indicate an interdependent relationship between quartz and mica, while negative correlations between Al$_2$O$_3$ and other major elements possibly reflect the importance of kaolinite in these samples. In comparison to Group I lithofacies, Group II turbidites are generally rich (>0.5 wt%) in sulphur (F5$_r$ is an exception) as the result of the formation of diagenetic pyrite (Cochran *et al.* 1989). The fact that the S–Fe$_2$O$_3^T$ correlation for Group II sediments closely parallels the stoichiometric pyrite line (Fig. 5) indicates that a proportion of Fe present in the majority of samples is contained within pyrite. Intersection of

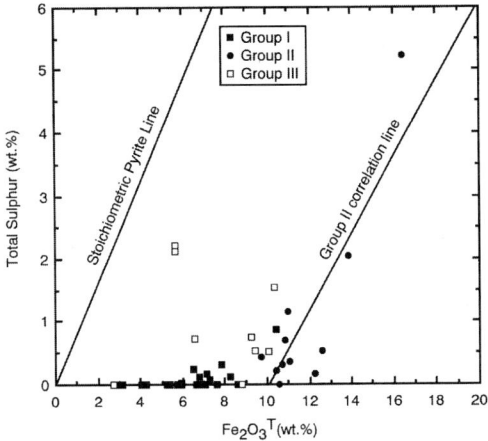

Fig. 5. Total sulphur–$Fe_2O_3^T$ diagram showing variations in Fe content of fan sediments with respect to S. The low S contents of Group I turbidites are consistent with low detrital/authigenic sulphate and sulphide abundances. In contrast, the high Fe concentration of Group II turbidites are in part dependent on pyrite content as indicated by the parallel trends of the Group II S–$Fe_2O_3^T$ correlation and the stoichiometric pyrite lines.

the regression with the $Fe_2O_3^T$ axis, however, shows that approximately 10 wt% $Fe_2O_3^T$ occurs in other (? detrital) phases. Although variable, the TOC content of Group II turbidites typically exceeds 1 wt% and provides a further important geochemical contrast with Group I sediments.

Group III. Carbonate-rich lithofacies are characterized by high CaO (7–40 wt%) and $Fe_2O_3^T/K_2O$, and low SiO_2/Al_2O_3. Group III turbidites consequently range from calcareous Fe-shales to limestones (Fig. 4). Although samples are variably diluted by biogenic carbonate, SandClass shows that the siliciclastic portions of these sediments have compositional affinities with Group II lithofacies, and this relationship is further supported by an overall similarity in major element geochemistry of Group III sediments when estimated on a CaO-free basis. TOC contents of carbonate-rich sediments are variable and occasionally higher than other distal fan facies. Although TOC data might possibly reflect an increased marine organic input associated with high biogenic carbonate contents, Rock-Eval pyrolysis data (Cochran *et al.* 1989) show that organic matter present in Group III sediments is dominated by terrestrial sources. TOC may be related to either (or both) high initial organic matter contents in source sediments or concentration of detrital organic matter by hydraulic factors during deposition.

Sulphur is only weakly correlated with other elements (Table 3c) and this indicates that S is probably present in a variety of sulphate and sulphide phases.

Trace element geochemistry

In order to rationalize variability in the trace element geochemistry of fan sediments, data from different lithofacies are compared by using

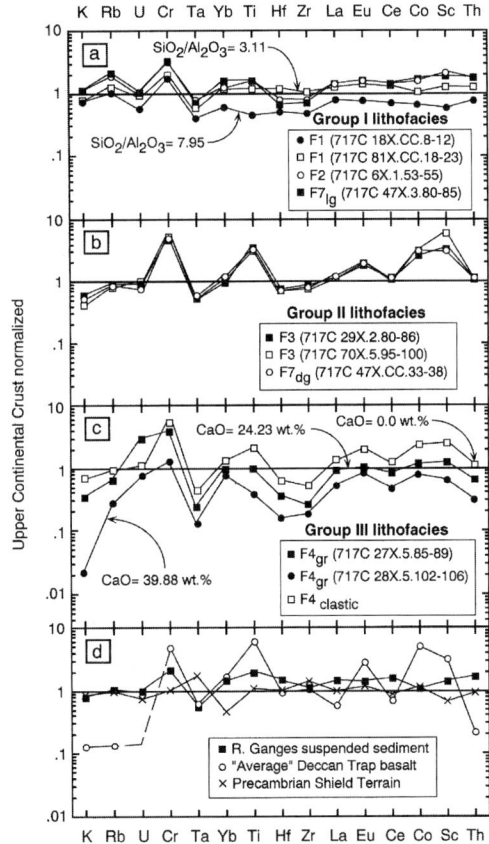

Fig. 6. Upper Continental Crust normalized trace element patterns (elements are ordered from left to right in terms of increasing ocean residence times). (**a**) Group I lithofacies. (**b**) Group II lithofacies. (**c**) Group III lithofacies (see text for details of methods used to estimate the composition of F4$_{clastic}$ [0. 0 wt% CaO]). (**d**) River Ganges suspended sediment (data from Martin & Meybeck 1979), 'average' Deccan Trap basalts (Ambenali, Mahabaleshwar, Panhala and Poladpur Formations (Lightfoot 1985)), Precambrian shield terrain (present-day Canadian Shield surface composition estimate of Shaw *et al.* (1986)). Upper Continental Crust values from Taylor & McLennan (1985). Zr concentrations in Bengal fan sediments are estimated assuming Zr = Hf × 39 (Condie 1991).

multielement plots (Figs 6 and 7) normalized to average Upper Continental Crust (Taylor & McLennan 1985) and chondrite (Briqueu *et al.* 1984).

Group I. Normalized trace element data (Fig. 6a) reveal relatively flat profiles corresponding to a bulk composition close to Upper Continental Crust. Chondrite-normalized rare earth element (REE) patterns (Figs 7a and 8) are similar to most

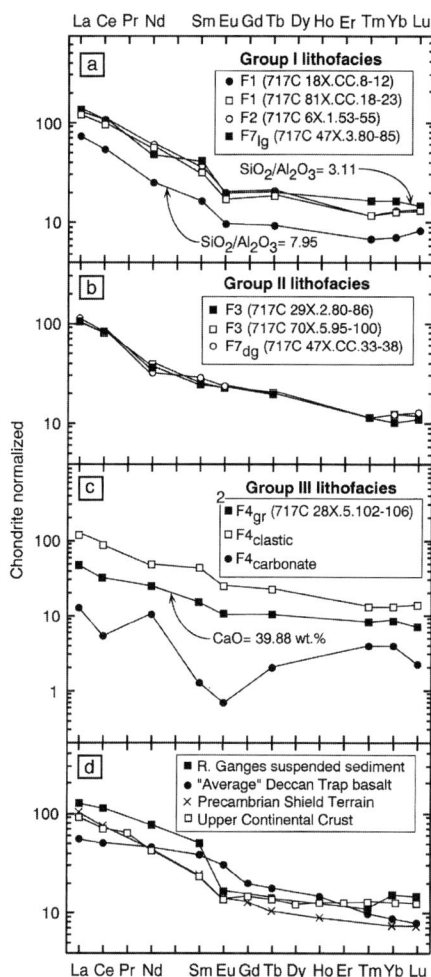

Fig. 7. Chondrite-normalized REE patterns. (**a**) Group I lithofacies. (**b**) Group II lithofacies. (**c**) Group III lithofacies (see text for details of methods used to estimate the REE composition of F4$_{clastic}$ and F4$_{carbonate}$). (**d**) River Ganges suspended sediment, Precambrian shield terrain and 'Average' Deccan Trap basalts and Upper Continental Crust (see Fig. 6 for details of data sources). Chondrite normalization values from Briqueu *et al.* (1984).

Phanerozoic clastic sediments (Taylor & McLennan 1985; McLennan 1989) and are distinguished by light rare earth element (LREE) enrichment (La$_N$/Yb$_N$ >7.5, La$_N$/Sm$_N$ >2.0), relatively flat heavy rare earth element (HREE) (Gd$_N$/Yb$_N$<2.0) and depleted Eu-anomalies (Eu/Eu* <0.85). Although the pattern of trace element and rare earth element behaviour between individual lithofacies is similar, total trace element and REE abundances decrease systematically with increasing SiO$_2$/Al$_2$O$_3$, consistent with the concentration of these elements in the fine-grained, sheet silicate-rich fraction of turbidite sediments (Taylor & McLennan 1985; McLennan *et al.* 1990).

Group II. In contrast to Group I turbidites, the majority of Group II samples (Figs 6b and 7b) are characterized by Co, Cr, Sc, Ti, Eu (Eu/Eu*>0.85) enrichment and K, Rb, Th depletion when compared to Upper Continental Crust. This is indicative of a higher mafic content than is typical of many Phanerozoic sediments (Taylor & McLennan 1985; McLennan 1989) and sedimentary successions deposited in a passive margin tectonic environment (McLennan *et al.* 1990). Group II sediments, however, exhibit LREE (La$_N$/Yb$_N$ >7.5, La$_N$/Sm$_N$ >2.0) patterns similar to those displayed by Group I turbidites (Fig. 8) and in this respect are similar in composition to Upper Continental Crust and average Phanerozoic clastic sediments (Taylor & McLennan 1985; McLennan 1989).

Group III. Carbonate-rich lithofacies exhibit differing total trace element concentrations and normalized patterns (Figs 6c and 7c) as a result of variations in the abundance of trace element-depleted biogenic carbonate. Estimates of the trace element and REE composition of clastic and carbonate end-members for F4$_{gr}$ samples, using a least-squares fit of trace element data against CaO (assuming bimodal mixing between terriginous (0 wt% CaO) and carbonate (56 wt% CaO) sediment), produce distinctly different geochemical signatures (Figs 6c and 7c). The resultant pattern for the terrigenous component reveals compositional features similar to those displayed by Group II sediments, although the depleted Eu/Eu* (Eu/Eu* = 0.72) is more consistent with Eu anomalies observed in Group I turbidites (Fig. 8). In contrast, the carbonate end-member is characterized by low total trace element concentrations and exhibits REE compositional features (depleted Ce anomaly, HREE enrichment) similar to those of marine bioclasts (cf. Elderfield *et al* 1981).

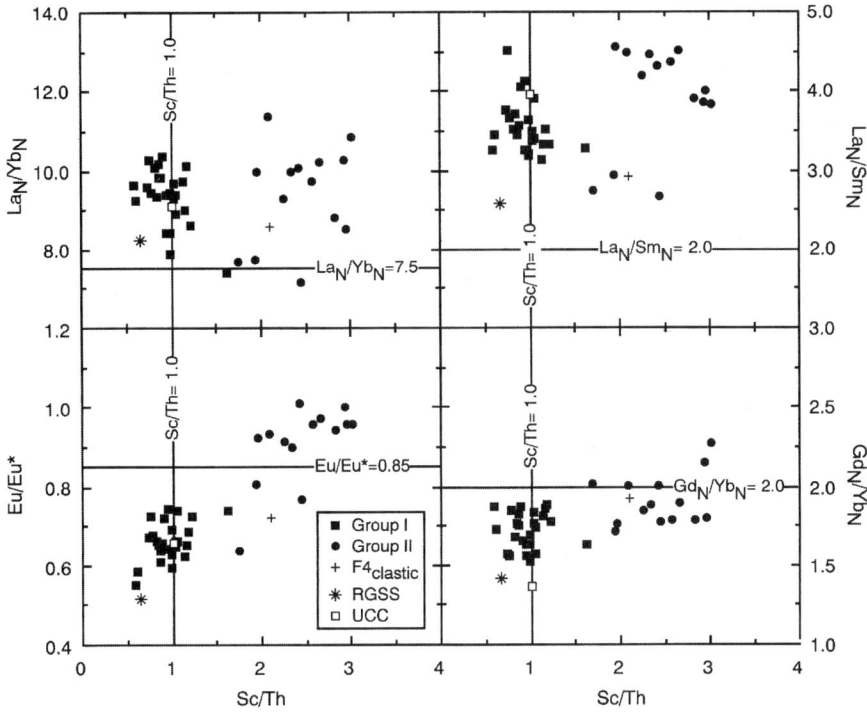

Fig. 8. La_N/Yb_N, La_N/Sm_N, Eu/Eu^*, Gd_N/Yb_N–Sc/Th diagrams showing the distribution of distal fan sediments, River Ganges suspended sediment (RGSS; Martin & Maybeck 1979) and Upper Continental Crust (UCC; Taylor & McLennan 1985). Gd_N estimated assuming $Gd_N = (Sm_N \times Tb_N^2)^{1/3}$, $Eu/Eu^* = Eu_N/(Sm_N \times Gd_N)^{1/2}$. All published data recalculated using the chondrite values of Briqueu *et al.* (1984).

Discussion

Source terrain weathering: mineralogical and geochemical evidence

Estimates of the extent of source terrain weathering based on the relative behaviour of specific groups of elements such as the alkali metals, alkaline earths and Th–U in resultant sediments have proven to provide reliable indicators of chemical weathering intensity (Nesbitt *et al.* 1980; Nesbitt & Young 1982, 1984; McLennan *et al.* 1990, 1993). For alkali and alkaline earth elements, Nesbitt *et al.* (1980) have shown that cations with large ionic radii (Rb, Cs) are generally fixed within the products of weathering profiles by formation of new phases and adsorption onto exchange sites of clay minerals, whereas cations with smaller radii (Na, Ca, Sr) are selectively leached during weathering and remain in solution. Similarly, differences in the solubility of U and Th during weathering offer a further means of assessing and comparing the degree of chemical weathering (McLennan *et al.* 1993) as a result of the increased

solubility and loss of U in oxidizing environments and the retention of Th (leading to increased Th/U) with progressive chemical weathering. The relative response of alkali and alkaline earth elements can be used as a guide to the extent of chemical weathering (Nesbitt & Young 1982) and this is expressed quantitatively as a Chemical Index of Alteration (CIA; see Fig. 9). Chemical weathering trends may also be monitored using the A–C*N*–K diagrams (see Fig. 10) of Nesbitt & Young (1989). These diagrams provide a convenient way of comparing the extent of source terrain weathering, assessing chemical weathering pathways and evaluating the bulk lithological and geochemical composition of source terrains.

Regardless of crustal source, differences in the mineralogy of distal fan sediments, particularly with respect to clay mineralogy and feldspar content, suggest a considerable variation in the relative intensity of chemical weathering prior to deposition of each of the lithofacies groupings. In general terms the high plagioclase feldspar, mica (including biotite), chlorite and amphibole content of Group I turbidites is consistent with rapid

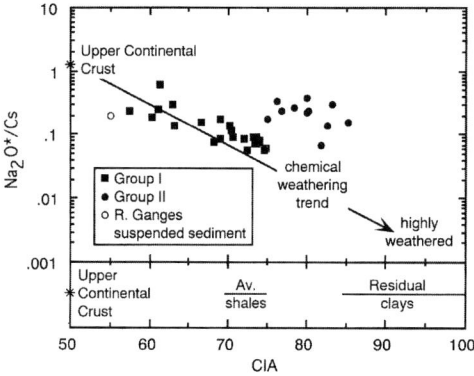

Fig. 9. Na_2O^*/Cs–CIA (Chemical Index of Alteration; Nesbitt & Young 1982) diagram for distal fan sediments. CIA = $Al_2O_3/(Al_2O_3 + CaO^* + Na_2O^* + K_2O) \times 100$, where: oxides are in molecular proportions; CaO^* is the CaO content minus CaO present in carbonates and apatite; Na_2O^* is Na_2O present in silicate minerals only (see text for details). XRD analysis shows that calcite and dolomite are the main carbonate phases present in Group I and Group II turbidites. CaO^* was calculated assuming that inorganic carbon (TIC) is divided equally between calcite and dolomite. Although CaO^* values are calculated with greater uncertainty, this compromise does not significantly affect interpretation. Values for Group III turbidites are not included due to high residual CaO^* contents after correction for the presence of carbonate minerals and apatite. River Ganges suspended sediment data from Martin & Meybeck (1979). Values for Upper Continental Crust, shales and residual clays are from Taylor & McLennan (1985).

erosion of sediments derived from a source region(s) dominated by mechanical weathering processes (Curtis 1990). In contrast, Group II and Group III turbidites contain significantly smaller quantities of feldspar and are characterized by high concentrations of smectite and kaolinite reflecting relatively intense chemical weathering of source terrains (Nesbitt & Young 1984; Curtis 1990).

Despite problems caused by the presence of both detrital (calcite, dolomite) and biogenic carbonates (calcite, aragonite), distal fan sediment data reveal good agreement between the extent of chemical weathering based on mineralogical observations and estimates of relative chemical weathering intensity using CIA (Fig. 9). Importantly, the separate trends defined by Group I and Group II turbidites in Na_2O^*/Cs–CIA space (Fig. 9) imply progressive chemical weathering of sources terrains with differing bulk compositions and this is supported by the pattern of geochemical behaviour observed from data plotted on an A-C*N*-K diagram (Fig. 10). Here the distribution of Group I turbidites corresponds with that expected for

progressive chemical weathering of rocks with a bulk composition similar to granite (although the progressive weathering trend displayed by these data may also reflect the effects of grain-size separation as K_2O-rich sheet silicates are partitioned into fine-grained lithofacies during deposition). In comparison, Group II turbidite data fall along a trend which is consistent with weathering of a source terrain with a bulk composition similar to tonalite–granodiorite (Fig. 10).

Similar conclusions regarding compositional differences in source terrain geochemistry may be drawn from U–Th data (Fig. 11). Although most samples exhibit Th/U ratios greater than upper crustal values, samples from Group I turbidites are characterized by higher Th/U ratios than those of Group II despite mineralogical and other geochemical evidence which suggests that Group I sediments are less intensely weathered. Because diagenetic processes may modify Th/U ratios it is not immediately certain whether authigenic U enrichment or differences in the composition of sediment source terrains are responsible for observed geochemical differences. Estimates of apparent authigenic U content (Myers & Wignall 1987) and U/Th ratios, however, indicate that both Group I and Group II turbidites were deposited under relatively oxic environmental conditions and that no significant diagenetic enrichment of U has occurred (see Jones & Manning (1994) for details). As a consequence the behaviour of Th–U in distal fan sediments is consistent with alkali and alkaline earth element data in supporting different bulk crustal compositions for the source terrains of Group I and Group II turbidites.

Crustal sources and sediment provenance: trace element geochemistry

Although trace elements may be partitioned by chemical weathering processes (Nesbitt 1979; Nesbitt *et al.* 1980; Marsh 1991), a number of studies (Taylor & McLennan 1985; McLennan 1989; McLennan *et al.* 1990, 1993; Condie 1991; Cullers 1994) have shown that certain elements with low solubilities do not remain in solution following weathering and are essentially recycled quantitatively as part of the detrital sediment load during erosion and transport. Consequently, although absolute abundances may change relative to source rock geochemistry as the result of the loss of highly soluble components (Na, Ca, Mg, Sr), elements such as Co, Cr, Hf, Sc, Th, REE are generally transported in ratios consistent with ratios present in primary crustal sources and may be used to infer the geochemical characteristics of sediment

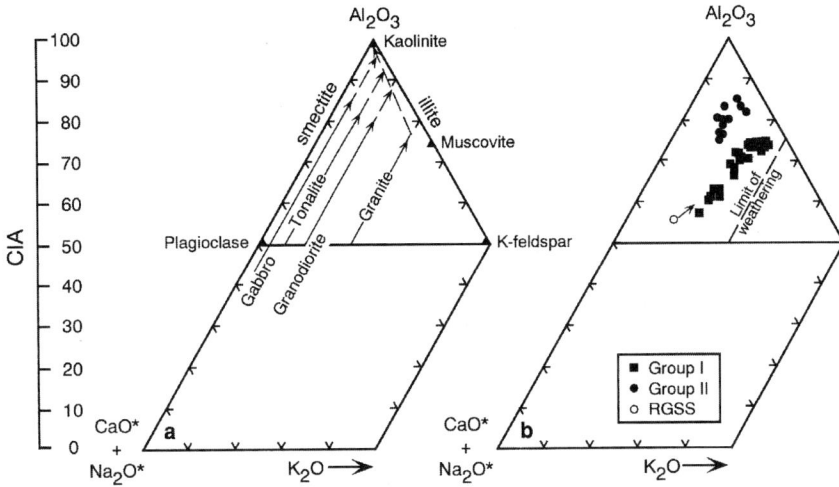

Fig. 10. A–C*N*–K diagrams (Nesbitt & Young 1989) illustrating: (a) ideal chemical weathering trends for common rock types (Nesbitt & Young 1989); and (b) distal fan sediments. Full arrows indicate predicted weathering paths; dashed arrows are extensions of these pathways to highly weathered, residual soils. River Ganges suspended sediment (RGSS) data are from Martin & Maybeck (1979). Note the relatively high CaO content (4.1 wt%) of RGSS has not been corrected for CaO present in carbonate minerals or apatite resulting in displacement towards CaO* (arrow indicates likely locus of RGSS). All values are in molar proportions (see Fig. 9 and Nesbitt & Young (1982) for details of CaO* calculations). Na$_2$O* is Na$_2$O present in silicate minerals only (see text for details).

source terrains and to assess the relative contribution of geochemically distinct crustal sources (Taylor & McLennan 1985; Condie & Wronkiewicz 1990). In addition, provenance-

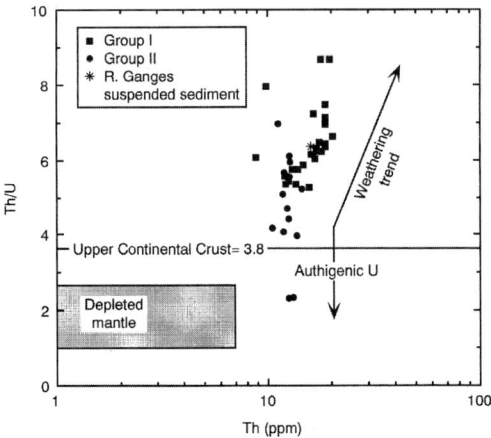

Fig. 11. Plot of Th/U–Th. Note that Group II samples have Th/U ratios similar to or less than those observed for Group I turbidites. River Ganges suspended sediment data are from Martin & Meybeck (1979). Upper Continental Crust, authigenic enrichment trend, weathering trend and depleted mantle compositions are from Taylor & McLennan (1985) and McLennan *et al.* (1993).

sensitive geochemical signatures based on a broad range of elements normalized against representative bulk crustal averages such as Upper Continental Crust (Taylor & McLennan 1985) can be used to discriminate between sediments derived from different source terrains and to identify the tectonic setting of sedimentary basins (McLennan *et al.* 1990, 1993). Both approaches are used to characterize the provenance of each of the major facies groups.

Group I. The relatively flat Upper Continental Crust normalized trace element signatures of Group I turbidites (Fig. 6a) indicate that these sediments are derived from a recycled orogenic (passive margin) provenance characterized by upper crustal rocks similar in bulk composition to granite (Taylor & McLennan 1985; McLennan *et al.* 1990, 1993). Specific features such as high SiO$_2$/Al$_2$O$_3$, marked depleted Eu anomalies (Eu/Eu* *c.* 0.65), LREE (La$_N$/Yb$_N$>7.5) and large ion lithophile element (Cs, K, Rb, Th) enrichment, and relative depletion in compatible and mafic elements (Co, Cr, Sc, Ti) are all consistent with a source terrain dominated by rocks of granitic composition. Together with the quartz–sodic plagioclase–mica–chlorite–amphibole mineralogy, the geochemical data are compatible with sediments sourced by mixed (meta)sedimentary–granitic lithologies similar to those present within the Himalayan mountain belt

(Bouquillon *et al.* 1990; Yokoyama *et al.* 1990; France-Lanord *et al.* 1993).

Similarities between the bulk geochemistry of the Himalayas and distal fan sediments are further strengthened by general comparison of Group I lithofacies with modern River Ganges suspended sediment (Figs 6d, 7d and 8). However, Group I turbidites are generally enriched in compatible elements (Co/Th, Cr/Th, Sc/Th) and have higher Eu/Eu* when compared with River Ganges suspended sediment (Figs 7 and 12). These features may indicate a greater contribution of basaltic crust to the Ganges–Brahamaputra sediment dispersal system during Neogene fan deposition compared to modern sediment supply (Figs 13 and 14).

Group II. In contrast to the bulk upper crustal provenance inferred from the geochemistry of Group I turbidites, Group II lithofacies exhibit mineralogical (high smectite content) and geochemical characteristics (Co, Cr, Fe, Ti, Sc enrichment; K, Rb, Th depletion; Sc/Th>1; Eu/Eu*>0.85; Figs 6b, 7b and 8) which suggest an important detrital input from basaltic crustal sources (Taylor & McLennan 1985; McLennan *et al.* 1990, 1993). This enrichment is shown clearly on Th–Hf–Co, La–Th–Sc and Co/Th–La/Sc diagrams (Figs 12 and 13) where individual samples plot towards basaltic compositions. Group II

turbidites, however, reveal features ($La_N/Yb_N > 7.5$, $Gd_N/Yb_N > 2.0$, low Sc/Th) and display mixing trends (Figs 12 and 13) which are incompatible with a purely basaltic sediment provenance such as the Deccan Trap basalts of central India (see Fig. 1), and this necessitates mixing of basaltic detritus with sediment derived from source lithologies characterized by LREE enrichment, HREE and compatible element depletion, and Eu/Eu*>0.85.

Although data clearly satisfy mixing between basaltic and grantic crustal sources for certain trace elements (Fig. 13), bimodal mixing of these components is unable to account for the high Eu/Eu* values which characterize Group II turbidites (see Fig. 8). An alternative bulk crustal source to granite which meets the requirements of LREE enrichment, HREE and compatible element depletion, and Eu/Eu*>0.85 may be found by substituting tonalite–trondhjemite–granodiorite (TTG; Condie 1993) for granite in mixing models (see Fig. 14 for details). Major crustal reservoirs with a bulk composition similar to TTG are most likely to occur within Precambrian Shield terrains where tonalitic rocks constitute a significant proportion of crustal volume (Condie 1993). Archaean rocks of the Indian Shield, in particular, provide a potentially important reservoir of tonalitic crust (Radhakrishna & Naqvi 1986;

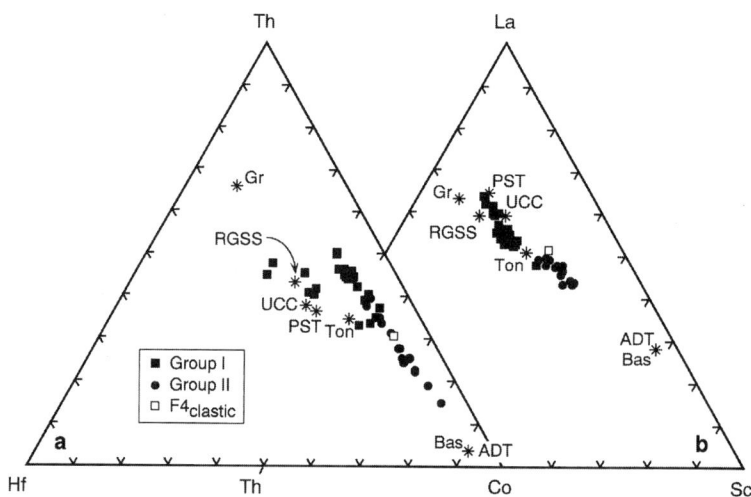

Fig. 12. Ternary diagrams (**a**) Th–Hf–Co and (**b**) La–Th–Sc showing the distribution of distal Bengal Fan sediments. Linear arrays indicate bimodal mixing. Displacement of some Group I lithofacies data towards the Hf apex implies a higher zircon content in these samples. A positive covariant relationship ($r = 0.673$) between Hf and SiO_2/Al_2O_3 suggests that high Hf contents are related to grain size partitioning and the concentration of zircon in coarser-grained turbidite deposits as opposed to recycling of old (meta)sedimentary rocks (McLennan *et al.* 1993). Data sources for River Ganges suspended sediment (RGSS), Precambrian shield terrain (PST), 'average' Deccan Traps and Upper Continental Crust (UCC) are given in Fig. 6. Gr, Phanerozoic granite; Ton, Phanerozoic tonalite–trondhjemite–granodiorite; Bas, Phanerozic basalt are taken from Condie (1993).

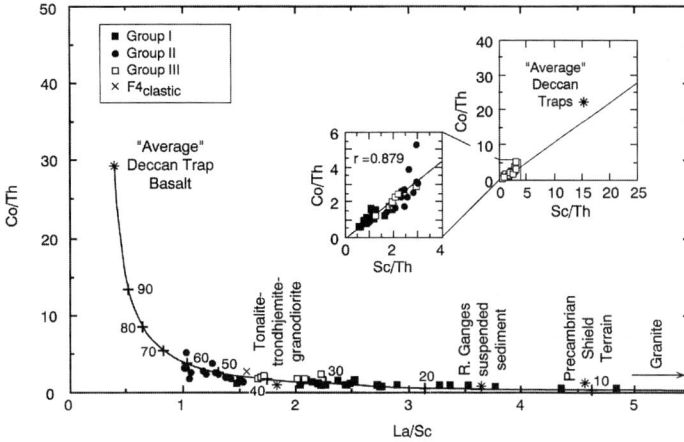

Fig. 13. Co/Th–La/Sc diagram showing the distribution of distal Bengal Fan sediments. An assessment of the possible proportionate contribution of sources to the different lithofacies groups has been made assuming basaltic and granitic end-members. Crosses (+) show mixing proportions in 10% increments. Basalt composition is an 'average' value calculated from Deccan Trap data reported in Lightfoot (1985). Phanerozoic granite is taken from Condie (1993). See Fig. 6 for all data sources except Phanerozoic tonalite–trondhjemite–granodiorite (TTG), for which data are taken from Condie (1993). Internal consistency of two-component mixing is supported by the linear correlation between Co/Th–Sc/Th (inset). The correlation coefficient is calculated using Bengal Fan data only.

Hansen *et al.* 1995) available as a source of sediment supply to the Bengal Fan (see also Einsele *et al.* 1996). Although estimates for the average geochemical composition of the Indian Shield are not currently available, the present-day 'Canadian

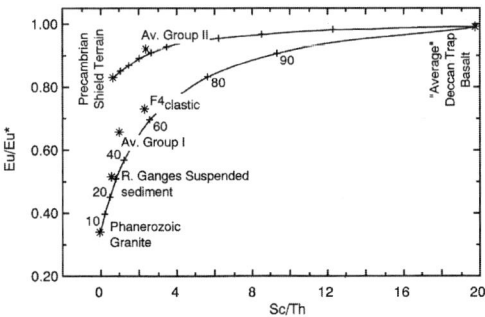

Fig. 14. Eu/Eu*–Sc/Th diagram illustrating the effects of bimodal mixing of sediment derived from tonalitic Precambrian Shield terrain and granite with Deccan Trap basalt. Crosses (+) show mixing proportions in 10% increments. Note that River Ganges suspended sediment and average Group I lithofacies plot close to the basalt–granite mixing line, while average Group II lithofacies plot close to the basalt–Precambrian shield terrain mixing line. Mean values for each lithofacies are used for clarity (distribution of all distal Bengal Fan data is shown in Fig. 8). See Fig. 6 for details of all data sources except Phanerozoic granite, for which data are taken from Condie (1993).

Precambrian Shield estimate' of Shaw *et al.* (1986) represents a possible proxy for Precambrian terrains world-wide and displays a number of features consistent with those expected if sediments equivalent in composition to Group II turbidites were produced by mixing of detritus derived from erosion of basaltic (Deccan Traps) and tonalitic (Precambrian Indian Shield) crust of the Indian subcontinent (see Figs 13 and 14). Bimodal mixing of sediments similar in composition to the Canadian Precambrian Shield estimate and Deccan Trap basalts are clearly capable of meeting many of the criteria demanded by trace element data.

Group III. Estimates of the trace element geochemistry of the clastic component (F4$_{clastic}$) of Group III lithofacies reveal strong compositional affinities with Group II turbidites. Similarities in the behaviour of the transition metals (Co, Cr, Fe, Ti, Sc enrichment) and large ion lithophile elements (K, Rb, Th depletion) suggest that carbonate-rich sediments were derived from the same Indian subcontinent provenance invoked as the source of Group II deposits. However, the clastic component of Group III turbidites is relatively depleted in Eu (Eu/Eu* *c.* 0.72) compared to most Group II samples and consequently requires an increased contribution from an Eu-depleted, differentiated granitic crustal source to account for this.

Stow *et al.* (1990) have previously inferred a Sri Lankan origin for Group III turbidites in order to account for the source of a bioclastic component (notably benthic foraminifera) of supposed shallow marine equatorial origin present in these sediments. A Sri Lankan source, dominated by late Archaean granulites (Newton & Hansen 1986), may be capable of accounting for the geochemistry of Group III turbidites. However, as with the Pre-cambrian rocks of southern India, it is difficult to estimate the likely average geochemical compo-sition of detritus eroded from Sri Lanka. A source terrain composition similar to the Precambrian Shield estimate of Shaw *et al.* (1986) fails to satisfy many of the geochemical characteristics displayed by the clastic component of Group III turbidites (depleted Eu/Eu*, low La/Sc, high Sc/Th and Cr/Th). The addition of a granitic crustal component with a depleted Eu anomaly may

modify Eu/Eu* (Fig. 14), but would also require a contribution from mafic sources to account for the relatively high Co, Cr, Sc contents of these sediments. At present, insufficient data are available to allow us to evaluate fully a Sri Lakan/southern India provenance for Group III turbidites on a geochemical basis.

Nd and Sr isotopic data

Nd isotopic data. Although France-Lanord *et al.* (1993) state that the Nd isotopic data obtained from fan deposits are not consistent with an Indian Shield or Deccan Trap provenance, comparisons between published isotopic data from Precambrian rocks of southern India (e.g. Peucat *et al.* 1989;

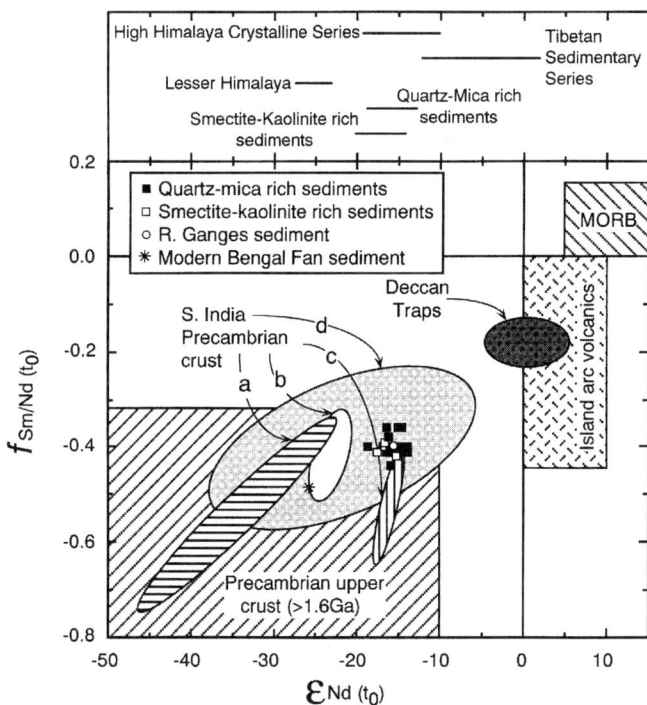

Fig. 15. Plot of f Sm/Nd$_{(t0)}$–εNd$_{(t0)}$ showing the Nd isotopic composition of distal fan sediments (quartz–mica and smectite–kaolinite turbidites of France Lanord *et al.* (1993) equate with Group I and Group II lithofacies respectively), River Ganges suspended sediment and Recent Bengal fan sediment relative to possible southern India Precambrian and Deccan Trap sources. The narrow range of εNd$_{(t0)}$ is consistent with the typical range of values observed in many Phanerozoic sediments (Goldstein 1988). The scatter of points suggests contributions from three or more isotopically distinct crustal sources (bimodal mixing should be characterized by a linear array). εNd(t$_0$) and f Sm/Nd were calculated using present-day values of 0.512636 and 0.325 for ^{143}Nd/^{144}Nd$_{CHUR}$ and Sm/Nd$_{CHUR}$ respectively (DePaolo 1988; McLennan & Hemming 1992). Data sources: Distal Bengal Fan sediments, Lesser Himalaya, High Himalayan Crystalline Series, Tibetan Sedimentary Series (Bouquillon *et al.* 1990; France-Lanord *et al.* 1993); modern Bengal Fan sediment (McLennan *et al*, 1990); River Ganges sediment (Goldstein *et al.* 1984); S. India Precambrian crust: (a) Peucat *et al.* (1989); (b) Paul *et al.* (1990); (c) Burton & O'Nions (1990); (d) Harris *et al.* (1994); Deccan Traps (Lightfoot 1985); fields for Precambrian upper crust, island arc volcanics and MORB from McLennan & Hemming (1992).

Burton & O'Nions 1990; Paul *et al.* 1990; Harris *et al.* 1994), the Deccan Traps (Lightfoot 1985) and the Bengal Fan (Fig. 15) suggest that observed $^{143}Nd/^{144}Nd$ values could be explained if turbidite sediments were either derived directly from the Indian Shield or involved mixing of Precambrian crust with material eroded from the Deccan Traps. The potential importance of an Indian subcontinent source to distal fan sedimentation is further supported by a single isotopic value from Recent sediments recovered from the western margin of the Bengal Fan (McLennan *et al.* 1990) which gives an $\varepsilon Nd_{(t0)}$ (−25.7) consistent with a Precambrian Indian subcontinent source (see Fig. 15).

The similarity between the geochemical and isotopic composition (see Fig. 15) of modern River Ganges suspended sediment and quartz–mica Group I turbidites overwhelmingly supports the

conclusion that these distal fan sediments were derived from erosion of the Himalayas. In contrast, interpretation of geochemical data obtained from smectite–kaolinite Group II and Group III turbidites indicates that these sediments were sourced by a mixture of detritus derived from the Precambrian Shield and the Deccan Trap basalts rather than granitic, upper crustal rocks exposed in the Himalayas. In order to test this hypothesis with respect to Nd isotopes we constructed a bimodal mixing model using published data from Precambrian Shield and Deccan Trap rocks.

Figure 16 illustrates the results of the model and demonstrates that mixing of Deccan Trap and Precambrian Shield-derived detritus could reasonably account for the range of isotopic values observed for smectite–kaolinite turbidites. Importantly, the mixing ratios calculated from Nd isotopic data are

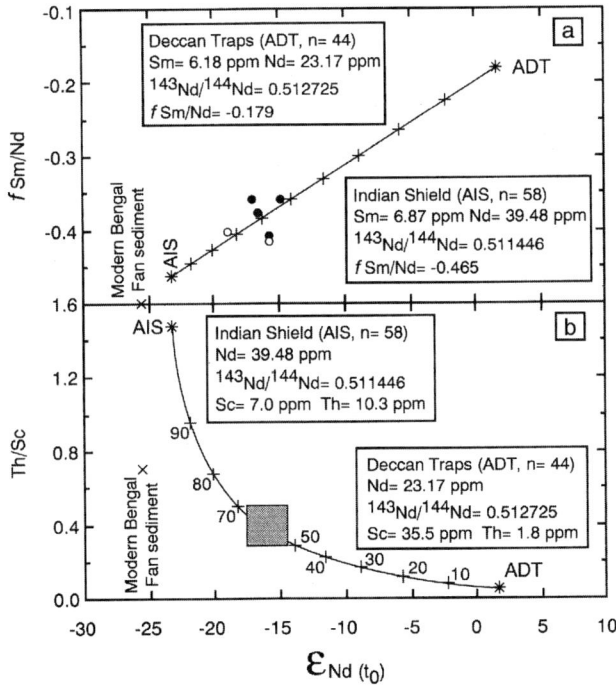

Fig. 16. (a) f Sm/Nd$_{(t0)}$–$\varepsilon Nd_{(t0)}$ and (b) Th/Sc–$\varepsilon Nd_{(t0)}$ mixing models for sediment derived from the Indian subcontinent. Data plotted in (a) are compositions of smectite–kaolinite turbidites (filled circles, F3; open circles, F4$_{gr}$) taken from Bouquillon *et al.* (1990) and France-Lanord *et al.* (1993). Shaded area in (b) shows the range of Th/Sc and $\varepsilon Nd_{(t0)}$ values for smectite–kaolinite (lithofacies F3) samples. The two-component mixing lines were calculated using an average value for Deccan Trap basalts calculated from the data of Lightfoot (1985; Mahabaleshwar, Panhala, Ambenali, Poladpur Formations) and an average value for the Indian shield calculated from the data of Peucat *et al.* (1989), Paul *et al.* (1990), Burton & O'Nions (1990) and Harris *et al.* (1994; not including samples 32D, 32F). Sc and Th composition of the Indian shield was assumed to be identical to the Canadian Precambrian Shield estimate of Shaw *et al.* (1986). Crosses (+) show mixing proportions in 10% increments. Recent Bengal Fan sediment data from McLennan *et al.* (1990). $\varepsilon Nd(t_0)$ and f Sm/Nd were calculated using present-day values defined in Fig. 15.

similar to those required to model successfully other features of the geochemistry of these Group II sediments (compare Figs. 13, 14, 16). For bimodal mixing a linear array of isotopic values would ideally be expected. In practice the small quantity of data (complete Sm, Nd and ^{143}Nd/^{144}Nd are only available for six samples) published for these turbidites define only a weak covariant trend and are insufficient in number to allow bimodal mixing to be evaluated fully. Furthermore, although two major source terrains (i.e. Deccan Traps and Indian Shield) are invoked to account for the geochemical composition of Group II turbidites, the diverse lithological make-up of the Indian Shield would indicate that a simple bimodal mixing model is unlikely to account precisely for the distribution of Nd isotopic data. Despite these uncertainties, it is sufficient to note that Nd isotope signatures preserved in distal fan sediments are not unique to a Himalyan source, and that mixing of detritus derived from the Indian Shield and the Deccan Traps can account for Nd isotopic data in samples where mineralogical and geochemical data are consistent with an Indian subcontinent provenance. Additional isotopic analyses of Group II and Group III lithofacies, together with modern sediment samples from Indian subcontinent rivers, are required to test this model further.

Sr isotopic data. As with Nd isotopic data, the Sr isotopic composition of distal fan sediments (particularly Group II turbidites) could be explained by mixing of young, Rb-depleted basaltic material (Lightfoot 1985) with radiogenic Precambrian crust (e.g. Taylor *et al.* 1984; Paul *et al.* 1990; Bhaskar Rao *et al.* 1992; Sarkar *et al.* 1993). The routine application of strontium isotopic data to provenance studies is, however, less reliable than ^{143}Nd/^{144}Nd because of the effects of differential chemical weathering on isotopic signatures preserved in clastic sediments (Goldstein 1988; Nelson & DePaolo 1988). Although Sr isotopes may provide a potentially unreliable indicator of sediment provenance, the data do allow us to test the 'variable weathering of a common source terrain' model proposed by France-Lanord *et al.* (1993) to account for differences in sediment mineralogy and similarities in ^{143}Nd/^{144}Nd.

Because chemical weathering processes result in the progressive (and mineralogically selective) loss of Sr relative to Rb with increased weathering (Taylor & McLennan 1985), Rb/Sr ratios of sediments are frequently modified to give values greater than those present in source rocks (Nesbitt *et al.* 1980; Goldstein 1988; Nelson & DePaolo, 1988). The effect of Sr removal is reflected in the young Sr mantle-depleted model ages (T_{DM}^{Sr}) relative to Nd (T_{DM}^{Nd}) model ages calculated for the

same sediment sample (Goldstein 1988; Nelson & DePaolo 1988). If the source of distal fan sediments remained constant, as suggested by France-Lanord *et al.* (1993), then it would be expected that T_{DM}^{Sr} ages should decrease systematically with progressive chemical weathering, while T_{DM}^{Nd} ages would remain constant due to the retention of Sm and Nd during weathering (Goldstein 1988; Nelson & DePaolo 1988; McLennan & Hemming 1992).

Although paired Sr and Nd data are only available for a small number of samples, it is clear that while Nd model ages fall within a narrow range, Sr model ages vary considerably (Fig. 17). Such behaviour is consistent with the effects of increased chemical weathering. However, the data show no systematic relationship between model age and smectite content (used as a measure of weathering intensity) which would indicate that the variation in sediment mineralogy was due to differences in the degree of weathering of a common crustal source (Fig. 17). In this respect Sr isotope values support the contention based on geochemical data that these sediments were derived from source terrains of differing bulk composition.

Changes in sediment provenance during fan deposition

Changes in sediment provenance with respect to time are conveniently represented by variations in the Cr/Th and Eu/Eu* (Fig. 18). The overall pattern of change in sediment provenance during the preceding *c.* 17.5 Ma is similar to that reported by Cochran (1990), Stow *et al.* (1990) and Amano & Taira (1992). Changes in sediment provenance with respect to time are summarized below (no geochemical data are available for recent deposits of <0.05 Ma):

0.05–0.73 Ma: quartz–mica turbidites derived from the Himalayan provenance;

0.73–3.40 Ma: smectite–kaolinite mud turbidites and bioclastic turbidites derived from the Indian subcontinent and Sri Lanka;

3.40–6.90 Ma: mixed provenance comprising turbidites sourced from the Indian subcontinent and the Himalayas.

6.90–17.5 Ma: quartz–mica turbidites derived from the Himalayas, with minor volumes of sediment sourced from the Indian subcontinent.

Although it is obvious that the bulk of sediment reaching the Bengal Fan has been sourced by erosion of the Himalayas (Curray & Moore 1971; Curray 1994; Johnson 1994; Einsele *et al.* 1996)

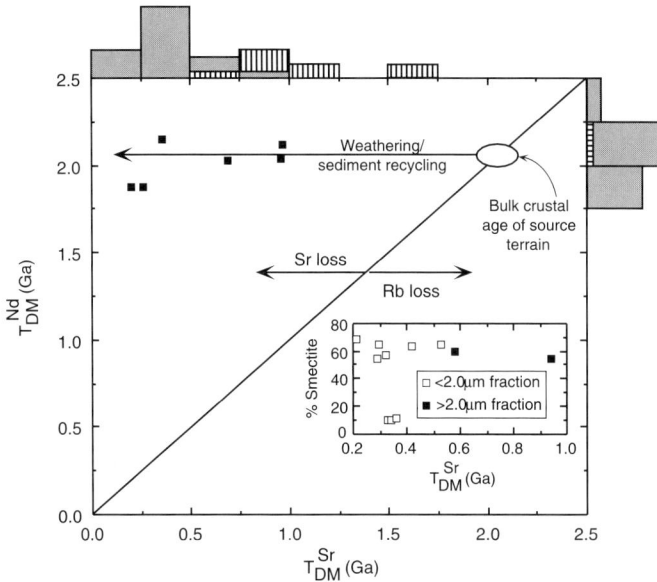

Fig. 17. Plot of T_{DM}^{Sr}–T_{DM}^{Nd} for Bengal fan sediments (all data from Bouquillon *et al.* 1990; France-Lanord *et al.* 1993). Samples were decarbonated with 10% acetic acid. Note that Sr mantle-depleted model ages are considerably younger than equivalent Nd ages consistent with Sr loss and Nd retention during weathering and transport of clastic detritus. Histograms show the mantle-depleted age distribution of all available (unpaired) isotopic data. Sr model ages show a wide spread of values with clay fraction samples (<2 μm; stippled infill) giving younger ages than coarser fraction samples (>2 μm; lined infill). Inset diagram shows the relationship between Sr model age and smectite content of individual fan samples. The data reveal no systematic relationships which would be consistent with progressive weathering of a common crustal source. Model ages were calculated using the depleted mantle parameters of Goldstein & Jacobsen (1988), and McLennan & Hemming (1992) for Sr and Nd respectively.

and transported *via* the Ganges–Brahamaputra dispersal system, it is not at all certain why the record of deposition on the distal fan should be interrupted by significant periods (up to 2.5 Ma) of turbidite sedimentation involving material derived predominantly from the Indian subcontinent. Assuming a continuous, high rate of terrigenous sediment transport to the Bay of Bengal (Copeland & Harrison (1990), although see Rea (1992) for a broader perspective of the relationship between Himalayan uplift and sediment supply to the northern Indian Ocean), it might be expected that sediment supply to the distal fan would be controlled predominantly by changes in relative sea level (Stow *et al.* 1983, 1990). Under these circumstances sediment supply to the Bengal Fan may be divided into: (1) periods of thick, relatively coarse-grained, quartz–mica turbidite sedimentation corresponding to low-stands associated with fluvial incision and efficient sediment bypassing of the Bengal shelf; and (2) periods of clay-rich, smectite–kaolinite turbidite deposition corresponding to high-stands when shorelines migrate

landwards allowing Himalayan-derived sediment to be stored on the Bengal shelf, while Indian subcontinent-derived sediment continued to be transported across the much narrower Indian continental margin.

Comparisons between the distal fan sedimentary record and eustatic sea level (Stow *et al.* 1990; Fig. 18) show that low-stand periods correlate only in a general sense with the occurrence of thick, coarser-grained Himalayan-derived turbidite sequences. Of particular importance is the relative absence of quartz–mica turbidites at the site of Leg 116 during the Pliocene and early Pleistocene when two periods of major sea-level fall occur. Evidence from modern Bengal shelf sediment dispersal patterns (Kuehl *et al.* 1989) suggests that efficient sediment bypassing of the shelf occurs even at relatively high, present-day sea levels. As a consequence Himalayan-derived sediment should continue to reach distal portions of the fan, even during relative high-stands, and this is supported by the deposition of Himalayan-derived turbidites on the distal fan during 3.4–6.9 Ma when estimated

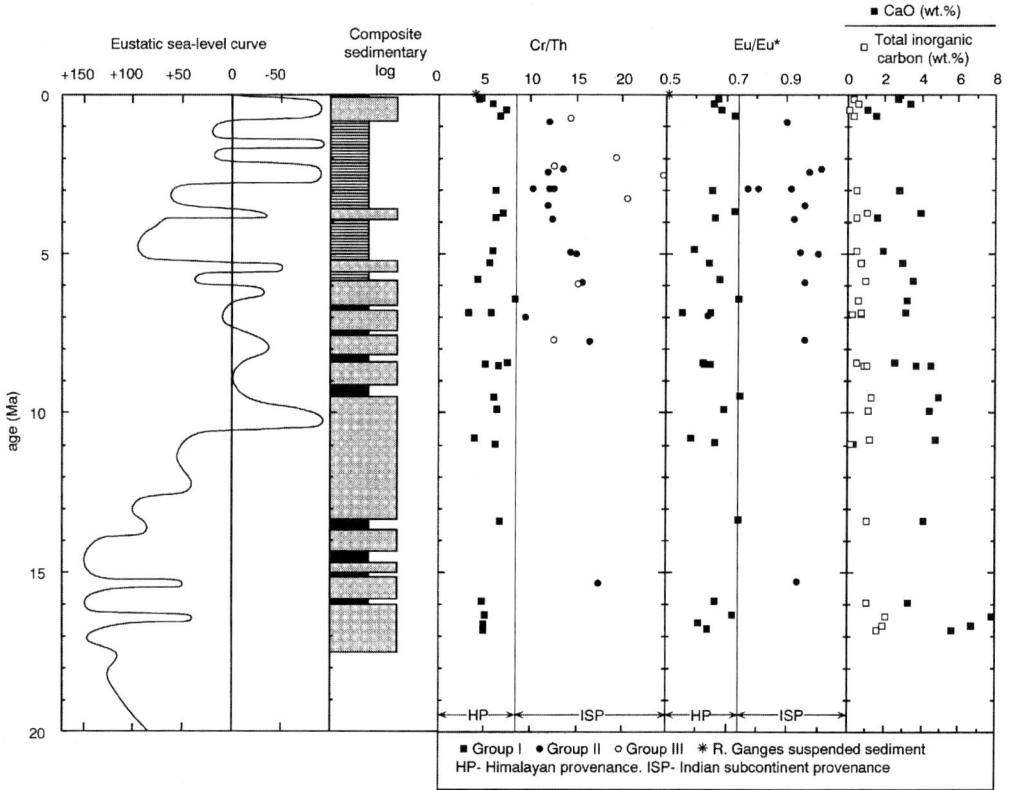

Fig. 18. Variation in sediment provenance with respect to time using Cr/Th (all data) and Eu/Eu* (from Group I and Group II turbidites only). Composite log of distal fan sedimentation and eustatic sea-level curve from Stow *et al.* (1990; symbols as for Fig. 2). River Ganges suspended sediment data from Martin & Maybeck (1979). CaO and TIC data from Group I turbidites only. The general increase in CaO and TIC with time suggests that the abundance of carbonate-containing source rocks (notably limestones and marbles present in the Tibetan Sedimentary Series and High Himalaya Crystalline Series) has decreased progressively since the Lower Miocene.

eustatic sea levels were over 50 m higher than the present-day.

Stow *et al.* (1990) have previously noted that changes in the position of major distributary fan channels, leading to a shift in sediment supply either to the currently abandoned easterly fan lobe of the Bengal Fan or the Nicobar Fan, could account for the occurrence of prolonged periods when little or no sediment derived from the Himalayan provenance reached distal fan sites of Leg 116. Assuming that Himalayan detritus was continuously reaching the Bay of Bengal submarine canyon, it would appear highly unlikely that relatively small-scale, medial to distal fan autocyclic sedimentary processes could account for significant periods of non-deposition and that regional-scale changes such as channel avulsion and lobe switching are required. If periodic fan-lobe abandonment did occur for significant lengths

of time then sediment supply to the distal fan would be relatively insensitive to sea-level change affecting the northern Bay of Bengal and would allow turbidites transported from the Indian subcontinental margin to accumulate in the absence of sediment derived from the Himalayas. Under these circumstances attempts to reconstruct major Himalayan tectonic and climatic events based on data obtained from the sedimentary record of the distal Bengal Fan may prove to be unreliable.

Conclusions

(1) Whole-rock mineralogical and geochemical investigation of samples recovered during ODP Leg 116 allow sediments to be resolved into three distinct groups consistent with major depositional lithofacies associations recognized in distal Bengal Fan cores.

(2) Estimates of the intensity of source-area chemical weathering based on CIA are consistent with the clay mineralogy of distal fan sediments. A–C*N*–K and Th/U–Th diagrams reveal important differences in the pattern of behaviour of quartz–mica and smectite–kaolinite turbidites with respect to changes in chemical weathering intensity. The most likely explanation for these differences is that they result from variations in the degree of chemical weathering of two different source terrains as opposed to differential weathering of a single source terrain of relatively uniform composition.

(3) Turbidites characterized by high concentrations of quartz and mica (Group I lithofacies) have trace element compositions consistent with erosion of a source terrain similar in bulk composition to Upper Continental Crust. Both mineralogical and geochemical signatures are compatible with a Himalayan source consisting of (meta) clastic sediments, granites and minor quantities of basalt.

(4) Organic-rich turbidites (Group II lithofacies), characterized by high concentrations of smectite and kaolinite, have trace element compositions which indicate a source containing a mixture of basaltic and tonalitic crust. Such a source is consistent with erosion of Deccan Trap basalts and Precambrian Shield terrains of the Indian subcontinent.

(5) The clastic component of carbonate-rich turbidites (Group III) exhibits many of the mineralogical and trace element characteristics displayed by Group II turbidites and indicates that a similar Indian subcontinent provenance is probable for these sediments. Although a Sri Lankan source has been proposed for these sediments on the basis of faunal content, geochemical data alone are insufficient to allow a Sri Lankan source to be reliably identified.

(6) Nd isotopic signatures published for smectite–kaolinite (Group II and Group III lithofacies) turbidites are not unique to a Himalayan source. Modelling of Nd isotopes shows that similar $\varepsilon Nd_{(t0)}$ values to those reported from the High Himalaya and distal fan sediments can be obtained by mixing of detritus derived from the Precambrian Shield and Deccan Trap basalts of the Indian subcontinent. Importantly, the mixing ratios obtained on the basis of Nd isotopic data are similar to those required to model successfully other features of the geochemistry of these sediments, and this further supports an Indian subcontinent provenance for these turbidites.

(7) Changes in the provenance of turbidites over the last c. 17.5 Ma indicate significant variation in the supply of Himalayan-derived sediment to the distal fan. The most likely causes of observed patterns of sediment supply are thought to be related to an interplay between sea-level change and distributary channel switching. As a consequence, attempts to reconstruct major Himalayan tectonic and climatic events based on data obtained from the sedimentary record of the distal Bengal Fan may be unreliable.

Appendix

Analytical results of 46 samples from Sites 717 and 718

Sample No. (ODP core ref)	Depth (mbsf)	Age (Ma)	Facies	SiO_2 (%)	Al_2O_3 (%)	TiO_2 (%)	Fe_2O_3T (%)	MgO (%)	CaO (%)	Na_2O (%)	K_2O (%)	MnO (%)	P_2O_5 (%)	LOI (%)	Total (%)	$Na_2O^{\dagger}$ (%)	Na_2O^{*} (%)	TOC (%)	TIC (%)
Hole 717C																			
2.2.35–40	15.35	0.16	1	70.60	11.24	0.54	4.20	2.09	2.73	2.13	2.63	0.06	0.10	3.83	100.15	0.42	1.71	0.15	0.32
6.1.53–55	46.53	0.29	2	57.48	15.99	0.76	5.92	3.30	3.40	2.25	3.79	0.09	0.14	6.91	100.03	0.47	1.78	0.46	0.55
10.1.110–112	75.61	0.47	2	54.08	18.37	0.86	8.60	3.15	1.08	2.17	3.63	0.09	0.13	7.89	100.05	0.96	1.21	0.66	0.05
18.CC.8–12	131.58	0.67	1	78.17	9.83	0.24	2.98	0.90	1.55	2.11	2.49	0.05	0.06	1.68	100.06	0.25	1.86	0.57	0.35
20.4.36–41	155.36	0.76	4gr	35.03	12.28	0.71	6.56	2.54	17.03	1.47	1.57	0.11	0.22	22.44	99.96	0.85	0.62	1.88	3.77
23.5.140–145	186.41	0.87	3	50.72	17.25	1.20	10.41	3.03	2.10	1.36	2.47	0.12	0.13	10.95	100.31	0.69	0.40	1.56	0.47
27.5.85–89	223.85	2.01	4gr	25.98	9.02	0.46	5.67	2.04	24.23	1.13	0.64	0.24	0.26	30.36	100.03	0.84	0.29	5.64	2.51
28.5.102–106	233.52	2.29	4gr	13.98	4.00	0.19	2.72	1.38	39.88	0.51	0.07	0.93	0.64	35.37	99.67	0.84	0.03	0.66	8.21
29.2.80–86	238.31	2.34	3	50.75	16.33	1.69	11.01	2.66	1.97	2.34	2.01	0.12	0.13	11.12	100.13	1.09	1.25	1.93	0.38
30.1.111–116	246.61	2.44	7dg	45.55	16.08	1.37	16.40	2.46	0.74	2.02	1.79	0.12	0.11	13.50	100.14	0.77	1.25	1.87	0.29
31.1.55–57	255.55	2.57	4gr	20.24	7.71	0.38	5.65	1.94	29.59	1.05	0.33	0.42	0.93	31.06	99.30	1.03	0.02	1.27	6.79
33.5.80–85	280.80	2.94	3	48.65	18.25	1.25	10.06	2.64	2.37	1.96	2.27	0.14	0.10	12.35	100.04	0.78	1.18	2.30	0.38
33.5.120–122	281.20	2.95	3	51.01	18.50	1.33	9.71	2.62	1.86	1.98	2.60	0.12	0.12	10.40	100.25	nd	nd	1.37	0.17
33.6.31–33	281.81	2.96	3	52.47	17.64	1.26	12.27	2.79	0.47	2.12	2.01	0.06	0.09	8.84	100.01	nd	nd	0.60	bdl
34.1.53–55	284.03	3.00	2	53.77	17.21	0.84	7.59	3.28	2.85	1.93	3.93	0.13	0.10	8.29	99.92	0.75	1.18	0.38	0.49
35.3.46–51	296.46	3.25	4gr	33.85	13.43	0.68	10.36	2.80	13.33	1.25	1.25	0.21	0.41	22.52	100.09	0.82	0.43	3.36	2.15
36.5.80–85	309.31	3.46	3	46.59	20.28	1.52	10.68	2.37	1.70	1.73	1.98	0.13	0.12	12.89	99.99	0.89	0.84	2.06	0.50
39.1.82–87	331.82	3.69	1	58.31	14.33	0.88	6.47	2.70	3.98	2.07	3.13	0.09	0.12	7.99	100.07	0.59	1.48	0.72	1.05
40.1.103–108	341.53	3.85	2	56.07	17.06	0.79	8.23	2.94	1.59	1.90	3.70	0.21	0.11	7.41	100.01	0.63	1.28	0.52	0.46
40.3.117–123	344.67	3.91	3	43.33	16.30	1.00	9.27	2.25	8.23	1.76	1.96	0.06	0.24	15.32	99.72	1.07	0.69	0.48	2.73
47.3.80–85	410.81	4.87	7lg	54.98	17.68	0.78	7.84	2.87	1.93	1.77	3.79	0.13	0.11	8.15	100.03	0.57	1.20	0.64	0.49
47.CC.33–38	416.11	4.94	7dg	48.91	18.91	1.61	10.82	2.24	1.78	1.84	1.81	0.09	0.10	11.90	100.01	0.82	1.02	2.18	0.30
48.2.80–86	418.81	4.98	2	50.86	17.75	1.47	12.59	2.46	0.55	2.00	1.93	0.05	0.11	10.40	100.17	0.77	1.24	1.61	bdl
50.1.11–16	435.61	5.27	2	56.59	16.22	0.76	6.76	2.90	2.99	1.89	3.97	0.15	0.11	7.63	99.97	0.61	1.28	0.43	0.75
53.1.104–108	465.04	5.81	1	70.06	10.37	0.54	4.04	1.97	3.56	1.82	2.54	0.07	0.10	5.30	100.37	0.42	1.40	0.17	0.95
54.5.126–131	480.76	5.90	3	49.12	19.26	1.50	11.03	2.44	0.74	1.97	1.56	0.07	0.09	12.46	100.24	0.63	1.34	2.32	0.20
55.CC.23–28	482.83	5.93	4gr	44.19	15.41	0.84	9.41	2.57	9.46	1.62	1.98	0.18	0.18	14.40	100.24	0.85	0.77	0.62	1.93
59.3.121–126	525.21	6.43	7lg	54.73	16.95	0.85	7.28	2.84	3.23	1.77	3.53	0.20	0.11	8.39	99.88	0.64	1.13	0.54	0.57
62.CC.11–15	559.46	6.84	1	74.75	8.99	0.53	3.14	1.45	3.19	1.75	2.09	0.06	0.09	4.09	100.14	0.31	1.44	0.10	0.70
63.2.42–47	560.92	6.86	2	53.34	18.54	0.83	10.39	3.19	0.71	1.29	4.53	0.10	0.10	6.72	99.73	0.42	0.87	0.30	0.24
64.1.53–55	569.03	6.96	7dg	48.37	15.81	0.94	10.02	2.58	6.76	1.49	2.56	0.09	0.11	11.36	100.09	0.78	0.71	0.40	1.10
70.5.63–69	632.13	7.71	4gr	41.95	12.99	0.93	8.82	2.35	13.73	1.46	1.45	0.09	0.11	16.32	100.20	0.85	0.61	1.03	0.65
70.5.95–100	632.45	7.72	3	46.73	17.12	1.57	13.84	2.28	2.69	1.87	1.39	0.13	0.09	12.49	100.20	0.82	1.05	1.30	0.58
79.6.91–96	719.41	8.41	2	54.90	17.87	0.79	7.12	3.12	2.57	1.73	3.56	0.18	0.13	8.16	100.13	0.54	1.19	0.69	0.50
81.CC.18–23	731.03	8.46	1	66.24	11.42	0.58	4.27	2.13	4.54	1.80	2.66	0.07	0.12	6.20	100.03	0.27	1.54	0.22	0.92
82.1.48–50	739.98	8.50	2	56.22	16.14	0.17	6.85	3.04	3.69	1.45	3.70	0.11	0.11	7.85	99.87	0.46	1.01	0.41	1.02
Hole 718C																			
46.1.53–55	437.33	9.49	2	55.57	15.14	0.71	5.74	3.33	4.90	1.57	3.42	0.13	0.11	8.96	99.59	0.56	1.01	0.23	1.33
50.1.60–63	475.41	9.89	2	53.07	16.84	0.75	6.65	3.52	4.41	1.55	3.75	0.11	0.14	9.03	99.82	0.62	0.93	0.24	1.13
59.2.87–92	562.67	10.80	1	71.91	8.93	0.51	2.72	1.76	4.79	1.72	1.94	0.04	0.12	5.66	100.11	0.36	1.36	0.01	1.18
65.2.53–55	619.33	10.96	1	62.08	16.71	0.76	7.04	2.52	0.31	1.49	3.91	0.05	0.11	4.79	99.78	0.46	1.03	0.01	0.16
72.2.46–51	685.76	13.38	1	62.43	12.95	0.62	5.21	2.82	4.13	1.40	2.95	0.08	0.11	7.27	100.03	0.38	1.07	0.21	1.06
79.1.53–55	750.83	15.35	5r	50.99	22.44	1.02	10.57	2.91	2.19	1.84	2.49	0.11	0.12	11.68	100.19	0.59	1.25	0.11	0.27
81.1.52–54	769.82	15.93	2	53.82	17.69	0.73	6.89	3.03	3.29	1.13	4.42	0.13	0.13	8.73	99.99	0.41	0.72	0.40	1.06
87.1.53–55	826.83	16.36	2	49.73	15.19	0.62	5.71	3.66	7.81	0.94	4.01	0.09	0.11	12.33	100.21	0.39	0.55	0.39	2.10
96.1.20–22	905.71	16.64	2	52.91	14.66	0.67	5.39	3.53	6.73	1.12	3.82	0.09	0.11	10.93	100.06	0.39	0.73	0.30	1.94
3R.2.15–17	953.75	16.80	2	51.37	16.31	0.67	5.92	3.72	5.64	1.11	4.41	0.09	0.12	10.61	99.98	0.42	0.69	0.46	1.59

Sample No. (ODP core ref)	S (%)	CIA	Co (ppm)	Cr (ppm)	Cs (ppm)	Hf (ppm)	Rb (ppm)	Sc (ppm)	Ta (ppm)	Th (ppm)	U (ppm)	La (ppm)	Ce (ppm)	Nd (ppm)	Sm (ppm)	Eu (ppm)	Tb (ppm)	Tm (ppm)	Yb (ppm)	Lu (ppm)
Hole 717C																				
2.2.35–40	0.01	57.4	11.8	56.0	7.2	5.6	162	9.23	1.18	12.49	2.29	32.5	65.8	26	5.12	1.00	0.62	0.36	2.12	0.37
6.1.53–55	0.03	63.1	21.4	110.0	13.0	4.4	207	16.76	1.68	18.75	2.96	41.6	90.7	36	6.90	1.38	0.97	0.36	2.65	0.46
10.1.110–112	0.01	69.0	28.0	139.0	14.0	3.8	195	22.43	1.62	18.86	2.70	45.7	97.6	37	7.68	1.57	1.02	0.36	2.82	0.43
18.CC.8–12	bdl	61.3	6.5	57.0	3.1	3.5	110	6.29	0.76	8.38	1.46	23.7	46.6	15	3.11	0.69	0.44	0.21	1.44	0.28
20.4.36–41	0.72	nd	17.2	117.0	4.1	2.7	95	16.43	0.76	8.42	5.42	27.8	56.0	24	5.37	1.06	0.65	0.29	1.96	0.34
23.5.140–145	0.22	81.8	29.3	153.0	5.9	3.5	134	29.29	1.28	12.54	2.15	38.4	84.5	26	5.10	1.55	0.98	0.34	2.40	0.40
27.5.85–89	2.22	nd	11.7	127.0	2.6	2.0	69	12.35	0.52	6.70	7.83	25.6	50.6	22	4.97	0.88	0.58	0.30	2.05	0.29
28.5.102–106	bdl	nd	7.9	40.0	1.8	0.9	29	6.95	0.27	3.20	2.12	15.5	28.0	13	2.88	0.74	0.49	0.25	1.70	0.24
29.2.80–86	1.15	76.3	32.7	163.0	3.8	4.3	97	29.00	1.17	11.99	2.49	34.2	72.1	22	4.70	1.62	0.93	0.35	2.12	0.38
30.1.111–116	5.22	80.1	40.4	125.0	3.3	3.3	79	27.65	0.98	10.43	2.28	34.8	65.6	21	4.57	1.50	0.87	0.31	2.13	0.31
31.1.55–57	2.12	nd	8.6	114.0	1.3	1.0	42	9.15	0.32	4.65	5.55	18.4	34.0	14	3.29	0.64	0.42	0.19	1.31	0.21
33.5.80–85	0.50	75.1	34.6	158.0	6.9	3.5	124	29.80	1.24	13.06	2.78	38.2	79.9	24	5.39	1.66	1.03	0.34	2.57	0.45
33.5.120–122	0.43	nd	27.9	149.0	6.8	3.6	124	27.94	1.29	14.49	2.85	38.5	79.1	35	5.79	1.86	1.01	0.50	3.10	0.50
33.6.31–33	0.16	nd	22.0	160.0	4.7	3.3	111	30.69	1.10	12.60	1.40	32.7	69.7	31	7.25	1.67	0.97	0.48	2.85	0.42
34.1.53–55	0.01	69.2	18.5	111.0	15.2	3.9	191	20.70	1.45	17.85	2.06	42.3	85.1	34	7.57	1.50	1.06	0.37	2.93	0.51
35.3.46–51	1.55	nd	19.1	177.0	3.1	2.3	81	17.04	0.58	8.34	3.42	29.2	58.9	20	5.00	1.11	0.65	0.29	2.17	0.34
36.5.80–85	0.32	85.2	31.2	163.0	5.4	4.1	101	34.80	1.22	13.54	2.01	42.0	78.3	27	5.70	1.80	1.02	0.34	2.70	0.46
39.1.82–87	0.23	70.2	21.9	97.0	10.9	5.7	159	16.93	1.49	13.80	2.41	35.3	78.5	33	6.32	1.40	0.89	0.40	2.56	0.45
40.1.103–108	0.12	73.6	21.1	115.0	13.8	4.5	197	19.15	1.44	18.60	2.62	42.0	87.5	34	7.37	1.44	0.95	0.35	2.83	0.52
40.3.117–123	0.74	83.4	19.8	154.0	5.7	2.9	109	24.00	1.05	12.29	3.06	35.9	69.8	20	4.68	1.43	0.84	0.29	2.25	0.38
47.3.80–85	0.31	74.0	18.5	110.0	14.5	3.8	238	18.67	1.50	18.87	2.93	42.3	87.2	29	7.82	1.35	0.95	0.50	3.35	0.48
47.CC.33–38	0.70	80.0	30.8	171.0	4.7	4.1	95	33.64	1.27	11.89	2.14	36.1	72.3	20	5.50	1.70	0.95	0.35	2.56	0.43
48.2.80–86	0.53	78.5	40.7	193.0	4.8	4.3	110	37.15	1.22	12.73	2.41	37.6	80.9	26	5.77	1.92	1.05	0.35	2.29	0.37
50.1.11–16	0.11	70.5	15.2	104.0	14.3	4.4	185	17.67	1.50	18.81	2.52	43.1	89.9	35	5.84	1.45	0.92	0.36	3.20	0.56
53.1.104–108	bdl	66.5	12.9	56.0	8.9	6.4	139	10.30	1.34	13.12	2.28	34.7	76.5	30	5.63	1.19	0.85	0.34	2.30	0.44
54.5.126–131	0.36	83.2	36.4	183.0	4.6	3.8	89	35.24	1.14	11.67	2.56	36.5	74.3	27	5.64	1.78	1.00	0.33	2.10	0.36
55.CC.23–28	0.53	nd	17.9	156.0	4.6	2.6	103	20.67	0.78	10.29	1.81	34.4	66.8	24	7.12	1.51	0.90	0.38	2.50	0.43
59.3.121–126	0.08	68.9	12.5	82.0	6.5	2.1	106	16.07	0.78	9.87	1.24	23.8	48.7	19	4.30	1.00	0.66	0.32	2.01	0.26
62.CC.11–15	bdl	61.0	8.9	48.0	5.8	9.0	118	8.62	1.31	14.79	2.52	41.7	90.8	35	7.61	1.23	0.95	0.35	2.70	0.51
63.2.42–47	0.86	74.5	19.8	113.0	15.7	5.0	250	20.20	1.52	19.75	2.28	45.3	91.8	36	7.70	1.49	1.03	0.36	2.92	0.52
64.1.53–55	bdl	nd	17.3	119.0	9.0	3.4	178	21.60	1.07	12.60	2.66	33.0	66.5	27	6.99	1.40	1.07	0.43	2.66	0.43
70.5.63–69	bdl	nd	21.8	113.0	4.6	3.2	82	22.58	0.86	9.17	1.60	33.2	63.1	26	7.38	1.79	1.19	0.45	3.10	0.47
70.5.95–100	2.04	80.3	58.9	185.0	4.5	4.1	89	32.83	1.19	11.16	2.77	34.0	69.6	24	5.03	1.64	0.98	0.36	2.50	0.41
79.6.91–96	0.16	71.9	22.7	156.0	13.8	4.5	177	19.82	1.36	17.44	2.70	42.5	85.7	33	8.05	1.41	0.89	0.34	2.73	0.51
81.CC.18–23	0.01	60.3	12.7	69.0	8.2	6.9	136	11.57	1.28	13.61	2.54	37.9	82.2	33	6.04	1.21	0.88	0.35	2.53	0.44
82.1.48–50	bdl	74.0	17.2	110.0	13.5	4.2	203	16.80	1.43	16.98	2.73	37.8	76.1	26	7.03	1.28	0.85	0.39	2.80	0.44
Hole 718C																				
46.1.53–55	bdl	73.2	16.5	103.0	11.1	4.2	192	15.89	1.43	16.85	2.80	39.0	80.9	33	5.62	1.30	0.84	0.35	2.59	0.40
50.1.60–63	bdl	73.5	18.5	115.0	12.9	4.1	209	17.57	1.48	17.94	2.89	39.2	83.3	34	6.39	1.32	0.84	0.38	2.60	0.51
59.2.87–92	bdl	63.0	7.0	48.0	4.6	6.3	101	7.39	1.03	12.05	2.26	32.2	65.4	27	5.55	0.96	0.71	0.36	2.18	0.40
65.2.53–55	bdl	73.8	26.4	102.0	12.6	4.4	237	17.37	1.44	16.50	2.28	41.1	85.3	35	7.19	1.39	0.89	0.39	2.74	0.55
72.2.46–51	bdl	70.4	20.0	81.0	9.2	4.2	165	12.71	1.23	12.16	2.18	32.0	66.1	27	4.87	1.10	0.69	0.39	2.25	0.40
79.1.53–55	bdl	76.9	21.4	218.0	5.4	3.2	86	25.91	1.05	12.48	2.05	36.8	80.1	25	4.87	1.49	0.85	0.31	2.82	0.35
81.1.52–54	bdl	74.8	16.0	97.0	12.2	4.2	231	16.73	1.51	20.14	3.04	46.0	95.1	40	7.77	1.45	0.87	0.42	2.86	0.53
87.1.53–55	bdl	72.3	14.9	83.0	9.6	3.8	208	14.71	1.24	16.11	2.63	40.0	83.6	33	5.86	1.25	0.76	0.40	2.41	0.46
96.1.20–22	bdl	73.3	14.2	78.0	8.4	4.1	188	13.54	1.29	17.75	2.99	39.2	80.7	32	6.76	1.21	0.86	0.40	2.49	0.49
3R.2.15–17	bdl	73.4	13.3	85.0	9.8	3.5	209	14.70	1.22	17.06	2.73	40.5	83.2	32	6.87	1.25	0.80	0.40	2.49	0.50

† Water-soluble fraction
* Silicate fraction

nd – not determined
bdl – below detection limits

References

AMANO, K. & TAIRA, A. 1992. Two-phase uplift of Higher Himalayas since 17Ma. *Geology*, **20**, 391–394.

AOKI, S., KOHYAMA, N. & ISHIZUKA, T. 1991. Sedimentary history and chemical characteristics of clay minerals in cores from the distal part of the Bengal fan. *Marine Geology*, **99**, 175–185.

ARGAST, S. & DONNELLY, T. W. 1987. The chemical discrimination of clastic sedimentary components. *Journal of Sedimentary Petrology*, **57**, 813–823.

BALSON, P. S. & STOW, D. A. V. 1990. Grain-size analysis: Leg 116, Bengal Fan. *In*: COCHRAN, J. R. & STOW, D. A. V. (eds) *Proceedings of the Ocean Drilling Program Leg 116, Scientific Results*. College Station, Texas, Ocean Drilling Program, 417–420.

BHASKAR RAO, Y. J., SIVARAMAN, T. V., PANTULU, G. V. C., GOPLAN, K. & NAQVI, S. M. 1992. Rb–Sr ages of late Archean metavolcanics and granites, Dhawar craton, south India and evidence for early Proterozoic thermotectonic event(s). *Precambrian Research*, **59**, 145–170.

BOUQUILLON, A., CHAMLEY, H. & FROHLICH, F. 1989. Sedimentation argileuse au Cenozoique superieur dans l'Ocean Indien nord-oriental. *Oceanologica Acta*, **12**, 133–147.

——, FRANCE-LANORD, C., MICHARD, A. & TIERCELIN, J.J. 1990. Sedimentology and isotopic chemistry of the Bengal fan sediments: the denudation of the Himalaya. *In*: COCHRAN, J. R. & STOW, D. A. V. (eds) *Proceedings of the Ocean Drilling Program Leg 116, Scientific Results*. College Station, Texas, Ocean Drilling Program, 43–58.

BRASS, G. W. & RAMAN, C. V. 1990. Clay mineralogy of sediments from the Bengal fan. *In*: COCHRAN, J. R. & STOW, D. A. V. (eds) *Proceedings of the Ocean Drilling Program Leg 116, Scientific Results*. College Station, Texas, Ocean Drilling Program, 35–41.

BRIQUEU, L., BOUGAULT, H. & JORON, J. L. 1984. Quantification of Nb, Ta, Ti and V anomalies in magmas associated with subduction zones: petrogenetic implications. *Earth and Planetary Science Letters*, **68**, 297–308.

BURTON, K. W. & O'NIONS, R. K. 1990. The timescale and mechanism of granulite formation at Kurunegala, Sri Lanka. *Contribributions to Mineralogy and Petrology*, **106**, 66–89.

COCHRAN, J. R. 1990. Himalayan uplift, sea-level, and the record of Bengal fan sedimentation at the ODP Leg 116 sites. *In*: COCHRAN, J. R. & STOW, D. A. V. (eds) *Proceedings of the Ocean Drilling Program Leg 116, Scientific Results*. College Station, Texas, Ocean Drilling Program, 397–414.

——, STOW, D. A. V. *ET AL.* 1989. *Proceedings of the Ocean Drilling Program Leg 116, Initial Reports*. College Station, Texas, Ocean Drilling Program.

——, ——, *ET AL.* 1990. *Proceedings of the Ocean Drilling Program Leg 116, Scientific Results*. College Station, Texas, Ocean Drilling Program.

CONDIE, K. C. 1991. Another look at rare earth elements in shales. *Geochimica et Cosmochimica Acta*, **55**, 2527–2531.

—— 1993. Chemical composition and evolution of the upper continental crust: contrasting results from surface samples and shales. *Chemical Geology*, **104**, 1–37.

—— & WRONKIEWICZ, D. J. 1990. The Cr/Th ratio in Precambrian pelites from the Kaapvaal Craton as an index of craton evolution. *Earth and Planetary Science Letters*, **97**, 256–267.

COPELAND, P. & HARRISON, T. M. 1990. Episodic rapid uplift in the Himalaya revealed by $^{40}Ar/^{39}Ar$ analysis of detrital K-feldspar and muscovite, Bengal fan. *Geology*, **18**, 354–357.

COX, K. G. 1989. The role of mantle plumes in the development of continental drainage patterns. *Nature*, **342**, 873–877.

CULLERS, R. L. 1994. The controls on the major and trace element variation of shales, siltstones, and sandstones of Pennsylvanian-Permian age from uplifted continental blocks in Colorado to platform sediment in Kansas, USA. *Geochimica et Cosmochimica Acta*, **58**, 4955–4972.

CURRAY, J. R. 1994. Sediment volume and mass beneath the Bay of Bengal. *Earth and Planetary Science Letters*, **125**, 371–383.

—— & MOORE, D. G. 1971. Growth of the Bengal deep-sea fan and denudation in the Himalayas. *Geological Society of America Bulletin*, **82**, 563–572.

CURTIS, C. D. 1990. Aspects of climatic influence on the clay mineralogy and geochemistry of soils, palaeosols and clastic sedimentary rocks. *Journal of the Geological Society, London*, **147**, 351–357.

DEPAOLO, D. J. 1988. Age dependence of the composition of continental crust: evidence from Nd isotopic variations in granitic rocks. *Earth and Planetary Science Letters*, **90**, 263–271.

DERRY, L. & FRANCE-LANORD, C. 1991. Chemical and physical erosion in the Himalaya: an isotopic and mineralogical view from the Bengal Fan. *Eos* (Transactions of the American Geophysical Union), **72**, 257.

——, GALY, A. & FRANCE-LANORD, C. 1993. Himalayan erosion versus sedimentation in Bengal Fan. Evidence from geochemical and stratigraphic study of ODP Leg 116. *Terra Nova*, **5** (abstract supplement 1), 161.

EINSELE, G. RATSCHBACHER, L. & WETZEL, A. 1996. The Himalayan–Bengal fan denudation-accumulation system during the past 20Ma. *Journal of Geology*, **104**, 163–184.

ELDERFIELD, H., HAWKESWORTH, C. J., GREAVES, M. J. & CALVERT, S. E. 1981. Rare earth element geochemistry of oceanic ferromanganese nodules and associated sediments. *Geochimica et Cosmochimica Acta*, **45**, 513–528.

EMMEL, F. J. & CURRAY, J. R. 1984. The Bengal submarine fan, northeastern Indian Ocean. *Geo-marine Letters*, **3**, 119–124.

FRANCE-LANORD, C., DERRY, L. & MICHARD, A. 1993. Evolution of the Himalaya since Miocene time: isotopic and sedimentalogical evidence from the Bengal Fan. *In*: TRELOAR, P. J. & SEARLE, M. P. (eds) *Himalayan Tectonics*. Geological Society, London, Special Publications, **74**, 603–621.

GOLDBERG, E. D. & GRIFFIN, J. J. 1970. The sediments of

the northern Indian Ocean. *Deep Sea Research*, **17**, 513–537.

GOLDSTEIN, S. L. 1988. Decoupled evolution of Nd and Sr isotopes in the continental crust and the mantle *Nature*, **336**, 733–738.

—— & JACOBSEN, S. B. 1988. Nd and Sr isotopic systematics of river water suspended material: implications for crustal evolution. *Earth and Planetary Science Letters*, **87**, 249–265.

——, O'NIONS, R. K. & HAMILTON, P. J. 1984. A Sm-Nd isotopic study of atmospheric dusts and particulates from major river systems. *Earth and Planetary Science Letters*, **70**, 221–236.

GOVINDARAJU K. 1989. Compilation of working values and sample descriptions for 272 geostandards. *Geostandards Newsletter*, **XIII** (special issue), 1–113.

HANSEN, E. C., NEWTON, R. C., JANARDHAR, A. S. & LINDENBERG, S. 1995. Differentiation of late Archean crust in the Eastern Dharwar Craton, Krishnagiri-Salem area, south India. *Journal of Geology*, **103**, 629–651.

HARRIS, N. B. W., SANTOSH, M. & TAYLOR, P. N. 1994. Crustal evolution in south India: constraints from Nd isotopes. *Journal of Geology*, **102**, 139–150.

HARVEY, P. K., TAYLOR, D. M., HENDRY, R. D. & BANCROFT, F. 1973. An accurate fusion method for analysis of rocks and chemically related materials by X-ray fluorescence spectrometry. *X-ray Spectrometry*, **2**, 33–44.

HERRON, M. M. 1988. Geochemical classification of terrigenous sands and shales from core or log data. *Journal of Sedimentary Petrology*, **58**, 820–829.

INGERSOLL, R. V. & SUCZEK, C. A. 1979. Petrology and provenance of Neogene sands from Nicobar and Bengal fans, DSDP Sites 211 and 218. *Journal of Sedimentary Petrology*, **49**, 1217–1228.

JOHNSON, M. R. W. 1994. Volume balance of erosional loss and sediment deposition related to Himalayan uplifts. *Journal of the Geological Society, London*, **151**, 217–220.

JONES, B. & MANNING, D. A. C. 1994. Comparison of geochemical indices used for the interpretation of palaeoredox conditions in ancient mudstones. *Chemical Geology*, **111**, 111–129.

KOLLA, V. & BISCAYE, P. 1973. Clay mineralogy and sedimentation in the eastern Indian Ocean. *Deep Sea Research*, **20**, 727–738.

KUEHL, S. A., HARIU, T. M. & MOORE, W. S. 1989. Shelf sedimentaion off the Ganges–Brahmaputra river systems: evidence for sediment bypassing to the Bengal fan. *Geology*, **17**, 1132–1135.

LIGHTFOOT, P. C. 1985. *Isotope and Trace Element Geochemistry of the South Deccan Lavas, India*. PhD Thesis, The Open University, UK.

MAHONEY, J. J. 1988. Deccan traps. *In*: MACDOUGALL, J. D. (ed.) *Continental Flood Basalts*. Kluwer, Dordrecht, 151–194.

MARSH, J. S. 1991. REE fractionation and Ce anomalies in weathered Karoo dolerite. *Chemical Geology*, **90**, 189–194.

MARTIN, J. M. & MEYBECK, M. 1979. Chemical composition of river-bourne particulates. *Marine Chemistry*, **7**, 193–206.

McLENNAN, S. M. 1989. Rare earth elements in sedimentary rocks: influence of provenance and sedimentary process. *Reviews in Mineralogy*, **21**, 169–200.

—— & HEMMING, S. 1992. Samarium/neodymium elemental and isotopic systematics in sedimentary rocks. *Geochimica et Comochimica Acta*, **56**, 887–898.

——, TAYLOR, S. R., McCULLOCH, M. T. & MAYNARD, J. B. 1990. Geochemical and Nd-Sr isotopic composition of deep-sea turbidites: crustal evolution and plate tectonic associations. *Geochimica et Cosmochimica Acta*, **54**, 2015–2050.

——, HEMMING, S., McDANIEL, D. K. & HANSON, G. N. 1993. Geochemical approaches to sedimentation, provenance, and tectonics. *In*: JOHNSON, M. J. & BASU, A. (eds) *Processes Controlling the Composition of Clastic Sediments*. Geological Society of America, Special Paper, **284**, 21–40.

MYERS, K. J. & WIGNALL, P. B. 1987. Understanding Jurassic organic-rich mudrocks- new concepts using gamma-ray spectrometry and palaeoecology: examples from the Kimmeridge Clay of Dorset and the Jet Rock of Yorkshire. *In*: LEGGETT, J. K. & ZUFFA, G. G. (eds) *Marine Clastic Sedimentology*. Graham & Trotman, London, 172–189.

NELSON, B. K. & DePAOLO, D. J. 1988. Comparison of isotopic and petrographic provenance indicators in sediments from Tertiary continental basins of New Mexico. *Journal of Sedimentary Petrology*, **58**, 348–357.

NESBITT, H. W. 1979. Mobility and fractionation of rare earth elements during weathering of a granodiorite. *Nature*, **279**, 206–210.

—— & YOUNG, G. M. 1982. Early Proterozoic climates and plate motions inferred from major element chemistry of lutites. *Nature*, **299**, 715–717.

—— & —— 1984. Prediction of some weathering trends of plutonic and volcanic rocks based on thermodynamic and kinetic considerations. *Geochimica et Cosmochimica Acta*, **48**, 1523–1534

—— & —— 1989. Formation and diagenesis of weathering profiles. *Journal of Geology*, **97**, 129–148.

——, MAKOVICS, G. & PRICE, R. C. 1980. Chemical processes affecting alkalis and alkaline earths during continental weathering. *Geochimica et Cosmochimica Acta*, **44**, 1659–1666.

NEWTON, R. C. & HANSEN, E. C. 1986. The South India-Sri Lanka high-grade terrain as a possible deep-crust section. *In*: DAWSON, J. B., CARSWELL, D. A., HALL, J. & WEDEPOHL, K. H. (eds) *The Nature of the Lower Continental Crust*. Geological Society, London, Special Publications, **24**, 297–307.

PAUL, D. K., RAY BARMAN, T., McHAUGHTON, N. J., FLETCHER, R., POTTS, P. J., RAMAKRISHNAN, M. & AUGUSTINE, P. F. 1990. Archean–Proterozoic evolution of Indian charnockites: isotopic and geochemical evidence from granulites of the Eastern Ghats belt. *Journal of Geology*, **98**, 253–263.

PEUCAT, J. J., VIDAL, P., BERNARD-GRIFFITHS, J. & CONDIE, K. C. 1989. Sr, Nd, and Pb isotopic systematics in the Archean low- to high-grade transition zone of southern India: syn-accretion vs post-accretion granulites. *Journal of Geology*, **97**, 537–549.

POTTS, P. J., TINDLE, A. G. & WEBB, P. C. 1992. *Geochemical Reference Material Compositions.* Whittles Publishing, Caithness.

RADHAKRISHNA, B. P. 1993. Neogene uplift and geomorphic rejuvenation of the Indian peninsula. *Current Science*, **64**, 787–793.

—— & NAQVI, S. M. 1986. Precambrian continental crust of India and its evolution. *Journal of Geology*, **94**, 145–167.

REA, D. K. 1992. Delivery of Himalayan sediment to the northern Indian Ocean and its relation to global climate, sea level, uplift and seawater strontium. *In*: DUNCAN, R. A., REA, D. K., KIDD, R. B., VON RAD, U. & WEISSEL, J. K. (eds) *Synthesis of Results from Scientific Drilling in the Indian Ocean.* American Geophysical Union, Geophysics Monograph **70**, 387–402.

SAGAR, W. W. & HALL, S. A. 1990. Magnetic properties of black mud turbidites from ODP Leg 116, distal Bengal fan, Indian Ocean. *In*: COCHRAN, J. R. & STOW, D. A. V. (eds) *Proceedings of the Ocean Drilling Program Leg 116, Scientific Results.* College Station, Texas, Ocean Drilling Program, 317–336.

SARKAR, G., CORFU, F., PAUL, D. K., MCNAUGHTON, N. J., GUPTA, S. N. & BISHUI, P. K. 1993. Early Archaean crust in Bastar Craton, central India – a geochemical and isotopic study. *Precambrian Research*, **62**, 127–137.

SHAW, D. M., CRAMER, J. J., HIGGINS, M. D. & TRUSCOTT, M. G. 1986. Composition of the Canadian Precambrian shield and the continental crust of the earth. *In*: DAWSON, J. B., CARSWELL, D. A., HALL, J. & WEDEPOHL, K. H. (eds) *The Nature of the Lower Continental Crust.* Geological Society, London, Special Publications, **24**, 275–282.

STOW, D. A. V., HOWELL, D. G. & NELSON, C. H. 1983. Sediment, tectonic and sea level controls on submarine fan and slope-apron turbidite systems. *Geo-Marine Letters*, **3**, 57–64.

——, AMANO, K., BALSON, P. S. *ET AL.* 1990. Sediment facies and processes on the distal Bengal fan. *In*: COCHRAN, J. R. & STOW, D. A. V. (eds) *Proceedings of the Ocean Drilling Program Leg 116, Scientific Results.* College Station, Texas, Ocean Drilling Program, 377–396.

TAYLOR, P. N., CHADWICK, B., MOORBATH, S., RAMAKRISHNAN, M. & VISWANATHA, M. N. 1984. Petrography, chemistry and isotopic age of Peninsular Gneiss, Dharwar acid volcanic rocks and the Chitradurga granite with special reference to the late Archaean evolution of the Karnataka Craton, southern India. *Precambrian Research*, **23**, 349–375.

TAYLOR, S. R. & MCLENNAN, S. M. 1985. *The Continental Crust: its composition and evolution.* Blackwell Scientific Publications, Oxford.

YOKOYAMA, K., AMANO, K., TAIRA, A. & SAITO, Y. 1990. Mineralogy of silts from the Bengal fan. *In*: COCHRAN, J. R. & STOW, D. A. V. (eds) *Proceedings of the Ocean Drilling Program Leg 116, Scientific Results.* College Station, Texas, Ocean Drilling Program, 59–73.

Structural, Tectonic and Sedimentary Issues

Structure and tectonics of intermediate-spread oceanic crust drilled at DSDP/ODP Holes 504B and 896A, Costa Rica Rift

YILDIRIM DILEK

Department of Geology, Miami University, Oxford, OH 45056, USA

Abstract: Deep Sea Drilling Project/Ocean Drilling Program (DSDP/ODP) Hole 504B, situated on the southern flank of the intermediate-spreading Costa Rica Rift in the eastern equatorial Pacific, penetrates 2.1 km into 5.9 Ma old oceanic crust and provides an *in situ* reference section for the physical and chemical structure of the upper oceanic lithosphere. Hole 896A, located on a basement high nearly 1 km south of Hole 504B, penetrates 290 m into the upper volcanic sequence, and together with Hole 504B provides an opportunity to examine variations in basement lithostratigraphy and structure. The sea-floor bathymetry in the vicinity of Holes 504B and 896A is defined by east–west trending asymmetric elongated ridges and troughs with wavelengths of 5–6 km and relief of 100–150 m. Hole 504B penetrates two major fault zones at about 800 metres below sea floor (mbsf) in the lower volcanic rocks and at 2111 mbsf in the lower dykes. The fault zone at 800 mbsf coincides with the lithological boundary between the volcanic sequence (layer 2B) above and the transition zone (pillow and massive lava flows, dykes) below, separates contrasting domains of magnetic properties, and marks a south-dipping (outward-facing) normal fault. The fault zone at 2111 mbsf in the bottom of the hole represents a dip-slip fault defined by closely spaced east-northeast striking microfractures with steep dips and steeply plunging lineations. The dyke margins that have been reoriented palaeomagnetically have strikes subparallel to the ridge axis of the Costa Rica Rift and steep dips towards both the north (inward-dipping) and the south (outward-dipping). The main deformation of the dyke complex has been fracturing and veining as suggested by the analyses of core samples and geophysical downhole measurements. Chlorite and/or actinolite veins represent extension fractures and form two orthogonal vein sets. E–W to ESE–WNW striking, steeply dipping extensional veins are commonly parallel to the dyke margins and to the orientation of the Costa Rica Rift axis; N–S to NNE–SSW striking veins are dyke-orthogonal, probably representing thermal contraction cracks. The apparent increase in grain size of the diabasic dyke rocks around 2 kmbsf coincides with an increase in abundance of actinolite at the expense of clinopyroxene and plagioclase, and a sharp decrease in the occurrence of chlorite in the core. These changes in the core mineralogy are accompanied by a steady increase in compressional wave velocity, which reaches a value of 6.8 km s^{-1} within the sheeted dyke complex nearly 1.4–1.6 km into the basement. The boundary between seismic layers 2 and 3 thus corresponds to changes in physical properties of the rocks, rather than a lithological boundary, and occurs over a finite depth interval within the sheeted dyke complex.

The upper oceanic crustal rocks recovered from Hole 896A include massive and pillow lava flows with interlayered breccias. An intensely fractured zone that coincides with sharp changes in acoustic velocity, electrical resistivity and magnetic properties in the volcanic strata occurs at 360 mbsf and marks a fault zone. The thickness of the volcanic strata, which have the same lithostratigraphy and magnetic properties, increases from Hole 896A in the south to Hole 504B in the north, suggesting off-axis ponding of lava(s) in the vicinity of Hole 504B. The sea-floor bathymetry defined by linear hills and intervening troughs in the vicinity of Sites 504 and 896 is interpreted to reflect a basement topography defined by fault-bounded asymmetric volcanic hills that developed within the crustal accretion zone in the spreading environment; it is inferred that the volcanic section drilled into Hole 896A represents one of these abyssal hills in the region.

The Deep Sea Drilling Project (DSDP) and Ocean Drilling Program (ODP) have made the investigation of the oceanic lithosphere a high-priority thematic objective and have played a major role in direct sampling of oceanic crust during the last several decades. In addition to the recovered core material, *in situ* measurements of a variety of geophysical properties through the use of wireline logs and borehole experiments have contributed significantly to our understanding of the internal

DILEK, Y. 1998. Structure and tectonics of intermediate-spread oceanic crust drilled at DSDP/ODP Holes 504B and 896A, Costa Rica Rift. *In*: CRAMP, A., MACLEOD, C. J., LEE, S. V. & JONES, E. J. W. (eds) *Geological Evolution of Ocean Basins: Results from the Ocean Drilling Program.* Geological Society, London, Special Publications, **131**, 179–197.

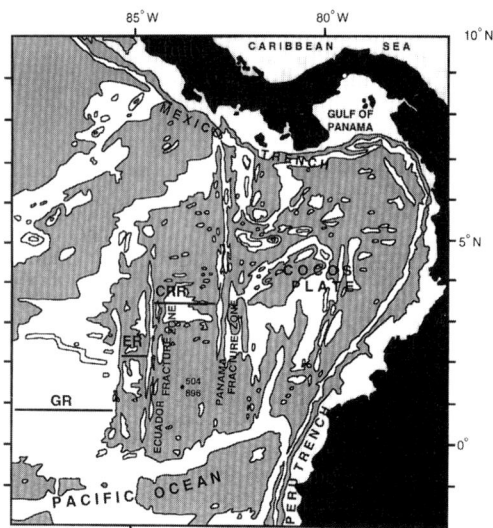

Fig. 1. Simplified tectonic map of the Panama Basin showing the location of DSDP/ODP Sites 504 and 896 south of the Costa Rica Rift and the sea-floor bathymetry (modified from Hobart *et al.* 1985). CRR = Costa Rica Rift, ER = Ecuador Rift, GR = Galapagos Rift. Dark grey contoured areas depict more than 2 km of depth below sea level.

structure and chemostratigraphy of oceanic crust. Recently, offset drilling through a series of relatively shallow, closely spaced holes has been used to do lateral correlation of lithologies in order to examine the nature and scale of heterogeneity in oceanic crust and to erect composite sections of the oceanic crust created in various spreading environments.

DSDP/ODP Hole 504B, located nearly 200 km south of the Costa Rica Rift in the eastern equatorial Pacific, has penetrated 2111 metres below the sea floor (mbsf) and over 1800 m in basement, and it represents the deepest reference section through *in situ* oceanic crust (Fig. 1). The borehole was started on DSDP Leg 69 in 1979, and ODP Leg 148 was the eighth leg to visit the site (Fig. 2). Drilling into 5.9 Ma, intermediate-spread (*c.* 36 mm a^{-1} half-rate) oceanic crust at Site 504 confirmed earlier models for a layered structure of the uppermost 2 km of oceanic crust and also provided the opportunity to conduct an extensive suite of *in situ* geophysical experiments, including permeability measurements in the sheeted dyke complex and the recording of continuous electrical resistivity data with the Formation Micro-Scanner^{TM} (FMS) and Dual Lateralog^{TM} tool

(DLL) of Schlumberger. The alteration history of oceanic rocks recovered from Hole 504B and their geochemical and geophysical features have been studied extensively from the beginning of the DSDP and ODP research (e.g. Honnorez *et al.* 1983; Alt *et al.* 1985, 1986, 1989; Anderson *et al.* 1985; Hobart *et al.* 1985; Newmark *et al.* 1985*a,b*; Becker *et al.* 1988; Morin *et al.* 1989, 1990; Pezard and Anderson, 1989; Pezard *et al.* 1992). However, systematic structural studies of the drill core and combined analyses of structures observed and documented from the Hole 504B core and borehole measurements have been limited and have only been conducted in recent years (i.e. Agar 1990, 1991; Dick *et al.* 1992; Alt *et al.* 1993; Agar & Marton 1995; Allerton *et al.* 1995; Tartarotti *et al.* 1995; Dilek *et al.* 1996*a*). Such structural studies are crucial to constrain the deformation mechanisms and histories of *in situ* oceanic crust and have the potential to provide significant information on the rheological properties of oceanic crust and its tectonic evolution.

The primary goal of Leg 148 was to deepen Hole 504B and to core through the dyke/gabbro and/or layer 2/3 boundary. Drilling was stopped abruptly, however, when the drill string became stuck at a fault zone in the bottom of the hole (2111.0 mbsf), and a contingency hole was drilled at Site 896 on a basement hill nearly 1 km south of Hole 504B (Fig. 3). Hole 896A penetrated to 469 mbsf (290 m into basement) into the upper volcanic sequence and has provided us with an opportunity to examine, via offset drilling, variations in basement lithostratigraphy, geochemistry and structure in order to understand accretion and magmatic processes at an intermediate-spreading mid-ocean ridge. These paired boreholes (Holes 504B and 896A) also provide sites to measure in the future the seismic and electrical tomography and the physical properties of the intervening crust.

This paper gives an account of the structure and tectonics of the upper and lower crust of intermediate-spread oceanic lithosphere drilled in Holes 504B and 896A. The first part of the paper presents new structural information and data combined with an alteration history derived from the core analyses and geophysical measurements associated with the Leg 148 studies (Holes 504B and 896A) and reviews the available data from previous studies at Site 504. Comparison of the lithostratigraphy and the mode and nature of structures observed in cores from Holes 504B and 896A is also included in this section. The second part of the paper presents a discussion of the tectonic interpretations with the aim of refining our ideas and models for the tectonic evolution of intermediate-spreading oceanic lithosphere in the eastern equatorial Pacific Ocean.

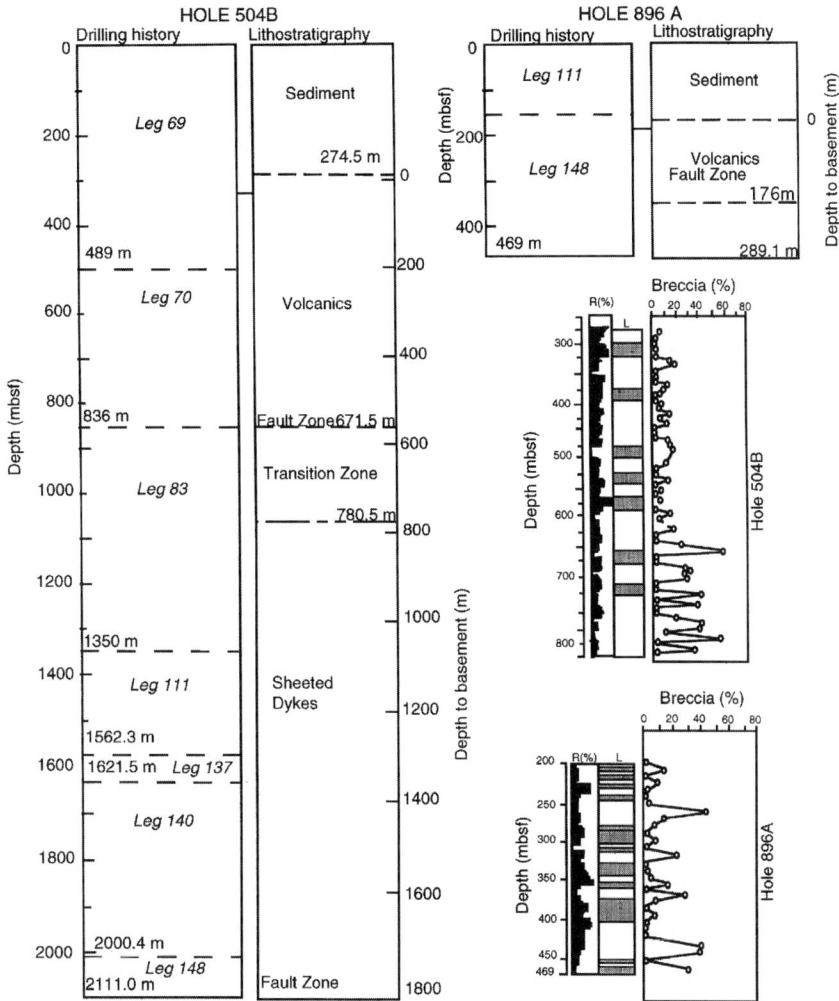

Fig. 2. Generalized lithostratigraphy and drilling histories of Holes 504B and 896A. Fault zones inferred from core observations and downhole bore measurements in Holes 504B and 896A are shown. Diagrams in the lower right depict the occurrence of breccias plotted as percentage of recovered core for each cored interval from Holes 504B (upper) and 896A (lower). Shaded zones represent massive lava units, and white zones show pillow lavas and thin flows. L = lithology, R = core recovery.

Structure of intermediate-spread oceanic crust at Holes 504B and 896A

Hole 504B

DSDP/ODP Hole 504B is situated 200 km south of the Costa Rica Rift and midway between the Ecuador and Panama fracture zones (Fig. 1) and remains the deepest basement hole in oceanic crust (Cann *et al.* 1983; Anderson *et al.* 1985; Becker *et al.* 1988; Becker *et al.* 1989; Dick *et al.* 1992; Alt *et al.* 1993). It penetrates 2.1 km into 5.9 Ma old

oceanic crust and provides a significant *in situ* reference section for the physical and chemical structure of the upper oceanic lithosphere. The lithostratigraphy in the hole includes (Adamson 1985), from top to bottom (Fig. 2): sediments (274.5 m); an upper zone of pillow lavas, pillow breccias, hyaloclastites, flows, and sills (571.5 m); a transition zone of pillow lavas, flows and dykes (209 m); and a lower zone of diabasic dyke rocks (1055.6 m). Thus, the drilled section penetrates geophysically defined oceanic layers 1, 2A, 2B and 2C to a depth of 2.1 km beneath the seafloor (Alt *et*

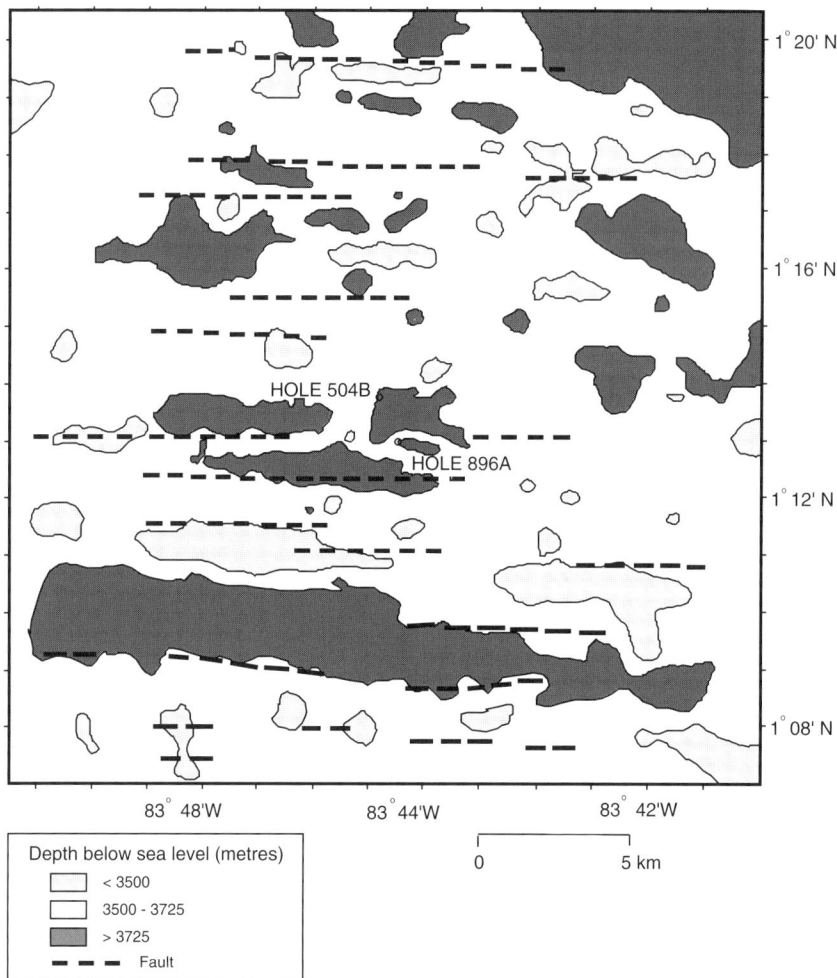

Fig. 3. Bathymetric map showing the fault-bounded abyssal hill fabric in the area around Sites 504 and 896 in the Panama Basin. Location and the areal extent of the faults are based on seismic reflection data from Kent *et al.* (1996) and Swift & Stephen (1995).

al. 1993). Although a significant increase in the grain size of diabasic rocks has been observed in the cores recovered on Legs 140 and 148, the rocks are still part of the sheeted dyke complex; the dyke/gabbro boundary has not yet been reached in the hole. The increased grain size of the diabasic rocks in the Leg 148 cores suggests higher temperatures and slower cooling rates near 2 km at depth during the evolution of oceanic crust (Alt *et al.* 1993) and may also point to the proximity of gabbros to the bottom of the hole at 2111.0 mbsf.

Fractures and vein systems are the most common features in cores from Hole 504B, suggesting that fracturing and brittle failure dominated the structural evolution of oceanic crust drilled at this site. Fractures in the upper extrusive sequence are mostly related to thermal cracking due to cooling and volume changes as a result of pervasive alteration of basalts (Agar 1991). A deformation zone occurs between 836 and 958.5 mbsf, where several discrete fault planes are spatially associated with cataclastic zones, isoclinal microfolds, and alignment of quartz and phyllosilicate minerals (Agar 1990, 1991). This deformation zone lies at the top of the lithological transition zone between the extrusive sequence above and the dyke complex below and thus coincides with a significant mechanical boundary in the basement (Fig. 2). A stockwork zone of more intense alteration occurs within this transition zone between 910 and 930 mbsf (Honnorez *et al.* 1985) and is possibly related to hydrothermal circulation facilitated by dilation during slip on shallow faults in the deformation zone (Agar & Marton 1995). Diabasic dyke rocks below the transition zone are relatively undeformed, and the structural evidence for brittle failure of the oceanic crust is limited to veins and microfractures, as observed in the cores recovered from Hole 504B prior to Leg 148. Observed microfracture networks in dyke rocks recovered during Leg 140 are interpreted, for example, to be related to thermal cracking along certain crystallographic planes (Agar & Marton 1995).

The massive, aphyric to phyric diabasic rocks in the lower sheeted dyke complex in Hole 504B mainly contain veins and fractures and/or microfractures with rare occurrences of igneous contacts and cataclastic zones. Millimetre-scale cataclastic zones in the drill core are spatially associated with veins and chilled dyke margins and are cemented by hydrothermal minerals. Rare (three) dyke margins have steep (76–88°) dips and are nearly parallel to the Costa Rica Rift axis. Heterogeneously distributed veins are composed mainly of actinolite, chlorite ± actinolite, and chlorite minerals and show two groups of orientations (Fig. 4). One group consists of veins that strike E–W to SSE–NNW with steep dips; these veins occur approximately parallel to the trend of the Costa Rica Rift spreading axis. The second group includes veins with a north-northeast strike that may represent dyke-orthogonal cooling fractures. Veins filled mainly with amphibole and chlorite formed as extension fractures and developed through syntaxial mineral overgrowth on host-rock clinopyroxene along the fracture walls and/or by a succession of crack-seal increments (Tartarotti *et al.* 1995; Dilek *et al.* 1996*a*). Although veins show evidence for internal deformation, such as crenulation of vein-filling mineral fibres and stair-shaped microfaults, they do not display any displacement and/or shearing along and across their boundaries. They are interpreted to have formed mostly as extensional mode I fractures (Dilek *et al.* 1996*a*).

Microfractures with lineated surfaces are abundant throughout the core from the lowest 110 m in the hole and are quite densely spaced in certain intervals at depth (e.g. 2026–2052 mbsf, 2103–2111 mbsf). The majority of the oriented microfractures have steep dips (>70°) and east-northeast strikes with subparallel, steeply plunging slickenline lineations and subhorizontal step structures; gently dipping microfractures display lineations with steep pitches (Dilek *et al.* 1996*a*). The steps on the fracture surfaces are smooth, and

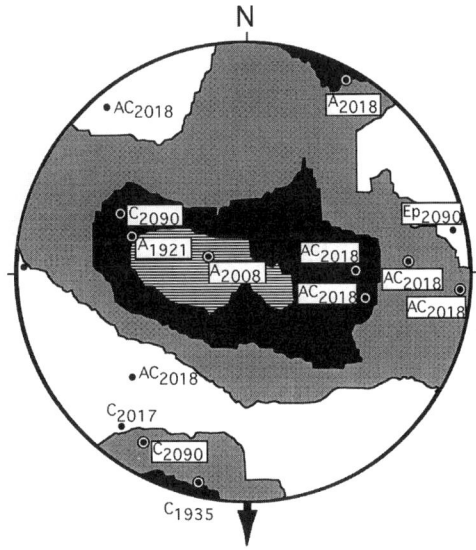

Fig. 4. Poles to veins from Leg 148 superimposed on a Kamb contour plot of Leg 140 vein data. A = actinolite vein; AC = actinolite/chlorite vein; C = chlorite vein; Ep = epidote vein. Numbers refer to depth (mbsf). Arrow is the palaeomagnetic reference direction used to correct the azimuths of the structural data. Contour interval (CI) = 2.0 sigma; N = 84 (after Dilek *et al.* 1996*a*).

their curvilinear traces are commonly orthogonal to the lineation. The sense of 'offset' along the steps is consistent on the core face and on parallel faces of the same core piece, and in general these lineated fracture surfaces are analogous to striated shear fractures in faulted rocks (Petit 1987). There is no evidence for displacement, shearing and/or mineralization along these microfracture surfaces in the core. The lack of any mineralization along them suggests a relatively young age, but the very small number of oriented samples and the large uncertainties in palaeomagnetically corrected azimuths do not allow comparison with the modern (Morin *et al.* 1990) or ridge-related stress fields.

Experimental and field studies in various rock types have shown that micro- and macrofractures become predominant in the vicinity of faults and that there is an abrupt decrease in fracture and crack density away from the faults (Kanaori *et al.* 1991; Scholz *et al.* 1993; Anders & Wiltschko 1994). This observation suggests that a wake of fractured rock (damage zone) develops adjacent to the fault and/or at the propagating fault tip as a zone of inelastic deformation when the fault is growing within the brittle field (Scholz *et al.* 1993). Systematic studies of microstructures in samples taken near thrust and normal faults also show that microfracturing can develop without block rotation, recrystallization, extensive granulation or crystal plasticity, and that maximum microfracture density is independent of slip (Anders & Wiltschko 1994). The micro-fractures observed in the Leg 148 core may, therefore, be related to fault zones (e.g. damage zone in vicinity of fault) and/or propagating fault tips, but evidence of shearing is virtually absent in the core. However, this could be an artefact of poor core recovery since fault rocks uncemented by secondary minerals are unlikely to have been recovered.

Formation MicroScanner (FMS) images depict a persistent fracture set with an average strike of 052° and a dip of 56° to the northwest between 1900 and 2000 mbsf in the hole (Dilek *et al.* 1996a). FMS produces electrical images of the borehole surface by mapping its conductivity and resistivity using an array of small, pad-mounted electrodes (Luthi & Banavar, 1988; Brewer *et al.* 1995; Pezard *et al.* 1996). The sampling rate of the FMS is about 2.5 mm giving a resolution on the order of 1 cm, and thus thin features (down to a few micrometres) that have a conductivity contrast to their host rock may be detected by the FMS. Mapping fractures on FMS images involves identifying traces that display a conductivity contrast to the surrounding rock. A best-fit sinusoid through these traces on the four images defines a planar feature with a dip angle (related to the amplitude of the sinusoid) and

dip direction (given by the lowermost point of the trace). The FMS-generated mean orientation of the persistent fracture set from Hole 504B is different from the orientation of the Costa Rica Rift axis and the modern horizontal principal stress directions (maximum horizontal compressive stress = $S_{Hmax} \sim N32°E$; Zoback *et al.* 1985). There are two zones of intense fracturing in the vicinity of 1930 and 2000 mbsf as deduced from FMS images (Fig. 5; Dilek *et al.* 1996a). The zone at 1930 mbsf is dominated by subhorizontal fractures, whereas the one at 2000 mbsf is dominated by vertical fractures. These zones display several domains with different dip directions.

Drilling in Hole 504B was stopped at 2111.0 mbsf when the drill string became stuck in a zone with a high penetration rate (7 m h^{-1} as opposed 1–2 m h^{-1} for the other dykes). When considered with the existence of highly fractured rocks near and at the bottom of the hole, this zone is interpreted as a fault zone at 2104–2111 mbsf (Alt *et al.* 1993; Dilek *et al.* 1996a), which may conceivably mark the boundary between the dykes above and gabbros below.

Alteration history and velocity structure

Alteration of the diabasic dyke rocks in Hole 504B is heterogeneous and shows evidence for increasing temperatures and metamorphic grade downward in

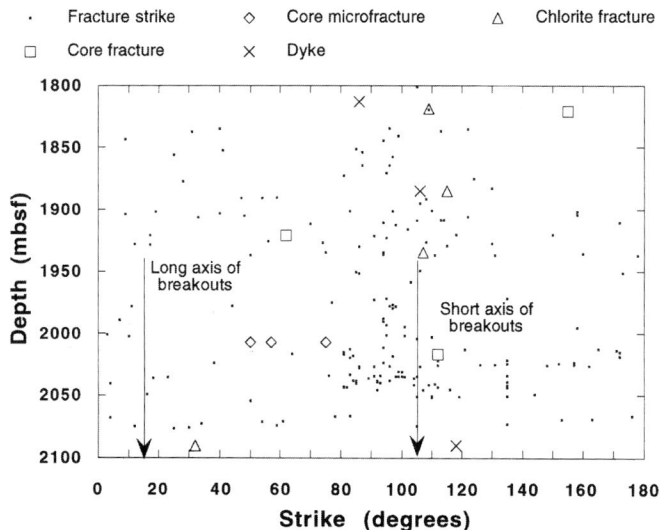

Fig. 5. A composite depth versus strike diagram showing the strikes of subvertical (>70°) FMS-depicted fractures together with oriented subvertical (>70°) fractures and microfractures observed in the drill core, refractured chlorite veins, and dyke margins. Strikes of the long and short axes of borehole breakouts are also shown for reference (after Dilek *et al.* 1996a).

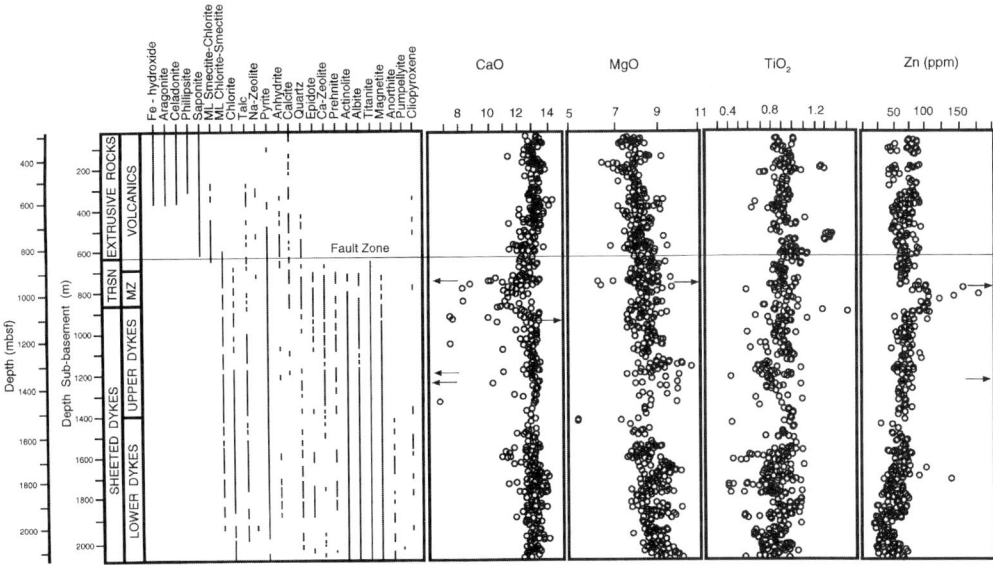

Fig. 6. Lithostratigraphy, distribution of secondary minerals, and selected whole-rock chemical compositions (in wt%) and zinc contents (in ppm) of rocks from Hole 504B (data are from Alt *et al.* 1986, 1989, 1995; Bach *et al.* 1996; Laverne *et al.* 1995, 1996; Zuleger *et al.* 1995). The fault zone at 836 mbsf marks the boundary between the volcanics and the transition zone (TRSN). ML = mixed layer; MZ = mineralized zone.

the dykes (Fig. 6; Alt *et al.* 1993). The diabases are affected by a ubiquitous slight to moderate background alteration (10–40% recrystallized) in centimetre- to decimetre-sized patches, and in centimetre-sized alteration halos around veins. The effects of increased hydrothermal alteration are reflected in the rock chemistry by slight enrichment of MgO and H_2O and depletion of CaO, TiO_2 and SiO_2 (Fig. 6); alteration patches are similarly significantly depleted in rare earth elements (REE), Zr, Nb, Ti, Cu(Zn) and S (Zuleger *et al.* 1995; Bach *et al.* 1996). Actinolite (-magnetite) replaces interstitial material, and olivine is totally replaced by talc + magnetite, chlorite + quartz, and actinolite (Alt *et al.* 1993). Clinopyroxene is slightly to totally replaced by actinolitic amphiboles, and plagioclase is partly replaced by secondary anorthite, chlorite, albite, actinolite and rare epidote. The proportion of amphibole increases with depth, and the range of amphibole compositions changes from mainly actinolitic in dykes at shallow depths to a range from actinolite to hornblende at depths more than 1600 mbsf; secondary anorthite occurs as replacement rims on plagioclase crystals below 1550 mbsf, and secondary clinopyroxene is present in the sections from Legs 111, 137, 140 and 148 (Alt *et al.* 1993, 1995; Laverne *et al.* 1995). Changes in the primary mineral compositions and the occurrence of these

secondary mineral phases indicate reactions of diabasic dyke rocks with hydrothermal fluids at temperatures of 350–500° C (Alt *et al.* 1989, 1995; Laverne *et al.* 1995; Sparks 1995).

The observed metamorphic gradient and the downhole alteration profile in Hole 504B have important implications for the velocity structure of oceanic crust (Detrick *et al.* 1994; Salisbury *et al.* 1996). Seismic experiments conducted in and around Hole 504B over the last 15 years have shown that the compressional wave velocity (V_p) increases rapidly with depth from 4.4–5 km s^{-1} at the top of basement to nearly 6.5 km s^{-1} at 1.2 km sub-basement (Fig. 7; Collins *et al.* 1989; Little & Stephen 1985; Detrick *et al.* 1994). The gradual increase of V_p continues slowly and reaches 6.7–6.8 km s^{-1} by 2.1 kmbsf at the base of Hole 504B. This velocity profile suggests that mean layer 3 velocities are reached within the sheeted dyke complex and that the layer 2/3 boundary occurs at around 1.4–1.6 kmbsf (Fig. 7). Recent seismic refraction experiments in the vicinity of Hole 504B have revealed good constraints on the shear wave velocity (V_s) of the crust that support the earlier findings in that the seismic layer 2/3 transition occurs within the sheeted dyke complex and nearly 600 m above the base of the hole (Collins *et al.* 1995). Laboratory measurements of V_p and V_s on water-saturated samples show that

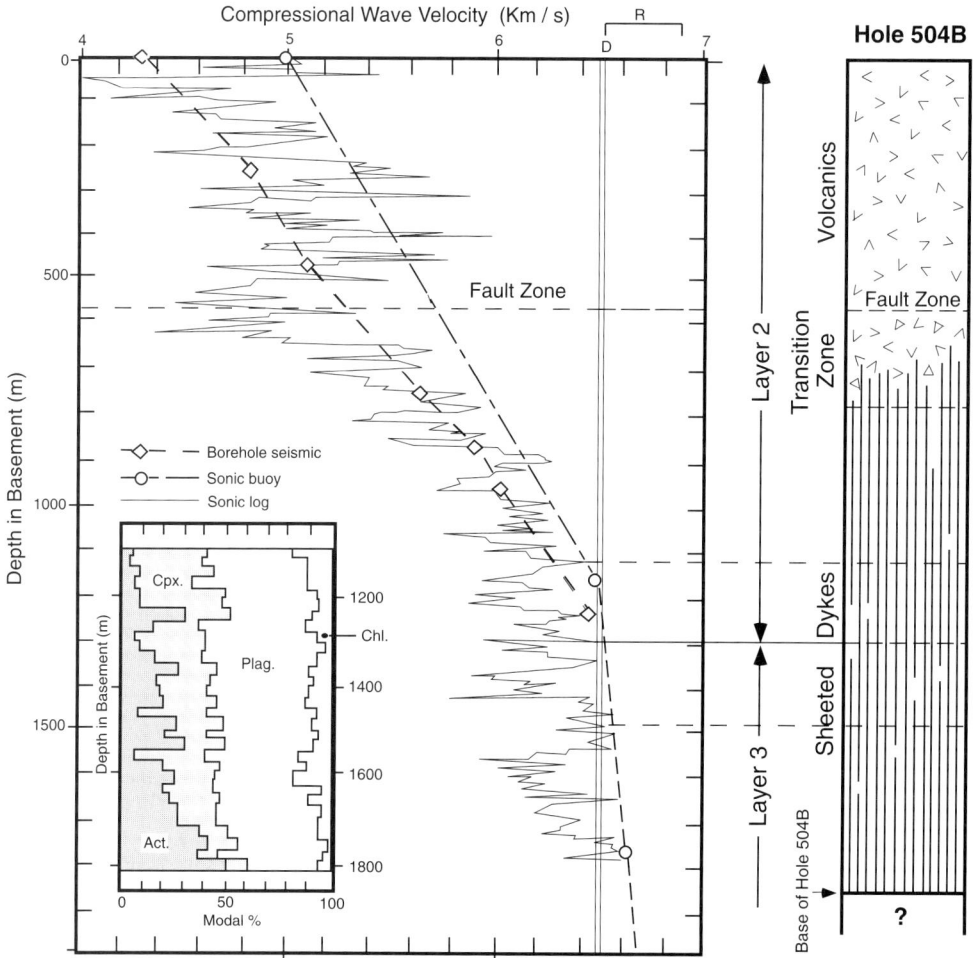

Fig. 7. Lithostratigraphy and compressional wave velocity (V_p) structure versus depth in Hole 504B (modified from Detrick *et al.* 1994). Bar labelled D marks the layer 2/3 boundary defined in terms of seismic velocity gradients, and bar labelled R shows layer 3 compressional wave velocity range (one standard deviation; from Salisbury *et al.* 1996). Hole 504B appears to have entered layer 3 within the sheeted dyke complex by either criterion. The inset diagram shows the modal composition versus depth for the diabasic sheeted dyke rocks as determined by point-count averages over 25 m intervals (after Salisbury *et al.* 1996). Act = actinolite, Chl = chlorite, Cpx = clinopyroxene, Plag = plagioclase. Chlorite is abundant in the upper sheeted dykes but decreases to <5% (by volume) by 1.8 km sub-basement, whereas actinolite increases at the expense of clinopyroxene from zero at the top to nearly 50% at the base of the hole. Classic layer 3 velocities are encountered at a depth where chlorite disappears with increasing metamorphic grade and pressure.

porosity becomes vanishingly small (<0.1%) at the bottom of the hole (Salisbury *et al.* 1996) in parallel with drastically reduced permeability (Alt *et al.* 1993), indicating that formation velocities at this level of the crust are not affected by porosity but are mainly controlled by mineralogy and temperature/ pressure conditions (Salisbury *et al.* 1996). This is

consistent with gradual decrease in chlorite and steady increase in actinolite content of the dyke rocks with depth since $V_p$cpx>$V_p$act>>$V_p$plag >>$V_p$chl (Fig. 7; Salisbury *et al.* 1996). Thus the layer 2/3 boundary as inferred from the seismic velocity structure appears to correspond to changes in physical properties over a depth interval within the sheeted dyke complex, rather than to the presence of a lithological change from dykes to gabbros.

Hole 896A

Hole 896A, located 1 km southeast of Hole 504B, is situated on a bathymetric high overlying a basement topographic high that coincides with a local heat-flow maximum (Fig. 3). It is a site of off-axis hydrothermal activity where low-temperature hydrothermal fluids are upwelling, as determined during a Leg 111 site survey (Becker *et al.* 1988). The basement rocks recovered at Hole 896A consist mainly of tholeiitic basalts, which include massive and pillow lava flows, volcanic breccias and minor dykes. Core observations suggest that pillow lavas constitute a major part (57% of the drilled section) of the core whereas massive lava flows and breccias together make up the remainder, with breccias constituting nearly 5% of the drilled section (Alt *et al.* 1993). However, the lithostratigraphy derived from qualitative visual interpretation of FMS images combined with descriptive use of DLL and caliper logs suggests much higher proportions of breccia (47%) with nearly equal proportions of pillow lavas (33%) and massive lava flows (34%; Brewer *et al.* 1995). The difference between the lithological and wireline log results is probably an artefact of drilling and logging procedures and core recovery, and indicates a potential underestimation of the breccias in the volcanic stratigraphy at Site 896. Breccias occur mainly in the upper part of the core and include hyaloclastites, angular basalt breccias (Fig. 8), and breccias with a 'jigsaw puzzle' fabric (Fig. 9). The main mechanisms of brecciation include fragmentation of glassy pillow rims, mass wasting on the sea floor, and hydraulic fracturing beneath the sea floor (Alt *et al.* 1993; Dilek *et al.* 1996b).

Veins are categorized in two groups on the basis of their internal structure (including fibrous and non-fibrous types) and fabric and mineral composition of vein-filling minerals (Dilek *et al.* 1996b; Tartarotti *et al.* 1996). Fibrous veins, composed mainly of carbonate + clay and carbonate minerals, are commonly younger than other vein types based on cross-cutting relationships, and they display shallow to moderate dip angles. Wall-perpendicular fibres in most of these veins imply pure extension that was accompanied by a succession of crack–seal increments. Non-fibrous veins are commonly filled with blocky carbonate/clay minerals, and/or Fe-oxyhydroxide, and in general they represent the early stages of veining in the upper ocean crust. Whereas some are spatially associated with chilled margins and rims of pillows,

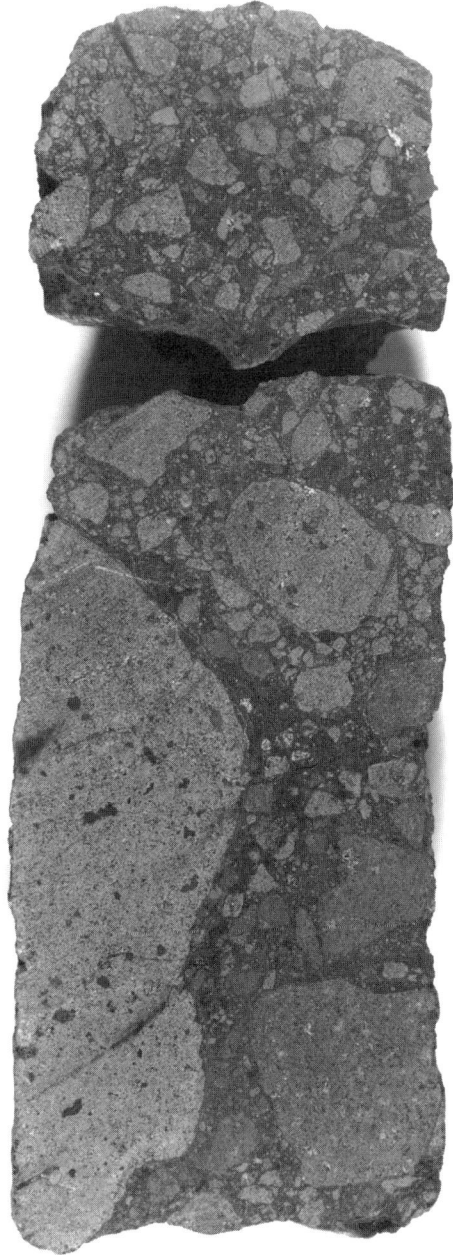

Fig. 8. Brecciated basalt with angular to subangular and poorly sorted clasts in a clay cement. Sample 148-896A-27R-2 (Pieces 5A and 5B). Maximum width of the core sample in the photo is *c.* 5.75 cm.

suggesting their formation during cooling of pillow lava flows, others (e.g. Fe-oxyhydroxide + clay and dark-green clay) are more common in massive lavas. These vein types, which display moderate to

Fig. 9. 'Jigsaw puzzle' fabric and hydraulic fracturing in an incipient breccia in a moderately phyric plagioclase–olivine basalt from Hole 896A. Sample 148-896A-27R-2 (Piece 7). Maximum width of the core sample in the photo is *c.* 5.75 cm.

steep dip angles, may represent steeply dipping cooling joints in massive lava flows.

Numerous vein faults occur in the core between 334 mbsf and 406 mbsf (Table 1). These vein faults have slickenfibres oblique to the vein walls, and the geometries of step structures suggest mostly oblique to reverse shearing along them. The palaeo-magnetically reoriented fibrous extension veins and faults in Sample 148-896A-16R-1 (Piece 9) all strike to the northwest with steep to vertical dips. These faults are oblique to the direction of extension along the Costa Rica Rift axis and nearly

perpendicular to the current maximum horizontal compressive stress ($S_{H_{max}}$ ~ N32°E) at crustal levels in the Panama Basin (Kent *et al.* 1996). The orientation and the nature of these faults are compatible with the focal mechanism solutions of seismic events from the eastern part of the Cocos Plate and borehole *in situ* stress estimates which collectively suggest that the current stress regime in the region is conducive to either reverse faulting on fault planes striking NW–SE or strike-slip faulting on faults striking nearly N–S (Kent *et al.* 1996). The two faults in Sample 148-896A-22R-3 (Piece 2) are normal faults dipping at 56° to the ESE and 90° to the NNE that strike subperpendicular and subparallel, respectively, to the Costa Rica Rift axis (Table 1).

Mapping of nearly 7700 features on images made with the Formation MicroScanner (FMS) has revealed the presence of several intensely fractured zones between 195.0 and 420.0 mbsf in the basement of Hole 896A (Larouzière *et al.* 1996). Sharp changes in acoustic velocity and electrical resistivity occur at 325 mbsf (3.5 km s^{-1}, 3.1 Ωm), 355 mbsf (3.8 km s^{-1}, 3.0 Ωm), and 385 mbsf (5.3 km s^{-1}, 12 Ωm) in the hole pointing to the existence of highly fractured intervals with thick-nesses ranging from 5 to 10 m, as detected both in the DLL and FMS images. The majority of fractures in these zones are subvertical. The fracture zone at 356 mbsf is of particular import-ance because it coincides with the lowest electrical resistivity value (*c.* 3.0 Ωm) recorded in Hole 896A with the DLL and separates fracture sets with two different strike directions (Larouzière *et al.* 1996). Subvertical to moderately dipping fractures above this zone have a mean azimuth of 145°N, whereas those below have a general orientation around 005°N. In addition, a significant change of the borehole shape from elliptical above this zone to cylindrical below suggests a shift in the modern stress field conditions. Based on these observations, Larouzière et al. (1996) have interpreted this intensely fractured interval at 356 mbsf as an active fault zone.

Measurements of bulk permeabilities by means of packer experiments in the basement penetrated by Hole 896A have shown that the upper 200 m of upper oceanic crust is quite permeable with average permeability values (on the order of 2×10^{-13} m^2) typical of the upper oceanic crust off-axis (Becker 1996). Below this depth and between 223 and 469 mbsf in the basement, the average permeability is on the order of $(1–2) \times 10^{-14}$ m^2. These measure-ments indicate that there is no sharp downhole reduction in permeability and that the basement is consistently permeable enough to support passive off-axis hydrothermal circulation. This perme-ability structure is compatible with the presence of

Table 1. *Summary of faults observed in drill core from Hole 896A, Leg 148*

Core, section	Piece	Depth (mbsf)	Dip		Fibres		Composition	Sense of shear*	Comments
			Direction	Angle (°)	Trend	Plunge (°)			
16R-1	9	334.7	36	88	306	5	Fe-OH/green clay		Fibres and polishing
16R-1	9	334.8	238	79	327	4	Fe-OH/green clay		Fibres and polishing
16R-1	9	335.1	220	87	131	23	Fe-OH/green clay	Oblique sinistral	Polished, no fibres
16R-3	5	337.7		48			Fe-OH/green clay		Oblique fibres
22R-3	2	385.6	117	56	117	40	Fe-OH/green clay	Normal	
22R-3	2	385.6	14	90	32	75	Light green clay	Normal	
22R-3	5	385.9		78		19	Light + dark green clay	Oblique sinistral	Fibres overgrowing vermicular clay
22R-3	8	386.2		41		41	Light + dark green clay	Reverse	
22R-3	8	386.2		78		78	Light + dark green clay	Reverse	
23R-2	16	394.3	92	61	58	56	Clay	Oblique reverse	
24R-1	9	403.1		39			Dark green clay	Reverse	
24R-2	4	403.7		72		71	Light green clay	Reverse	
24R-2	10	404.1	162	72			Dark green clay		Cataclastic?
24R-2	10	404.2	99	32			Clay		Cataclastic?
24R-2	11	404.3		32			Clay		Cataclastic?
24R-3	12	405.8		79		14	Dark green clay	Oblique strike-slip	Three directions of fibres
						48	Dark green clay	Oblique normal	Same as surface above
						49	Dark green clay	Oblique reverse	Same as surface above
25R-1	11	413.7		25			Clay/carbonate		

* Determined from steps or obliquity of fibres

highly fractured intervals and a fault zone in the basement, as inferred from FMS electrical images.

Magnetic properties of the basement in Hole 896A show a variation with depth (Alt *et al.* 1993) and significant changes at about 330 mbsf and at 360 mbsf (Allerton *et al.* 1996). Both of these breaks correspond approximately to the tops of massive flow units and separate the basement into three sections that have different magnetic properties (Fig. 10). The magnetic susceptibilities of the rock samples increase below 330 mbsf, and again below 360 mbsf, with an increase in variability, and these changes do not correspond to the petrological changes observed in the core (Allerton *et al.* 1996).

The presence of highly scattered magnetic inclinations below 360 mbsf and a clear linear magnetization with subhorizontal inclinations (mean inclination = $-7.5° \pm 10.2°$) above it points to a major break at this zone that nearly coincides with the fault zone at 356 mbsf inferred from the FMS data.

Comparison of lithostratigraphy and structures between Holes 504B and 896A

The volcanic sequence penetrated by Hole 504B contains 1287.8 m of oceanic basalts and includes

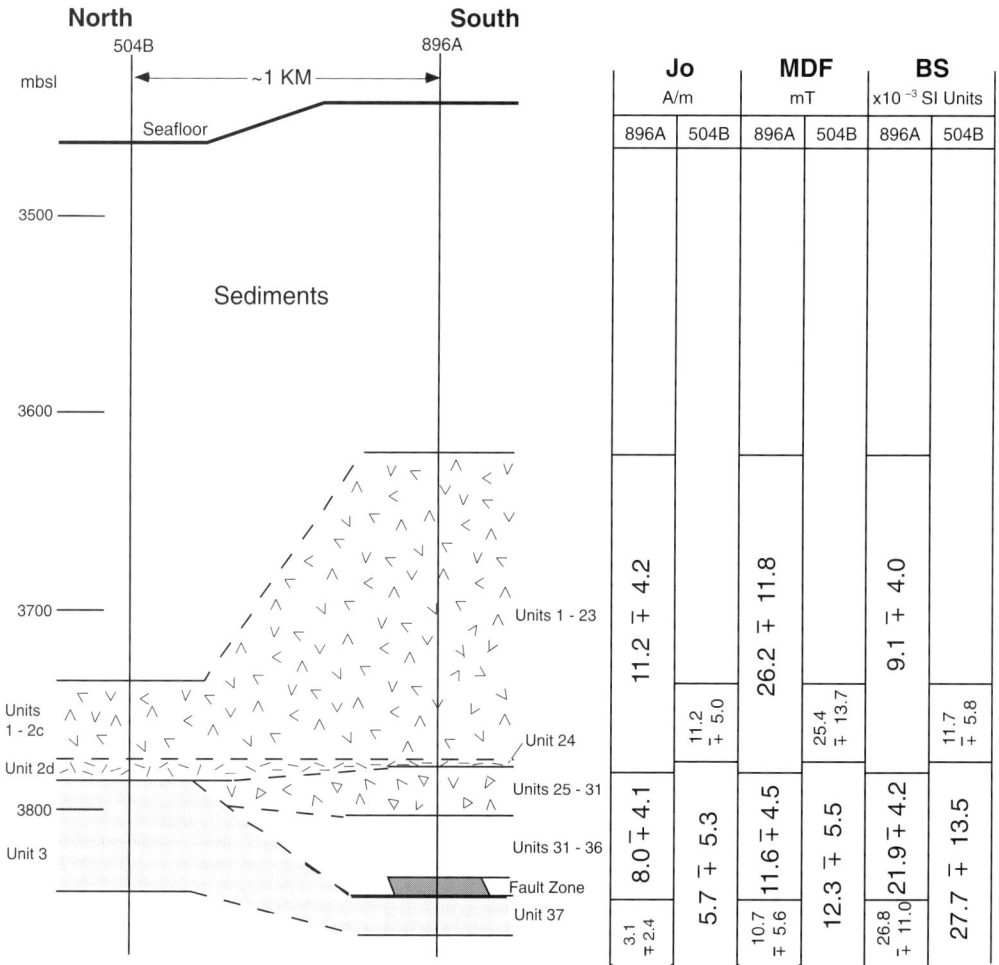

Fig. 10. Schematic diagram on the left showing the correlation of lithostratigraphic units in the extrusive sequence between Hole 504B and Hole 896A based on magnetic properties listed on the right (data from Allerton *et al.* 1996). See text for discussion. J_0 = natural remanent magnetism (NRM) intensity, MDF = median destructive field, BS = bulk susceptibility.

three major lithostratigraphic zones (Anderson *et al.* 1982; Adamson 1985; Becker *et al.* 1989). These zones are: (1) 571.5 m of predominantly pillow lavas and thin flows, intercalated with pillow breccias, hyaloclastites, massive basalts, and localized flow breccias and tectonic breccias (274.5 to 846 mbsf); (2) a 209 m thick transition zone composed of pillow lavas, thin flows, massive basalts and dykes (846 to 1055 mbsf); and (3) 1045 m of sheeted dykes and massive basalts (1055 to 2100.0 mbsf). The rocks are predominantly fine- to medium-grained, aphyric to highly phyric, tholeiitic basalts which are altered to some degree (Becker *et al.* 1989). Alteration in Hole 504B basalts shows three zones with distinct assemblages of secondary minerals (Honnorez *et al.* 1985; Alt *et al.* 1986): (1) the upper pillow alteration zone (320 m of the pillow lavas) characterized by oxidative alteration; (2) the deeper extrusive section (320–624 m) and the topmost 52 m of the lithological transition zone, characterized by smectite and pyrite; and (3) the combined transition zone and dyke section (624–1075.5 m) character- ized by greenschist-facies minerals.

The type and occurrence of major alteration minerals associated with bulk rock and vein hydrothermal alteration observed in Hole 896A are similar to those documented in the upper pillow alteration zone in the Hole 504B core (Alt *et al.* 1993; Laverne *et al.* 1996). Thus, the cored sub- basement (total of 290 m) in Hole 896A is analogous to the upper part (320 m) of the extrusive sequence in Hole 504B in terms of alteration characteristics. However, the mode and nature of hydrothermal veins observed in these holes show some differences. In general, there exists a greater number of veins cross-cutting both pillow and massive lava units in Hole 896A compared to Hole 504B, and the veins are commonly thicker (for example, 54 carbonate + clay mineral veins >2 mm wide in the Hole 896A core versus 17 veins >2 mm wide in the Hole 504B core), particularly in the uppermost part of the volcanic sequence in Hole 896A. Nearly 98% of veins in the upper volcanic section in Hole 504B and 90% in Hole 896A are less than 1 mm wide. The late-stage fibrous veins filled with carbonate minerals (particularly aragonite) also seem to be more abundant in the Hole 896A core.

There are two major differences in the litho- stratigraphy of Holes 504B and 896A (Alt *et al.* 1993): massive lava units intercalated with pillow lavas are more abundant in the cores from Hole 896A. They may represent single or multiple eruptions of magma (Allerton *et al.* 1996). Such massive and sheet lava flows are widespread within the rift valleys of intermediate- and fast-spreading ridges and are interpreted to represent brief but voluminous eruptions of episodic nature (Ballard *et al.* 1979). Breccias and hyaloclastites occur also more commonly in the cores of Hole 896A. They are particularly abundant in the upper part (top 370 mbsf) of the core and are commonly spatially associated with glassy rinds of pillow lavas. Extreme enrichments of Mg, alkalis and CO_2 in these breccia zones indicate that development and the existence of breccias produced an increased permeability and supported intense geochemical changes in the basement (Teagle *et al.* 1996).

Correlations of the magnetic properties of the volcanic rocks from Holes 504B and 896A suggest that the base of the upper extrusive sequence corresponds to the massive lava flow (Unit 24 composed of moderately phyric plagioclase– olivine basalt) in Hole 896A and to the massive lava flow (Unit 2D) in Hole 504B (Fig. 10; Allerton *et al.* 1996). The apparent dip of the upper surface of these massive lava flows between the two holes is about 2°, indicating minimal tilt. The cor- responding lava flow unit above this horizon is much thicker (nearly three to four times as thick) in Hole 504B (Unit 2d) than in Hole 896A (Unit 24), suggesting a possibility of off-axis ponding of lava(s) in the vicinity of Hole 504B.

Tectonic interpretations and discussion

Sites 504 and 896 occur on the south flank of the Costa Rica Rift, where the basement topography is defined by linear east–west trending ridges and troughs (Fig. 3) interpreted to represent asymmetric (steeper to the north), tilted fault blocks (Fig. 11; Langseth *et al.* 1988; Kent *et al.* 1996). Hole 504B is located on the southern flank of one of these asymmetric fault blocks and north of two salient ridges, whereas Hole 896A is situated about 1 km to the southeast near the crest of a small hill that lies near the centre of a 4-km-wide trough running east– west through the area (Figs 3 and 11; Langseth *et al.* 1988). Recent seismic surveys in this region reveal the existence of distinct offsets in sediment reflectors overlying the volcanic basement, and these offsets are commonly spatially associated with basement escarpments with a relief of 100 m or more (Kent *et al.* 1996). The scarps correspond to faults in the basement which are mostly active as indicated by their upward continuation through the sedimentary strata and to the sea floor. The offsets of sediment reflectors show that the downdropped side is generally to the north, towards the Costa Rica spreading axis, pointing to the existence of north-dipping normal faults in the basement (Fig. 11). Apparent decrease in fault displacements upsection in the sedimentary strata suggests a growth fault nature of some of these normal faults and possible change in their geometry (e.g.,

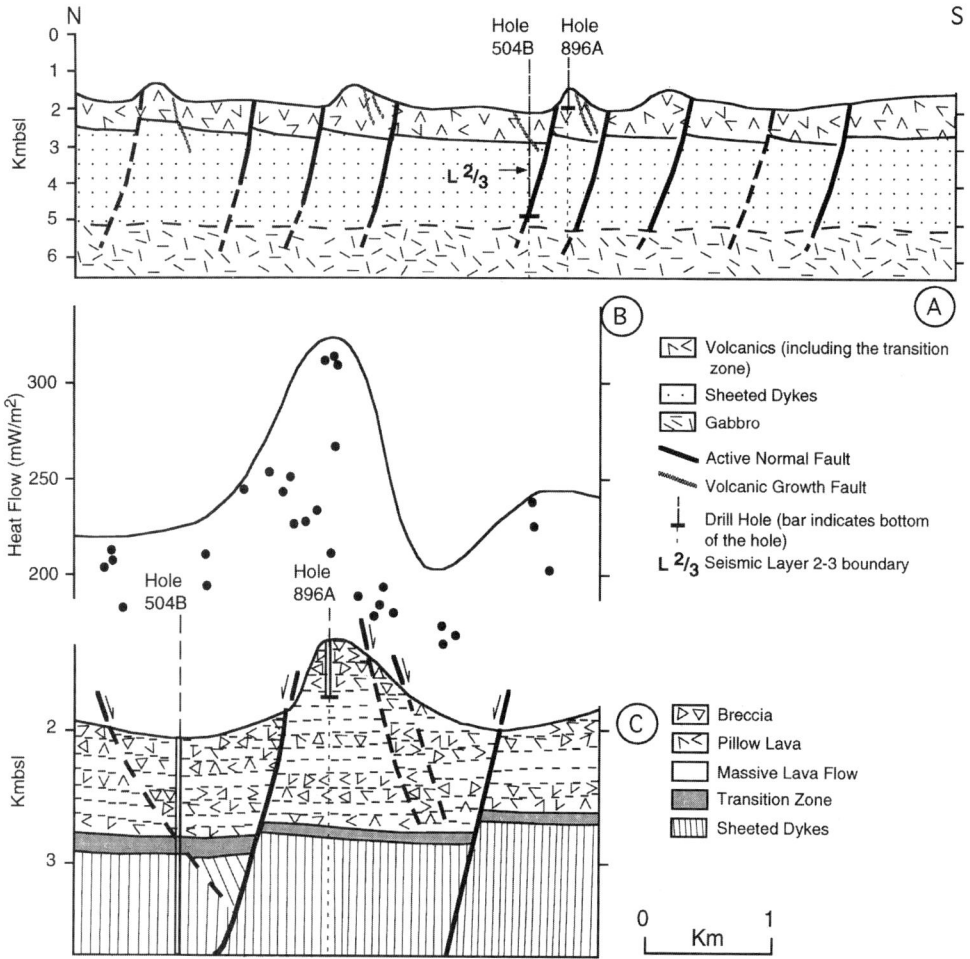

Fig. 11. (**A**) North–south trending geological cross-section across Holes 504B and 896A in the region shown in Fig. 3. The location and geometry of the active normal faults are determined based on seismic reflection data and profiles from Kent *et al.* (1996) and Swift & Stephen (1995). Depth to the dyke–gabbro boundary is based on average thickness of sheeted dyke complex in modern oceanic crust and is consistent with the apparent increase in grain size of diabasic rocks in the bottom of Hole 504B. Occurrence of the faults at *c.* 836 mbsf and 2111 mbsf (based on core observations) coincides with the south-dipping volcanic growth fault and the north-dipping active normal fault, respectively, on this diagram. (**B**) Heat flow data and maximum heat flow curve along a north–south trending profile across Holes 504B and 896A (from Langseth *et al.* 1988) corresponding to the geological cross-section below. (**C**) Geological cross-section across Holes 504B and 896A showing the volcanic stratigraphy in the extrusive sequence, transition zone between the volcanic rocks above and the sheeted dykes below, and the upper part of the sheeted dyke complex. South-dipping volcanic growth faults are represented by thick dashed lines. Development of the asymmetric abyssal hill fabric in this region is interpreted to have developed via synthetic (dipping towards the rift axis) and antithetic (dipping away from the rift axis) faulting and synchronous volcanism in the flanking tectonic province of the intermediate-spreading oceanic lithosphere along the Costa Rica Rift.

lessening dip angles) with depth. An east–west striking, north-dipping normal fault occurs about 1 km south of Hole 504B and passes within 100 m of Hole 896A (Fig. 11; Swift & Stephen 1995; Kent *et al.* 1996). The sediments and the sea floor are offset by this fault indicating activity on it during the last 100 000 years. Hole 504B is thus located on the hanging wall of this active normal fault, whereas Hole 896A is situated on its footwall. Extrapolation of fault dip, as imaged in reflection profiles, into the basement yields a dip angle of 70±5° for this north-dipping active fault, which can be estimated to intersect Hole 504B at a depth of c. 4.8 kmbsl assuming a planar geometry (Fig. 11).

Hole 504B, located on the hanging wall of the north-dipping active fault, penetrates two probable faults at about 800 mbsf in the lower volcanic rocks and at 2111 mbsf in the lower dykes. The inferred fault zone at 800 mbsf coincides with the lithological boundary between the lower part of the volcanic section (layer 2B) and the transition zone (containing dykes, pillow lavas, and flows within the interval of 840–958.5 mbsf) and occurs 50 m above the most deformed section of the Hole 504B core from the extrusive sequence (Fig. 11C; Agar 1991). Reduced recovery (to <10%; Fig. 2) and the highest penetration rates during drilling at 800 mbsf are consistent with the existence of a strongly fractured interval at this depth (Adamson 1985). Possible tectonic tilting due to faulting at this level is suggested by the occurrence of a magnetic property boundary separating contrasting domains of natural remanent magnetization and stable magnetization inclinations (Kinoshita et al. 1989). Pariso & Johnson (1989) interpreted this fault as a south-dipping normal fault based on the tilt direction of the pillow lavas towards the spreading axis of the Costa Rica Rift. This fault geometry and the tilt direction of the pillow lavas are compatible with observations and models from Iceland and the Bay of Islands ophiolite, where lavas dip towards the ridge axis and dykes dip in the opposite direction, away from the ridge axis (Bodvarson & Walker 1964; Rosencrantz 1982). Karson (1987) suggested that this tilting mechanism may result from differential subsidence of lavas and dykes of the crust as it moves laterally away from a spreading centre with a large magma supply. It involves slip on steeply outward (away from the spreading axis) dipping faults near the axis that become locked as they move away from the spreading centre, and rapid subsidence near the axis results from axial loading by extrusives. Dykes with subvertical dips (76–84°) to the south are documented in Hole 504B core, supporting this model (Dilek et al. 1996a). Although the occurrence of south-dipping faults has not been reported from the discontinuous core record, their existence in the volcanic sequence and the dyke complex cannot be ruled out.

Correlation of the lithostratigraphy in Holes 504B and 896A based on magnetic properties shows that the inclinations of the upper volcanic sequences in both holes are very close to the stable reference inclination, indicating that these sequences have not been tilted significantly (Allerton et al. 1996). Therefore, the tilting of the lower volcanic sequence in Hole 504B must have taken place prior to the extrusion of the upper volcanic sequence and possibly within and/or near the crustal accretion zone in the spreading axis environment.

The second fault zone encountered at Hole 504B occurs in the interval of 2104–2111 mbsf and is represented by highly fractured rocks (Alt et al. 1993; Dilek et al. 1996a). Closely spaced east-northeast striking microfractures with steep dips (>70°) and steeply plunging lineations in the core suggest that this is a dip-slip fault zone. The FMS images also point to the existence of a fault zone characterized by densely populated subvertical fracture sets at 2000 mbsf (Dilek et al. 1996a). There is no direct evidence for the dip direction of this fault zone; however, it may correspond to the active fault on the sea floor nearly 1 km south of the drill site that dips 70±5° northwards beneath the hole (Fig. 11A; Swift & Stephen 1995).

The alteration history of the volcanic and dyke rocks from Hole 504B shows that the upper volcanic section was affected by low-temperature (<100°C) oxidizing alteration at high seawater/rock ratios (c. 100), whereas the lower volcanic section underwent low-temperature alteration at lower seawater/rock ratios (c. 10; Alt et al. 1995). The greenschist facies metamorphism of the rocks in the underlying transition zone and sheeted dykes is attributed to higher temperatures (250–380°C) downward in the oceanic crust. The formation of the stockwork-like sulphide mineralization in the transition zone is a result of the mixing of hot hydrothermal fluids (>250°C) upwelling along fracture systems in the dyke complex with large volumes of cooler seawater circulating in the permeable volcanic rocks beneath the spreading axis (Alt et al. 1995). As the crust moved away from the Costa Rica Rift axis upon continued sea floor spreading, rocks were recharged with seawater and more evolved (Mg-depleted, Ca-enriched) off-axis hydrothermal fluids were circulated in the crust, resulting in the formation of anhydrite and laumontite–prehnite veins.

The presence of abundant veins and breccias in the core suggests that the basaltic basement in Hole 896A was affected by interaction with seawater at high pore fluid pressures. The existence of crack–seal veins and vein cross-cutting relations in drill cores indicates that sealed fractures have been reopened or new fractures have been produced as fluid pressures have increased intermittently as a result of deposition of minerals in veins and hence sealing of fractures. This phenomenon is probably related to the presence of higher heat flow and upwelling fluids at the site overlying a basement topographic high on the footwall of the active fault dipping north towards Hole 504B (Fig. 11B). The uniform permeability structure in the basement of

Hole 896A has similar upper crustal permeability values as in Hole 504B and other upper crustal holes on ridge flanks (Becker 1996). Commonly, faults on ridge flanks provide discrete high-permeability zones that facilitate focused fluid flow. The intensely fractured interval at 356 mbsf in Hole 896A may correspond to such a pathway for focused fluid flow in the basement. However, extensive breccia zones and fractured pillow lavas have also been probably equally important in producing horizontal high-permeability zones to provide pathways for fluid flow.

Recent studies of fine-scale variations in ridge morphology and characteristics of mid-ocean ridge faulting based on bathymetric and side-scan sonar measurements show the existence of both inward-dipping (dipping towards the spreading axis) and outward-dipping (dipping away from the spreading axis) normal faults and fault-bounded abyssal hills within the crustal accretion zone of the spreading environments (Carbotte & Macdonald 1994; Macdonald et al. 1996). Observations associated with these studies suggest that: the occurrence of inward- and outward-dipping faults is dependent on spreading rate; there is a monotonic increase in the occurrence of outward-dipping faults with increased spreading rate; and inward-dipping faults continue to be active for a longer time than outward-dipping faults. Outward-dipping faults in the 'flanking tectonic province' of a fast-spreading ridge axis develop as volcanic growth faults and together with the inward-dipping faults result in ponding of syntectonic lava flows within fault-bounded asymmetric linear volcanic hills. These features form abyssal hills, which are typically 10–20 km long, 2–5 km wide, and 50–300 m high on the Pacific Ocean floor (Macdonald et al. 1996). It is inferred that abyssal hills develop as bathy-metric lows and fault-bounded grabens in the flanking tectonic province of the ridge axis, lengthen and link with each other via crack propagation as the lithosphere is stretched (Macdonald et al. 1996). Continuation of this process through time forms linear arrays of subparallel abyssal hills within the axial high which in turn creates a basement topography defined by linear hills and intervening troughs. The area encompassing Sites 504 and 896 on the south flank of the Costa Rica Rift has a basement topography similar to the one depicted in this abyssal hill model. The basement high at which Hole 896A is located may possibly represent a fault-bounded abyssal hill (Fig. 11C), which is at least 10 km long in the east–west direction (Kent et al. 1996). This interpretation is consistent with the substantial increase in lava thickness of the upper extrusive sequence from south to north (from Hole 896A to 504B) as predicted in the abyssal hill development model. The existence of east-west elongated ridges and troughs with wavelengths of 5–6 km and relief of 100–150 m, as established by the recent seismic surveys in the area, is compatible with bathymetric observations along some sections of the fast-spreading East Pacific Rise. These observations may thus suggest that active normal faulting associated with crustal extension and brittle deformation can continue on the ridge flanks outside of the primary crustal accretion zone along the rift axis.

Funding for this research was provided by grants from the Joint Oceanographic Institutions–United States Science Advisory Committee (JOI–USSAC) to Y. Dilek. Invaluable discussions with the shipboard scientists on Leg 148 and the collaborative studies on the geology of Holes 504B and 896A with S. Allerton, P. Pezard, G. Harper, P. Tartarotti and D. Vanko have been most helpful in the analysis and synthesis of the data and observations presented here. M. Salisbury, S. Swift, S. Allerton, D. Teagle, H.-U. Worm, R. Wilkens, C. Laverne and W. Bach kindly made their manuscripts and data available for this study. Constructive reviews by C.J. MacLeod and an anonymous referee improved the paper.

References

ADAMSON, A. C. 1985. Basement lithostratigraphy, Deep Sea Drilling Project Hole 504B. *In*: ANDERSON, R. N., HONNOREZ, J., BECKER, K. ET AL. *Initial Reports of the Deep Sea Drilling Project*, **83**. US Govt Printing Office, Washington, 121–128.

AGAR, S. M. 1990. Fracture evolution in the upper oceanic crust: evidence from DSDP Hole 504B. *In*: KNIPE, R. J. & RUTTER, E. H. (eds) *Deformation Mechanisms, Rheology and Tectonics*. Geological Society, London, Special Publications, **54**, 41–50.

—— 1991. Microstructural evolution of a deformation zone in the upper ocean crust: evidence from DSDP Hole 504B. *Journal of Geodynamics*, **13**, 119–140.

—— & MARTON, F. C. 1995. Microstructural controls on strain localization in ocean crust diabases: evidence from Hole 504B. *In*: ERZINGER, J., BECKER, K., DICK, H. J. B. & STOKKING, L. B. (eds) *Proceedings of the Ocean Drilling Program, Scientific Results*, **137/140**, 219–229.

ALLERTON, S., MCNEILL, A. W., STOKKING, L. B., PARISO, J. E., TARTTAROTTI, P., MARTON, F. C. & PERTSEV, N. N. 1995. Structures and magnetic fabrics from the lower sheeted dike complex of Hole 504B reoriented using stable magnetic remanence. *In*: ERZINGER, J., BECKER, K., DICK, H. J. B. & STOKKING, G. L. B. (eds) *Proceedings of the Ocean Drilling Program, Scientific Results*, **137/140**, 245–251.

——, WORM, H.-U. & STOKKING, L. 1996. Paleomagnetic and rock magnetic properties of Hole 896A. *In*: ALT,

J. C., KINOSHITA, H., STOKKING, L. B. & MICHAEL, P. J. (eds) *Proceedings of the Ocean Drilling Program, Scientific Results*, **148**, 217–226.

ALT, J. C., LAVERNE, C. & MUEHLENBACHS, K. 1985. Alteration of the upper oceanic crust: Mineralogy and processes in Deep Sea Drilling Project Hole 504B. *In*: ANDERSON, R. N., HONNOREZ, J., BECKER, K. *ET AL. Initial Reports of the Deep Sea Drilling Project, * **83**. US Government Printing Office, Washington, 217–248.

——, HONNOREZ, J., LAVERNE, C. & EMMERMANN, R. 1986. Hydrothermal alteration of a 1-km section through the upper oceanic crust, DSDP Hole 504B: The mineralogy, chemistry and evolution of basalt-seawater interactions. *Journal of Geophysical Research*, **91**, 10 309–10 335.

——, ANDERSON, T. F., BONNELL, L., & MUEHLENBACHS, K. 1989. Mineralogy, chemistry, and stable isotopic compositions of hydrothermally altered sheeted dykes: ODP Hole 504B, Leg 111. *In*: BECKER, K., SAKAI, H. *ET AL.* (eds) *Proceedings of the Ocean Drilling Program, Scientific Results*, **111**, 27–40.

——, KINOSHITA, H., STOKKING, L. B. *ET AL.* 1993. *Proceedings of the Ocean Drilling Program, Initial Reports*, **148**. College Station, TX (Ocean Drilling Program).

——, ZULEGER, E. & ERZINGER, J. 1995. Mineralogy and stable isotopic compositions of the hydrothermally altered lower sheeted dike complex, Hole 504B, Leg 140. *In*: ERZINGER, J., BECKER, K., DICK, H. J. B. & STOKKING, L. B. (eds) *Proceedings of the Ocean Drilling Program, Scientific Results*, **137/140**, 155–166.

ANDERS, M. H. & WILTSCHKO, D. V. 1994. Microfracturing, paleostress and the growth of faults. *Journal of Structural Geology*, **16**, 795–815.

ANDERSON, R. N., HONNOREZ, J., BECKER, K. *ET AL.* 1982. DSDP Hole 504B, the first reference section over 1 km through oceanic crust. *Nature*, **300**, 589–594.

——, ——, ——. *ET AL.* 1985. *Initial Reports of the Deep Sea Drilling Project*, **83**. US Government Printing Office, Washington.

BACH, W., ERZINGER, J., ALT, J. C. & TEAGLE, D. A. H. 1996. Chemistry of the lower sheeted dike complex, Hole 504B (Leg 148): influence of magmatic differentiation and hydrothermal alteration. *In*: ALT, J. C., KINOSHITA, H., STOKKING, L. B. & MICHAEL, P. J. (eds) *Proceedings of the Ocean Drilling Program, Scientific Results*, **148**, 39–55.

BALLARD, R. D., HOLCOMB, R. T. & VAN ANDEL, T. H. 1979. The Galapagos Rift at 86°W: 3. Sheet flows, collapse pits, and lava lakes of the rift valley. *Journal of Geophysical Research*, **84**, 5407–5422.

BECKER, K. 1996. Permeability measurements in Hole 896A, Leg 148, and implications for the lateral variability of permeability in young upper oceanic crust. *In*: ALT, J. C., KINOSHITA, H., STOKKING, L. B. & MICHAEL, P. J. (eds) *Proceedings of the Ocean Drilling Program, Scientific Results*, **148**, 353–363.

——, SAKAI H. *ET AL.* 1988. *Proceedings of the Ocean Drilling Program, Scientific Results*, **111**. College Station, TX (Ocean Drilling Program).

—— *ET AL.* 1989. Drilling deep into young oceanic crust, Hole 504B, Costa Rica Rift. *Reviews of Geophysics*, **27**, 79–101.

BODVARSSON, G. & WALKER, G. P. L. 1964. Crustal drift in Iceland: *Geophysical Journal, Royal Astronomical Society*, **9**, 285–300.

BREWER, T., LOWELL, M., HARVEY, P. & WILLIAMSON, G. 1995. Stratigraphy of the ocean crust in ODP Hole 896A from FMS images. *Scientific Drilling*, **5**, 87–92.

CANN, J. R., LANGSETH, M. G. *ET AL.* 1983. *Initial Reports of the Deep Sea Drilling Project*, **69**. US Govt Printing Office, Washington.

CARBOTTE, S. & MACDONALD, K. C. 1994. Comparison of seafloor tectonic fabric at intermediate, fast, and super fast spreading ridges: Influence of spreading rate, plate motions, and ridge segmentation on fault patterns. *Journal of Geophysical Research*, **99**, 13 609–13 631.

COLLINS J. A., PURDY, M. G. & BROCHER, T. M. 1989. Seismic velocity structure at Deep Sea Drilling Project Site 504B, Panama Basin: Evidence for thin oceanic crust. *Journal of Geophysical Research*, **94**, 9283–9302.

——, DETRICK, R. S., STEPHEN, R. A., KENT, G. M. & SWIFT, S. A. 1995. Hole 504B Seismic experiment: New constraints on the depth to the seismic layer 2/layer 3 boundary. *EOS (Transactions of the American Geophysical Union)*, **76**, F616.

DETRICK, R. S., COLLINS, J., STEPHEN, R. & SWIFT, S. 1994. *In situ* evidence for the nature of the seismic layer 2/3 boundary in oceanic crust. *Nature*, **370**, 288–290.

DICK, H. J. B., ERZINGER, J. A., STOKKING, L. B. *ET AL.* 1992. *Proceedings of the Ocean Drilling Program, Initial Reports*, **140**. College Station, TX (Ocean Drilling Program).

DILEK, Y., HARPER, G. D., PEZARD, P. & TARTAROTTI, P. 1996a. Structure of the sheeted dike complex in Hole 504B (Leg 148). *In*: ALT, J. C., KINOSHITA, H., STOKKING, L. B. & MICHAEL, P. J. (eds) *Proceedings of the Ocean Drilling Program, Scientific Results*, **148**, 229–243.

——, HARPER, G. D., WALKER, J., ALLERTON, S. & TARTAROTTI, P. 1996b. Structure of upper layer 2 in Hole 896A. *In*: ALT, J. C., KINOSHITA, H., STOKKING, L. B. & MICHAEL, P. J. (eds) *Proceedings of the Ocean Drilling Program, Scientific Results*, **148**, 261–279.

HOBART, M. A., LANGSETH, M. G. & ANDERSON, R. N. 1985. A geophysical and geothermal survey on the south flank of the Costa Rica Rift: Sites 504B and 505. *In*: ANDERSON, R. N., HONNOREZ, J., BECKER, K. *ET AL. Initial Reports of the Deep Sea Drilling Project*, **83**. US Govt Printing Office, Washington, 379–404.

HONNOREZ, J., LAVERNE, C., HUBBERTEN, H., EMMERMANN, R. & MUEHLENBACHS, K. 1983. Alteration processes of Layer 2 basalts from Deep Sea Drilling Project Hole 504B, Costa Rica Rift. *In*: CANN, J. R., LANGSETH, M. G., HONNOREZ, J., VON HERZEN, R. P., WHITE, W. M. *ET AL. Initial Reports of the Deep Sea Drilling Project*, **83**. US Government Printing Office, Washington, 509–546.

——, ALT, J. C., HONNOREZ-GUERSTEIN, B.-M., LAVERNE, C., MUEHLENBACHS, K., RUIZ, J. & SALTZMAN, E. 1985. Stockwork-like sulfide mineralization in young oceanic crust: Deep Sea Drilling Project Hole 504B. *In*: ANDERSON, R. N., HONNOREZ, J., BECKER, K. *et al. Initial Reports of the Deep Sea Drilling Project*, **83**. US Government Printing Office, Washington, 263–245.

KANAORI, Y., YAIRI, K. & ISHIDA, T. 1991. Grain boundary microcracking of granitic rocks from the northeastern region of the Atotsugawa fault, central Japan: SEM backscattered electron images. *Engineering Geology*, **30**, 221–235.

KARSON, J. A. 1987. Factors controlling the orientation of dykes in ophiolites and oceanic crust. *In*: HALLS, H. C. & FAHRIG, W. F. (eds) *Mafic Dyke Swarms*. Geological Association of Canada, Special Paper **34**, 229–241.

KENT, G. M., SWIFT, S. A., DETRICK, R. S., COLLINS, J. A. & STEPHEN, R. A. 1996. Evidence for active normal faulting on 5.9 Ma crust near Hole 504B on the southern flank of the Costa Rica rift. *Geology*, **24**, 83–86.

KINOSHITA, H., FRUTA, T. & PARSIO, J. 1989. Downhole magnetic field measurements and paleomagnetism, Hole 504B, Costa Rica Ridge. *In*: BECKER, K., SAKAI H. *ET AL.* (eds) *Proceedings of the Ocean Drilling Program, Scientific Results*, **111**, 147–156.

LANGSETH, M. G., MOTTL, M. J., HOBART, M. A. & FISHER, A. 1988. The distribution of geothermal and geochemical gradients near Site 501/504: implications for hydrothermal circulation in the oceanic crust. *In*: BECKER, K., SAKAI, H. *ET AL. Proceedings of the Ocean Drilling Program, Scientific Results*, **111**, 23–32.

LAROUZIERE, F. D., PEZARD, P., AYADI, M. & BECKER, K. 1996. Downhole measurements and electrical images in ODP Hole 896A, Costa Rica Rift. *In*: ALT, J. C., KINOSHITA, H., STOKKING, L. B., & MICHAEL, P. J. (eds) *Proceedings of the Ocean Drilling Program, Scientific Results*, **148**, 375–388.

LAVERNE, C., VANKO, D. A., TARTAROTTI, P. & ALT, J. C. 1995. Chemistry and geothermometry of secondary minerals from the deep sheeted dike complex, Hole 504B. *In*: ERZINGER, J., BECKER, K., DICK, H. J. B. & STOKKING, L. B. (eds) *Proceedings of the Ocean Drilling Program, Scientific Results*, **137/140**, 167–189.

——, BELAROUCHI, A. & HONNOREZ, J. 1996. Alteration mineralogy and chemistry of the upper oceanic crust from ODP Hole 896A, Costa Rica Rift. *In*: ALT, J. C., KINOSHITA, H., STOKKING, L. B. & MICHAEL, P. J. (eds) *Proceedings of the Ocean Drilling Program, Scientific Results*, **148**, 151–170.

LITTLE, S. A. & STEPHEN, R. A. 1985. Costa Rica Rift borehole seismic experiment, Deep Sea Drilling Project Hole 504B, Leg 92. *In*: ANDERSON, R. N., HONNOREZ, J., BECKER K. *ET AL.* (eds) *Initial Reports of the Deep Sea Drilling Project*, **83**. US Government Printing Office, Washington, 517-528.

LUTHI, S. M. & BANAVAR, J. R. 1988. Application of borehole images to three-dimensional geometric modeling of eolian sandstone reservoirs, Permian Rotliegende, North Sea. *AAPG Bulletin*, **72**, 1074–1089.

MACDONALD, K. C., FOX, P. J., ALEXANDER, R. T., POCKALNY, R. & GENTE, P. 1996. Volcanic growth

faults and the origin of Pacific abyssal hills. *Nature*, **380**, 125–129.

MORIN, R. H., ANDERSON, R. N. & BARTON, C. A. 1989. Analysis and interpretation of the borehole televiewer log: Information on the state of stress and the lithostratigraphy at Hole 504B. *In*: BECKER, K., SAKAI, H. *ET AL. Proceeding of the Ocean Drilling Program, Scientific Results*, **111**, 109–118.

——, NEWMARK, R. L., BARTON, C. A. & AANDERSON, R. N. 1990. State of lithospheric stress and borehole stability at Deep Sea Drilling Project Site 504B, Eastern Equatorial Pacific. *Journal of Geophysical Research*, **95**, 9293–9303.

NEWMARK, R. L., ANDERSON, R. N., MOOS, D. & ZOBACK, M. D. 1985*a*. Sonic and ultrasonic logging of Hole 504B and its implications for the structure, porosity and stress regime of the upper 1 km of the oceanic crust. *In*: ANDERSON, R. N., HONNOREZ, J., BECKER, K. *ET AL.* (eds) *Initial Reports of the Deep Sea Drilling Project*, **83**. US Government Printing Office, Washington, 479-510.

——, ZOBACK, M. D. & ANDERSON, R. N. 1985*b*. Orientation of *in situ* stresses near the Costa Rica Rift and Peru - Chile trench: DSDP Hole 504B. *In*: ANDERSON, R. N., HONNOREZ, J., BECKER, K. *ET AL. Initial Reports of the Deep Sea Drilling Project*, **83**. US Government Printing Office, Washington, 511–515.

PARISO, J. E. & JOHNSON, H. P. 1989. Magnetic properties and oxide petrography of the sheeted dike complex in Hole 504B: *In*: BECKER, K., SAKAI, H. *ET AL.* (eds) *Proceedings of the Ocean Drilling Program, Scientific Results*, **111**, 159–167.

PETIT, J. P. 1987. Criteria for the sense of movement on fault surfaces in brittle rocks. *Journal of Structural Geology*, **9**, 597–608.

PEZARD, P. A. & ANDERSON, R. N. 1989. Morphology and alteration of the upper oceanic crust from *in situ* electrical experiments in DSDP/ODP Hole 504B. *In*: BECKER, K., SAKAI, H. *ET AL.* (eds) *Proceedings of the Ocean Drilling Program, Scientific Results*, **111**, 133–146.

——, ——, RYAN, W. B. F., BECKER, K., ALT, J. C. & GENTE, P. 1992. Accretion, structure and hydrology of intermediate spreading-rate oceanic crust from drillhole experiments and seafloor observations. *Marine Geophysical Research*, **14**, 93–123.

——, BECKER, K., REVIL, A., AYADI, M. & HARVEY, P. K. 1996. Fractures, porosity, and stress in the dolerites of Hole 504B, Costa Rica Rift. *In*: ALT, J. C., KINOSHITA, H., STOKKING, L. B. & MICHAEL, P. J. (eds) *Proceedings of the Ocean Drilling Program, Scientific Results*, **148**, 317–329.

ROSENCRANTZ, E. 1982. Formation of uppermost oceanic crust. *Tectonics*, **1**, 471–494.

SALISBURY, M. H., CHRISTENSEN, N. I. & WILKENS, R. H. 1996. Nature of the Layer2/3 transition from a comparison of laboratory and logging velocities and petrology at the base of Hole 504B. *In*: ALT, J. C., KINOSHITA, H., STOKKING, L. B. & MICHAEL, P. J. (eds) *Proceedings of the Ocean Drilling Program, Scientific Results*, **148**, 409–414.

SCHOLZ, C. H., DAWERS, N. H., YU, J.-Z., & ANDERS, M. H. 1993. Fault growth and fault scaling laws:

preliminary results. *Journal of Geophysical Research*, **98**, 21 951–21 961.

SPARKS, J. W. 1995. Geochemistry of the lower sheeted dike complex, Hole 504B, Leg 140. *In*: ERZINGER, J., BECKER, K., DICK, H. J. B. & STOKKING, L. B. (eds) *Proceedings of the Ocean Drilling Program, Scientific Results*, **137/140**, 81–97.

SWIFT, S. A. & STEPHEN, R. A. 1995. Hole 504B Seismic experiment: Single channel seismic survey reveals recent faulting and basement relief near crustal boreholes. *EOS (Transactions of the American Geophysical Union)*, **76**, F616.

TARTAROTTI, P., ALLERTON, S. & LAVERNE, C. 1995. Vein formation mechanisms in the sheeted dike complex from Hole 504B. *In*: ERZINGER, J., BECKER, K., DICK, H. J. B. & STOKKING, L. B. (eds) *Proceedings of the Ocean Drilling Program, Scientific Results*, **137/140**, 231–241.

——, VANKO, D. A., HARPER, G. D. & DILEK, Y. 1996. Crack-seal veins in upper layer 2 in Hole 896A. *In*: ALT, J. C., KINOSHITA, H., STOKKING, L. B. & MICHAEL, P. J. (eds) *Proceedings of the Ocean Drilling Program, Scientific Results*, **148**, 281–288.

TEAGLE, D. A. H., ALT, J. C., BACH, W., HALLIDAY, A. N. & ERZINGER, J. 1996. Alteration of upper ocean crust in a ridge flank hydrothermal up-flow zone: mineral, chemical, and isotopic constraints from Hole 896A. *In*: ALT, J. C., KINOSHITA, H., STOKKING, L. B. & MICHAEL, P. J. (eds) *Proceedings of the Ocean Drilling Program, Scientific Results*, **148**, 119–150.

ZOBACK, M. D., MOOS, D., MASLIN, L. & ANDERSON, R. N. 1985. Well borehole breakouts and in situ stress. *Journal of Geophysical Research*, **90**, 5523–5530.

ZULEGER, E., ALT, J. C. & ERZINGER, J. 1995. Primary and secondary variations in major and trace element geochemistry of the lower sheeted dike complex: Hole 504B, Leg 140. *In*: ERZINGER, J., BECKER, K., DICK, H. J. B. & STOKKING, L. B. (eds) *Proceedings of the Ocean Drilling Program, Scientific Results*, **137/140**, 65–80.

Sediment accumulation rates from Deep Tow profiler records and DSDP Leg 70 cores over the Galapagos spreading centre

NEIL C. MITCHELL

*Department of Geological Sciences, University of Durham, South Road,
Durham DH1 3LE, UK*

*Present address: Department of Earth Sciences, Oxford University, Parks Road,
Oxford OX1 3PR, UK*

Abstract: Variations in pelagic input rates over abyssal hill areas cannot be inferred easily from sedimentation rate variations between cores because of redistribution by downslope gravity processes and bottom currents. Some spatial averaging or adjustment for sediment transport is required. Relatively accurate mean sedimentation rates may be obtained, however, by regressing sediment thickness with distance from a mid-ocean ridge spreading centre, where the regression averages out the variations due to local redistribution. This is shown using sediment profiler records from the Galapagos spreading centre (SC) at 1° N in the eastern Pacific. Physical property data from Deep Sea Drilling Project (DSDP) Leg 70 are used to correct sediment thicknesses for compaction and to convert sedimentation rates to mass accumulation rates (MARs), which are *c.* 2 g cm^{-2} ka^{-1} in this region. These high MARs are due to enhanced equatorial productivity of pelagic organisms, which is also reflected in a *c.* 6% higher MAR for the ridge flank closest to the equator, corresponding to a rate of change of MAR with latitude of *c.* 16% per degree. The equatorial high productivity zone in the Panama Basin lies further south than in the central Pacific; peak sedimentation along the Ocean Drilling Program (ODP) Leg 138 transect (110° W) occurs near 1° N so has nearly zero gradient at this latitude, but has comparable gradient to the Galapagos SC north of 2° N. The zone of peak enhancement at 86° W in the Panama Basin may therefore be 1° or more further south than at 110° W. Some further sedimentation characteristics of the Galapagos spreading centre are also described, such as a scaling of thickness variability and the possibility of dating sea floor using sediment thickness.

High rates of sedimentation occur beneath oceanic zones of upwelling, nutrient-rich deep waters (e.g. Broecker & Peng 1982). The deposited sediments are commonly carbonates, though highly siliceous sediments also occur, particularly in waters that are strongly carbonate-undersaturated causing dissolution of the carbonate component. Studying the form of enhanced deposition across these upwelling regions, however, is often not straightforward. Mass accumulation rates (MARs) vary between sediment cores either because of downslope gravity-driven transport and sediment redistribution or because of variable sedimentation rates caused by bottom currents (e.g. Berger & Stax 1994). Sedimentation rates are greatly affected in abyssal hill areas, which is important because significant proportions of these areas lie above the sedimentary lysocline and are therefore important contributors to the carbonate 'sink'. Since cores are commonly taken in areas between abyssal hills to avoid core damage, global estimates of total

carbonate accumulation may be too high, and consequently the ability of pelagic sedimentation to remove carbon (e.g. Broecker & Peng 1982) may be overstated. This paper reconsiders a simple geophysical method for determining long-term average sedimentation rates from sediment thickness variations across mid-ocean ridges (Keen & Manchester 1970; van Andel *et al.* 1973).

MARs at the Galapagos spreading centre (SC) in the Panama Basin are determined here using high-resolution sediment profiler records. The basin (Fig. 1a) is enclosed by the American continents and the Cocos and Carnegie Ridges, with bottom waters supplied through sills in the ridges and along the Ecuador trench (Lonsdale 1977*c*). The Galapagos SC (east–west ridge at 1° N in Fig. 1a) is relatively distant from these inlets so bottom currents are generally low (Detrick *et al.* 1974; Lonsdale 1977*a*,*b*). The equatorial zone of high productivity in the basin lies at *c.* 2° S over the Carnegie Ridge, further south than elsewhere in the

MITCHELL, N. C. 1998. Sediment accumulation rates from Deep Tow profiler records and DSDP Leg 70 cores over the Galapagos spreading centre. *In*: CRAMP, A., MACLEOD, C. J., LEE, S. V. & JONES, E. J. W. (eds) *Geological Evolution of Ocean Basins: Results from the Ocean Drilling Program*. Geological Society, London, Special Publications, **131**, 199–209.

199

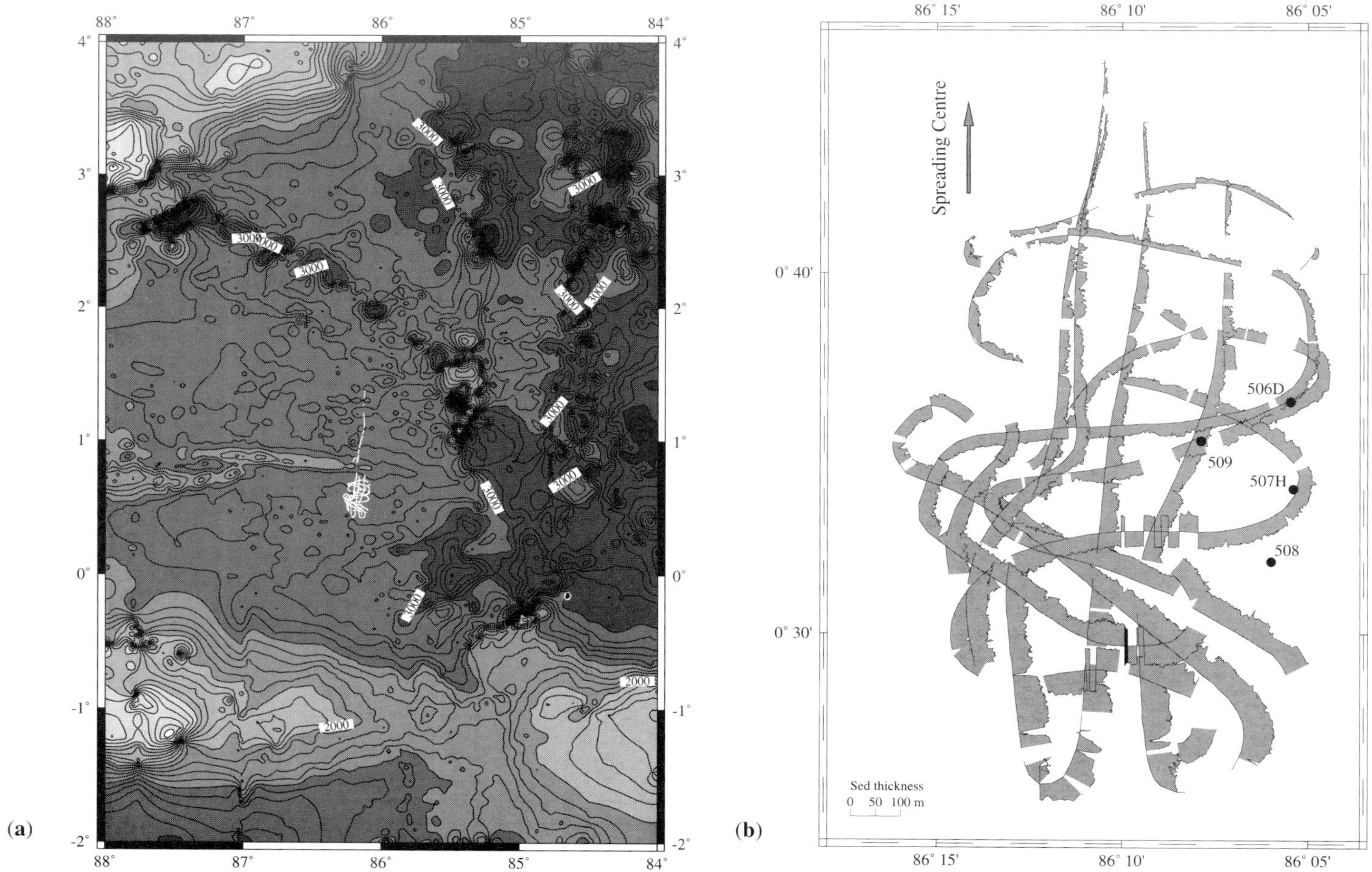

Fig. 1. (a) Location of the Deep Tow profiler data (white lines) over the Galapagos spreading centre, which is the east–west ridge at 1° N. The survey area lies in the western Panama Basin, encircled by the Carnegie and Cocos Ridges (at the bottom and upper left of the map respectively). (b) Sediment thicknesses plotted perpendicular to Deep Tow tracks together with DSDP Leg 70 sites.

Pacific due to the interaction of the equatorial divergence with the Peru Current (Moore *et al.* 1973). Sediment in cores recovered from the spreading centre is a siliceous–calcareous ooze (Klitgord & Mudie 1974; Lonsdale 1977*b*; Honnorez *et al.* 1983). The sedimentary lysocline lies at 2750–3000 m (Thunell *et al.* 1981), marginally below the depth of the spreading centre, so these MARs should be more representative of primary input rates than at deeper sites.

The profiler data are also used to explore the possibility of sea-floor dating using these accurate MARs. Dating sea floor from sediment thickness data could provide a complementary method to magnetic anomaly and isotope geochronology (Duncan & Hogan 1994; Goldstein *et al.* 1994) because it could provide a wider coverage than is possible with bottom sampling and it may provide an intermediate age range (50 ka to 10 Ma is possibly achievable). Dating has already been attempted semi-quantitatively using thicknesses measured from Deep Tow profiler records, for example to infer off-axis volcanism (Lonsdale 1977*d*) and locate overlapping spreading centres (Lonsdale 1989). This present study attempts to quantify the age precision using data from the Galapagos spreading centre and identify some of the assumptions and problems with this method.

Deep Tow data preparation

The profiler records were collected with the Scripps Deep Tow (Spiess & Tyce 1973; Spiess & Lonsdale 1982) during the 1972 Southtow-6 expedition (Klitgord & Mudie 1974). The topography of the sediment surface and basement were digitized from the records, linearly interpolated, differenced to produce sediment thicknesses and merged with the Deep Tow navigation (Fig. 1b). Thicknesses were corrected for compaction (Mitchell 1995) so they effectively represent thickness at the surface porosity (*c.* 85%), and then multiplied by the mean

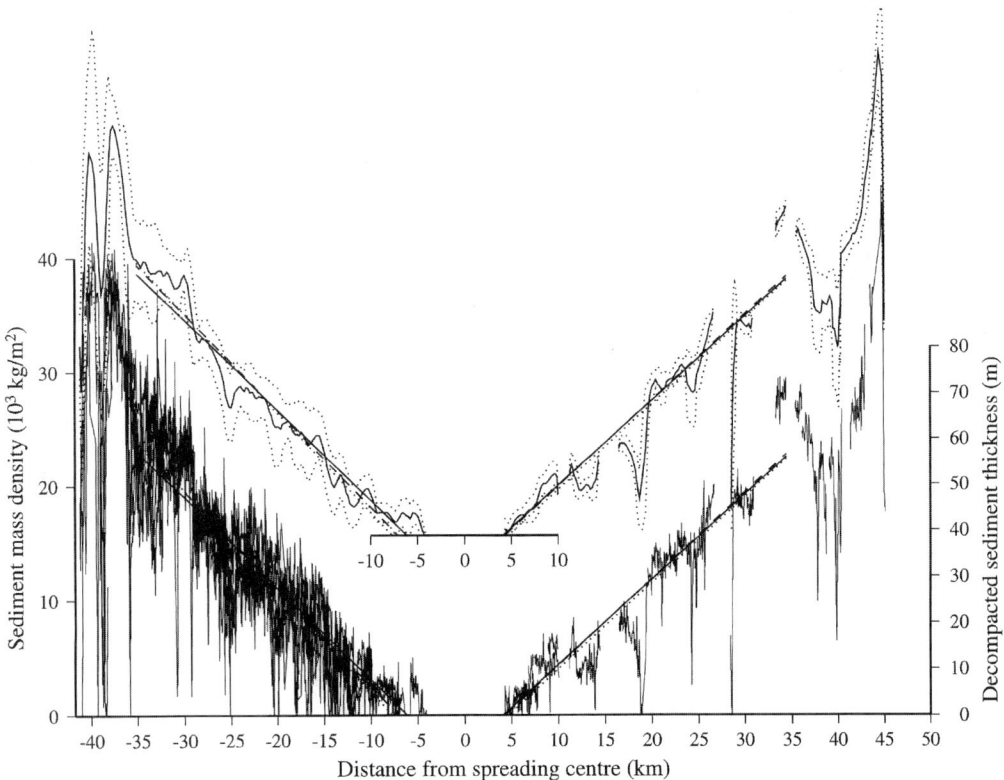

Fig. 2. Sediment thickness and dry bulk mass density over the Galapagos SC. The lines show regressions by minimizing the L1 norm (continuous line), squared error (dotted) and median of squared errors (LMS, dashed). The continuous and dotted curves above the graph show a 1 km running mean and standard deviation offset by 39 m (regression lines superimposed).

dry bulk density (0.403 g cm^{-3} from data in Karato & Becker (1983)) to produce accumulated dry mass density in Fig. 2. The data show generally increasing sediment cover with distance from the spreading centre, with fluctuations due to sediment redistribution, in particular thinning around fault scarps (Mitchell 1996). On detailed examination (Mitchell 1995), there are some areas of near-constant sediment thickness (e.g. at 20–25 km) that are probably due to large volcanic flows, which were erupted when those areas were close to the spreading centre. These have left regions of constant age and hence constant thickness. The gap between the two curves at the centre represents the extent of axial volcanism and also an effect of the resolution threshold of the profiler (Mitchell 1995).

Sedimentation rates

The average mass accumulation rate for each ridge flank is obtained by combining the slope of the data in Fig. 2 with the sea-floor spreading rate obtained by regressing Deep Tow magnetic anomaly identifications (Fig. 3). This accumulation rate also represents the supply of material from the water column, minus the effects of dissolution, if the sediment is transported only short distances (less than 10 km) when it reaches the sea floor. If, on the

other hand, some material is resuspended by erosion around fault scarps and is removed from the survey area by bottom currents (Klitgord & Mudie 1974), the supply of material is perhaps better characterized by the rate at which the sediment cover increases in the undisturbed areas between scarps. To characterize the accumulation rate for the first case, sediment mass density was regressed on distance from the spreading centre by minimizing the L1 norm (continuous line in Fig. 2). Combining the sea-floor spreading rates (Fig. 3) with the slopes of the L1 regressions, the accumulation rates are 2.17 and 2.30 g cm^{-2} ka^{-1} for the north and south flanks respectively (decompacted sedimentation rates of 54 and 57 m Ma^{-1} at 85% surface porosity). The difference in accumulation rate (c. 6%) occurs over a lengthscale of c. 40 km (the distance between the regression centres) and represents a gradient of MAR of 16% per degree of latitude.

To characterize the MAR for the second case, the data were regressed by minimizing the median of squared errors, which locates the central trend in the data without being biased by outliers (Rousseeuw & Leroy 1987). The eroded areas around scarps cover a minor proportion of the area so this should characterize the MAR of areas between scarps. The MAR estimates are only slightly higher than those computed from the L1 regression; 2.21 and 2.46 g cm^{-2} ka^{-1} for the north and south flanks respectively.

The standard errors of these rates are difficult to estimate. The errors for least-squares regressions are 1.1% (south flank) and 3.2% (north), and the error for the MAR difference is 3.3%; however the least-squares regression is biased by outliers so these are only upper bounds on the L1 and LMS regression errors. Assuming also a 1% sea-floor spreading rate error for each flank and 0.5% mean sediment density error, these maximal errors are 1.5% (south flank) and 3.4% (north) and net gradient errors are 3.6%. The L1 results are preferred because of the low bottom currents in this area. The rates are broadly in accord with a compilation by Lyle (1992), which included late Pleistocene–Holocene rates of 40–70 m Ma^{-1} for this region. Other estimates include an average Holocene rate of 52 m Ma^{-1} from C-14 at 0° 52.3'N, 86° 07.7'W (Pederson 1983) and 50 m Ma^{-1} from oxygen isotope stratigraphy at DSDP Site 506D (Lalou et al. 1983).

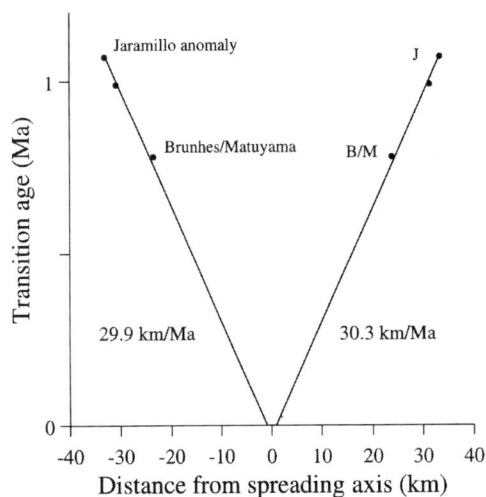

Fig. 3. Sea-floor spreading rates for the Galapagos spreading centre obtained by constrained least-squares regressions of magnetization transitions identified by Klitgord & Mudie (1974). The identifications were made at the mid-points of transitions so the constraining points were chosen at 900 m from the spreading axis (half the average transition width). Transition dates are from recent timescale revisions (Cande & Kent, 1992, 1995).

Characterization of variability lengthscales and sea floor dating

Decompacted sediment thicknesses, combined with the sedimentation rate information, can potentially be used to date sea floor, though thickness needs to be averaged over a significant area to compensate for sediment redistribution. The size of area required for this averaging was estimated by assuming that the data in Fig. 2 would form a straight line if there were no redistribution. Since the sea floor does not age exactly linearly with distance from the spreading centre but increases in discrete steps due to large volcanic flows (Mitchell 1995) and sedimentation rates probably fluctuate with time, the results are only upper bounds. The lengthscale is found by comparing averaged thicknesses to those predicted by the L1 regression line and determining the averaging lengthscale which produces an acceptable error. Sediment transport is assumed to cause a simple stochastic thickness variation, which has a standard deviation that scales linearly with the systematic trend in Fig. 2. The next section shows that this is also not strictly correct although this assumption simplifies matters.

Sediment thicknesses were first averaged within a window of fixed across-axis width, which was run over the tracks in Fig. 1. Each average was subtracted from the mean value predicted by the L1 regression, and normalized by dividing the

anomaly by the predicted value. The variability was characterized by computing the root-mean-square (r.m.s.) anomaly for each window size over the data set, to produce the upper curve in Fig. 4. The graph shows that the variation decreases relatively slowly with increasing window size from an initial value of 0.29 for the raw data. The relatively slow decrease is partly due to systematic variations in sediment thickness along the abyssal hill fabrics, probably due to variable extents of volcanic flows (Mitchell 1995), which is not reduced by along-track averaging. The lower curve in Fig. 2 shows that variability is reduced more quickly when data are averaged between tracks before binning in the across-axis direction, e.g. r.m.s. error is below 0.1 when the averaging lengthscale is greater than 2 km, compared to 8 km with only along-track averaging. We can use these values to predict the age precision expected from averaged thicknesses. From where the graphs in Fig. 4 fall below 0.12, sea-floor age should be predicted at the 95% confidence level to within 0.24 of the true value if the data are averaged over $c.$ 1.5 km (2-D averaging) and 5 km (1-D averaging).

Scaling of thickness variability

Scaling relationships between thickness mean and standard deviation (SD) may perhaps be more useful for dating sea floor because it may be possible to obtain SD and the scaling relationship from routinely collected reconnaissance side-scan sonar, as described below. The scaling relationship also characterizes the sedimentation since thickness variability will be low where sediment is mostly draping but will be high if sediment is strongly redistributed.

Side-scan sonar backscatter data from the Southeast Indian Ridge suggest that there may be a simple linear scaling between SD and mean sediment thickness (Fig. 5a from 6.5 kHz GLORIA data in Mitchell (1993)). This conclusion was based on the assumption that, due to its low frequency, the sonar is capable of penetrating the thin cover of pelagic sediments and sensing the underlying volcanic basement. Variations in the signal level compared to that of bare basalt (the attenuation) could then be interpreted as due to variations in sediment thickness. If correct, this method would be useful because sediment thickness variability could be determined remotely using low-frequency swath-mapping sonars and, coupled with scaling relationships, these could be used to estimate mean thickness and hence sea-floor age (this of course relies on an assumption of statistical uniformity).

The thickness mean and standard deviation were computed within 10 km bins across the Galapagos

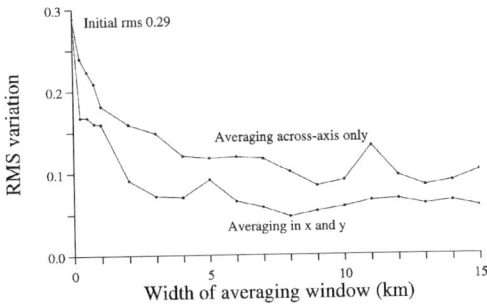

Fig. 4. The variability of averaged sediment thicknesses compared to the linear trend in Fig. 2 as a function of the width of the averaging window. Thickness anomalies were obtained by subtracting the average thickness from that predicted by the LMS regression (dashed line in Fig. 2) and normalized by dividing by the predicted mean. The variability was then characterized by calculating the root-mean-square (r.m.s.) anomaly over the profile. The upper curve shows the result of averaging along individual tracks in the across-axis direction only. The lower curve shows the result of averaging all the track data falling within fixed across-axis bins, i.e. effectively also averaging the data in the axis-parallel direction (data only from the southern flank). With 2-D averaging, the r.m.s. error is reduced below 0.1 when the averaging window is greater than 2 km. The calculations used data from within 10–35 km of the spreading centre.

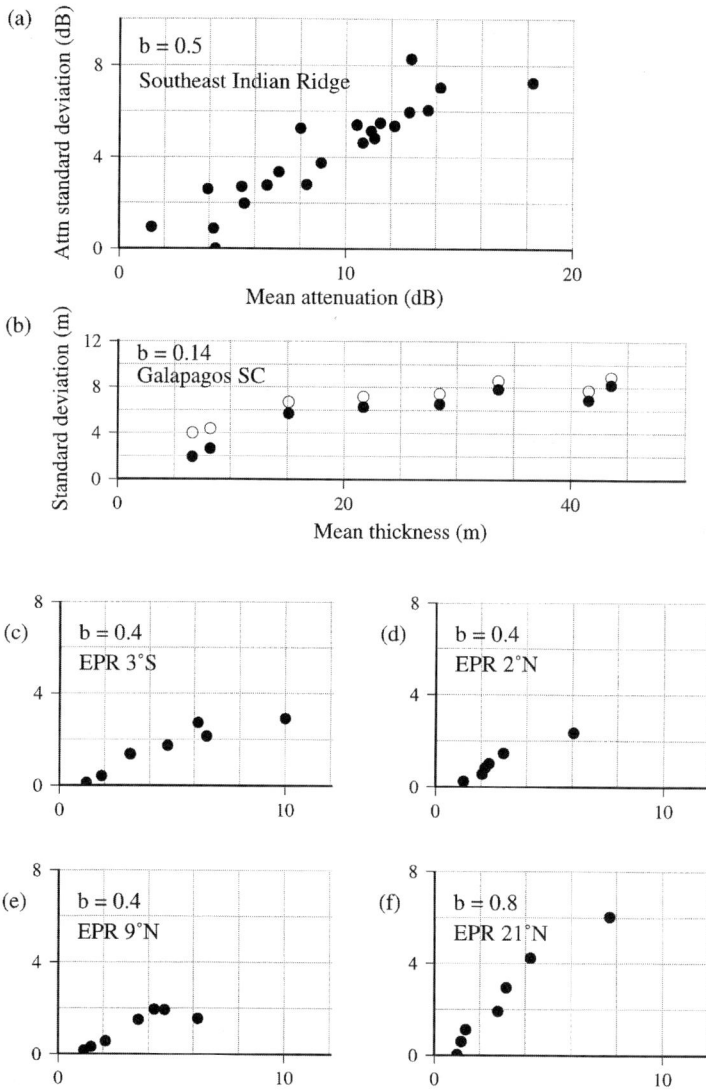

Fig. 5. Scaling of sediment thickness variability with mean thickness for various spreading centres. The slopes of the graphs (*b*) were computed by least-squares regression to characterize the sedimentation. In areas where sediments predominantly drape the basement, such as the Galapagos SC, the graphs have low regression slopes, while areas with more redistributed sediment are represented by higher slopes. (**a**) Variability of acoustic attenuation produced from GLORIA side-scan sonar backscatter data from within 4.5 km of the Southeast Indian Ridge spreading centre (Mitchell 1993) at 25.5° S. It was proposed in Mitchell (1993) that the sediment thickness is proportional to the attenuation in decibels so the correlation between variability and mean attenuation suggests scaling between variability and mean thickness at a rate of 0.5. (Mean attenuations were measured with respect to a signal level of -38 dB for the data in Mitchell (1993).) (**b**) The standard deviation and mean thickness for the Galapagos SC data of Fig. 2 (open symbols) computed within 10 km wide bins, excluding data from the central 8 km. The filled circles show the statistics adjusted for possible overestimation of variability due to digitizing errors (assuming an additive error variance of 12 m^2) and the slope *b* was computed from the adjusted data. Thicknesses were not corrected for compaction. Statistics computed from (**c**) Fig. 6a, (**d**) Fig. 6b, (**e**) from Fig. 6c and (**f**) Fig. 6d.

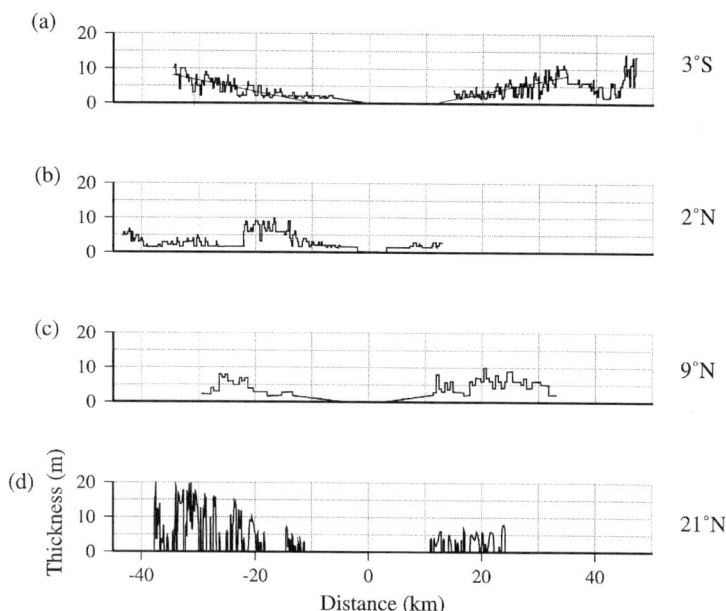

Fig. 6. Sediment thickness with distance from spreading centres digitized from (**a**) figure 3 of Lonsdale (1977*d*), (**b**) figure 15 of Lonsdale (1989), (**c**) figure 14 of Lonsdale & Spiess (1980) and (**d**) figure 2 of Larson & Spiess (1969). These authors used slightly different interpretation criteria which may affect the statistics (e.g. maximum thickness in 200 m sections (Lonsdale 1977*d*, 1989) and maximum thickness in 500 m sections (Lonsdale & Spiess 1980)). The lines in (**a**) are least-squares regressions.

SC and plotted with the open symbols in Fig. 5b. The filled circles show the data adjusted for digitizing error. In contrast to Fig. 5a, the variability increases non-linearly with the mean, showing a rapid initial increase followed by a more gentle rise. The initial increase may reflect initial redistribution occurring on young sea floor to fill local depressions, while the lower slope shows that sediment is then largely draping (Klitgord & Mudie 1974; Lonsdale 1977*b*). This draping implies that the primary input rate is higher than the average magnitude of lateral transport rates.

Figure 5c–f show statistics for East Pacific Rise sediments, calculated from the data in Fig. 6a–d. Slightly different interpretation criteria were used to produce Fig. 6a–d and the thickness range is small so the statistics should be interpreted with caution. The rate of variability increase, 0.4, is higher than for the Galapagos SC and reflects the lower sedimentation rates, but is similar to the Indian Ocean data (indirectly supporting the attenuation model (Mitchell 1993) since the Indian Ocean site was also an area of normal pelagic activity). The lack of an initial rise in SD in these graphs, except for Fig. 5b, is partly due to the low data resolution.

Discussion

Sediment accumulation rate

Figure 7 shows the gradient of MAR with latitude at the Galapagos SC compared to MAR variation at 110° W (rescaled to the Galapagos results). Sedimentation in the two areas can be compared despite their different water depths (4000 m at 110° W, 2700 m at the Galapagos SC) because both sites lie slightly above their local carbonate lysoclines (4000–4500 m at 110° W (Snoeckx & Rea 1994), 2900 m at the Galapagos (Lyle *et al.* 1988)). Figure 7 was produced from seismic two-way time through sediments (Mayer *et al.* 1992*a*) combined with ODP Leg 138 density and velocity data (Mayer *et al.* 1992*b*) to produce a mass density profile. The profile was corrected for broadening due to sedimentation on the northwards drifting Pacific Plate (Berger & Winterer 1974; van Andel *et al.* 1975; Lancelot 1978). Fluctuations in the profile are artefacts of this correction (details will be presented elsewhere) so the Galapagos SC data should be compared to the dashed curve, which is a Gaussian curve fitted to the corrected data. The graph shows that the MAR gradient for the

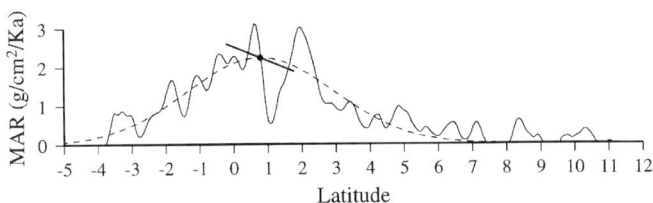

Fig. 7. Comparison of the MAR variation at 110° W with the MAR gradient of the Galapagos SC (oblique line). The irregular continuous line shows the average latitude variation of MAR computed from sediment thicknesses derived from seismic reflection records (Mayer *et al.* 1992*a*, figure 4) corrected for broadening caused by sedimentation on the northward migrating Pacific Plate (method to be presented elsewhere). Because the correction method increases variability due to sediment redistribution, the centre and width of equatorial enhancement are characterized by fitting the Gaussian curve (dashed line). The Gaussian curve and 110° W profile have been rescaled to match the Galapagos SC rates. The Galapagos SC gradient is much higher than the Gaussian at the same latitude, suggesting that the centre of primary input near 85° W is further south than at 110° W.

Galapagos SC is higher than for the 110° W at the same latitude, but would match the 110° W curve if translated by 100 km or more. Indeed, if we assume that equatorial enhancement in the Panama Basin and at 110° W have similar profile shapes, the distance to its centre can be directly estimated from $S'\sigma^2$, where S' is the proportional gradient of MAR (0.16 per degree) and σ is the Gaussian parameter (4.71° at 110° W). These figures imply an enhancement centre at 2.7° S, which matches the present productivity maximum (Moore *et al.* 1973). Although this match is largely fortuitous because the MAR gradient at the Galapagos SC is only marginally resolved and the equal shape assumption is not verified, it nevertheless demonstrates that accurate MAR gradient measurements are potentially more useful than MARs alone.

Regressions were computed for Fig. 6a for further illustration. Assuming a spreading half-rate of 68 km Ma^{-1} (DeMets *et al.* 1989), the 500 ka mean sedimentation rate is 21 m Ma^{-1}. Further assuming a dry bulk density of 0.659 g cm^{-3} (mean surface density within 4.2° of the equator (Murray *et al.* 1993)), the MAR is 1.4 g cm^{-2} ka^{-1}. This is 40% greater than the 0.963 g cm^{-2} ka^{-1} MAR at 3° S, 135–150° W (Murray *et al.* 1993) and 55% greater than 0.9 g cm^{-2} ka^{-1} at 3° S, 110° W (ODP Site 848) computed from data in Shackleton & Shipboard Scientific Party (1992) (the timescales are 12 ka and 1 Ma respectively). The difference is probably largely due to the different water depths of the East Pacific Rise (EPR) compared to the other sites but a further interesting possibility is that the zone of peak equatorial productivity may be deflected southwards further west than previously thought. This could be tested by collecting high-resolution sediment thickness profiles at various latitudes along the EPR.

Sea-floor dating

A study using sea-floor dating from sediment cover would require an assessment of sediment transport lengthscales for that area although the Galapagos SC results can be used to illustrate how age precision might be estimated in practice, using some interpretations of the EPR data in Fig. 6. (This is merely an illustration because thickness variability is greater over the EPR and the assumption of statistical uniformity is not verified.) Lonsdale (1977*d*) suggested that off-axis volcanism produced the thin sediment at 40 km in Fig. 6a, and (Lonsdale 1989) that thin sediment at –30 km and thick sediment at –15 km in Fig. 6b are due to a fossil axial zone and inter-rift plateau of an overlapping spreading centre. In both cases, thinning occurs over greater than 10 km, so this is used as the averaging lengthscale. The mean thickness at 35–45 km in Fig. 6a is 4.8 m and the predicted mean thickness at –35 to –45 km is 8.7 m (extrapolated from 6.6 m at –25 to –35 km), so the thickness anomaly is 3.9 m. To estimate the age anomaly, we assume a sedimentation rate of 21 m Ma^{-1} from the regressions of Fig. 6a and a 95% confidence interval of ±0.2 from the upper line in Fig. 4. By simple propagation of errors, the lava flow age anomaly is 190±50 ka. For the overlapping spreading centre in Fig. 6b, the mean thickness at –40 to –20 km is 2.7 m and at –22 to –12 km is 6.3 m, so the thickness contrast is 3.3 m. Repeating the previous calculation but assuming a 50% higher sedimentation rate due to the different latitude, the age contrast is 110±30 ka and the age of the inter-rift plateau is 210±40 ka, a lower value than estimated by Lonsdale (1989). These precision estimates do not include sedimentation rate or digitizing errors, but suggest that, even accounting for the greater sediment variability over the EPR, thicknesses from profiler records are capable of resolving age information that is useful for tectonic or volcanic interpretations. This procedure could also be improved by taking account of temporal fluctuations in sedimentation rates if suitable cores were available. This would in practice involve mapping the thicknesses to age using an MAR–age curve derived from cores and then processing these model ages as done for thicknesses above. Hiatuses in the sedimentary record due to variable bottom currents could also potentially be incorporated using core information. Hiatuses should not affect the method greatly provided that the MAR–age curve includes them. However, local effects such as non-deposition around fault scarps may have different spatial extents depending on the intensity of bottom currents and their directions, so other methods would be needed if studying at a high resolution. This might be possible using core information together with data from a modern 'CHIRP' sediment profiler if capable of resolving the sediment stratigraphy and hence structures due to variable deposition.

Conclusions

High-resolution sediment profiler records across spreading centres provide accurate estimates of sedimentation rates and, by comparing adjacent ridge flanks, the gradient of sedimentation rates. Provided that bottom currents do not remove sediment preferentially from one ridge flank, this may provide a way of assessing geographic variations in long-term primary input rate; variations along the East Pacific Rise where it crosses the equator may provide an interesting application. Mass accumulation rates over the Galapagos SC are 2.17 and 2.30 g cm^{-2} ka^{-1} for the north and south flanks respectively, which correspond to a north–south gradient of MAR of 16% per degree of latitude. Although this gradient is not strongly resolved because of limited data from the north flank, the gradient is relatively high for this latitude compared to the ODP Leg 138 equatorial transect at 110° W, suggesting that the centre of upwelling and enhanced sedimentation in this part of the Panama Basin is 1° or more further south than at 110° W, as shown by productivity data (Moore *et al.* 1973).

Sediment thickness data are potentially useful for characterizing the age of near-ridge basement if the data are averaged over a sufficient area to compensate for the redistribution. The averaging lengthscale was assessed assuming that the effect of sediment redistribution can be modelled as a simple stochastic variation. For the Galapagos SC, stacking a series of ridge-axis profiles and then averaging with a cell size of 10 km reduces the 95% confidence interval to less than a factor of ±0.1 of the thickness that would exist in the absence of redistribution. The standard deviation and mean thickness appear to show a rough scaling relationship and their ratio (SD/mean) characterizes the general nature of the sedimentation: high ratios occur where the sediment is strongly redistributed and low ratios occur where sediment is draping the basement topography. The Galapagos SC has low ratios (0.14 compared to 0.4 for the EPR) reflecting the high sedimentation rate over the ridge compared to redistribution rates.

F.N. Spiess of the Marine Physical Laboratory, Scripps Institution of Oceanography, is gratefully acknowledged for permission to copy the Deep Tow data used in this study. Figures were produced using the GMT software system developed by P. Wessel and W. Smith (Smith & Wessel 1990; Wessel & Smith 1991). I am also grateful to M. Audet for help with the decompaction correction. This work was supported by a Research Fellowship from the Natural Environment Research Council (GT5/92/GS/5). The British Antarctic Survey is also thanked for access to computer facilities while some of this work was done at sea on *RRS James Clark Ross*.

References

BERGER, W. H. & STAX, R. 1994. Neogene carbonate stratigraphy of Ontong Java Plateau (western equatorial Pacific): three unexpected findings. *Terra Nova*, **6**, 520–534.

—— & WINTERER, E. L. 1974. Plate *Stratigraphy and the Fluctuating Carbonate Line*. International Association of Sedimentologists, Special Publications, **1**, 11–48.

BROECKER, W. S. & PENG, T.-H. 1982. *Tracers in the Sea*. Lamont-Doherty Geological Observatory, Columbia University, Palisades, New York.

CANDE, S. C. & KENT, D. V. 1992. A new geomagnetic polarity time scale for the late Cretaceous and Cenozoic. *Journal of Geophysical Research*, **97**, 13 917–13 951.

—— & —— 1995. Revised calibration of the geomagnetic polarity timescale for the late Cretaceous and Cenozoic. *Journal of Geophysical Research*, **100**, 6093–6095.

DEMETS, C., GORDON, R. G., ARGUS, D. F. & STEIN, S. 1989. Current plate motions. *Geophysical Journal International*, **101**, 425–478.

DETRICK, R. S., WILLIAMS, D. L., MUDIE, J. D. & SCLATER, J. G. 1974. The Galapagos Spreading Centre: bottom-water temperatures and the significance of geothermal heating. *Geophysical Journal of the Royal Astronomical Society*, **38**, 627–637.

DUNCAN, R. A. & HOGAN, L. G. 1994. Radiometric dating of young MORB using the ^{40}Ar–^{39}Ar incremental heating method. *Geophysical Research Letters*, **21**, 1927–1930.

GOLDSTEIN, S. J., PERFIT, M. R., BATIZA, R., FORNARI, D.

& MURRELL, M. T. 1994. Off-axis volcanism at the East Pacific Rise detected by uranium-series dating of basalts. *Nature*, **367**, 157–159.

HONNOREZ, J., VON HERZEN, R. P. & BORELLA, P. E. 1983. Introduction, principal results, and explanatory notes, Deep Sea Drilling Project Leg 70. *In*: HONNOREZ, J. & VON HERZEN, R. P. (eds) *Initial Reports of the Deep Sea Drilling Project*, **70**. US Government Printing Office, Washington, 5–30.

KARATO, S. & BECKER, K. 1983. Physical properties of sediments from the Galapagos region and their implications for hydrothermal convection. *In*: HONNOREZ, J. & VON HERZEN, R. P. (eds) *Initial reports of the Deep Sea Drilling Project*, **70**. US Government Printing Office, Washington, 355–368.

KEEN, M. J. & MANCHESTER, K. S. 1970. The Mid-Atlantic Ridge near 45°N. X. Sediment distribution and thickness from seismic reflection profiling. *Canadian Journal of Earth Sciences*, **7**, 733–747.

KLITGORD, K. D. & MUDIE, J. D. 1974. The Galapagos spreading centre: a near-bottom geophysical survey. *Geophysical Journal of the Royal Astronomical Society*, **38**, 563–586.

LALOU, C., BRICHET, E., LECLAIRE, H. & DUPLESSY, J.-C. 1983. Uranium series disequilibrium and isotope stratigraphy in hydrothermal mound samples from DSDP sites 506–509, Leg 70, and site 424, Leg 54: an attempt at chronology. *In*: HONNOREZ, J. & VON HERZEN, R. P. (eds) *Initial Reports of the Deep Sea Drilling Project*, **70**. US Government Printing Office, Washington, 303–314.

LANCELOT, Y. 1978. *Relations entre evolution sedimentaire et tectonique de la plaque Pacifique depuis le Cretace inferieur*. Societe Geologique de France Memoire, **134**.

LARSON, R. L. & SPIESS, F. N. 1969. East Pacific Rise crest: a near-bottom geophysical profile. *Science*, **163**, 68–71.

LONSDALE, P. 1977a. Abyssal pahoehoe with lava coils at the Galapagos rift. *Geology*, **5**, 147–152.

—— 1977b. Deep-tow observations at the mounds abyssal hudrothermal field, Galapagos Rift. *Earth and Planetary Science Letters*, **36**, 92–110.

—— 1977c. Inflow of bottom water to the Panama Basin. *Deep-Sea Research*, **24**, 1065–1101.

—— 1977d. Structural geomorphology of a fast-spreading rise crest: the East Pacific Rise near 3° 25'S. *Marine Geophysical Research*, **3**, 251–293.

—— 1989. The rise flank trails left by migrating offsets of the Equatorial East Pacific Rise axis. *Journal of Geophysical Research*, **94**, 713–743.

—— & SPIESS, F. N. 1980. Deep-tow observations at the East Pacific Rise, 8° 45'N, and some interpretations. *In*: ROSENDAHL, B. R. & HEKINIAN, R. (eds) *Initial Reports of the Deep Sea Drilling Project*, **54**. US Government Printing Office, Washington, 43–62.

LYLE, M. 1992. Composition maps of surface sediments of the eastern tropical Pacific Ocean. *In*: MAYER, L., PISIAS, N. & JANECEK, T. (eds) *Proceedings of the Ocean Drilling Program, Initial Reports*, **138**. Ocean Drilling Program, College Station, Texas, 101–115.

——, MURRAY, D. W., FINNEY, B. P., DYMOND, J., ROBBINS, J. M. & BROOKSFORCE, K. 1988. The record of Late Pleistocene biogenic sedimentation in the eastern tropical Pacific Ocean. *Paleoceanography*, **3**, 39–59.

MAYER, L. A., PISIAS, N. G. & JANECEK, T. R. 1992a. Introduction. *In*: MAYER, L., PISIAS, N. & JANECEK, T. (eds) *Proceedings of the Ocean Drilling Program, Initial Reports*, **138**. Ocean Drilling Program, College Station, Texas, 5–12.

——, —— & —— 1992b. *Proceedings of the Ocean Drilling Program, Initial Reports*, **138**. Ocean Drilling Program, College Station, Texas.

MITCHELL, N. C. 1993. A model for attenuation of backscatter due to sediment accumulations and its application to determine sediment thickness with GLORIA sidescan sonar. *Journal of Geophysical Research*, **98**, 22 477–22 493.

—— 1995. Characterising the extent of volcanism at the Galapagos Spreading Centre using Deep Tow profiler records. *Earth and Planetary Science Letters*, **134**, 459–472.

—— 1996. Creep in pelagic sediments and potential for morphologic dating of marine fault scarps. *Geophysical Research Letters*, **23**, 483–486.

MOORE, T. C., HEATH, G. R. & KOWSMANN, R. O. 1973. Biogenic sediments of the Panama Basin. *Journal of Geology*, **81**, 458-472.

MURRAY, R. W., LEINEN, M. & ISERN, A. R. 1993. Biogenic flux of Al to sediment in the central equatorial Pacific Ocean: evidence for increased productivity during glacial periods. *Paleoceanography*, **8**, 651–670.

PEDERSON, T. F. 1983. Increased productivity in the eastern equatorial Pacific during the last glacial maximum (19,000 to 14,000 yr B.P.). *Geology*, **11**, 16–19.

ROUSSEEUW, P. J. & LEROY, A. M. 1987. *Robust Regression and Outlier Detection*. John Wiley and Sons, New York.

SHACKLETON, N. J. & SHIPBOARD SCIENTIFIC PARTY. 1992. Sedimentation rates: toward a GRAPE density stratigraphy for Leg 138 carbonate sections. *In*: MAYER, L., PISIAS, N. & JANECEK, T. (eds) *Proceedings of the Ocean Drilling Program, Initial Reports*, **138**. Ocean Drilling Program, College Station, Texas, 87–91.

SMITH, W. H. F. & WESSEL, P. 1990. Gridding with continuous curvature splines in tension. *Geophysics*, **55**, 293–305.

SNOECKX, H. & REA, D. K. 1994. Late Quaternary $CaCO_3$ stratigraphy of the eastern equatorial Pacific. *Paleoceanography*, **9**, 341–351.

SPIESS, F. N. & LONSDALE, P. F. 1982. Deep Tow rise crest exploration techniques. *Marine Technology Society Journal*, **16**, 67–74.

—— & TYCE, R. C. 1973. *Marine Physical Laboratory Deep Tow Instrumentation System*. University of California, San Diego, Marine Physical Laboratory of the Scripps Institution of Oceanography, SIO reference **73–4**.

THUNELL, R. C., KEIR, R. S. & HONJO, S. 1981. Carbonate dissolution: An in situ study in the Panama Basin. *Science*, **212**, 659–661.

VAN ANDEL, T. H., REA, D. K., VON HERZEN, R. P. & HOSKINS, H. 1973. Ascension fracture zone,

Ascension Island, and the Mid-Atlantic Ridge. *Bulletin of the Geological Society of America*, **84**, 1527–1546.

—— HEATH, G. R. & MOORE, T. C. 1975. *Cenozoic Tectonics, Sedimentation, and Paleoceanography of the Central Equatorial Pacific*. Geological Society of America Memoir, **143**.

WESSEL, P. & SMITH, W. H. F. 1991. Free software helps map and display data. *EOS (Transactions of the American Geophysical Union)*, **72**, 441.

Sedimentary evidence relating to the tectonic evolution of the Lau Basin, SW Pacific, from ODP Sites 834–839 (ODP Leg 135)

R. G. ROTHWELL

Southampton Oceanography Centre, Empress Dock, Southampton SO14 3ZH, UK

Abstract: Six sites were drilled during Ocean Drilling Program Leg 135 in the western Lau backarc basin in the southwest Pacific (Sites 834–839). These sites are all located in basins within a horst and graben terrain and form an approximate transect across the rifted arc basement onto crust considered to have been formed at the East Lau Spreading Centre. The sedimentary sequences recovered from these backarc sites range in age from the late Miocene to the Holocene; they consist primarily of a lower succession of volcaniclastic sediment gravity-flow deposits interbedded with hemipelagic clayey nannofossil oozes and nannofossil clays, overlain by a distinctive upper succession of hydrothermally stained hemipelagic, and locally redeposited, clayey nannofossil oozes. The volcaniclastic sediment gravity-flow deposits are predominantly massive, proximal, vitric gravels, sands and silts, that are mainly locally derived from adjacent basement ridges and intrabasin seamount volcanoes.

At Site 835, which was drilled in a small extensional sub-basin in the oldest part of the Lau Basin, the upper clayey nannofossil ooze sequence is anomalously thick and rigorous sedimentological analysis shows that much of this sequence is redeposited. Thick clayey nannofossil ooze turbidite muds are identified that closely resemble the enclosing clayey nannofossil ooze hemipelagites. These thick turbidite muds are associated with mudclast conglomerates, interpreted as muddy debris-flow deposits, and a number of coherent rafted blocks of older hemipelagic sediment. These allochthonous deposits testify to several episodes of instability in the sub-basin that may be related to large-scale tectonic activity caused by the southward passage of ocean-ridge propagator tips past the latitude of the drillsite. Episodes of increased sediment deposition, due to increased frequency of turbidite emplacement, are evident at all of the Lau Basin backarc drillsites, and these correlate moderately well with the closest approach of the propagating ridge tip to each sub-basin. The research presented illustrates how turbidites, once identified, can be used to decipher complex tectonic histories, and demonstrates the importance of local tectonic controls on sediment redeposition in backarc basins.

Thick structureless muds are a common lithology drilled at many DSDP/ODP (Deep Sea Drilling Project/Ocean Drilling Program) drillsites, especially those drilled on sediment fans, deep-sea basins and continental rises. Such muds are commonly redeposited in origin and great care is needed in interpreting sequences containing thick muds in order to distinguish between autochthonous and allochthonous deposits. Correct identification of deep sea muds as turbidites, hemipelagic deposits or contourites is essential if sedimentary columns are to be correctly interpreted and inferences made on depositional history. Once turbidites, or groups of turbidites, are identified in a sediment sequence, it may be possible to use their occurrence to infer episodes of instability, which may be correlatable with known or inferred tectonic, volcanic or other events. Turbidites can therefore be useful tools in deciphering tectonic or volcanic histories in certain geological settings.

Turbidites tend to show an organized sequence of sedimentary structures, whose development commonly depends on proximality–distality relationships. Bouma (1962) demonstrated such a consistent ordered set of sedimentary structures, which records waning of the flow, and proposed a now universally accepted structural scheme. In Bouma's scheme, all fine material was classified as e division mud. Kuenen (1964) divided the e division into e_t (turbidite mud) and e_p (pelagic/hemipelagic mud) to distinguish between mud deposited by the turbidity current and the overlying mud deposited by pelagic and hemipelagic processes. Van der Lingen (1969) and Hesse (1975) used the letter e for turbidite mud and f for pelagic/hemipelagic mud. Piper (1978) subdivided turbidite mud (e) into three structural divisions: e_1 (laminated mud), e_2 (graded mud) and e_3 (ungraded mud), and retained the f notation for overlying hemipelagic or pelagic sediment. The boundary

ROTHWELL, R. G. 1998. Sedimentary evidence relating to the tectonic evolution of the Lau Basin, SW Pacific, from ODP Sites 834–839 (ODP Leg 135). *In*: CRAMP, A., MACLEOD, C. J., LEE, S. V. & JONES, E. J. W. (eds) *Geological Evolution of Ocean Basins: Results from the Ocean Drilling Program*. Geological Society, London, Special Publications, **131**, 211–229.

between turbidite mud (e) and overlying hemipelagic or pelagic mud (f) is typically bioturbated
or gradational.

If turbidites show marked differences in colour
to the enclosing hemipelagic sediments and if the
hemipelagic sediments are moderately to heavily
bioturbated, it may be relatively easy to visually
distinguish between turbidite and hemipelagic
muds (cf. Weaver & Kuijpers 1983; Weaver &
Rothwell 1987; Rothwell *et al.* 1992). However, in
instances where turbidites and hemipelagites are
very similar in colour and bioturbation of the

sediments is low, it can, in practice, be difficult to
determine the boundaries between e and f division
muds. Many authors (e.g. Kelts & Arthur 1981;
Brunner & Normark 1985; Wetzel 1986; Brunner &
Ledbetter 1987; ODP Leg 107 Shipboard Scientific
Party 1987; Pilkey 1987; and many others) have
discussed, or noted difficulty in identifying
turbidite muds and/or establishing the true thicknesses of turbidites (and hence the proportion that
turbidite muds make up of particular sediment
columns). Kelling & Stanley (1976) suggested that
many so-called hemipelagic muds are actually

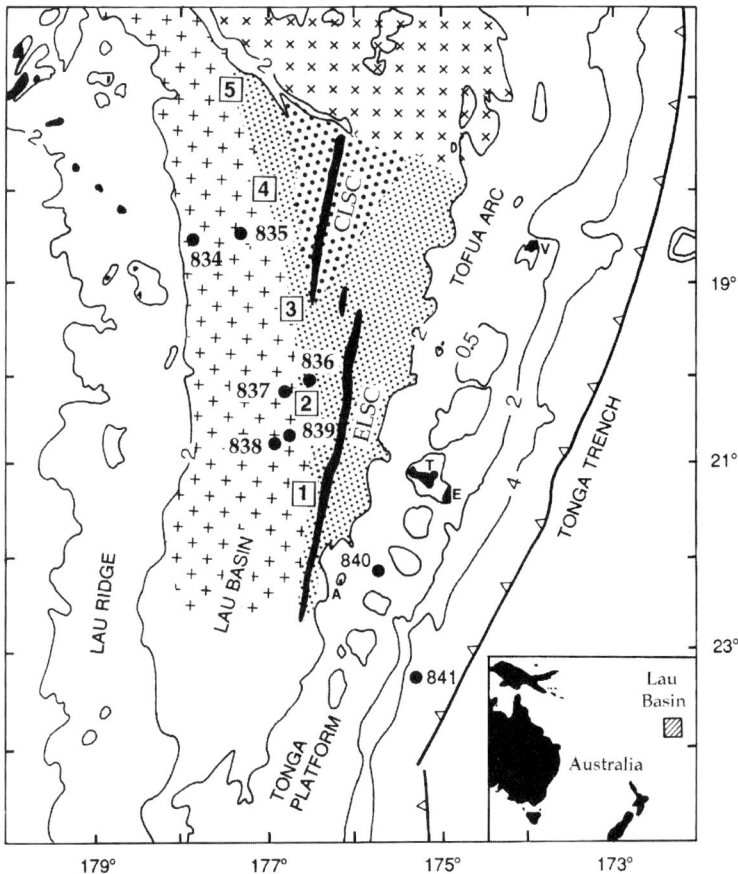

Fig. 1. Map showing location of the Lau Basin, the location of the ODP Leg 135 drillsites, and crustal types with the
backarc basin, overlain on a regional bathymetry (water depth is in kilometres). The + symbol indicates attenuated
and rifted arc crust (horst and graben terrain) which floors the western Lau Basin. × symbol indicates attenuated crust
in the northern Lau Basin which may be equivalent to the western basin terrain (shown by + symbol). Lighter, more
closely spaced stipple indicates the distribution of oceanic crust generated at the East Lau Spreading Centre (ELSC),
while the darker, coarser stipple indicates oceanic crust generated at the Central Lau Spreading Centre (CLSC).
Numbers in boxes are ages in Ma at which the ELSC propagator would have reached that latitude on its progression
southward, assuming constant spreading and propagation rates. Islands on the Tonga Platform shown are Ata (A),
Tongatapu (T), 'Eua (E), and Vavau (V). Locations of CLSC and ELSC are also shown (after Parson & Hawkins
1994; base map from Bøe 1994).

turbidite muds on the basis of structures, composition and texture.

Biogenic turbidites can be particularly difficult to distinguish from hemipelagites, as their composition, colour and sometimes the biostratigraphic age of the contained microfossils may be very similar to the enclosing hemipelagic sediments. This is especially true in regions where the organic carbon content of the sediment is low and hence there is little bioturbation of the hemipelagic sediments to distinguish them from turbidite muds. As Kelts & Arthur (1981) point out, thick turbidite muds have not been widely recognized in sediment sequences from DSDP sites. This is undoubtedly because they are often difficult to distinguish in disturbed intervals and because many sequences of hemipelagic mud are monotonous and may be of little interest to shipboard scientists. Any turbidite muds present may remain unrecognized.

This paper discusses data obtained during Leg 135 of the Ocean Drilling Program from the Lau backarc basin in the southwest Pacific. During this leg, six sites (ODP Sites 834–839, Fig. 1) were drilled in the western Lau Basin. It was expected that drilling would recover a sequence dominated by volcaniclastic and hemipelagic sediments. This is what largely was recovered, but some sediment sequences contained thick mud turbidites that closely resembled, both in colour and composition, the enclosing hemipelagic deposits. These beds were largely unrecognized during shipboard description of the sediment cores. However, once identified, they provided key indicators of episodes of local instability, possibly related to the approach and passage of ocean-ridge propagator tips past the latitude of the drillsites. This research illustrates how turbidites, once identified, can be used to decipher complex tectonic histories, and demonstrates the importance of local tectonic controls on sediment redeposition in backarc basins.

The Lau Basin

The Lau Basin is a southerly tapering, actively spreading, backarc basin at the convergent boundary of the Indo-Australian and Pacific Plates, situated about 330 to 700 km east of Fiji. The basin has an area of 250×10^3 km^2 and is bounded to the west by the Lau Ridge, a remnant extinct arc, and to the east by the Tonga Ridge/Tofua Arc, an active island arc (Fig. 1).

Generation of backarc oceanic crust has been through two southward propagating ocean ridges which form the main morphotectonic elements within the central Lau Basin: the Central Lau Spreading Centre (CLSC) and the East Lau Spreading Centre (ELSC; Parson et al. 1990; Fig. 1). The CLSC is presently an active spreading ridge which is propagating southward into the basin at the expense of the ELSC, approximately along

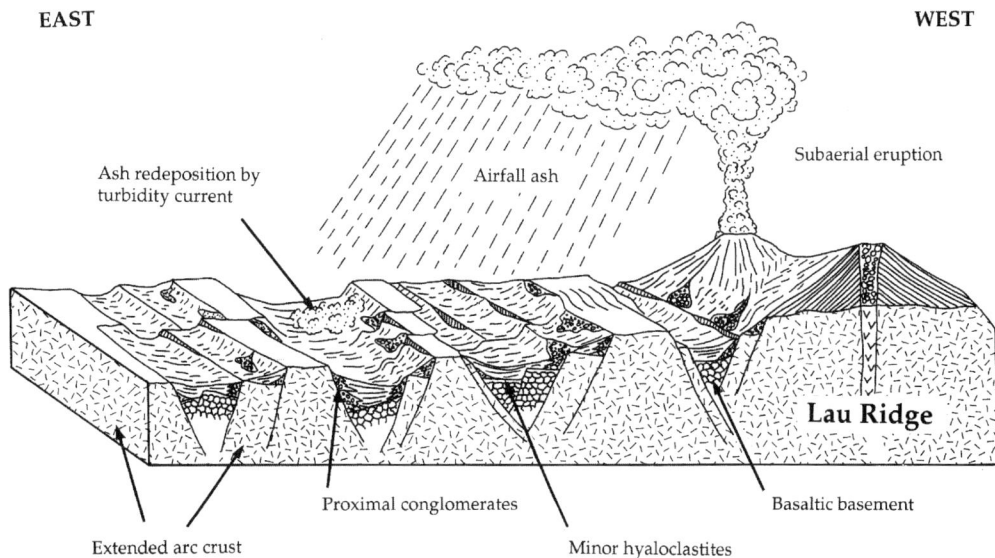

Fig. 2. Schematic cross-section across the western Lau Basin during the late Pliocene, showing the horst and graben terrain, in which proximal volcanic conglomerates and breccias, hyaloclastites and vitric sediment gravity flows were deposited. These form the main sediments of lithostratigraphic Units II and III in the Lau Basin (after Clift & Dixon 1994).

176° 30' W longitude (Fig. 1). The propagator tip is located at around 19°16′ S, 176°32′ W. The ELSC lies 50 km to the east of the CLSC and extends southward for over 180 km from near 19°20′ S at longitude 175°55′ W to about 21° S where it merges with the Valu Fa Ridge (Fig. 1).

The floor of the Lau Basin consists of two principal crustal types: normal mid-oceanic ridge (MORB-like) basalts generated through backarc spreading, and rifted island-arc basaltic andesitic to dacitic crust of Oligocene to Pliocene age (Parson *et al.* 1992; Parson & Hawkins 1994; Fig. 1). The rifted island-arc crust forms a terrain of irregular horst and graben (Fig. 2) which was produced through rifting of a late Palaeogene to early Neogene volcanic island-arc chain (Karig 1970; Parson & Hawkins 1994). Rifted island-arc crust forms the floor of much of the western part of the

Lau Basin, west of 177° W longitude (Parson *et al.* 1992).

Leg 135 of the Ocean Drilling Program drilled six sites (Sites 834 to 839, Fig. 1), forming an approximate transect across the rifted arc basement, that floors the western Lau Basin, onto crust considered to have formed at the ELSC. The sediment sequence at each site was drilled using the advanced hydraulic piston corer (APC) with the goal of obtaining complete and relatively undisturbed sediment sections. Although the main drilling objectives from these sites were to sample the igneous basement rocks and to determine the age of the beginning of opening of the backarc basin, an important objective was to obtain sedimentary, palaeontological and palaeomagnetic records of the basin fill.

All the Leg 135 backarc sites are located within

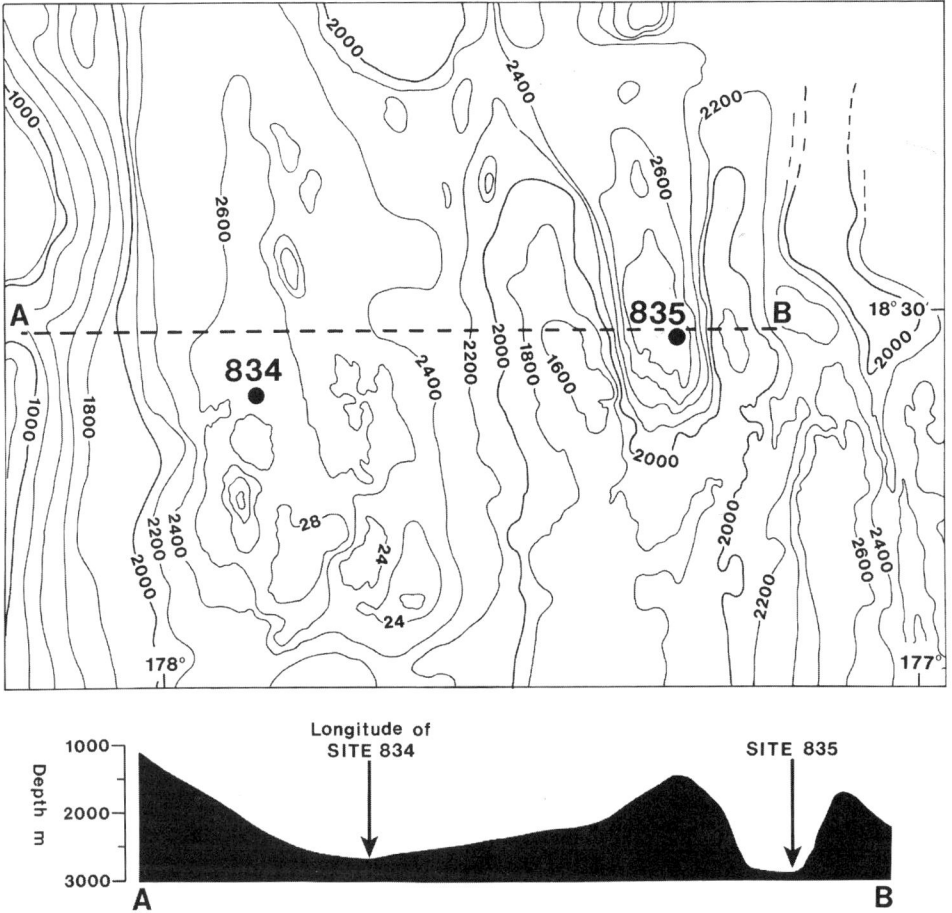

Fig. 3. Local bathymetry around the two sub-basins containing ODP Sites 834 and 835. Contour interval 200 m. Bathymetric profile along the line A–B is shown below (vertical exaggeration of A–B profile × 8).

small, elongate, partially sediment-filled graben within the horst and graben terrain. Sites 834 and 835 were drilled in two separate, but adjacent, sub-basins, 48 km apart, between 100–200 km east of the Lau Ridge (Figs 1 and 3); Sites 836–839 were drilled about 200 km to the southeast of Sites 834 and 835, in three fault-bounded sub-basins on the western flank of the ELSC (Fig. 1). These four sites are less than 90 km apart. Sites 838 and 839 were drilled within the same sub-basin. The data collected during the drilling leg are presented by Parson *et al.* (1992).

The sedimentary sequence

The sediment sequences recovered from the backarc sites range from the late Miocene to Holocene in age and a two-fold subdivision of the sedimentary sequence can be recognized: a lower sequence of volcaniclastic sediment gravity-flow deposits interbedded with hemipelagic clayey nannofossil oozes and nannofossil clays, overlain by an upper sequence of hemipelagic clayey nan-nofossil ooze, locally containing calcareous tur-bidite beds. The volcaniclastic sediment gravity-flow deposits that make up much of the lower sequence are predominantly massive proximal, vitric sands, silts and sometimes gravels (Parson *et al.* 1992), that are mainly locally derived from adjacent basement ridges and intrabasin seamount volcanoes. The thickness and age of the lower volcaniclastic-dominated sequence and the over-lying hemipelagic clayey nannofossil ooze sequence, however, varies substantially across the western Lau Basin.

Sites 834 and 835 were drilled in the oldest part of the Lau Basin in two sub-basins near the Lau Ridge (Figs 1, 2 and 3) and the thickest sediment columns were recovered at these two sites. A continuous sediment sequence overlying basalts was recovered at both sites, and both show a similar sequence, consisting of clayey nannofossil oozes and mixed sediments interbedded with epiclastic vitric sands and silts (Fig. 4). The vitric sands and silts are largely restricted to the deeper part of the sediment column and generally show an increase in thickness and frequency downhole (ODP Leg 135 Shipboard Scientific Party 1992*a,b*).

The upper part of the sediment column at both sites consists of a distinctive sequence of brown, iron and manganese oxyhydroxide-stained, clayey nannofossil ooze (40–70% $CaCO_3$, but commonly 60–70% $CaCO_3$; ODP Leg 135 Shipboard Scientific Party 1992*a,b*), which is commonly structureless in appearance. This sequence forms a distinct lithostratigraphic unit, termed Unit I at both sites, and is 130 m thick at Site 835, compared to only 42 m in thickness at Site 834 (Fig. 4). The iron

and manganese oxyhydroxides that give the sedi-ments a distinctive reddish-brown appearance are hydrothermally derived, and these oxyhydroxides are widely distributed throughout the sediments of the Lau Basin (Cronan *et al.* 1986; Hodkinson *et al.* 1986; Reich *et al.* 1990). Epiclastic vitric sands and silts are not present within Unit I, although it does contain several thin megascopic and disrupted ash layers that represent discrete air-fall pyroclastic events. The data presented and discussed in this paper are largely restricted to lithostratigraphic Unit I at both drillsites.

Lithologic Unit I at Site 835 contains several medium to very thick beds of matrix-supported mudclast conglomerate. These are interpreted as muddy debris-flow deposits (ODP Leg 135 Shipboard Scientific Party 1992*b*). The total thickness of the mudclast conglomerates is at least 20 m. Such beds are not present at Site 834. Shipboard micropalaeontological analysis also identified a redeposited block, some 15 m thick, within the lower Pleistocene sediments of Unit I at Site 835. This interval contains a few thin layers of mudclast conglomerate, but consists mainly of structureless mud that closely resembles the background Pleistocene hemipelagic deposits, although abundant discoasters within this interval betray its allochthonous nature (ODP Leg 135 Shipboard Scientific Party 1992*b*).

Palaeomagnetic and biostratigraphic data suggest that at Site 834, Unit I was deposited since 2.3 Ma (ODP Leg 135 Shipboard Scientific Party 1992*a*). During this time, 42 m of sediments were deposited, of which about 7 m are redeposited nannofossil oozes and foraminiferal sands (cal-careous turbidites). These contain many shallow-water-derived benthic foraminifers (e.g. *Calcarina spengleri*, *Sorites marginalis*) as well as *Halimeda* and abundant pteropod fragments (ODP Leg 135 Shipboard Scientific Party 1992*a*), suggesting that some of these turbidites are probably derived from carbonate reefs built up on the eroded Lau Ridge. As these beds may be regarded as being deposited by geologically instantaneous events, the remaining 35 m of sediment was deposited since 2.3 Ma, giving an average sedimentation rate of 1.52 cm ka^{-1}. This is a relatively low sedimentation rate consistent with wholly hemipelagic sedi-mentation. Reich *et al.* (1990) calculated a similar value for an average hemipelagic sedimentation rate (1.5 cm ka^{-1}) for the Lau Basin from studies of piston and gravity cores.

At Site 835, shipboard interpretation of the geomagnetic and biostratigraphic data suggests that deposition of lithologic Unit I represents a time interval of perhaps 2.9 Ma. During this time 130 m of sediment was deposited, of which 20 m consists of muddy debris-flow deposits, and a further 15 m

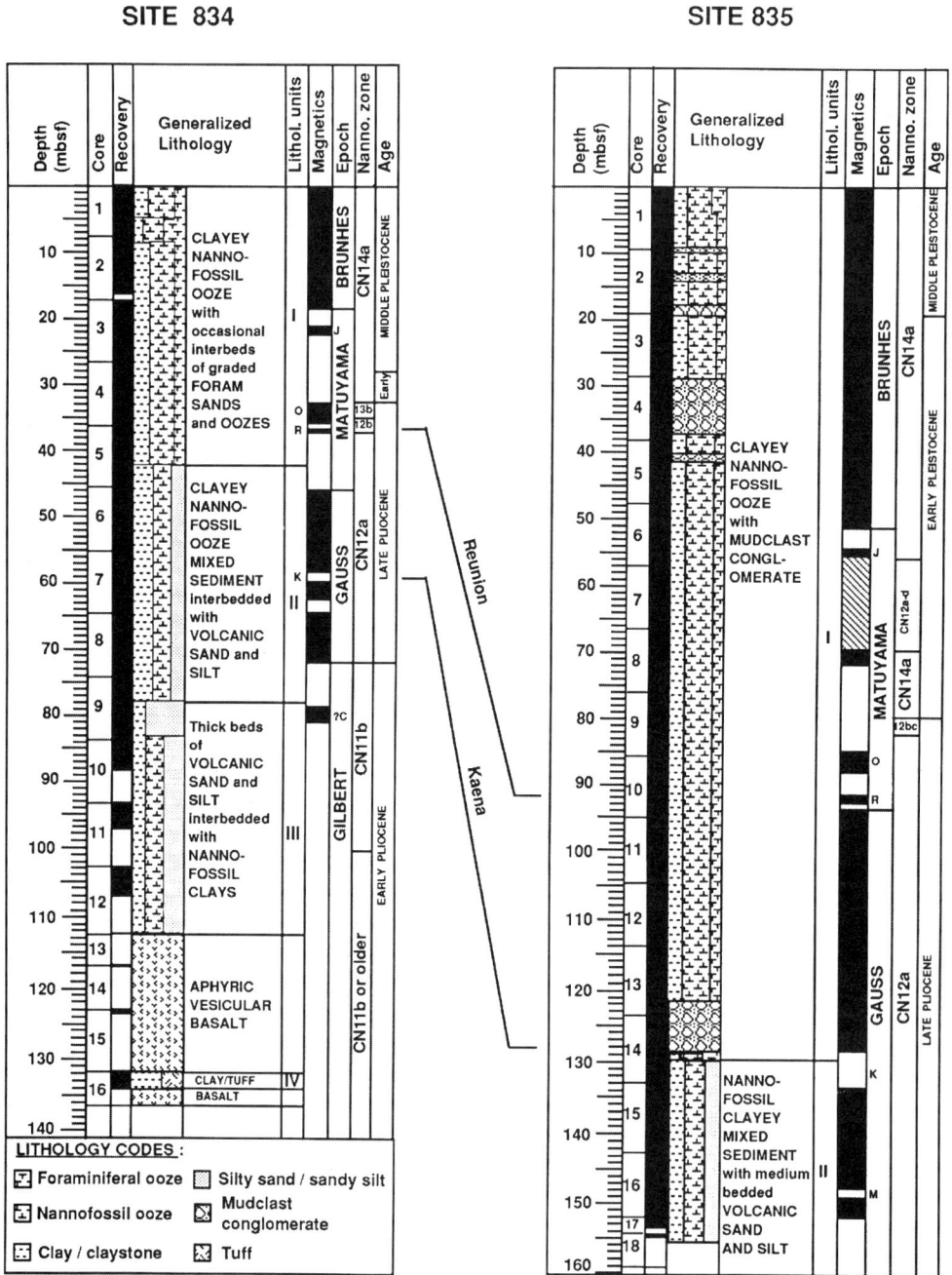

Fig. 4. Downhole logs of the sediment sequence recovered at Sites 834 and 835, showing generalized lithology determined through shipboard logging with shipboard palaeomagnetic and biostratigraphic logs alongside. Note the anomalously thick clayey nannofossil ooze sequence in lithostratigraphic Unit I at Site 835 compared to lithostratigraphic Unit I at Site 834.

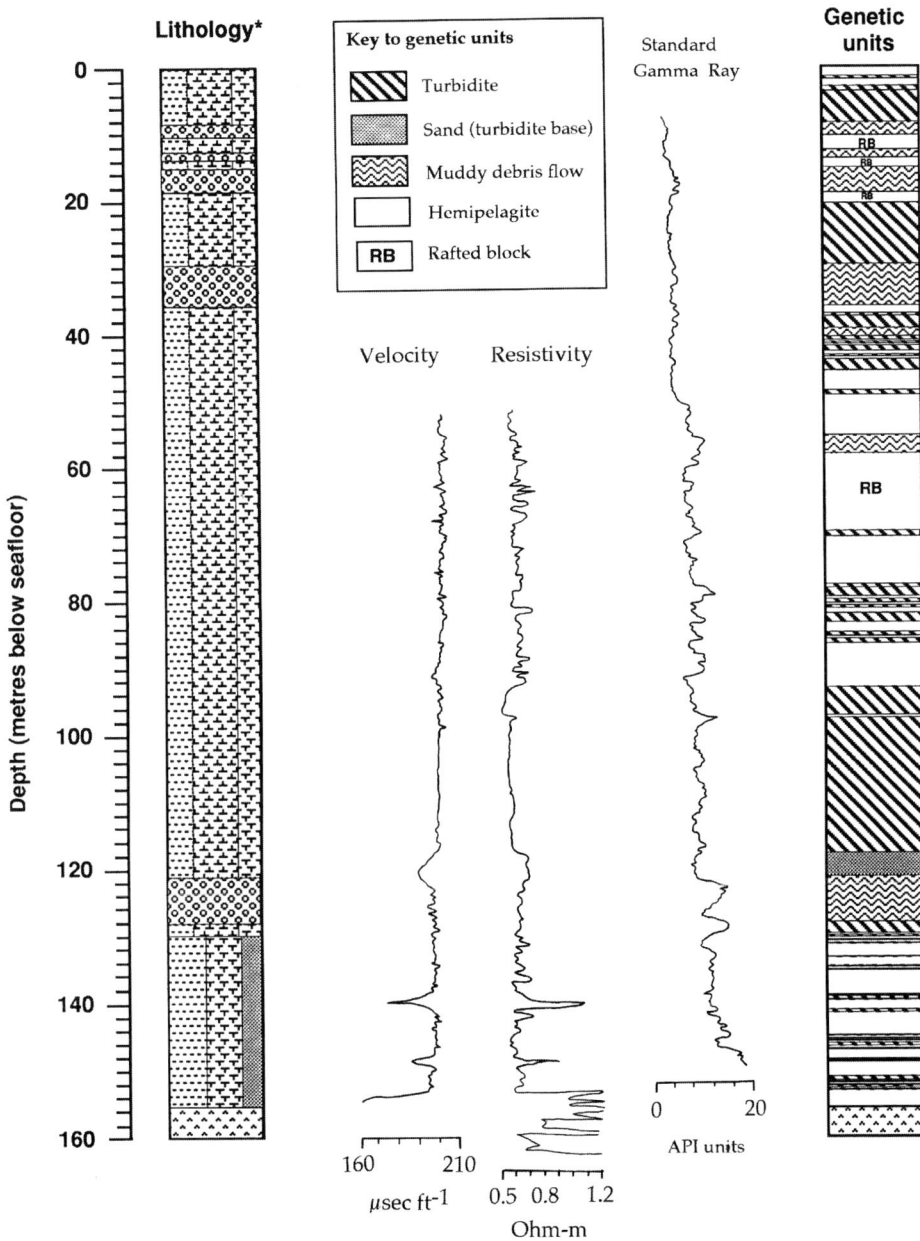

Fig. 5. Shipboard lithological log for Site 835 (*for lithology symbols, see Fig. 4), with downhole velocity, resistivity and standard gamma ray wireline logs alongside. Note the variability shown in the log signatures, especially over the visually 'monotonous' interval of clayey nannofossil ooze between 50 and 120 mbsf. Division of the stratigraphic column into genetic units, using the sedimentological criteria discussed in the text, is shown to the right, showing that the sediment column drilled at Site 835 represents a complex depositional sequence. Note that variations in the wireline logs correlate moderately well with genetic unit intervals and boundaries.

represents a redeposited Pliocene block distinguished on micropalaeontological evidence. Assuming that these deposits resulted from geologically instantaneous events, then the remaining 95 m of clayey nannofossil ooze reflects an average sedimentation rate of 3.2 cm ka^{-1}.

Considering that Sites 835 and 834 were drilled in two adjacent graben, only 50 km apart, it seems legitimate to assume a relatively constant average hemipelagic accumulation rate at both sites. If we consider that the lower sedimentation rate (1.5 cm ka^{-1}) is closer to the true average hemipelagic sedimentation rate, then lithologic Unit I at Site 835 must contain nearly 90 m of additional redeposited material compared to Site 834. The mudclast conglomerate and the redeposited block account for only 30 to 35 m of this. Therefore the remaining extra thickness must be due to the presence of redeposited clayey nannofossil ooze (turbidites), and these closely resemble the background hemipelagic ooze deposits in both colour and composition. An average hemipelagic sedimentation rate of 1.5 cm ka^{-1} implies that

perhaps only 44 m of the remaining 95 m of clayey nannofossil oozes in Unit I at Site 835 are true hemipelagic deposits. Therefore, 51 m of the remaining clayey nannofossil ooze at Site 835 are probably clayey nannofossil ooze turbidites rather than hemipelagites.

The wireline logs obtained at Site 835 (Fig. 5) show clear variability and suggest an obscure, but real, structure to the visually rather monotonous clayey nannofossil ooze sequence of lithological Unit I. At Site 835, the bottom of the drill pipe was held at 48 metres below sea floor (mbsf) while logging uphole. Therefore few wireline log data were recovered from the part of the sediment sequence over the interval 0–48 mbsf. Clear variability is seen in most of the parameters measured throughout both lithological Units I and II. For example, the resistivity log shows numerous peaks of high resistivity in lithologic Unit II (130–155 mbsf) suggesting the presence of numerous discrete layers of higher porosity and water content, and these correlate with vitric sandy turbidites. However, clear variation in resistivity,

Table 1. *Characteristics of clayey nannofossil ooze turbidites and hemipelagites at Site 835*

	Hemipelagites	Turbidites
Colour	Reddish brown to orange brown colours that show subtle variation in hue and chroma over short segments of core.	Reddish brown to orange brown colours of consistent hue and chroma, with some over long intervals.
Bioturbation	Some light mottling due to bioturbation may occur.	Sediments appear homogeneous; bioturbation occurs only near the tops of units.
Foraminiferal distribution	Occur scattered throughout sediment. Sediment appears speckled under a hand lens.	Large foraminifers rare or absent, except near base of units. Sediment appears featureless under hand lens.
Texture	Cut core surface commonly slightly pitted.	Cut core surface usually smooth.
Grain-size characteristics	Grain-size frequency distribution curves variable over short segments of core. Contain variable sand/coarse silt modes. Variable sorting values.	Grain-size frequency distribution curves highly consistent over depositional intervals. Characterized by persistent restricted coccolith modes; sand/coarse silt modes absent. Restricted sorting values.
CaCO$_3$/C$_{org.}$	Variable over short core segments.	Consistent values over depositional intervals which can be several to tens of metres in thickness.
Micropalaeontology	Discoasters absent from Pleistocene hemipelagites.	May contain a variable proportion of discoasters.
Dropstones	Weathered pumice clasts common.	Absent.
Geochemical analysis	Chemistry variable	Geochemical profiles show consistent chemistry over depositional intervals. Commonly clear differences in element concentrations to the enclosing hemipelagites.

although to a lesser extent, is also apparent within Unit I, although this is superficially 'monotonous' clayey nannofossil ooze. Particularly striking is a 20 m thick interval of remarkably constant values within the clayey nannofossil ooze sequence between 98 and 118 mbsf (Fig. 5). Between 52 and 98 mbsf the resistivity profile shows a clear repeated 'peakiness' on a scale of $0.2 \, \Omega \, m^{-1}$.

Post-cruise re-examination of the cores from Unit I at Sites 834 and 835 suggested that the autochthonous and allochthonous clayey nannofossil oozes could be distinguished on a number of subtle sedimentological criteria (Table 1, Figs 6 and 7), and that four distinct types of sediment could be distinguished within Unit I at Site 835 (Table 2):

1. discrete beds of mud-clast conglomerate;
2. coherent blocks of clayey nannofossil ooze characterized by abundant scattered foraminifera and containing isolated pumice dropstones; commonly bioturbated and showing diffuse variation in colour hue and chroma; within the Pleistocene section that contain discoasters;
3. intervals of clayey nannofossil ooze characterized by abundant scattered foraminifera and containing isolated pumice dropstones; commonly bioturbated and showing diffuse variation in colour hue and chroma that in the Pleistocene section do not contain discoasters;
4. homogeneous intervals of clayey nannofossil ooze, in which the foraminiferal content is 1% or less and of restricted size, and pumice dropstones are absent; bioturbated only towards their upper boundaries, otherwise characteristically structureless; of uniform colour and frequently overlie graded and massive foraminiferal sands.

Discussion

The beds of mud-clast conglomerate identified within lithostratigraphic Unit I at Site 835 were interpreted as muddy debris-flow deposits by the

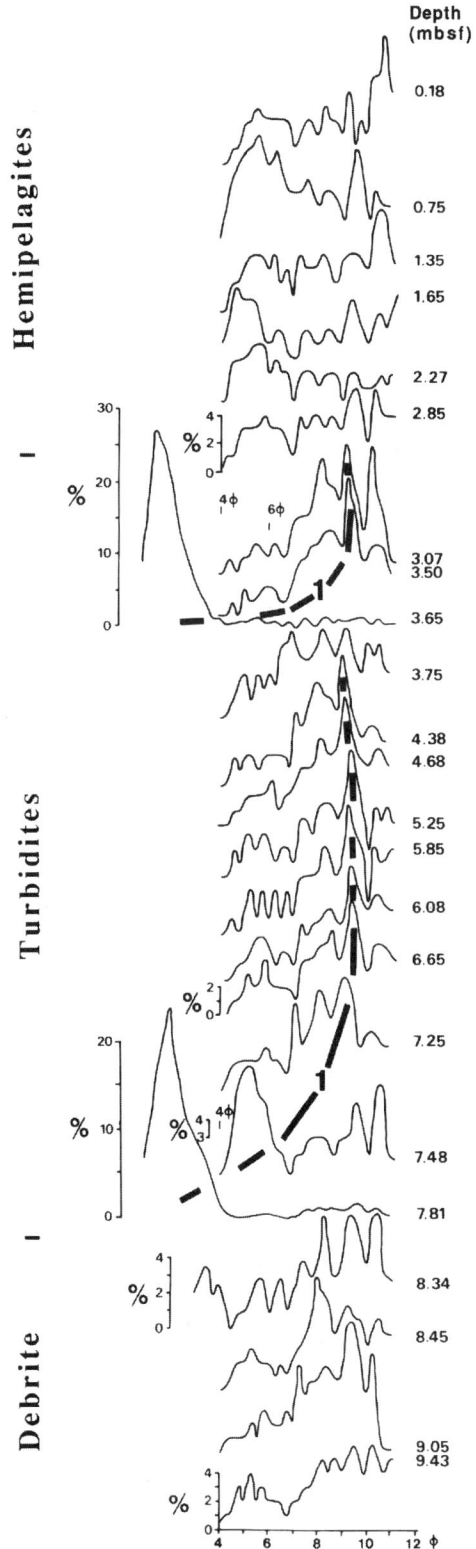

Fig. 6. Stacked grain-size frequency distribution curves through Core 135-835A-1H (0–9.5 mbsf). Note the variable sand/coarse silt modes in those samples taken from hemipelagic intervals (0.18–2.85 mbsf); compared to those taken from units interpreted as turbidites (3.07–7.81 mbsf), which show consistent frequency distribution curves over depositional intervals and are characterized by persistent, restricted coccolith modes and lack pronounced sand/coarse silt modes, except near their bases. Samples from 8.34 to 9.43 mbsf are from the debris-flow matrix. Primary mode (1) in individual turbidites is indicated by thick broken line.

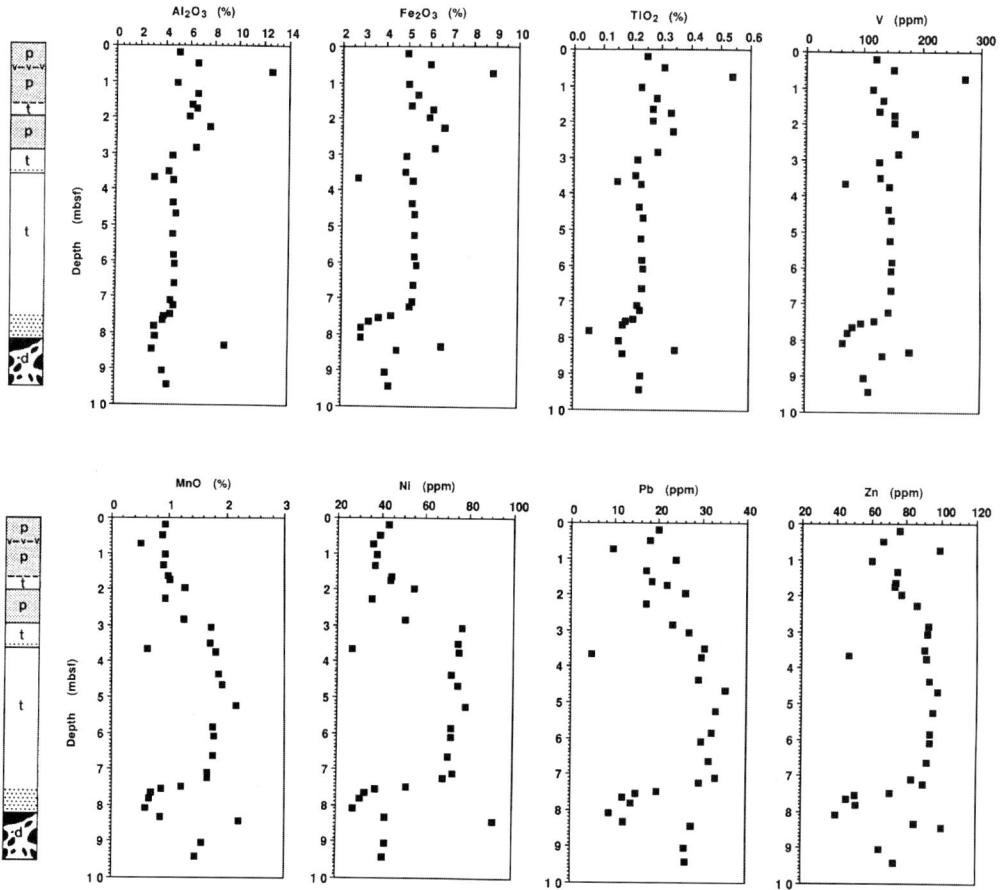

Fig. 7. Graphic log of Core 135-835A-1H (0–9.5 mbsf), showing a sequence of clayey nannofossil ooze hemipelagites (p) and turbidites (t), overlying a debrite (d) with selected downcore geochemical profiles plotted alongside. A thin pyroclastic ash layer is indicated by v in the graphic log. Note the variability in the hemipelagic intervals compared to the turbidite muds, which show consistent values throughout and are enriched in Mn, Ni, Pb and Zn. The core has been chosen for illustration as it contains the main sediment types recorded at Site 835. Deeper cores show similar variation in geochemistry depending on genetic unit.

shipboard sedimentologists (Parson *et al.* 1992) from their petrographic and textural characteristics (see Table 2). Smear slide samples from some of the mud-clasts from beds in the Pleistocene section contain discoasters showing them to originate from a pre-Pleistocene sequence. Around some clasts, flow deformation structures are clearly visible in the clayey nannofossil ooze matrix, supporting the shipboard interpretation. Fortuin *et al.* (1992) describe similar Neogene mud-clast-rich conglomeratic debris flows from the island of Sumba, which lies west of Timor, north of the Java Trench in Indonesia. Sumba is regarded as a recently emergent uplifted element of the present forearc

basins now separating the Lombok Basin of the Sundra Arc from the Savu Basin of the Banda Arc (Karig *et al.* 1987; Van Weering *et al.* 1989). Fortuin *et al.* (1992) consider that the mud-clast-rich conglomerates of Sumba and their associated turbidites and slide deposits resulted from tectonically induced oversteepening. The deposits found at Site 835 may have a similar origin (see discussion below).

Within the Pleistocene section at Site 835, several intervals of clayey nannofossil ooze were identified, which are characterized by abundant scattered foraminifera and contain both isolated, often weathered, pumice dropstones and dis-

Table 2. *Characteristics, occurrence and origin of sediment types identified at Sites 834 and 835*

Sediment type	Visual characteristics	Occurrence	Probable origin
Mud-clast conglomerates	Typically matrix-supported, with rounded to angular clasts and blocks of clayey and foraminifer-rich nannofossil ooze, or more rarely pumice, enclosed in a matrix of clayey nannofossil ooze. Clasts are usually rounded and may show a subvertical alignment within the core. Some beds show crude grading.	Only occur at Site 835, where they occur as medium to very thick beds up to 7 m thick.	Allochthonous, probably muddy debris flow deposits.
Discoaster-bearing blocks of clayey nannofossil ooze within the Pleistocene section	Intervals contain abundant scattered foraminifera and contain isolated, often weathered, pumice dropstones. Commonly bioturbated and show diffuse variations in colour hue and chroma. Discoasters tend to be well preserved.	Only occur at Site 835, where they occur as medium to very thick beds up to 15 m thick. Commonly overlie mud-clast conglomerates.	Clearly allochthonous, probably coherent rafted blocks of Pliocene hemipelagic sediments.
Clayey nannofossil ooze containing abundant scattered foraminifera	Contain isolated pumice dropstones, commonly bioturbated and show diffuse variation in colour hue and chroma over short intervals of core. Lack discoasters when within the Pleistocene section.	Occur at both Sites 834 and 835, and may occur as very thick beds up to 9.8 m thick at Site 834 and up to 4.8 m thick at Site 835.	Autochthonous, probably represent hemipelagic accumulation.
Structureless clayey nannofossil ooze	Homogeneous intervals in which the foraminifera content is 1% or less and restricted to juveniles or test fragments of restricted size range. Pumice dropstones absent. Only bioturbated towards the upper boundaries of individual beds. Show uniform colour, even over several metres. May contain discoasters, which are often fragmented or corroded. Commonly overlie graded and massive foraminiferal sands.	Occur at both Sites 834 and 835. Thinly- to medium bedded at Site 834 (beds are up to 2.6 m thick), but occur as very thick beds (up to 24.4 m thick) at Site 835.	Allochthonous, probably turbidite muds, foraminifers removed by gravitative settling.

coasters. These intervals are commonly bioturbated and show diffuse variation in colour hue and chroma. Because discoasters became extinct in the late Pliocene, these intervals clearly must be allochthonous. However, these intervals show no evidence for gravitative sorting or deformation structures, showing that emplacement could not have been by any type of sediment gravity flow. If it were not for their discoaster content, then these intervals could be interpreted as hemipelagic deposits, the scatter of foraminifers being due to pelagic settling through the water column, and the diffuse variations in colour value, hue and chroma explained as reflecting temporal variations in the hydrothermal input to the sediments. Pervasive bioturbation is also a characteristic of hemipelagic sediments. These intervals, which may be up to 15 m thick, commonly overlie mud-clast con-

glomerate horizons. The most likely explanation for the origin of these deposits is that they represent rafted blocks of older, pre-Pleistocene, hemipelagic material, which have their origin in slope failures on the steep basin walls which bound the graben in which Site 835 was drilled. Some of these blocks appear to have been emplaced by being carried into the basin as coherent slabs by muddy debris flows. Their absence from the sequence drilled at Site 834 may be explained by the fact that the basin in which this site was drilled is surrounded by relatively gentle slopes (see Fig. 3) and hence oversteepening leading to catastrophic slope failure is less likely.

Within the Pleistocene section at both Sites 834 and 835, there are intervals of clayey nannofossil ooze containing abundant scattered foraminifera but lacking discoasters. These contain isolated pumice dropstones, are frequently bioturbated, and show diffuse variation in colour shade over short intervals of core. Deposits of similar character also occur within the Pliocene section at both sites, but these contain discoasters. These intervals of clayey nannofossil ooze containing abundant scattered foraminifera in the Pliocene and Pleistocene sequence are generally of much greater thickness and continuity at Site 834 (where they occur as beds up to 9.8 m thick) than at Site 835 (maximum continuous thickness 4.8 m). Samples taken from these intervals show variable grain-size frequency distribution curves and sorting values and contain variable sand and coarse silt modes (see Fig. 6). Calcium carbonate, organic carbon and major, minor and trace element geochemistry (see Fig. 7) are variable over short segments within these intervals. These intervals are clearly autochthonous and represent hemipelagic accumulation. The scatter of foraminifera showing a wide range of grain size (hence the variable grain-size characteristics and presence of sand and coarse silt modes), reflects pelagic settling of foraminifer tests, whether individually or incorporated into faecal pellets, through the water column. This forms the normal hemipelagic/pelagic 'rain' of material from the surface or near-surface waters to the sea floor. Accumulation rates of hemipelagic/pelagic sediments are low (1–3 cm ka^{-1} are commonly quoted figures; e.g. Kennett 1982). Therefore, the variability in calcium carbonate and organic carbon, seen over short segments of core within these intervals, reflects the variability in these parameters on 10^3–10^4 year timescales. Variability in the geochemistry, like the variation in the colour staining, however, reflects temporal variations in the hydrothermal input to the sediments, which is especially marked in the Lau Basin sediments due to the relative proximity of active spreading centres.

Intervals of homogeneous clayey nannofossil ooze, often of considerable thickness, occur at both Sites 834 and 835. Within these intervals the foraminiferal content is 1% or less, but the beds commonly overlie massive or graded foraminiferal sands of variable thickness. These beds are clearly allochthonous and are identified as calcareous turbidites, the foraminifers having been removed from the mud by gravitative settling. They, like the rafted blocks identified above, commonly overlie beds of mud-clast conglomerate. Pumice dropstones are also absent from the structureless mud intervals, although they may occur within the basal foraminiferal sands. The turbidite muds, in contrast to the hemipelagic intervals, tend to show a uniform colour, calcium carbonate and organic carbon content which may be constant and consistent over several metres of core. This probably reflects homogenization of the original material within the

turbidity current during flow, resulting in consistency over depositional intervals. Grain-size analyses (Fig. 6) show the muds to be largely ungraded and therefore they comprise the E$_3$ structural division of Piper (1978), although they commonly show normal grading in their lower parts and grade into, initially muddy, foraminiferal sands. Major, minor and trace element analyses of samples from the homogeneous muds show them to be commonly enriched in Mn, Ni, Pb, Zn, Cr and P (Fig. 7). This probably reflects derivation from source areas on the flanks of the sub-basins where the hemipelagic sediments may be enriched in hydrothermally derived and/or volcanically derived elements compared to the hemipelagic sediments accumulating in the sub-basins.

Clearly the initially visually 'monotonous' intervals of brown clayey nannofossil ooze recovered and logged during drilling at Site 835, are complex sequences comprising rafted blocks, thick turbidite muds, muddy debris-flow deposits and hemipelagic accumulation (Table 2). The allochthonous deposits are commonly closely associated suggesting that they form parts of larger slide complexes. Identification of allochthonous and autochthonous beds at Sites 834 and 835 allows the sediment sequences to be divided into genetic units (Figs 8 and 9) and, as a result, the depositional history for each site can now be determined and described. Revised sedimentation curves can now be drawn up for each site by subtracting the allochthonous units (considered to have been deposited by geologically instantaneous events) and summing the remaining autochthonous (hemipelagic) intervals and plotting thickness versus occurrence of biostratigraphic and magnetic datums that appear within hemipelagic intervals (Figs 10 and 11). The revised sedimentation rate curves for both Sites 834 and 835 show similar average hemipelagic sedimentation rates of 1.46 and 1.44 cm ka^{-1} respectively. These values closely agree with the average hemipelagic sedimentation rate of 1.5 cm ka^{-1} calculated by Reich et al. (1990) for the Lau Basin.

The genetic logs for both Sites 834 and 835 show that allochthonous beds tend to be grouped and this is interpreted as reflecting episodes of local instability during the site's history. For instance, at Site 834 there has been a marked increase in emplacement of allochthonous deposits since 200 ka. Before this, until 2.3 Ma, deposition had been mainly hemipelagic (Fig. 8). The sediment sequence at Site 835 is dominated by allochthonous deposits. These form the main sediment type deposited during the Brunhes geomagnetic epoch, making up 80% of the thickness of sediment deposited during this period. Based on the occurrence of allochthonous deposits, four main episodes of instability can be identified at Site 835. These are

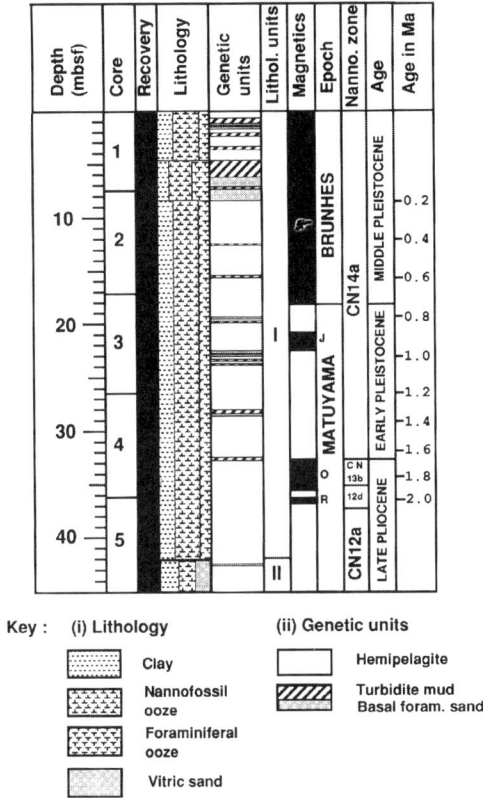

Fig. 8. Genetic log through the late Pliocene–
Pleistocene sequence at Site 834 showing the occurrence
of clayey nannofossil ooze hemipelagites and turbidites
in the sequence, with magnetostratigraphic and
biostratigraphic logs alongside. Age (in Ma) is estimated
primarily from calculations using the average
hemipelagic sedimentation rate (1.5 cm ka^{-1}) constrained
by shipboard palaeomagnetic dating.

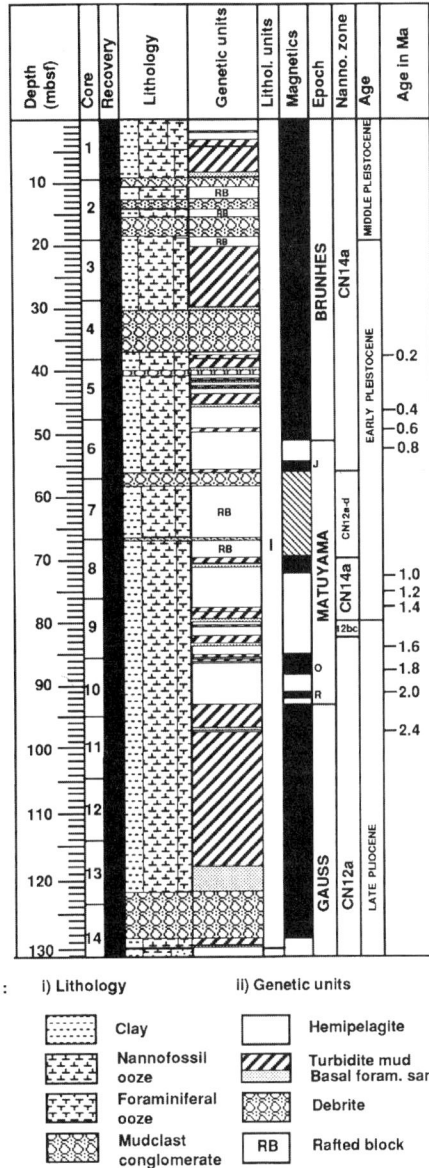

Fig. 9. Genetic log through the late Pliocene–
Pleistocene sequence at Site 835 showing the occurrence
of debrites, turbidites, hemipelagites and rafted blocks
(identified from micropalaeontological evidence) in the
sequence, with magnetostratigraphic and biostratigraphic
logs alongside. Age (in Ma) is estimated primarily from
calculations using the average hemipelagic
sedimentation rate (1.5 cm ka^{-1}) constrained by
shipboard palaeomagnetic dating.

estimated as having occurred from 0 to 0.4, 0.9 to
1.0, 1.4 to 1.7 and 2.1 to 2.9 Ma (see Figs 9 and 12).
These involved catastrophic slope failures with the
emplacement of thick muddy debris flows that
commonly carried large rafted blocks, and thick
mud turbidites.

The occurrence of groups of turbidites can
therefore be used to infer episodes of instability,
which may be correlatable with known or inferred
tectonic or volcanic events. Parson *et al.* (1994)
consider volcanic and tectonic controls on
sedimentation in the Lau Basin. They conclude that
although Lau Ridge volcanism is likely to have
made a significant contribution to the adjacent
basins in the early to middle Pliocene (i.e. within
lithostratigraphic Unit II), the effects of recent

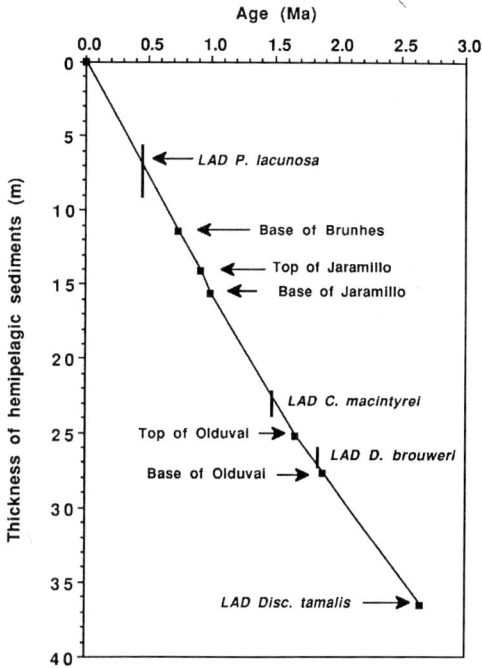

Fig. 10. Sedimentation rate curve for hemipelagic sediments at Site 834. Curve is derived by subtracting redeposited beds from the sediment sequence and summing the remaining hemipelagic intervals and plotting thickness vs. occurrence of biostratigraphic and palaeomagnetic datums that appear within hemipelagic intervals. Curve shows an average hemipelagic sedimentation rate of 1.46 cm ka^{-1} for Site 834.

off-axis volcanism are more difficult to assess and in the late Pliocene to Pleistocene (i.e. within lithostratigraphic Unit I) tectonic controls on sedimentation may have been more important. Parson & Hawkins (1994) estimate that the propagating ridge that formed the East Lau Spreading Centre passed the latitude of Sites 834 and 835 at around 3.6 Ma and that the propagator that formed the Central Lau Spreading Centre passed the latitude of the two sites at about 0.5 to 0.6 Ma. Sedimentation at Site 834 at 3.6 Ma was dominated by deposition of thick epiclastic volcanic sands and silts, which are interbedded with hemipelagic nannofossil clays and mixed sediments (ODP Leg 135 Shipboard Scientific Party 1992a). The passing of the East Lau propagator occurred close to the transition between lithologic Units II and III at Site 834. The boundary between lithologic Units II and III marks a perceived decrease in the thickness and frequency of volcaniclastic turbidites (ODP Leg 135

Shipboard Scientific Party 1992a). However, close to this boundary, an episode of increased sandy turbidite deposition is marked by two 'packets' of stacked turbidites, with several turbidites emplaced directly one on top of one another, with little or no intervening hemipelagic deposition between the individual turbidites. The first of these occurs at about 3.5 Ma, and the second at about 3.9 Ma. The increase in frequency and 'sandiness' of volcaniclastic turbidites at these times may possibly correlate with increased instability due to the approach and passing of the East Lau propagator. The oceanic crust that forms the floor of the basin in which Site 835 was drilled is about 3.6 Ma old (ODP Leg 135 Shipboard Scientific Party 1992b).

Between 2.4 Ma and 200 ka, a marked decrease in frequency of turbidite emplacement is seen at Site 834, with turbidites becoming very rare and when they do occur, they are typically thinly bedded. This interval corresponds with a period of predominantly hemipelagic deposition and may

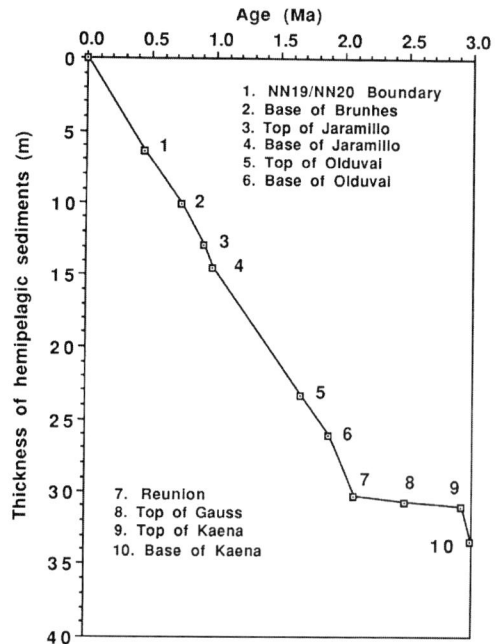

Fig. 11. Sedimentation rate curve for hemipelagic sediments at Site 835. Curve is derived by subtracting redeposited beds from the sediment sequence and summing the remaining hemipelagic intervals and plotting thickness vs. occurrence of biostratigraphic and palaeomagnetic datums that appear within hemipelagic intervals. Curve shows an average hemipelagic sedimentation rate of 1.44 cm ka^{-1} for Site 835. Note the prominent hiatus evident between 2.1–2.9 Ma.

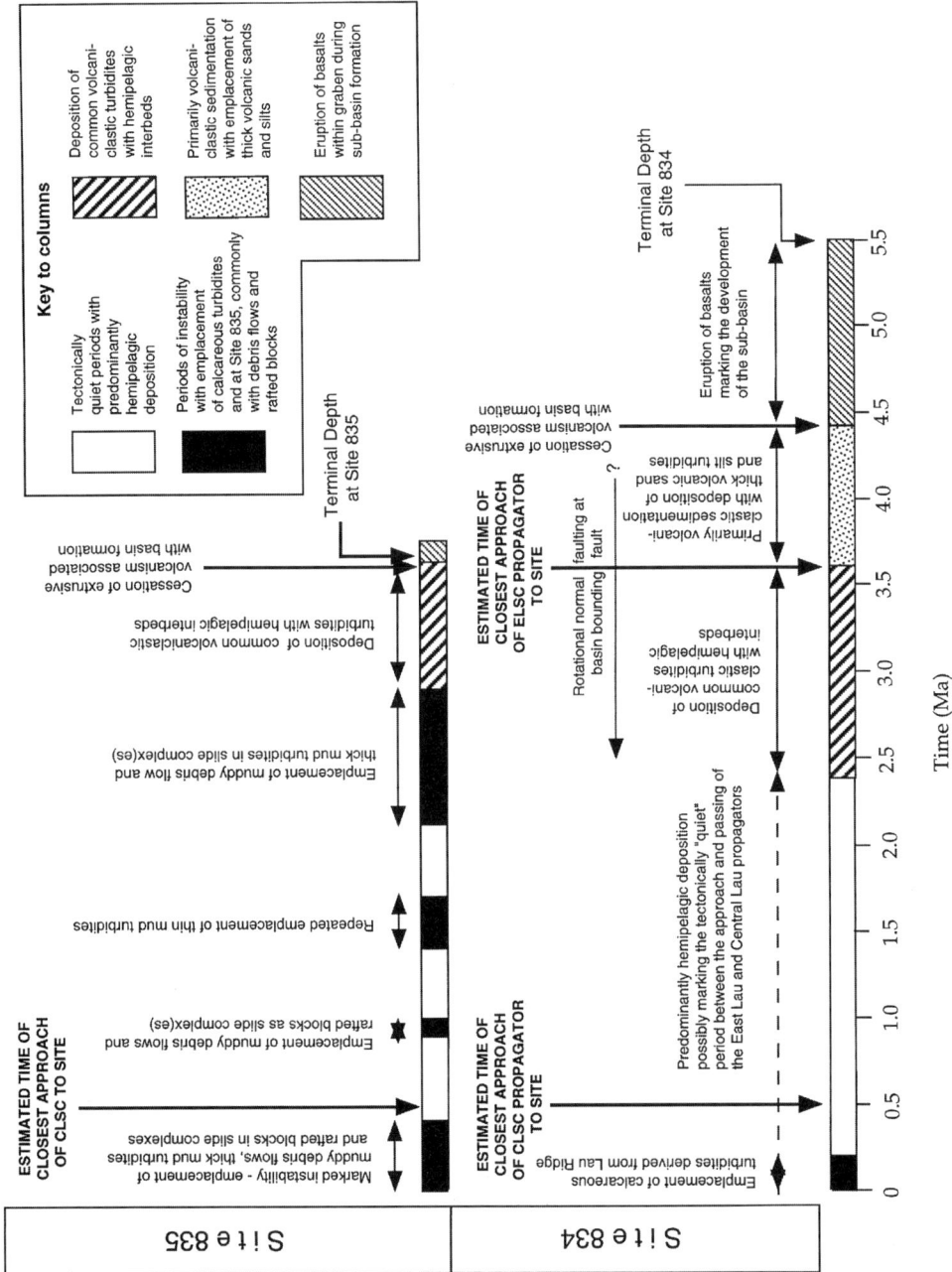

Fig. 12. Summary showing the geological history of Sites 834 and 835, in terms of periods of primarily hemipelagic accumulation and periods dominated by allochthonous sedimentation. Time estimates for the closest approach of the Lau Basin ocean-ridge propagators to the latitude of the drillsites are also shown (from Parson & Hawkins 1994) and for Site 834, the timing of the most intense rotational normal faulting on the basin-bounding faults (from Parson et al. 1994).

mark the tectonically 'quiet' period between the approach and passing of the East Lau and Central Lau propagators (Fig. 12). Although the CLSC propagator is considered to have made its closest approach to the latitude of Site 834 at around 0.5–0.6 Ma (Parson & Hawkins 1994; see Fig. 12), there may be some latitude in both the estimated ages for the sediment sequence given in this paper, which are first-order estimates based on calculations using average hemipelagic sedimentation rates constrained by shipboard palaeomagnetic dating, and the dates given for southward propagation of the propagating ridge tip which assume constant spreading and propagation rates. This may explain this discrepancy. Indeed, propagation of the ELSC appears to have taken place in an intermittent fashion, advancing southward in a stepwise, rather than smooth, mode; its southward propagation is probably controlled by the pre-existing rifted fabric of the western basin (Parson & Hawkins 1994).

Deposition since 3 Ma at Site 835, in contrast to Site 834, has been dominated by redeposited sediments, and four main episodes of instability can be identified at this site. These are estimated to have occurred from 0 to 0.4, 0.9 to 1.0, 1.4 to 1.7, and 2.1 to 2.9 Ma and involved catastrophic slope failures with the emplacement of thick muddy debris flows, that often carried large rafted blocks and thick mud turbidites. The reason for these periods of instability are as yet unknown, but the most recent episode (0–0.4 Ma) may possibly relate to tectonic activity and crustal readjustment following the passing of the Central Lau propagator at around 0.5 Ma.

Turbidites and other sediment gravity-flow deposits are common in the sediment columns recovered at the other backarc sites drilled in the Lau Basin and allochthonous bed groupings may relate to the tectonic evolution of the Lau Basin. Figure 13 shows generalized lithologic columns drilled at the Lau Basin backarc sites, presented with age as the vertical scale. The ages of the sedimentary horizons are calculated from consideration of hemipelagic accumulation rates (taken as 1.5 cm ka^{-1}), constrained by shipboard biostratigraphic and palaeomagnetic data. Times of emplacement of allochthonous units are shown in black in the 'Turbidites' column. The predicted ages of the closest approach of the Central and Eastern Lau ocean ridge propagators to the latitude of the drillsites, as estimated by Parson & Hawkins (1994) and Parson (pers. comm.), are also shown. The time of approach and passing of the ELSC (the only propagator to pass all backarc sites; see Fig. 1) in relation to the latitude of each drillsite correlates moderately well with times of increased deposition of allochthonous sediments at the respective drillsites. The increased advective sediment flux to

the sub-basins at these times possibly reflects instability caused by tectonic activity and crustal readjustment relating to the approach and passing of the propagator. Parson et al. (1994) note that the most intense episodes of rotational normal faulting on basin-bounding faults appears to straddle the time of the closest approach of the ELSC propagator to each sub-basin, suggesting that the transition from rifting to sea-floor spreading is marked by increased extension of the rifted arc crust and then a decrease as extension is taken up by the spreading centre. Hodkinson & Cronan (1994) also record general correspondence between increased hydrothermal input to the sediments at individual drillsites and the approach and passing of the ELSC propagator as it moved southward through the Lau Basin, although at each site a time lag between these events is apparent, suggesting that a period of time occurs before the hydrothermal discharge from the newly formed ridge becomes established and maximum hydrothermal discharge into adjacent sediments is recorded. Recently, Taylor et al. (1995) have suggested different spreading rates for the Lau Basin backarc propagators to those proposed by Parson & Hawkins (1994) on analysis of magnetic data. However, timescales are such that the correlation between the approach and passing of the propagating ridge tips and instability at the drillsites, discussed above, is still likely to hold. Figure 13 shows that episodes of instability, marked by increased emplacement of turbidites and debris flows, straddle the times of the closest approach of the propagating ridge tips as estimated by Parson & Hawkins (1994). This may be expected in that tectonism and crustal readjustment associated with the passage of a propagating ridge tip may correlate with a period of instability at the drillsite. This period may start with the approach of the propagator and perhaps increase in intensity as the propagator gets closer, and end well after (but perhaps with diminishing intensity) the propagating ridge tip has passed the latitude of the drillsite. Data presented by Parson & Hawkins (1994) suggest that the ridge tips propagated southward at a rate of approximately 100 km in 1 Ma. This suggests that periods of instability affecting the graben due to the approach and passing of the propagators may have been surprisingly long.

The pattern of allochthonous sediment deposition at each of Lau Basin backarc sites may provide a signature of the tectonic evolution of the basin. Turbidites, once identified, can be an important tool for deciphering tectonic histories, and in the Lau Basin, at least, local tectonism, driven by the mechanism of sea-floor spreading, may form the main control on local sediment redeposition.

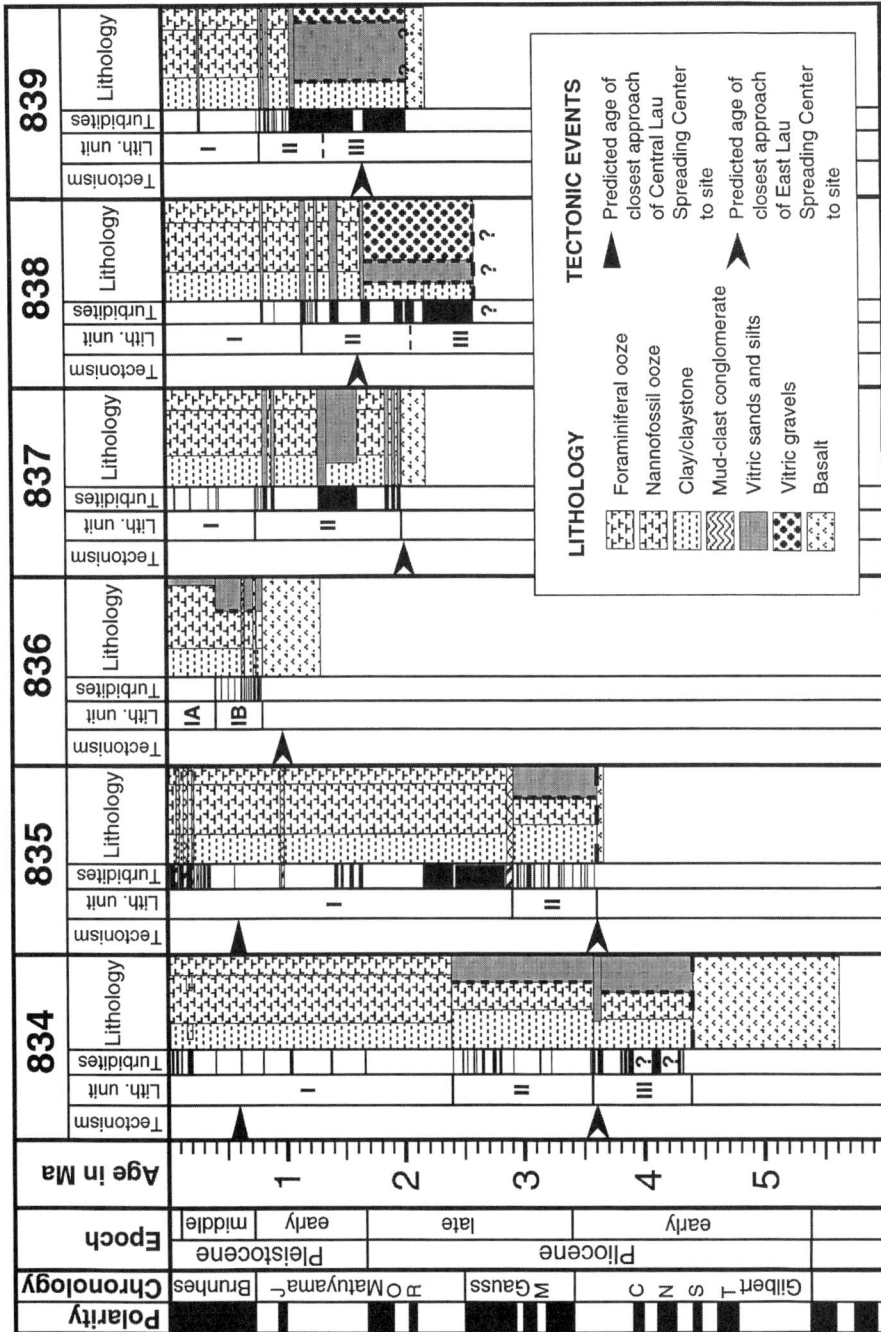

Fig. 13. Generalized lithological columns for the Lau Basin backarc sites, presented with age equivalence (in Ma) as the vertical scale. Sediment lithology is shown as percentages on a horizontal scale of constituent sediment types (for example, the dominant sediment type of lithologic Unit I is clayey nannofossil ooze, made up of up to 30% clay and up to 10% foraminiferal ooze with the remainder of the sediment consisting of calcareous nannofossils). The age equivalence for sedimentary horizons was calculated from consideration of average hemipelagic sedimentation rates (considered to be about 1.5 cm ka^{-1}) constrained by shipboard biostratigraphic and palaeomagnetic data. Times of emplacement of allochthonous beds within the sedimentary succession at each site are shown in black in the 'Turbidites' column. These were identified on compositional and textural criteria for the volcaniclastic beds and on the criteria discussed in the text for calcareous mud turbidites. The lithologic units identified by Parson et al. (1992) and the predicted ages of the closest approach of the Central and East Lau propagators to the sites are also shown. Thick dashed vertical lines indicate that the lithologies shown are interbedded. Note the broad correlation of periods of instability marked by increased sediment redeposition and the closest approach of the ocean ridge propagators to the latitudes of the drillsites.

Conclusions

1. Allochthonous pelagic sediments (calcareous turbidites) and other sediment gravity-flow deposits make a significant contribution to the basin fill of sub-basins within the Lau Basin.
2. Allochthonous pelagic sediments can be distinguished from autochthonous pelagic sediments on several sedimentological criteria. These include variation in colour hue and chroma, presence or absence of bioturbation, presence or absence of scattered foraminifera, grain-size characteristics, variability in calcium carbonate content, presence or absence of pumice clasts and micropalaeontology.
3. Allochthonous and autochthonous pelagic sediments are also geochemically distinct, with allochthonous units being commonly enriched in Mn, Ni, Pb, Zn, Cr and P.
4. Identification of allochthonous beds within sediment sequences allows us to infer episodes of instability, and thus seek correlation with known or inferred volcanic and tectonic events.
5. At the Lau Basin backarc sites, periods of increased sedimentary activity, as recorded by influx of turbiditic sediments or slumping, correlate moderately well with the closest approach of ocean-ridge propagators to each sub-basin. Therefore, local sediment instability close to the sites may relate to tectonic activity and crustal readjustment due to the passage of the southward propagating ridge tips past the latitude of individual sub-basins.

I would like to thank C. Gravestock and N. Higgs for preparing samples and running the ICP-AES analyses. I thank K. Lyons for her help in drafting some of the figures. I am very grateful to C. MacLeod, P. Weaver, R. Hodkinson, C. Pratt, M. Styzen, L. Parson, P. Lineberger and L. Carter for helpful discussion during the post-cruise analysis of the Lau Basin data. I thank C. Mato, J. Miller and the staff at the ODP Gulf Coast Repository, College Station, Texas, for their assistance in sampling Leg 135 sediment cores. I also express my thanks to the shipboard scientific party, the ODP technicians and the SEDCO drilling crew of Leg 135 during which the data were collected. Thanks are owed to R. Kidd and R. Hiscott for reviewing the manuscript.

References

BØE, R. 1994. Nature and record of Late Miocene mass-flow deposits from the Lau-Tonga forearc basin, Tongan Platform (Hole 840B). *Proceedings of the Ocean Drilling Program, Scientific Results,* **135**. Ocean Drilling Program, College Station, TX, 87–100.

BOUMA, A. H. 1962. *Sedimentology of some Flysch Deposits*. Elsevier, Amsterdam.

BRUNNER, C. A. & LEDBETTER, M. T. 1987. Sedimentological and micro paleontological detection of turbidite muds in hemipelagic sequences : an example from the late Pleistocene levee of Monterey Fan, central California continental margin. *Marine Micropaleontology*, **12**, 223–239.

—— & NORMARK, W. R. 1985. Biostratigraphic implications for turbidite depositional processes on the Monterey Deep Sea Fan, central California. *Journal of Sedimentary Petrology*, **55**, 495–505.

CLIFT, P. D. & DIXON, J. E. 1994. Variations in arc volcanism and sedimentation related to rifting of the Lau Basin (Southwest Pacific). *Proceedings of the Ocean Drilling Program, Scientific Results,* **135**. Ocean Drilling Program, College Station, TX, 23–45.

CRONAN, D. S., HODKINSON, R. A., HARKNESS, D. D., MOORBY, S. A. & GLASBY, G. P. 1986. Accumulation rates of hydrothermal metalliferous sediments in the Lau Basin, S.W. Pacific. *Geo-Marine Letters*, **6**, 51–56.

FORTUIN, A. R., ROEP, TH. B., SUMOSUSASTRO, P. A., VAN WEERING, T. C. E. & VAN DER WERFF, W. 1992. Slumping and sliding in Miocene and Recent developing arc basins, onshore and offshore Sumba (Indonesia). *Marine Geology*, **108**, 345–363.

HESSE, R. 1975. Turbiditic and non-turbiditic mudstone of Cretaceous flysch sections of the East Alps and other basins. *Sedimentology*, **22**, 387–416.

HODKINSON, R. A. & CRONAN, D. S. 1994. Variability in the hydrothermal component of the sedimentary sequence in the Lau Backarc Basin (Sites 834–839). *Proceedings of the Ocean Drilling Program, Scientific Results*, **135**. Ocean Drilling Program, College Station, TX, 75–86.

——, ——, GLASBY, G. P. & MOORBY, S. A. 1986. Geochemistry of marine sediments from the Lau Basin, Havre Trough and Tonga-Kermadec Ridge. *New Zealand Journal of Geology and Geophysics*, **29**, 335–344.

KARIG, D. E. 1970. Ridges and basins of the Tonga-Kermadec island arc system. *Journal of Geophysical Research*, **75**, 239–254.

——, BARBER, A. J., CHARLTON, T. R., KLEMPERER, S. & HUSSONG, D. M. 1987. Nature and distribution of deformation across the Banda Arc-Australian collision zone at Timor. *Bulletin of the Geological Society of America*, **98**, 18–32.

KELLING, G. & STANLEY, D. J. 1976. Sedimentation in canyon, slope and base-of-slope environments. *In*: STANLEY, D. J. & SWIFT, D. J. P. (eds) *Marine Sediment Transport and Environmental Management*. John Wiley & Sons, New York, 379–435,

KELTS, K. & ARTHUR, M. A. 1981. Turbidites after ten years of Deep Sea Drilling – wringing out the mop. *In*: WARME, J .E., DOUGLAS, R. G. & WINTERER, E. L. (eds) *The Deep Sea Drilling Project : A Decade Of Progress*. Society of Economic Paleontologists and Mineralogists, Special Publications, **32**, 91–127.

KENNETT, J. 1982. *Marine Geology*. Prentice Hall, Englewood Cliffs, NJ.

KUENEN, P. H. 1964. The shell pavement below oceanic turbidites. *Marine Geology*, **2**, 236–246.

ODP LEG 107 SHIPBOARD SCIENTIFIC PARTY. 1987. Site 650 : Marsili Basin. *In*: *Proceedings of the Ocean Drilling Program, Initial Reports*, **107**. Ocean Drilling Program, College Station, TX, 129–286.

ODP LEG 135 SHIPBOARD SCIENTIFIC PARTY. 1992a. Site 834. *Proceedings of the Ocean Drilling Program, Initial Reports*, **135**. Ocean Drilling Program, College Station, TX, 85–180.

—— 1992b. Site 835. *Proceedings of the Ocean Drilling Program, Initial Reports*, **135**. Ocean Drilling Program, College Station, TX, 181–245.

PARSON, L. M. & HAWKINS, J. W. 1994. Two stage ridge propagation and the geological history of the Lau backarc Basin. *Proceedings of the Ocean Drilling Program, Scientific Results,* **135**. Ocean Drilling Program, College Station, TX, 819–828.

——, PEARCE, J. A., MURTON, B. J. *ET AL*. 1990. Role of ridge jumps and ridge propagation in the tectonic evolution of the Lau backarc Basin, southwest Pacific. *Geology*, **18**, 470–473.

——, HAWKINS, J. W., ALLAN, J. *ET AL*. 1992. *Proceedings of the Ocean Drilling Program, Initial Reports*, **135**. Ocean Drilling Program, College Station, TX, (2 volumes).

——, ROTHWELL, R. G. & MACLEOD, C. J. 1994. Tectonics and sedimentation in the Lau Basin (Southwest Pacific). *Proceedings of the Ocean Drilling Program, Scientific Results,* **135**. Ocean Drilling Program, College Station, TX, 9–21.

PILKEY, O. H. 1987. Sedimentology of basin plains. *In*: WEAVER, P. P. E. & THOMSON, J. (eds) *Geology and Geochemistry of Abyssal Plains*. Geological Society, London, Special Publications, **31**, 1–12.

PIPER, D. J. W. 1978. Turbidite muds and silts on deep sea fans and abyssal plains. *In*: STANLEY D. J. & KELLING, G. (eds) *Sedimentation in Submarine Canyons, Fans and Trenches*. Hutchinson & Ross, Stroudsburg, PA, 163–176.

REICH, V., MARCHIG, V., SUNKEL, G. & WEISS, W. 1990. Hydrothermal and volcanic input in sediments of the Lau backarc Basin, SW Pacific. *Marine Mining*, **9**, 183–203.

ROTHWELL, R. G., PEARCE, T. J. & WEAVER, P. P. E. 1992. Late Quaternary evolution of the Madeira Abyssal Plain, Canary Basin, NE Atlantic. *Basin Research*, **4**, 103–131.

TAYLOR, B., ZELLMER, K., MARTINEZ, F. & GOODLIFFE, A. M. 1995. Tectonic evolution of the Lau Basin. *Eos*, **76**(46), T32C–11.

VAN DER LINGEN, G. J. 1969. The turbidite problem. *New Zealand Journal of Geology and Geophysics*, **12**, 7–50.

VAN WEERING, T. C. E., KUSNIDA, D., TJOKROSAPOETRO, S., LUBIS, S. & KRIDOHARTO, P. 1989. Slumping, sliding and the occurrence of acoustic voids in recent and subrecent sediments of the Savu forearc Basin (Indonesia). *Netherlands Journal of Sea Research*, **24**, 415–430.

WEAVER, P .P. E. & KUIJPERS, A. 1983. Climatic control of turbidite deposition on the Madeira Abyssal Plain. *Nature*, **306**, 360–363.

—— & ROTHWELL, R. G. 1987. Sedimentation on the Madeira Abyssal Plain over the last 300,000 years. *In:* WEAVER, P. P. E. & THOMSON, J. (eds) *Geology and Geochemistry of Abyssal Plains*. Geological Society, London, Special Publications, **31**, 71–86.

WETZEL, A. 1986. Anisotrophy and modes of deposition of pelitic Mississippi Fan deposits. *Initial Reports of the Deep Sea Drilling Project*, **96**. US Government Printing Office, Washington, DC, **96**, 811–817.

Hydrothermal inputs at ODP Sites 836, 837, 838 and 839 in relation to Eastern Lau Spreading Centre propagation in the Lau Basin, southwest Pacific

RICHARD A. HODKINSON & DAVID S. CRONAN

Department of Geology, Imperial College of Science, Technology & Medicine,
London SW7 2BP, UK

Abstract: Lau Basin Ocean Drilling Program (ODP) Sites 836, 837, 838 and 839 were drilled in basins west of the axial rift zone of the Eastern Lau Spreading Centre (ELSC), south of the Central Lau Spreading Centre (CLSC). Based on the models of basin evolution, Site 836 is thought to be located on ELSC-generated crust while Site 838 is located on pre-existing extended arc terrain to the west of ELSC-generated crust. Sites 837 and 839 are postulated to be close to the boundary separating ELSC-generated and pre-existing crust.

Sediments recovered at all sites comprise late Pliocene to Pleistocene clayey nannofossil ooze containing sparse calcareous turbidites overlying a thick sequence of redeposited volcaniclastic sediments interbedded with hemipelagic clayey nannofossil oozes. All calcareous ooze sections contain hydrothermal ferromanganese oxides.

The hydrothermal flux to the sediments at each site has been assessed using non-detrital Mn+Fe accumulation rates for recalculated, hemipelagic sediment intervals alone in order to remove the effect of sediment redeposition. Site 836 shows a typical mid-ocean ridge up-hole pattern of decreasing accumulation rates with increasing crustal accretion, as expected from its predicted location on ELSC-generated crust. Site 837 shows a similar pattern suggesting its location on ELSC-generated rather than pre-existing western Lau Basin attenuated crust. At Site 839 non-detrital Mn+Fe accumulation rates show an increase up-hole from basement to a peak before decreasing towards surface, reflecting increasing hydrothermal flux to the sediments with the southward propagation of the ELSC past this site and suggesting its location on pre-existing crust west of ELSC-generated crust. For the time-equivalent sediment section at Site 838, non-detrital Mn+Fe accumulation rate patterns are similar to those at Site 839. Although a period of low non-detrital Mn+Fe accumulation rate reflecting lower plume fallout flux prior to ELSC propagation past the site is not seen at the base of the sediment section at Site 838 (as the hole did not penetrate to basement), the accumulation rate patterns seen are not at variance with its postulated location on pre-existing attenuated crust west of ELSC-generated crust.

Mean non-detrital Mn+Fe accumulation rate values are highest at Site 836 with lower mean values at Sites 837, 838 and 839. These values are within the range of those for East Pacific Rise sediments but lower than those for ELSC-propagation-associated sediments at Sites 834 and 835 in the northwestern part of the basin, suggesting that hydrothermal plume fallout associated with the southern, more recently generated portion, of the ELSC was lower than that of the older portion in the north.

The Lau Basin forms part of the Lau Basin, Havre Trough, Tonga–Kermadec Ridge arc–backarc complex, located at the convergent boundary of the Pacific and Indo-Australian Plates (Fig. 1). The basin is relatively shallow (2000 to 3000 m water depth) and lies between the islands and atoll reefs of the Lau Ridge in the west and the volcanic islands, atoll reefs and uplifted carbonate platforms of the Tonga Ridge and the active Tofua Arc in the east.

Originally thought to have formed by a single spreading event initiated some 3 to 5 Ma, recent studies have indicated that the basin has had a significantly more complex tectonic evolution. Different models of basin evolution are described and discussed in Parson *et al.* (1990, 1992*a*), Hawkins *et al.* (1994) and Parson & Hawkins (1994) (and references therein). Based on a compilation of bathymetric, magnetic and side-scan sonar data and the results of drilling during Ocean Drilling Program (ODP) Leg 135, the Parson & Hawkins (1994) model of tectonic evolution defines morphotectonic domains and basin evolution as follows.

HODKINSON, R. A. & CRONAN, D. S. 1998. Hydrothermal inputs at ODP Sites 836, 837, 838 and 839 in relation to Eastern Lau Spreading Centre propagation in the Lau Basin, southwest Pacific. *In*: CRAMP, A., MACLEOD, C. J., LEE, S. V. & JONES, E. J. W. (eds) *Geological Evolution of Ocean Basins: Results from the Ocean Drilling Program.* Geological Society, London, Special Publications, **131**, 231-242.

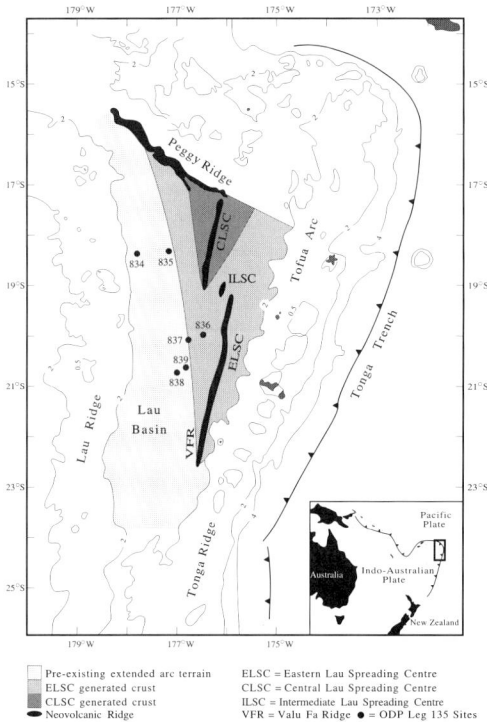

Pre-existing extended arc terrain	ELSC = Eastern Lau Spreading Centre
ELSC generated crust	CLSC = Central Lau Spreading Centre
CLSC generated crust	ILSC = Intermediate Lau Spreading Centre
Neovolcanic Ridge	VFR = Valu Fa Ridge ● = ODP Leg 135 Sites

Fig. 1. Location, simplified bathymetry, simplified morphotectonic domains and major tectonic features of the central Lau Basin and location of ODP Leg 135 Sites (after Parson & Hawkins 1994). Isobaths in kilometres.

The western part of the Central Lau Basin (Fig. 1) comprises a region of attenuated, heterogeneous crust that probably represents a mixture of relict arc fragments, ephemeral rift grabens and volcanic constructs derived from easterly migrating arc magmatism. The area shows an irregular topography with a complex pattern of discontinuous ridges and basins, locally showing a north–south trend. Basins are partly sediment-filled and are separated by uplifted, elongate basement ridges and highs which have only thin sediment cover, forming a 'horst and graben' terrain.

The central part of the Lau Basin contains two major morphotectonic features: the Eastern Lau Spreading Centre (ELSC) and Central Lau Spreading Centre (CLSC) (Fig. 1). The ELSC, lying close to the Tofua Arc, and becoming more proximal to the arc in the south, extends for over 180 km from 19°20′S to around 21°S, where it merges with the Valu Fa Ridge, and extends southwards to 23°S where its propagating rift tip is located. The ELSC is postulated to have propagated

southwards, commencing at around 5.5 Ma, from the southeast end of a structure that became the Peggy Ridge, into the horst and graben dominated extensional terrain derived from the fragmented Lau/Tonga composite arc. Propagation is thought to have taken place in an intermittent manner, probably controlled by the pre-existing rifted fabric of the western Lau Basin. The ELSC has generated the majority of crust seen in the eastern most central Lau Basin (Fig. 1).

The CLSC is a gently curving axial ridge, concave to the east, extending from around 17°10′S to around 19°20′S. It is postulated that the CLSC commenced southward propagation from the southeastern limit of the Peggy Ridge at between 1.2 and 1.5 Ma. The continued CLSC propagation into ELSC-generated crust has occurred at the expense of the ELSC, with spreading being transferred via a relay spreading segment (the Intermediate Lau Spreading Centre, ILSC), forming a wedge-shaped area of crust (Fig. 1). As the CLSC has yet to propagate to the latitude of Sites 836, 837, 838 and 839 (Fig. 1), the hydrothermal input to the sediments at these sites mainly will reflect hydrothermal plume fallout from the ELSC, rather than from the CLSC.

Most recently, Taylor *et al.* (1995) and Taylor (pers. comm. 1996), using a compilation of bathymetric, magnetic and geological data, have suggested a more complex model of Lau Basin evolution, based on a three-phase opening history of the basin. In summary, they propose that crustal accretion began between the Lau and Tonga Ridges at 6 Ma by magma intrusion and extensional faulting. At 4 Ma, focused accretion began on the ELSC at 16°S, with it propagating into the rifted volcanic terrain to 20°S by 2 Ma. At 2 Ma, the ELSC rotated 20° clockwise, subdivided into left-stepping overlapping segments and propagated faster. At the same time, the CLSC was initiated and propagated southwards, replacing the ELSC and transferring Anomaly 2 and younger crust eastwards.

Morphotectonic settings of sites

Sites 836, 837, 838 and 839 (Fig. 1) were all drilled in fault-bounded sub-basins west of the ELSC within an area of about 150 × 50 km (Parson *et al.* 1992*b*). Sites 838 and 839 are located within the same sub-basin and are separated by a flexure that forms a low ridge. All but Site 838 reached basement.

Based on the nature of basement recovered during ODP Leg 135 and their model of basin evolution (see above), the Parson & Hawkins (1994) interpretation of the morphotectonic settings of Sites 836, 837, 838 and 839 is summarized below. Qualifications concerning the timing of events such as propagator passage and basement ages, consequent on the most recent data from Taylor (Taylor *et al.* 1995; Taylor pers. comm. 1996), are also given where appropriate.

Site 836

Site 836 (Fig. 1) is situated less than 50 km west of the ELSC and is located on ELSC-generated crust with an estimated age of either 0.9 Ma (Parson & Hawkins 1994) or 0.99–1.07 Ma (Taylor *et al.* 1995; Taylor pers. comm. 1996). Oldest sediments, however, are dated at 0.7 Ma and Parson & Hawkins (1994) speculate that the sedimentation history at the site may have been interrupted as a result of restricted sediment supply and/or sediment reworking and scouring due to currents.

Site 837

Site 837 is situated about 70 km west of the ELSC (Fig. 1) on the eastern margin of a basin (postulated to delineate the western pseudofault of the ELSC) which is thought to form an extension of crust generated at the ELSC (Parson & Hawkins 1994). Basement geochemical data, however, are equivocal, with Pb isotope data supporting an ELSC-generated basement but major and trace element data suggesting a western Lau Basin attenuated crust basement. According to Parson & Hawkins (1994) the estimated data of the closest passage of the southwardly propagating ELSC to Site 837 is 2.0 Ma, corresponding exactly to the age of the youngest basement recovered at the site. Taylor *et al.* (1995) and Taylor (pers. comm. 1996) suggest a basement age of 1.77–1.95 Ma.

Site 838

Site 838 is located about 90 km west of the ELSC (Fig. 1) on the western flank of an irregular-shaped ridge in a region dominated by rift graben and horst structures and is thus thought to be sited on western Lau Basin attenuated crust. Although basement was not reached during drilling, seismic data collected on Leg 135 indicate a recognizable acoustic basement. Extrapolation of sedimentation rate data from the recovered sediment section (17.7 cm ka^{-1}; Parson *et al.* 1992b) suggests that basement could be as old as 4 Ma (Parson & Hawkins 1994). Their estimated date for the closest passage of the southwardly propagating ELSC to Site 838 is around 1.65 Ma, while extrapolation of dates derived from the magnetic anomaly data of Taylor *et al.* (1995) and Taylor (pers. comm. 1996) suggest 1.22 Ma.

Site 839

Site 839 is located about 75 km west of the ELSC (Fig. 1) at the western margin of a basin in a region of sea floor ascribed to western Lau Basin attenuated crust. Youngest basement rocks have an estimated age of around 1.8 to 1.9 Ma, whilst the predicted date for closest passage of the southwardly propagating ELSC to Site 839 is 1.7 Ma (Parson & Hawkins 1994). Extrapolation of dates derived from the magnetic anomaly data of Taylor *et al.* (1995) and Taylor (pers. comm. 1996) suggest a date of 1.36 Ma for closest passage of the propagating ELSC.

Hydrothermal sedimentation in the Lau Basin

Studies of recent sediments (surface deposits and short cores) in the central Lau Basin have shown the sediments to comprise mixtures of volcaniclastic detritus and biogenic carbonate, with a superimposed hydrothermal component derived from hydrothermal discharge at the ELSC and CLSC (Cronan 1983; Cronan *et al.* 1984, 1986; Hodkinson *et al.* 1986; Riech 1990; Riech *et al.* 1990; Walter *et al.* 1990; Hodkinson & Cronan 1991). Possible off-axis hydrothermalism has, however, also been postulated (Cronan *et al.* 1986; Hodkinson & Cronan 1991; Parson *et al.* 1994a).

Consequent on the ODP Leg 135 drilling in the Lau Basin (Parson *et al.* 1992b; 1994b) the long-term temporal variability in the flux of the hydrothermal component of the sediments during the evolution of the basin, particularly with respect to ELSC and CLSC propagation, can be evaluated. Hydrothermal input to the sites occurs in the form of Mn and Fe oxides, with associated coprecipitated elements, and there is a relationship between the record of hydrothermal plume fallout flux at each site and spreading centre propagation (Hodkinson & Cronan 1994).

Assessment of the hydrothermal input to Sites 834 and 835 using metal accumulation rate data (Hodkinson & Cronan 1995) has shown that variation in hydrothermal flux to the sediment section at each site reflects the hydrothermal activity associated with ELSC and CLSC propagation into the basin, with possible modification of plume dispersion by palaeocurrents and basin topography. In this paper we present results of geochemical studies on sediments from ODP Sites 836, 837, 838 and 839, with particular reference to variation in the hydrothermal flux to the sediments at these sites in relation to ELSC propagation, and a comparison of the flux between these sites, Sites 834 and 835 to the north and the East Pacific Rise (EPR).

Analytical methods and data processing

Prior to chemical analysis, sediment samples ($n = 267$) were air-dried at room temperature over silica gel and finely ground using an agate mortar and pestle. Bulk chemical analyses were performed by inductively coupled plasma atomic emission spectrometry (ICP–AES) after total digestion with a mixture of HNO_3, $HClO_4$ and HF acids and subsequent leaching with 2 ml of 4M HCl and 8 ml of 0.3M HCl to give a final 1M HCl solution (Thompson & Walsh 1989). Accuracy was checked with the use of certified international reference materials, along with in-house reference materials. No systematic errors were observed and accuracy was better than ±5% relative for the elements studied here. Analytical precision was checked with the use of duplicate samples and was found to be better than ±3% relative.

To assess the geochemical partitioning of elements, a series of selective leaches was carried out on some samples ($n = 35$ for leaches 1 and 2, $n = 267$ for leach 3), based on methods described by Chester & Hughes (1967) as subsequently modified by Cronan (1976). The leaches used provide a means of assessing with which phase of the sediments individual element enrichments, particularly the hydrothermal enrichments, are associated. These phases are operationally defined as follows:

(1) carbonates, interstitial water evaporates, loosely adsorbed phases and some amorphous Fe oxides (leached using 25% acetic acid);
(2) as (1) plus reducible ferromanganese oxides and associated elements (leached using a combined acid-reducing agent comprising 1M hydroxylammonium chloride in 25% acetic acid);
(3) as (2) plus more crystalline Fe oxides, clays, weathered detrital phases and partial sulphide dissolution (leached using hot, 50% HCl);
(4) resistant detrital phases (by subtraction of (3) from bulk concentrations).

Subsequent to filtration through 0.45 µm cellulose nitrate filter papers, decomposition of leachates using HNO_3 and leaching of residues in 1M HCl (for leaches 1 and 2, above), leachate compositions were determined by ICP–AES techniques. For leach 3, composition determinations were similarly made after filtration and direct dilution of the leachate to 1M HCl. For all leaches, analytical precision was found to be better than ±4% relative for all elements studied here. The weight percentage of individual samples soluble in each leach was also determined by weighing the dried filter residues for each leach.

To assess interelement associations in the non-detrital, carbonate-free fraction of the sediments, Factor Analysis was performed on geochemical data obtained on the HCl-soluble, carbonate-free sediment phase. By using such data, the effects of two major sources of variance in the bulk uncorrected data – biogenic carbonate and volcaniclastic detritus – have been removed. The effect of these two dominant and antipathetically varying sediment components is to mask variations in the hydrothermal component of the sediments, which is the subject of this paper. Their removal allows a closer assessment of the variability in the hydrothermal sediment phase to be made. Factor Analysis itself is a technique which enables a large number of variables (in this case, chemical elements) to be grouped into a lesser number of Factors, each of which is a linear function (transformation) of the element concentrations and is based on the element associations inherent in a suitably conditioned correlation matrix (Howarth 1983). The relative influence of each Factor (in this case a hydrothermal oxide Factor) on individual samples is then determined by means of Factor Scores, enabling the relative input of hydrothermal oxide plume fallout throughout the sediment section at each site be assessed. Factor Analysis has been widely used in the study of metalliferous sediments from other regions in order to identify the nature and abundance of the hydrothermal phases in them (e.g. Leinen & Pisias 1984; Murphy et al. 1991; McMurtry et al. 1991).

Combined, non-detrital Mn+Fe accumulation rates in the sediments have been calculated using the equation:

$$\text{Accumulation Rate } (\mu g\ cm^{-2}\ ka^{-1}) = M \times \rho \times S$$

where M = concentration of Mn+Fe (ppm) in the HCl-soluble sediment phase; ρ = dry, *in situ* sediment density (g cm^{-3}); and S = Sedimentation rate (cm ka^{-1}).

Dry, *in situ* density values were calculated from shipboard physical property measurements for both sites (Parson et al. 1992b). Sedimentation rates are based on the recalculated hemipelagic sediment thicknesses for each site, after subtraction of allochthonous sediments.

Sedimentation at sites

Figure 2 shows the generalized lithologic columns for Sites 836, 837, 838 and 839 (after Rothwell et al. 1994a), with age (Ma) as a vertical scale. Sediment ages have been determined from shipboard biostratigraphic and palaeomagnetic data (Parson et al. 1992b).

All sites show a similar late Pliocene to Pleistocene sediment sequence of hemipelagic

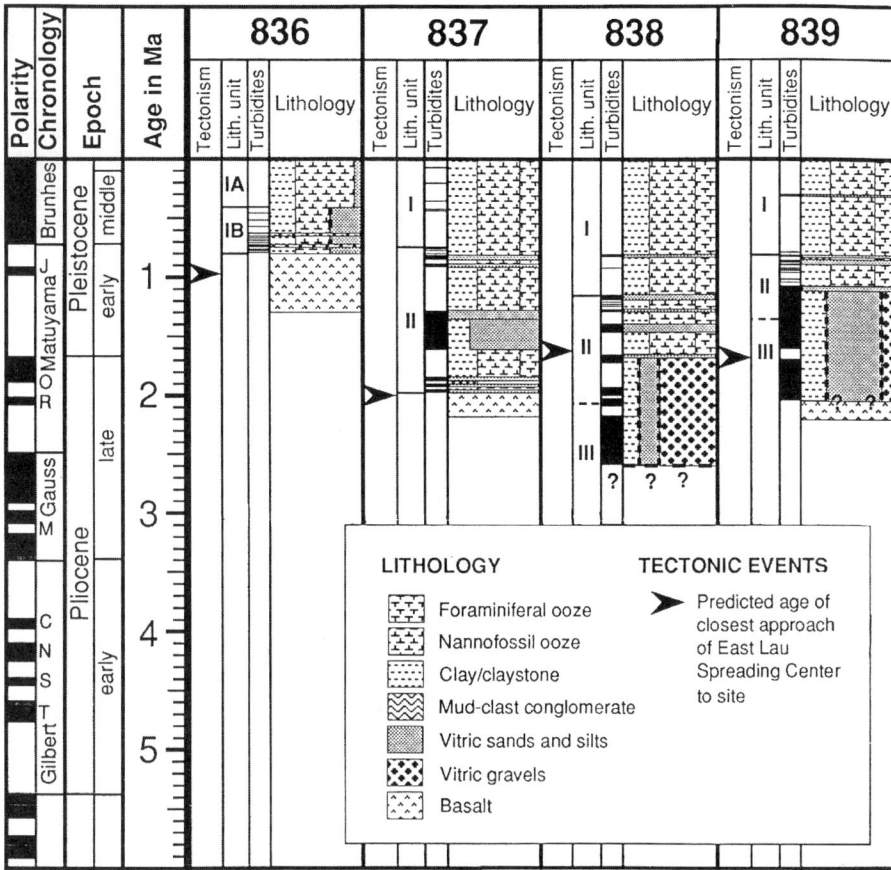

Fig. 2. Simplified lithologic columns for Sites 836, 837, 838 and 839 with age (Ma) as a vertical scale (after Rothwell *et al.* 1994*a*). Allochthonous beds are shown in the 'Turbidites' column. Thick dashed lines indicate that the lithologies shown are interbedded (see Parson *et al.* 1992*b*). Predicted ages of the closest approach of the ELSC to each site are shown (see Parson & Hawkins 1994). Recovered sediment sequences are 20 m at Site 836, 82 m at Site 837, 103 m at Site 838 and 215 m at Site 839. Recalculation to subtract the redeposited intervals would reduce these thicknesses to 16 m at Site 836, 36 m at Site 837, 32 m at Site 838 and 37 m at Site 839 (see text).

clayey nannofossil oozes containing rare volcani-clastic horizons and minor calcareous turbidites, overlying thick sequences of volcaniclastic silts, sands and pumiceous gravels with clayey nannofossil ooze and mixed sediment interbeds (Rothwell *et al.* 1994*a*).

A significant proportion of the sedimentary section at all the sites contains both redeposited clayey nannofossil ooze intervals and/or intervals of epiclastic volcanic ash deposition (Rothwell *et al.* 1994*a*). To assess the downhole variability in hydrothermal inputs to the hemipelagic sediment intervals alone, the thickness of the sediment section at each site has been recalculated by subtracting the redeposited intervals. This reduces

the thickness of the sediment section at Site 836 from 20 m to 16 m, at Site 837 from 82 m to 36 m, at Site 838 from 103 m to 32 m and at Site 839 from 215 m to 37 m. Subsequently, the depth below sea floor of individual samples used in this work has been recalculated. This enables a more accurate assessment of the temporal variability in hydro-thermal input at each site to be made. All depths referred to are quoted as recalculated depths in metres below sea floor (mbsf*).

Shipboard smear slide analysis of sediments at Sites 836, 837, 838 and 839 (Parson *et al.* 1992*b*) show that Fe and Mn oxyhydroxides are abundant throughout the hemipelagic sediment sections, suggesting that they have been continuously in receipt of hydrothermal input. These oxides pervasively stain the sediments red/brown (see Parson *et al.* 1992*b*). X-ray diffraction analysis carried out on selected samples from the sites shows no crystalline Fe or Mn oxide, Fe-silicate or sulphide mineral phases. However, it suggests that a significant X-ray-amorphous ferromanganese oxide component occurs in the sediments.

Sediment geochemistry

Factor Analysis of HCl-soluble, carbonate-corrected geochemical data from Sites 836, 837, 838 and 839 (Hodkinson & Cronan 1994) shows an association of Mn, Fe, Co, Ni, Cu, Zn, Pb, V and P in the sediments at all sites, characteristic of a hydrothermal oxide (plume fallout) phase comprising hydrothermally derived Mn and Fe oxides with associated coprecipitated elements.

In order to confirm that the hydrothermal metal enrichments in the sediments from Sites 836, 837, 838 and 839 studied here do occur as a result of oxide plume fallout, selective chemical leach analysis has been carried out on sediment samples from all four sites. These samples ($n = 35$) were chosen from throughout the hemipelagic sediment sequence at Sites 836, 837, 838 and 839 and, based on the results of Factor Analysis, encompass the entire range of relative hydrothermal inputs to these sites. The relative hydrothermal inputs have been determined by Factor Scores calculated from Factor Loadings for the hydrothermal oxide Factor. Up to Factor Score 20th, 40th, 60th, 80th and 100th percentile divisions have been termed very low, low, medium, high and very high hydrothermal inputs respectively. Figure 3 shows the mean geochemical partitioning of Mn and Fe for sediments from Sites 836, 837, 838 and 839 based on these hydrothermal input divisions.

Mean bulk concentrations of both Mn and Fe, as expected, show an increase in the sediments as the relative hydrothermal input increases (Fig. 3). Selective leach data show that Mn occurs principally in the acid-reducible phase, and thus as an amorphous oxide, and that the increase in mean bulk concentration with increasing relative hydrothermal input occurs due to increased Mn concentrations in this form. In the case of Fe, although a significant amount is associated with insoluble and HCl-soluble phases (detrital and altered detrital respectively), the increasing mean bulk concentrations with increasing relative hydrothermal input occur, like Mn, due to increased concentrations in the reducible (oxide-associated) phase. No hydrothermal sulphide or silicate enrichments were found, confirming that hydrothermal input occurs entirely as a result of oxide plume fallout.

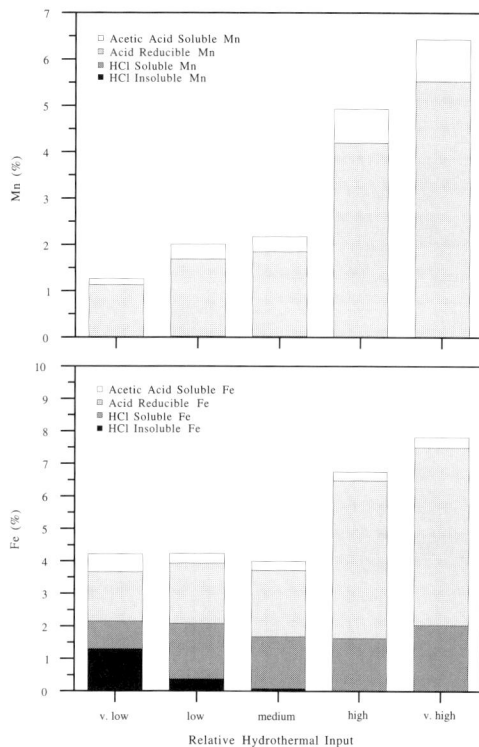

Fig. 3. Mean geochemical partitioning of Mn and Fe at Sites 836, 837, 838 and 839. Samples ($n = 35$) have been assigned a 'relative hydrothermal input' (very low, low, medium, high and very high), based on the 20th, 40th, 60th and 80th percentile divisions of Factor Scores for the hydrothermal oxide Factor (see also Hodkinson & Cronan 1994, 1995). For each 'relative hydrothermal input' group, the overall height of the bar represents the mean bulk concentration. Shading within the bar represents the mean proportions soluble in each of the four selective leaches (see 'Analytical Methods and Data Processing' section for the sediment phases dissolved by each leach). For Mn, no samples show any significant HCl-soluble or HCl-insoluble, non-oxidic Mn. For Fe, samples assigned to 'high' and 'very high' hydrothermal input groups show no significant HCl-insoluble (detrital) component. Significant amounts of HCl-soluble Fe (principally altered detrital phases) occur in all hydrothermal influence groups.

Sediment records of hydrothermal plume fallout and non-detrital metal accumulation rates

The lithological and geochemical studies of sediments recovered from Sites 836, 837, 838 and 839 (see above) indicate that hydrothermal inputs occur throughout the entire hemipelagic sedimentary intervals as they do at Sites 834 and 835 to the north (Hodkinson & Cronan 1994, 1995). Only in volcaniclastic sediment intervals, comprising predominantly pyroclastic and epiclastic deposits that were rapidly sedimented (Rothwell *et al.* 1994*a*), is evidence of hydrothermal plume fallout absent. While individual hydrothermal discharge vents associated with a spreading ridge can be short-lived on a geological timescale and the discharges from them can exhibit significant temporal variability (e.g. Baker 1994; Von Damm *et al.* 1995), the sedimentary record in the Lau Basin would suggest that hydrothermal discharge associated with the spreading centres has occurred there throughout the period of basin history recorded by the hemipelagic sediment intervals recovered. Furthermore, studies on both the ELSC and CLSC (e.g. Malahoff & Falloon 1991; Fouquet *et al.* 1993) have found evidence for active and recently active hydrothermal systems close to the propagating tips of both spreading centres, indicating that hydrothermal systems evolve soon after ridge propagation, although studies on the Valu Fa Ridge by Fouquet *et al.* (1993) suggest that the most actively discharging vents are located on the older, more evolved parts of the ridge.

In order to assess the general trends in temporal variation of the hydrothermal flux to the sediments at Sites 836, 837, 838 and 839, in particular their patterns with regard to the postulated positions of the sites with respect to crustal province and ELSC evolution, non-detrital metal accumulation rates for Mn+Fe have been used (see 'Analytical methods and data processing' section). These are plotted for the hemipelagic, recalculated sediment interval in each core (Figs 4 to 7). Likely ages of prominent

peaks in accumulation rates are shown (Ma), based on shipboard age data applied to the recalculated, hemipelagic sediment section at each site (see above).

Studies of hydrothermal plumes associated with currently active systems indicate that hydrothermal discharge and dispersal can be highly variable over short time periods, with factors such as local geology, topography, currents and oceanographic conditions all being important in this regard (e.g. Klinkhammer & Hudson 1986; Dymond & Roth 1988; German *et al.* 1995). However, the hydrothermal plume fallout signature in sediments represents a record built up over a period of time. In the case of the sediments studied here, the sampling interval of 2 cm represents an average 1000–2000 year period (based on a typical sedimentation rate of 1–2 cm ka^{-1} for the hemipelagic sediment intervals; Rothwell *et al.* 1994*b*). Over such a period, much of the short-term temporal variability in hydrothermal discharge and periods of preferential plume dispersion due to varying current activity will be 'averaged out'. The resulting record in the sediment section will be an *integrated* one. Thus, metal accumulation rates in the sediment section provide a valuable tool to assess temporal variability in spreading-centre hydrothermal activity.

SITE 836

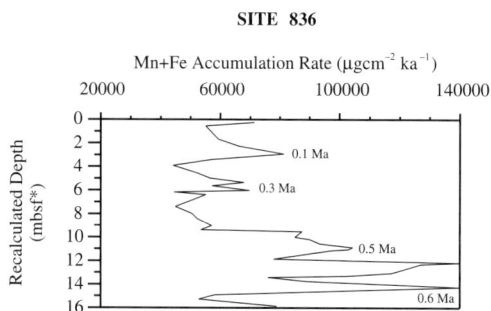

Fig. 4. Plot of non-detrital Mn+Fe accumulation rates versus recalculated depth (mbsf*) at Site 836. Estimates of basement age are 0.9 Ma (Parson & Hawkins 1994) and 0.99–1.07 Ma (Taylor *et al.* 1995; Taylor pers. comm. 1996) and oldest sediments at 16 mbsf* are 0.7 Ma (Parson & Hawkins 1994). Ages (Ma) of prominent peaks in accumulation rates are indicated (see text).

SITE 837

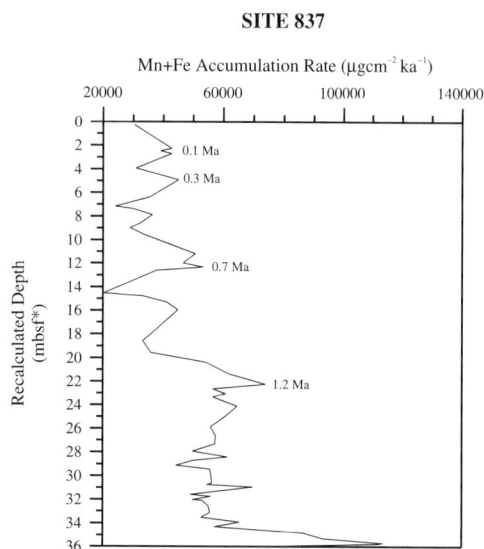

Fig. 5. Plot of non-detrital Mn+Fe accumulation rates versus recalculated depth (mbsf*) at Site 837. Estimates of basement age (at 36 mbsf*) are 2.0 Ma (Parson & Hawkins 1994) and 1.77–1.95 Ma (Taylor *et al.* 1995; Taylor pers. comm. 1996). Ages (Ma) of prominent peaks in accumulation rates are indicated (see text).

SITE 839

Mn+Fe Accumulation Rate (μgcm^{-2} ka^{-1})

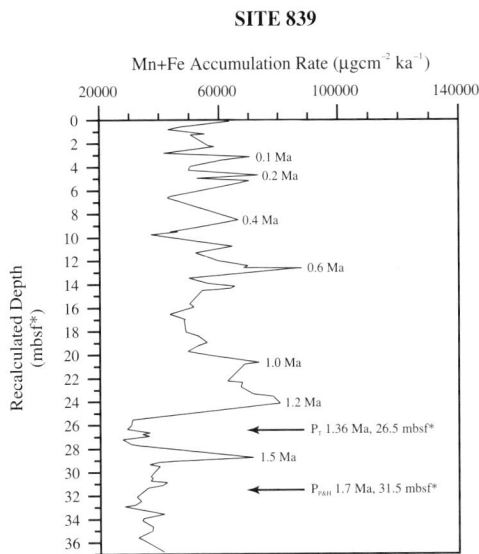

Fig. 6. Plot of non-detrital Mn+Fe accumulation rates versus recalculated depth at Site 839. $P_{P\&H}$ = age of closest passage of southward propagating ELSC (1.7 Ma), based on Parson & Hawkins (1994) model of basin evolution. P_T = age of closest passage of southward propagating ELSC (1.36 Ma), based on Taylor *et al.* (1995; Taylor pers. comm. 1996) model of basin evolution. Ages (Ma) of prominent peaks in accumulation rates are indicated (see text).

SITE 838

Mn+Fe Accumulation Rate (μgcm^{-2} ka^{-1})

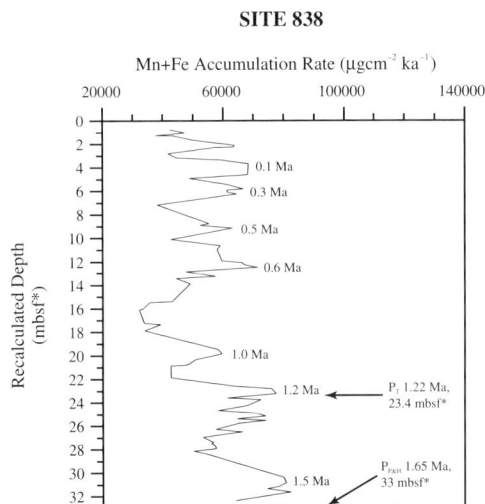

Fig. 7. Plot of non-detrital Mn+Fe accumulation rates versus recalculated depth at Site 838. $P_{P\&H}$ = age of closest passage of southward propagating ELSC (1.65 Ma), based on Parson & Hawkins (1994) model of basin evolution. P_T = age of closest passage of southward propagating ELSC (1.22 Ma), based on Taylor *et al.* (1995; Taylor pers. comm. 1996) model of basin evolution. Ages (Ma) of prominent peaks in accumulation rates are indicated (see text).

Site 836. Site 836 shows a general trend of decreasing hydrothermal flux to the sediment section up-hole from peaks just above basement at 0.55 and 0.6 Ma (Fig. 4). Up-hole, some local variability does occur, with peaks in non-detrital accumulation rates at 0.3 and 0.1 Ma. With the exception of sediments immediately overlying basement, such a trend is similar to that typically seen on mid-ocean ridges (MORs), such as the EPR, where the hydrothermal flux to the sediments decreases up-hole as continued crustal accretion increases the distance between hydrothermal discharge at the spreading ridge and the site of plume fallout. At Site 836 the low non-detrital Mn+Fe accumulation rates immediately above basement may reflect the discrepancy between the proposed basement age (0.9 Ma (Parson & Hawkins 1994), 0.99–1.07 Ma (Taylor *et al.* 1995; Taylor pers. comm. 1996)) and the oldest sediments recovered (0.7 Ma), postulated to result from delayed sediment deposition at this site (Parson & Hawkins 1994). The observed overall trend is, however, consistent with the location of Site 836 on ELSC spreading-generated crust.

Site 837. Site 837 (Fig. 5) shows maximum non-

detrital Mn+Fe accumulation rates in sediments immediately overlying basement. Up-hole, accumulation rates show a general decrease towards surface with some local variability, including peaks in input at 1.2, 0.7, 0.3 and 0.1 Ma. Such a pattern, like that at Site 836, is typical of that seen on MORs. The results would thus suggest that Site 837 is located on, or very close to, ELSC-generated crust, like Site 836.

Site 839. In contrast to Sites 836 and 837, accumulation rates at Site 839 show an increase up-hole from basement to an interval of high input, followed by a slight decrease towards surface (Fig. 6).

Just above basement, non-detrital Mn+Fe accumulation rates are low and fairly constant before a sharp, isolated peak in input occurs at around 1.5 Ma, followed by another at around 1.2 Ma and a subsidiary peak at 1.0 Ma, forming a broad section of enhanced accumulation rate input (Fig. 6). Above this, accumulation rate trends show much more variability with pronounced peaks at 0.6, 0.4, 0.2 and 0.1 Ma, although the general overall trend is one of a decrease in accumulation rates towards the surface.

The accumulation rate trend observed at Site 839

is similar to that previously described from Sites 834 and 835 in the north of the basin (Hodkinson & Cronan 1995) and is interpreted as an increase in hydrothermal input to the site as the ELSC propagates southwards into pre-existing crust, resulting in the hydrothermal discharge from it becoming more proximal to the site. Based on these previous studies, it seems probable that the combined peaks in input at 1.2 and 1.0 Ma reflect the principal hydrothermal plume fallout signature associated with the propagation of the ELSC past Site 839. The sharp peak seen at around 1.5 Ma is likely to be due to a short-lived increase in plume fallout to the sediments as a result of a specific local event.

From the models of basin evolution proposed by Parson & Hawkins (1994) and Taylor *et al.* (1995; Taylor pers. comm. 1996) estimated ages of closest passage of the southward propagating ELSC to Site 839 are 1.7 Ma and 1.36 Ma respectively. Significantly, the combined period of peak hydrothermal input recorded at the Site (1.0 to 1.2 Ma) is not coincident with either estimate for the timing of the predicted closest passage of the propagating ELSC past it, but instead occurs in sediments either some 0.5–0.7 Ma or 0.16–0.36 Ma later. Either model for the timing of propagator passage would suggest a 'time lag' between closest propagator pass and maximum hydrothermal flux to the sediments. This 'time lag' may be due to a delay between ridge formation and either the initiation of hydrothermal systems on the newly formed ridge and/or their full development, as suggested by Hodkinson & Cronan (1995) for the propagation of the ELSC past Sites 834 and 835 further north in the basin.

The peak in hydrothermal input at 1.0 to 1.2 Ma is followed by a general pattern of decreasing hydrothermal input to the site as continued crustal accretion at the newly formed ELSC increases the distance between it and the hydrothermal vents in a manner similar to that seen at Sites 834 and 835 to the north (Hodkinson & Cronan 1995). These observations confirm that Site 839 is located on pre-existing western Lau Basin attenuated crust rather than on ELSC-generated crust.

Site 838. Site 838 is located southwest of Site 839 and thus further from the ELSC-generated crust–western Lau Basin attenuated crust boundary. Its observed pattern of accumulation rates, which like Site 839 show significant local variability, shows a general decrease up-hole from the base of the sediment section (Fig. 7). The recalculated sediment section (≤1.6 Ma) at Site 838 (which did not penetrate to basement) is shorter than that at Site 839 (≤2.0 Ma) but, where accumulation rate patterns for sediment sections younger than about 1.6 Ma are available at both sites, a generally

similar pattern of peaks in input is seen. Towards the base of the section at Site 838, peaks in input occur at around 1.5 Ma and 1.2 Ma, followed by a subsidiary peak at around 1.0 Ma, all time-equivalent to those seen at Site 839. Had Site 838 penetrated to basement its lowermost sediments might also be expected to mirror those of Site 839. However, the peaks show different relative intensities and form compared to those at Site 839. For example, the peak at 1.5 Ma is broader and of greater magnitude than that at 1.2 Ma (compared to Site 839) and a greater degree of separation between the peaks at 1.0 and 1.2 Ma is seen compared with Site 839. Such differences between the two sites may reflect local factors affecting patterns of plume dispersion and fallout, such as local sub-basin topography.

In relation to the timing of propagator passage past Site 838, the Parson & Hawkins (1994) and Taylor *et al.* (1995; Taylor pers. comm. 1996) models give ages of 1.65 Ma and 1.22 Ma respectively. If, as at Site 839, the peaks in hydrothermal input to the sediment section at 1.2 to 1.0 Ma record maximum hydrothermal input associated with propagator passage, a 'time lag' is again indicated. In the case of the Parson & Hawkins (1994) model this would be 0.45–0.65 Ma, while for the Taylor model (Taylor *et al.* 1995; Taylor pers. comm. 1996) it would be only 0.02–0.22 Ma.

In more recent sediments, again like Site 839, accumulation rate patterns show much more local variability, with pronounced peaks at 0.6, 0.5, 0.3 and 0.1 Ma, although the general overall trend is one of a slight decrease in accumulation rates towards surface, particularly in sediments younger than about 0.1 Ma.

The fact that Site 838 did not penetrate to basement means that the hydrothermal record cannot be used to confirm the location of the site on western Lau Basin attenuated crust. However, it does show a pattern of non-detrital Mn+Fe accumulation rates consistent with that seen in time-equivalent sections at Site 839 which is not at variance with its location on western Lau Basin attenuated crust.

Intersite variation in metal accumulation rates and comparison with other regions

In addition to assessing hydrothermal flux variability in relation to ELSC evolution, element accumulation rates provide a means of quantifying the intensity of the hydrothermal flux to sediments at each site and comparing relative fluxes between the Lau Basin ODP sites, recent sediments in the basin and hydrothermal sediments from MOR settings such as the EPR. Table 1 gives mean and

Table 1. *Mn and Fe accumulation rates (μg cm⁻² ka⁻¹) for the non-detrital (HCl-soluble) sediment phase at ODP Leg 135 Sites 836, 837, 838 and 839 and comparative rates from Sites 834 and 835 and other areas of known hydrothermal plume fall-out*

Site/Location		Mn	Fe
Site 836	Mean	21 600	51 500
	Max.	66 700	119 800
Site 837	Mean	13 200	38 400
	Max.	47 100	71 700
Site 838	Mean	17 800	39 300
	Max.	31 700	59 400
Site 839	Mean	13 100	37 500
	Max.	30 300	58 400
Site 834* (ELSC)	Mean	82 500	97 500
	Max.	114 900	124 300
Site 835* (ELSC)	Mean	47 400	69 700
	Max.	54 700	81 100
Site 834* (CLSC)	Mean	11 700	24 300
	Max.	17 900	34 300
Site 835* (CLSC)	Mean	11 000	25 700
	Max.	23 900	45 400
Recent Lau Basin(CLSC)[†]		4 700–15 000	10 500–48 500
EPR Crestc[‡]		5 800–28 000	11 000–82 000
EPR Leg 92[§]		130–35 100	900–102 900
EPR (Recent Ridge Proximal)[∥]		36 000–86 000	120 000–240 000

*Hodkinson & Cronan (1995)
[†]Cronan *et al.* (1986)
[‡]Dymond & Veeh (1975)
[§]Barrett *et al.* (1987)
[∥]Lyle *et al.* (1986)

maximum non-detrital Mn and Fe accumulation rates for Sites 836, 837, 838 and 839 studied here, along with comparative data for Sites 834 and 835 further north in the basin, and for recent sediments (short cores) proximal to the CLSC and the EPR.

In comparison with MOR environments, non-detrital Mn and Fe accumulation rates at Sites 836, 837, 838 and 839 fall within the range of values reported for hydrothermal sediments derived from the EPR.

Mean and maximum non-detrital Mn and Fe accumulation rates are similar at Sites 837, 838 and 839, but markedly higher at Site 836 (Table 1). Such a pattern is in accord with Site 836 being nearer to the source of hydrothermal discharge at the ELSC than Sites 837, 838 and 839 which are located further to the west (Fig. 1).

In comparison with non-detrital Mn and Fe accumulation rates for older (around 3.2 Ma) hydrothermal sediments associated with ELSC propagation past Sites 834 and 835 in the north of the basin, mean values are lower at Site 836 and considerably lower at Sites 837, 838 and 839 (Table 1). Such data suggest that the hydrothermal flux

associated with the ELSC during its earlier propagation in the north of the basin was significantly higher than during its more recent propagation further south.

In comparison with younger (≤1.0 Ma) sediments from Sites 834 and 835 associated with the propagation of the CLSC past Sites 834 and 835 and recent (surface) Lau Basin sediments from the latitude of Sites 834 and 835, both of which have received plume fall-out from the CLSC, both mean and maximum non-detrital metal accumulation rates at Sites 836, 837, 838 and 839 are higher, particularly those at Site 836 (Table 1). This supports previous work on hydrothermal sediments associated with the Lau Basin spreading ridges (Riech *et al.* 1990; Hodkinson & Cronan 1995) that suggests a higher plume fallout flux from the ELSC, both present and past, than from the CLSC.

Summary and conclusions

Hydrothermal inputs to Lau Basin ODP sediment sections at Sites 836, 837, 838 and 839 have been assessed with respect to crustal province and proposed models of tectonic evolution of the Lau Basin. Non-detrital Mn+Fe accumulation rates at these sites have been compared with those at Sites 834 and 835 further north in the basin and on the EPR. The results are as follows:

(1) Sites 836 and 837 show a pattern of non-detrital Mn+Fe accumulation rates that generally decrease up-hole from basement, consistent with their location on ELSC-generated crust.

(2) Non-detrital Mn+Fe accumulation rates at Site 839 increase upwards from basement and then decrease, confirming its suggested location on western Lau Basin attenuated crust. An observed 'time lag' between ELSC propagation past the site and the peak in hydrothermal input may be due to a delay between ridge formation and either the initiation of hydrothermal systems on the ridge and/or their full development. Such a pattern is similar to that previously described from Sites 834 and 835 in the north of the basin.

(3) Where time-equivalent sediment sections have been recovered at Sites 838 and 839, accumulation rate patterns are similar, and the lack of recovered sediments down to basement at Site 838 probably accounts for the lack of an observed period of reduced hydrothermal input in the oldest sediments there. Available non-detrital Mn+Fe accumulation rate patterns are, however, not at variance with the location of Site 838 on western Lau Basin attenuated crust.

(4) Mean non-detrital Mn and Fe accumulation rates show that the hydrothermal flux to Site 836 was higher than that at Sites 837, 838 and 839. This probably reflects the closer proximity of this site to the ELSC than the other sites. Accumulation rates are within the range of those reported on the EPR.

(5) Hydrothermal flux to the sediments at Sites 836, 837, 838 and 839 is lower than that associated with the propagation of the ELSC past the latitude of Sites 834 and 835 further north. Such data suggest that the hydrothermal flux from the ELSC was significantly greater during the early phase of its southward propagation than during its more recent history.

(6) Non-detrital metal accumulation rates at Sites 836, 837, 838 and 839 are higher than in sediments ≤ 1.0 Ma associated with CLSC propagation past Sites 834 and 835 and recent Lau Basin sediments receiving plume fallout from the CLSC. This supports previous studies that indicated a higher plume fallout flux from the ELSC than from the CLSC.

This work was funded by NERC (UK), under ODP Special Topic Research Grant GST/02/442 to D.S.C. We thank Mrs K. St Clair-Gribble for help with chemical analysis, and L. M. Parson and R. G. Rothwell (Southampton Oceanography Centre) for their helpful discussions during the research grant. Thanks are owed to the master, crew and technical staff of the *JOIDES Resolution* for their efforts during Leg 135, in which R.A.H. participated, and to the Leg 135 Scientific Party.

References

BAKER, E. T. 1994. A 6-year time series of hydrothermal plumes over the Cleft segment of the Juan de Fuca Ridge. *Journal of Geophysical Research*, **99**, (B3), 4889–4904.

BARRETT, T. J., TAYLOR, P. N. & LUGOWSKI, J. 1987. Metalliferous sediments from DSDP Leg 92: The East Pacific Rise transect. *Geochimica et Cosmochimica Acta*, **51**, 2241–2253.

CHESTER, R. & HUGHES, M. J. 1967. A chemical technique for the separation of ferromanganese minerals, carbonate minerals and adsorbed trace elements from pelagic sediments. *Chemical Geology*, **2**, 249–262.

CRONAN, D. S. 1976. Basal metalliferous sediments from the eastern Pacific. *Bulletin of the Geological Society of America*, **87**, 928–934.

—— 1983. Metalliferous sediments in the CCOP/SOPAC region of the Southwest Pacific, with particular reference to geochemical exploration for the deposits. *CCOP/SOPAC Technical Bulletin*, **4**.

——, MOORBY, S. A., GLASBY, G. P., KNEDLER, K. E., THOMPSON, J. & HODKINSON, R. A. 1984. Hydrothermal and volcaniclastic sedimentation on the Tonga-Kermadec Ridge and its adjacent marginal basins. *In*: KOKELAAR, B. P. & HOWELLS,

M. F. (eds) *Marginal Basin Geology*. Geological Society, London, Special Publications, **16**, 137–149.

——, HODKINSON, R. A., HARKNESS, D. D., MOORBY, S. A. & GLASBY, G. P. 1986. Accumulation rates of hydrothermal metalliferous sediments in the Lau Basin, S.W. Pacific. *Geo-Marine Letters*, **6**, 51–56.

DYMOND, J. & ROTH, S. 1988. Plume dispersed hydrothermal particles: A time-series record of settling flux from the Endeavor Ridge using moored sensors. *Geochimica et Cosmochimica Acta*, **52**, 2525–2536.

—— & VEEH, H. H. 1975. Metal accumulation values in the southeast Pacific and the origin of metalliferous sediments. *Earth and Planetary Science Letters*, **28**, 13–22.

FOUQUET, Y., VON STACKELBERG, U., CHARLOU, J. L., ERZINGER, J., HERZIG, P. M., MÜHE, R. & WIEDICKE, M. 1993. Metallogenesis in back-arc environments: The Lau Basin example. *Economic Geology*, **88**, 2154–2181.

GERMAN, C. R., BAKER, E. T. & KLINKHAMMER, G. 1995. Regional setting of hydrothermal activity. *In*: PARSON, L. M., WALKER, C. L. & DIXON, D. R. (eds) *Hydrothermal Vents and Processes*. Geological Society, London, Special Publications, **87**, 3–15.

HAWKINS, J. W., PARSON, L. M. & ALLAN, J. F. 1994. Introduction to the scientific results of Leg 135: Lau Basin–Tonga Ridge drilling transect. *Proceedings, ODP, Scientific Results*, **135**, 3–5.

HODKINSON, R. A. & CRONAN, D. S. 1991. Geochemistry of recent hydrothermal sediments in relation to tectonic environment in the Lau Basin, southwest Pacific. *Marine Geology*, **98**, 353–366.

—— & —— 1994. Variability in the hydrothermal component of the sedimentary sequence in the Lau back-arc Basin (Sites 834–839). *Proceedings, ODP, Scientific Results*, **135**, 75–86.

—— & —— 1995. Hydrothermal sedimentation at ODP Sites 834 and 835 in relation to crustal evolution of the Lau Backarc Basin. *In*: PARSON, L. M., WALKER, C. L. & DIXON, D. R. (eds) H*ydrothermal Vents and Processes*. Geological Society, London, Special Publications, **87**, 231–248.

——, ——, GLASBY, G. P. & MOORBY, S. A. 1986. Geochemistry of marine sediments from the Lau Basin, Havre Trough, and Tonga-Kermadec Ridge. *New Zealand Journal of Geology and Geophysics*, **29**, 335–344.

HOWARTH, R. J. 1983. *Handbook of Exploration Geochemistry (Vol. 2): Statistics and Data Analysis in Geochemical Prospecting*. Elsevier, New York.

KLINKHAMMER, G. & HUDSON, A. 1986. Dispersal patterns for hydrothermal plumes in the South Pacific using manganese as a tracer. *Earth and Planetary Science Letters*, **79**, 241–249.

LEINEN, M. & PISIAS, N. G. 1984. An objective technique for determining end-member compositions and for partitioning according to their sources. *Geochimica et Cosmochimica Acta*, **48**, 47–62.

LYLE, M. W., OWEN, R. M. & LEINEN, M. 1986. History of hydrothermal sedimentation at the EPR, 19°S. *Initial Reports of the DSDP*, **92**, 585–596.

MALAHOFF, A. & FALLOON, T. 1991. *Preliminary Report of the Akademik Mstislav Keldush/Mir Cruise 1990. Lau Basin Leg. (May 7–21)*. Suva, South Pacific Applied Geoscience Commission, unpublished cruise report **137**.

McMURTRY, G. M., DE CARLO, E. H. & KIM, K. H. 1991. Accumulation rates, chemical partitioning, and Q-mode factor analysis of metalliferous sediments from the North Fiji Basin. Marine Geology, **98**, 271–295.

MURPHY, E., McMURTRY, G. M., KIM, K. H. & DE CARLO, E. H. 1991. Geochemistry and geochronology of a hydrothermal ferromanganese deposit from the North Fiji Basin. *Marine Geology*, **98**, 297–312.

PARSON, L. M. & HAWKINS, J. W. 1994. Two-stage ridge propagation and the geological history of the Lau backarc basin. *Proceedings, ODP, Scientific Results*, **135**, 819–828.

——, PEARCE, J. A., MURTON, B. J. & HODKINSON, R. A. 1990. Role of ridge jumps and ridge propagation in the tectonic evolution of the Lau back-arc basin, southwest Pacific. *Geology*, **18**, 470–473.

——, HAWKINS, J. W. & HUNTER, P. M. 1992a. Morphotectonics of the Lau Basin seafloor – Implications for the opening history of backarc basins. *Proceedings, ODP, Initial Reports*, **135**, 81–82.

——, ——, ALLAN, J. ET AL. 1992b. *Proceedings, ODP, Initial Reports*, **135**.

——, ROTHWELL, R. G. & MACLEOD, C. J. 1994a. Tectonics and sedimentation in the Lau Basin (Southwest Pacific). *Proceedings, ODP, Scientific Results*, **135**, 9–22.

——, HAWKINS, J. W., ALLAN, J. ET AL. 1994b. *Proceedings, ODP, Scientific Results*, **135**.

RIECH, V. 1990. Calcareous ooze, volcanic ash, and metalliferous sediments in the Quaternary of the Lau and North Fiji Basins. *Geologische Jahrbuch*, **D92**, 109–162.

——, MARCHIG, V., SUNKEL, G. & WEISS, W. 1990. Hydrothermal and volcanic input in sediments of the Lau Back-Arc Basin, SW Pacific. *Marine Mining*, **9**, 183–203.

ROTHWELL, R. G., BEDNARZ, U., BØE, R., CLIFT, P., HODKINSON, R. A., LEDBETTER, J. K., PRATT, C. E. & SOAKAI, S. 1994a. Sedimentation and sedimentary processes in the Lau Backarc Basin: Results from ODP Leg 135. *Proceedings, ODP, Scientific Results*, **135**, 829–842.

——, WEAVER, P. P. E., HODKINSON, R. A., PRATT, C. E., STYZEN, M. J. & HIGGS, N. C. 1994b. Clayey nannofossil ooze turbidites and hemipelagites at Sites 834 and 835 (Lau Basin, SW Pacific). *Proceedings, ODP, Scientific Results*, **135**, 101–130.

TAYLOR, B., ZELLMER, K., MARTINEZ, F. & GOODLIFFE, A. M. 1995. Tectonic evolution of the Lau Basin. *EOS: Transactions, American Geophysical Union*, **76**(46), 595 (Abst.).

THOMPSON, M. & WALSH, J. N. 1989. *The Handbook of Inductively Coupled Plasma Spectrometry* (2nd edn). Blackie, Glasgow, 156–160.

VON DAMM, K. L., OOSTING, S. E., KOZLOWSKI, R., BUTTERMORE, L. G., COLODNER, D. C., EDMONDS, H. N., EDMOND, J. M. & GREBMEIER, J. M. 1995. Evolution of East Pacific Rise hydrothermal vent fluids following a volcanic eruption. *Nature*, **375**, 47–50.

WALTER, P., STOFFERS, P., GLASBY, G. P. & MARCHIG, V. 1990. Major and trace element geochemistry of Lau Basin sediments. *Geologische Jahrbuch, Reihe D*, **92**, 163–188.

Collision-related break-up of a carbonate platform (Eratosthenes Seamount) and mud volcanism on the Mediterranean Ridge: preliminary synthesis and implications of tectonic results of ODP Leg 160 in the Eastern Mediterranean Sea

A. H. F. ROBERTSON[1], K.-C. EMEIS, C. RICHTER,

M.-M. BLANC-VALLERON, I. BOULOUBASSI, H-J. BRUMSACK, A. CRAMP, G. J. DI STEFANO,
R. FLECKER, E. FRANKEL, M. W. HOWELL, T. R. JANECEK, M.-J. JURADO, A. E. S. KEMP,
I. KOIZUMI, A. KOPF, C. O. MAJOR, Y. MART, D. F. C. PRIBNOW, A. RABAUTE, A. P. ROBERTS,
J. RULLKÖTTER, T. SAKAMOTO, S. SPEZZAFERRI, T. S. STAERKER, J. S. STONER, B. M. WHITING &
J. M. WOODSIDE

[1] *Department of Geology and Geophysics, University of Edinburgh, West Mains Road,
Edinburgh EH9 3JW, UK*

Abstract: Drilling of the Eratosthenes Seamount south of Cyprus documented incipient collision of the African and Eurasian plates. The oldest sediments recovered, mid?-Cretaceous shallow-water limestones, are overlain by Upper Cretaceous to Lower Oligocene pelagic carbonates, with several hiatuses. Following uplift, a carbonate platform was established in the Miocene; Eratosthenes was then below eustatic sea level during the Messinian desiccation crisis. The platform subsided to bathyal depths during the Lower Pliocene, associated with localized breccia deposition. Further subsidence occurred in Late Pliocene–early Quaternary, coeval with strong surface uplift of southern Cyprus. Subsidence and break-up of Eratosthenes was achieved by a combination of flexural loading and normal faulting. In addition, the Milano and Napoli mud volcanoes were drilled on the northern flank of the Mediterranean Ridge accretionary complex, south of Crete. A mainly extrusive, sedimentary origin is indicated. Multiple debris flows include clasts of sandstone and limestone of at least partly Miocene age. Both mud volcanoes are dated as >1 Ma old and have been active episodically. Hydrocarbon gas is associated with both mud volcanoes, while methane hydrates (clathrates) exist locally at Milano. The driving force of mud volcanism is overpressuring caused by incipient plate collision. Messinian evaporites may have acted as a localized seal. Material escaped through a zone of backthrusting against rigid Cretan crust to the north.

This paper's aim is to present a preliminary summary and synthesis of the tectonic results of drilling in the Eastern Mediterranean Sea during Ocean Drilling Program (ODP) Leg 160 (April, May, 1995), and to integrate these results within a regional tectonic context. Some of the results have implications for fundamental processes, including the preservation of crustal fragments in subduction/accretion complexes, and the role of mud volcanism in accretionary prisms.

The drilling during Leg 160 investigated two tectonic aspects. The first was concerned with the process of break-up of a carbonate platform, the Eratosthenes Seamount, at a subducting plate boundary, represented by the Cyprus active margin

(Sites 965–968; Fig 1). The second objective concerned the origin of two mud volcanoes, the Milano (Site 970) and Napoli (Site 971) mud domes, located on the northern part of the Mediterranean Ridge accretionary complex, south of Crete (Fig. 1).

The Eastern Mediterranean Sea is an ideal location for study of crustal processes related to the collision of continental plates. The northern, passive margin of the African plate (i.e. Gondwana) was rifted in the Early Mesozoic, giving rise to an irregular pattern of embayments and promontories. Continental fragments were rifted from Gondwana and then drifted northwards into the Tethys Ocean. The opposing, northern margin of Neotethys (i.e.

ROBERTSON, A. H. F., EMEIS, K.-C., RICHTER, C. *ET AL.* 1998. Collision-related break-up of a carbonate platform (Eratosthenes Seamount) and mud volcanism on the Mediterranean Ridge: preliminary synthesis and implications of tectonic results of ODP Leg 160 in the Eastern Mediterranean Sea. *In*: CRAMP, A., MACLEOD, C. J., LEE, S. V. & JONES, E. J. W. (eds) *Geological Evolution of Ocean Basins: Results from the Ocean Drilling Program.* Geological Society, London, Special Publications, **131**, 243–271.

the southern margin of Eurasia) was initially located well to the north, extending from northern Greece through the Pontides of northern Turkey to the Caucasus. Africa and Eurasia converged in the Late Mesozoic–Early Tertiary, as a result of northward subduction. As northward subduction progressed, a number of continental fragments (previously rifted from Gondwana) were accreted to the active southern margin of Eurasia. The overall subduction front migrated southwards with time. By the Miocene the front was located in the vicinity of the present convergent plate boundary that extends across the Eastern Mediterranean Sea south of Crete, and then south of Cyprus, to connect with the Tethys suture zone further east (Fig. 1). At the present time, the Eastern Mediterranean Sea can be considered as a final remnant of the southerly Mesozoic Neotethys that is now in its final stages of closure associated with diachronous collision of the African and Eurasian plates.

Collisional emplacement of a carbonate platform: the Eratosthenes Seamount

The Eratosthenes Seamount is located in the easternmost Mediterranean Sea south of Cyprus and north of the Nile cone, with the passive margin

of the Levant to the east (Fig. 1). The term 'seamount' has given rise to some confusion in the past as to many it implies an origin mainly as extrusive igneous rocks, whereas drilling during Leg 160 has shown that at least the upper part is composed mainly of limestones without igneous rocks. For this reason the term 'Eratosthenes platform' is preferred here.

Drilling during Leg 160 followed on from the earlier results obtained during Deep Sea Drilling Project (DSDP) Leg 13 (Ryan *et al.* 1973) and Leg 42 (Hsü *et al.* 1978). In particular, drilling of the Florence Rise, west of Cyprus, during Leg 42 revealed a deep-sea Plio-Quaternary succession underlain by Late Miocene sediments, including evaporites. However, at that stage of research the plate tectonic setting of the easternmost Mediterranean remained obscure, prior to obtaining more extensive seismic reflection and refraction data.

The easternmost Mediterranean is a part of the wider Eastern Mediterranean Sea, formed by rifting of the northern margin of Gondwana in the Triassic (Garfunkel & Derin 1984). The Eratosthenes platform is generally envisaged as a carbonate platform constructed on rifted continental crust. Oceanic crust was formed in the easternmost Mediterranean by the Late Triassic and is repre-

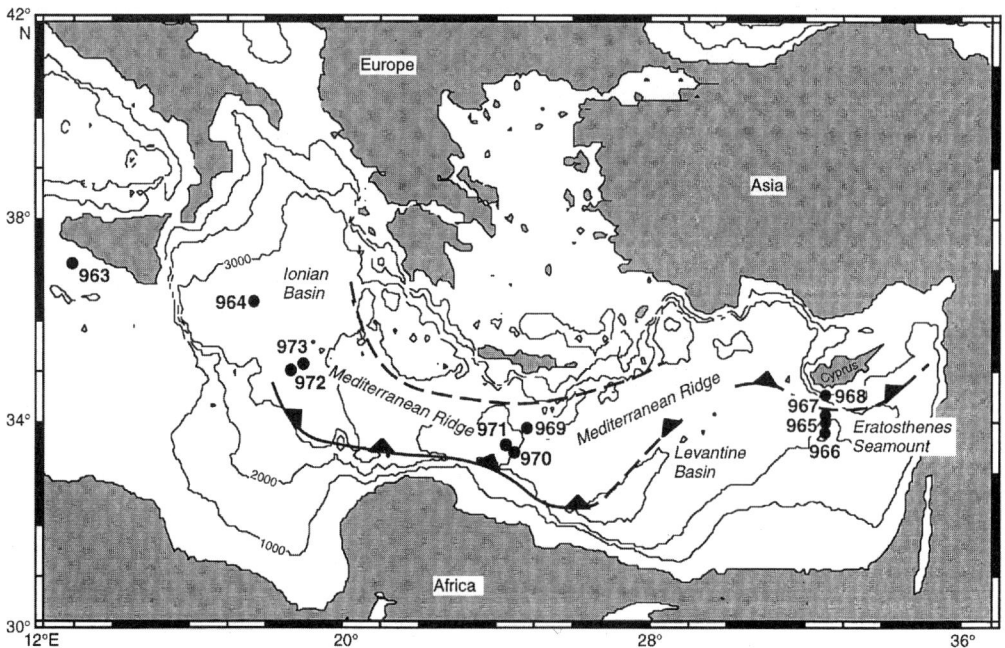

Fig. 1. Outline map of the Mediterranean region showing the main tectonic features and the locations of the sites drilled during Leg 160, specifically the Eratosthenes platform and the mud volcanoes south of Crete.

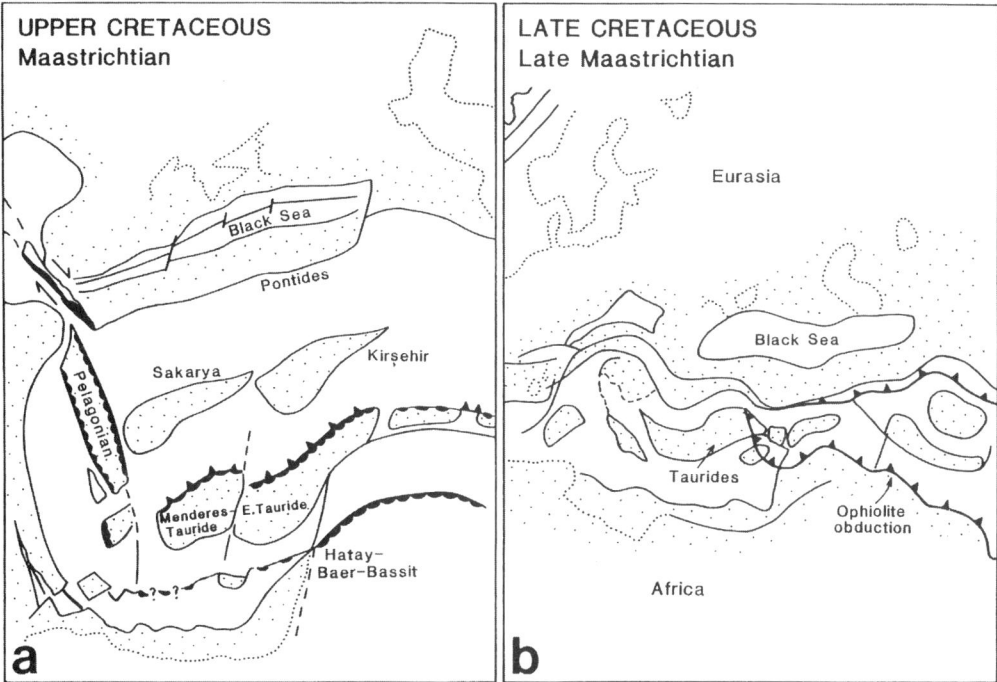

Fig. 2. Plate tectonic sketch maps showing alternative reconstructions of the Eastern Mediterranean during the Upper Cretaceous: (**a**) with the Eastern Mediterranean as a southerly Neotethyan Ocean basin, bordered by microcontinents rifted from Gondwana to the north. Ophiolites formed above subduction zones and were emplaced onto the Arabian shelf to the east and onto microcontinents. The Cyprus ophiolite represents part of the southerly Neotethys that was not obducted in the Upper Cretaceous but remained within the ocean; it is only now being emplaced related to collision with the continental fragment represented by the Eratosthenes platform (from Robertson & Dixon 1984). (**b**) In this case an ocean basin is envisaged to have formed in the Eastern Mediterranean, but all the ophiolites, including the Troodos, are seen as having been thrust from far to the north (simplified from Dercourt *et al.* 1993). Recent data, including the Leg 160 results favour model (a).

sented by 'accreted' fragments in SW Cyprus (Mamonia Complex), SW Turkey (Antalya Complex) and northern Syria (Baer-Bassit) (e.g. Robertson *et al.* 1991*a*). The easternmost Mediterranean oceanic basin formed the most southerly of a number of oceanic stands, separated by continental slivers rifted from Gondwana, including the Tauride carbonate platforms of southern Turkey. In the Cretaceous, relative motion of the African and Eurasian plates became convergent, coupled with opening of the south Atlantic (Livermore & Smith 1984). In response to regional plate convergence, subduction was initiated within the southerly Eastern Mediterranean ocean basin (and more northerly basins also), leading to genesis of the Troodos ophiolite by spreading above a subduction zone (see Robertson & Xenophonotos 1993; Fig. 2a). The subduction zone probably dipped northward and was located south of Cyprus. During the Late Cretaceous–Early Eocene, much of

Cyprus underwent anticlockwise rotation as a discrete microplate (Clube & Robertson 1986; Morris *et al.* 1990). Northward subduction of remaining oceanic crust in the easternmost Mediterranean to the south of Cyprus probably began at the beginning of the Miocene and the North African plate, including the Eratosthenes platform, began its final northward drift towards Cyprus (Dewey & Şengör 1979; Dercourt *et al.* 1993). During this time, southern Cyprus was located on the leading edge of what was, by then, effectively part of the Eurasian plate.

The Eratosthenes platform is a large subrectangular, elevated feature in the Mediterranean Sea south of Cyprus. Geophysical data indicate the presence of a regional magnetic anomaly, that is more extensive than the Eratosthenes platform. Previous work had suggested that the 'seamount' was in the process of collision with the Cyprus active margin to the north (Ben-Avraham *et al.*

1976; Robertson 1990; Robertson *et al.* 1991*b*; Woodside 1991; Kempler 1993), but it was the TREDMAR cruise that produced the key seismic evidence of collision and underthrusting (Limonov *et al.* 1994; Robertson *et al.* 1994, 1995*a,b*; Fig. 3). There is some evidence of collapse and under-thrusting also to the south, beneath the Levantine Basin (Limonov *et al.* 1994). Dredging results also suggested that Eratosthenes includes some lime-stone (Krasheninnikov *et al.* 1994) and this was confirmed by drilling during Leg 160.

Any collision-related tectonic hypothesis for the emplacement of the Eratosthenes platform could only be tested by drilling: the test had to involve only relatively shallow penetration (i.e. a few hundred metres), as time and safety considerations precluded deeper penetration. Key questions were the nature of the Eratosthenes platform, and the timing and processes of its subsidence related to collision with the Cyprus active margin to the north.

Suture zones in Greece, Turkey and other Tethyan areas include numerous deformed car-bonate platforms. These platforms commonly overlie rifted continental crust, although some may overlie oceanic crust (i.e. volcanic seamounts). Examples include the Mesozoic Tauride carbonate platforms of southern Turkey and counterparts in Greece. An important question is how these platforms came to be incorporated into the Tethyan orogenic collage. Drilling of the Eratosthenes platform also had the potential to shed light on this problem.

Summary of drilling results

A transect of four holes was drilled, from south to north: one on the northern crestal area of the Eratosthenes platform (Site 966), one on the upper northern flanks (Site 965), one at the base of the northern slope (Site 967), and one at the base of the opposing slope of the Cyprus active margin (Site 968) (Fig. 4). Neither time nor site survey data allowed drilling of the interesting southern margin of the Eratosthenes platform and related basin bordering the Levantine Basin. Details of the recovery at each of the four sites occupied are given in Emeis *et al.* (1996) and Robertson *et al.* (1995*a*). The main results are summarized here in terms of four time slices: (i) Mid-Cretaceous–Early Tertiary; (ii) Messinian (latest Miocene); (iii) Early Pliocene; and (iv) Plio-Pleistocene, Fig. 5.

Mid-Cretaceous to Early Tertiary

The evolution of the Eratosthenes platform, as recorded by drilling (at Site 967), commenced with the accumulation of shallow-water carbonates that are assumed to be of mid-Cretaceous age, based on micropalaeontological data and lithological cor-relation with similar limestones onshore in the Levant to the east (Mart *et al.* 1997). This unit is overlain by Late Cretaceous deep-water carbonates of Coniacian to Maastrichtian age, dated by cal-careous nannofossils and planktonic foraminifera, followed by a hiatus until Early Tertiary time. Initial shipboard results suggested that these limestones were Middle Eocene in age. However, post-cruise studies indicate that Late Eocene to Early Oligocene planktonic foraminifera are present, mixed with abundant Middle Eocene foraminifera (S. Spezzaferri, pers. comm. 1996). Rare Palaeocene planktonic foraminifera are also present. Similar pelagic carbonates with reworked Middle Eocene foraminifera were recovered further south, at Site 966. Sedimentary structures and the nature of benthic foraminifera indicate a bathyal, relatively deep-water environment, with pelagic

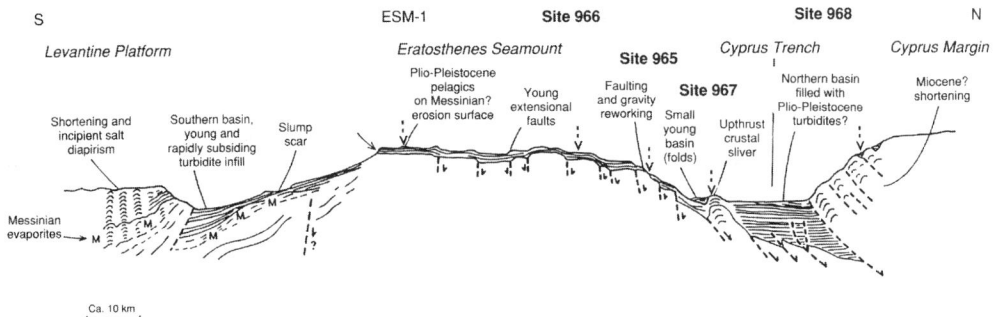

Fig. 3. Tectonic setting of the Eratosthenes platform as envisaged based on the 1993 TREDMAR site survey data (Limonov *et al.* 1994; Robertson *et al.* 1995*a*).

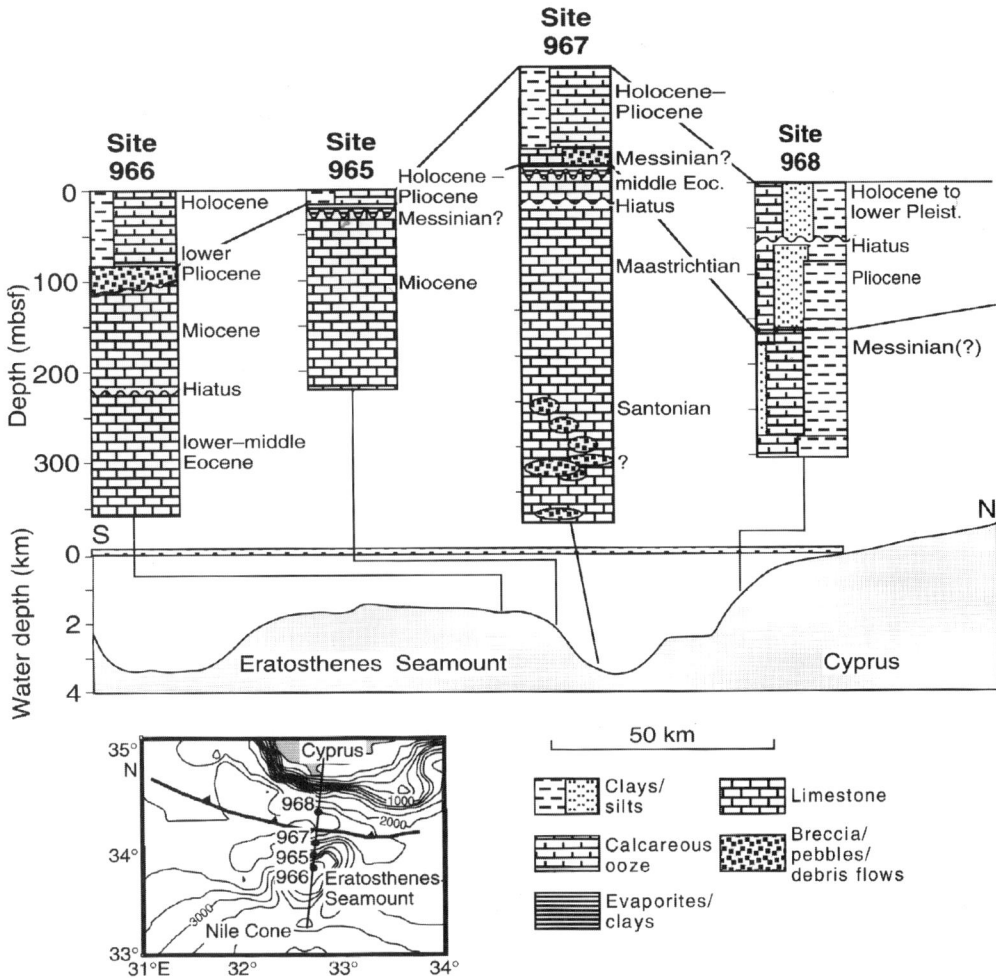

Fig. 4. Summary of the sequences recovered from the Eratosthenes Seamount drilled during Leg 160 (see text for explanation).

carbonate deposition and little terrigenous sediment or redeposited carbonates. A possible explanation of the mixed faunas is that extensive reworking took place on a submerged carbonate platform.

After the Late Eocene–Early Oligocene there was a hiatus; Eratosthenes was then tectonically uplifted at least 1 km, based on palaeo-depth estimates, followed by deposition of shallow-water carbonates. Similar carbonates were deposited at the two most southerly sites (Sites 965 and 966). The limestones are provisionally dated as Miocene, based on a benthic foraminiferal assemblage (I. Premoli-Silva, pers. comm. 1995). These limestones record alternations of relatively high- and relatively low-energy settings, typical of lagoonal and near-reef environ-

ments. Similar Miocene shallow-water and reef-related carbonates are known in other areas of the Eastern Mediterranean including Israel, southern Turkey and Cyprus (e.g. Flecker et al. 1995; Follows & Robertson 1990; Follows et al. 1996).

At the northernmost of the three Eratosthenes sites (Site 967), Upper Cretaceous–Middle Eocene pelagic carbonates are overlain by Messinian fine-grained sediments, up to several tens of metres thick. These sediments were only recognized and dated during post-cruise studies (S. Spezzaferri, pers. comm. 1996). This interval includes ostracods (i.e. *Cyprideis pannomica, Ammonia tepida, A. Beccarii*) and some other shallow-water benthic foraminifera (including *Elphidium* sp.). These

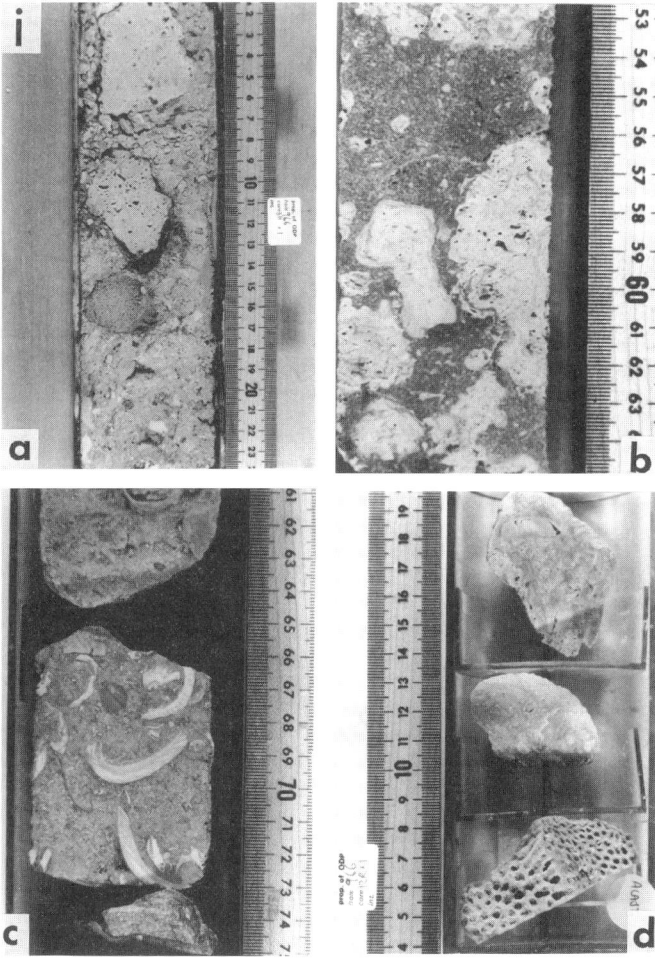

Fig. 5. Photographs of cores from sites S to N across the Eratosthenes platform and on the Cyprus margin (all photos taken by A. H. F. Robertson). (**i**) Crest of the Eratosthenes platform (Site 966). (**a**) Early Pliocene matrix-supported breccias, Hole 966X, Core 3H, Section 1, 2–24 cm. (**b**) Calcareous algal oncolites, Hole 966R, Core 8H, Section 1, 53–65 cm. (**c**) Large bivalves in a bioclastic matrix, Hole 966R, Core 10R, Section 2, 61–75 cm. (**d**) Fragments of bioclastic limestone and coral (lower), Hole 966R, Core 12R, Section 1, 4–20 cm. (b–d are Early Miocene? in age).

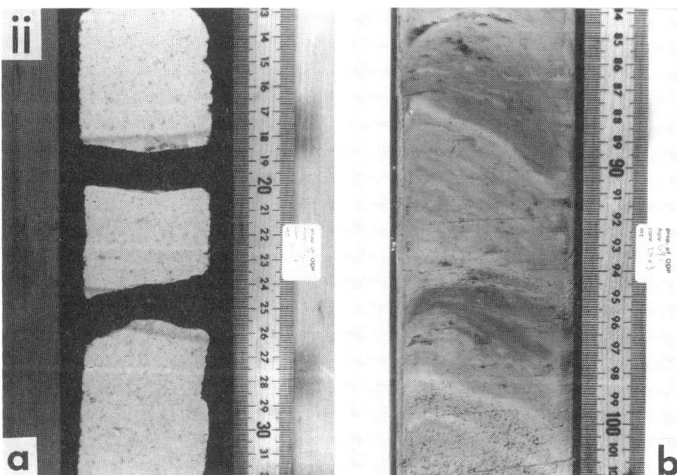

Fig. 5. (cont.) (**ii**) Upper northern flanks of Eratosthenes platform (Site 965). (**a**) Coarse bioclastic limestone composed of shallow-water bioclasts, Hole 955R, Core 20R, Section 1, 13–32 cm (Miocene?). (**b**) Slumped Early Pliocene deep-sea mud, Hole 955H, Core 2H, Section 3, 84–102 cm.

Fig. 5. (cont.) (**iii**) Lower northern flanks of Eratosthenes platform (Site 967). (**a**) Reworked pelagic carbonate (upper), with an infilled neptunian fissure (lower), Hole 967E, Core 6R, Section 3, 4–20 cm. (**b**) Normal faults in pelagic limestone, Hole 967E, Core 14R, Section 1, 103–118 cm. (**c**) Laminated shallow-water carbonate of mid-Cretaceous? age, Hole 967E, Core 41R Section 1, 5-16 cm. (**d**) Tectonically brecciated limestone near the maximum depth drilled (upper two pieces), Hole 967E, Core 46R, Section 1, 3–18 cm.

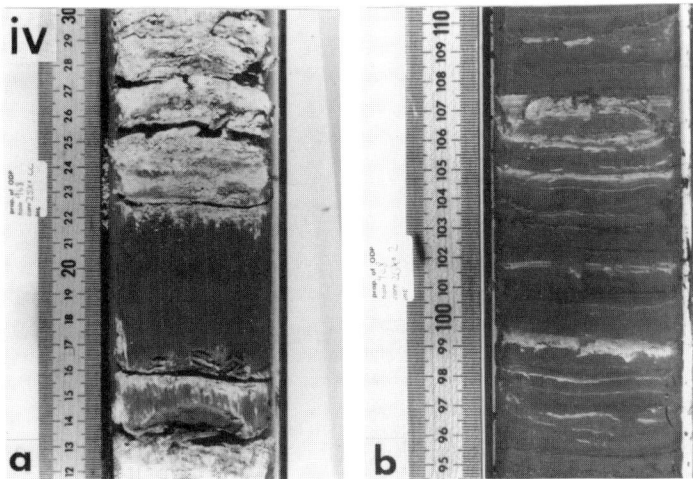

Fig. 5. (cont.) (**iv**) Lower Cyprus slope. (**a**) Detrital sand-sized gypsum (white) within dark silty clays; Hole 9678, Core 25X, core catcher, 12–30 cm. (**b**) Laminated calcareous muds without coeval biota, pale layers are detrital fine-grained gypsum, Hole 968A, Core 26X, Section 2, 95–110 cm.

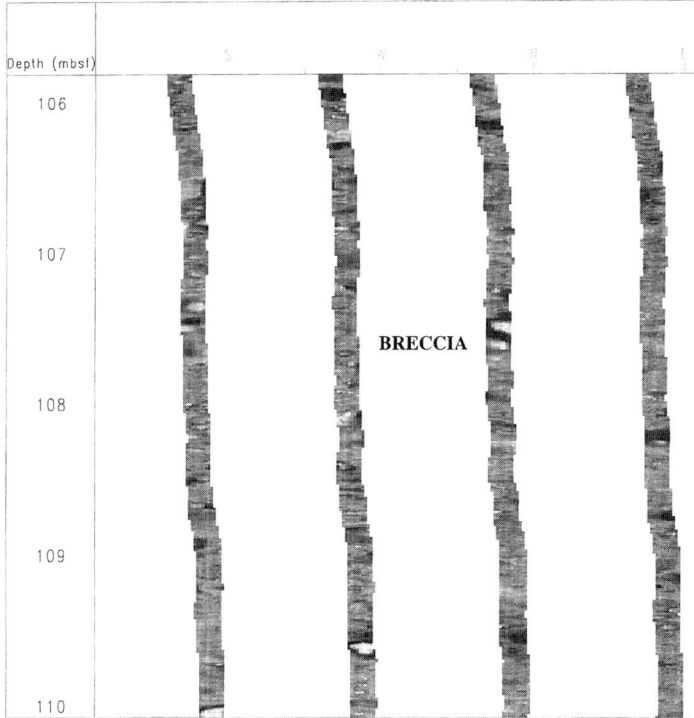

(a)

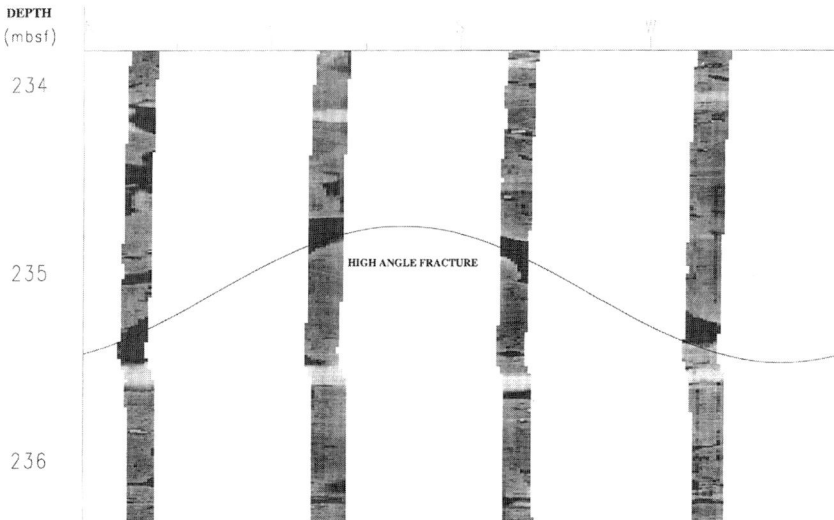

(b)

Fig. 6. Formation MicroScanner (FMS) images. (**a**) FMS image of Breccia Unit at Site 966. The 'lumpy' appearance of the record is interpreted to indicate the presence of clasts. The pale lenses are interpreted as limestone clasts, while the grey lenses are mainly hemipelagic sediments. (**b**) High-angle fault at Site 967. High-angle faults observed in the cores are of extensional type. (**c**) Evidence for the presence of evaporites (i.e. gypsum) at Site 967, that were not recovered. Note the distinctive bright image, in contrast to the darker, more argillaceous, sediments above and below. See text for explanation (M.-J. Jurado, unpublished data).

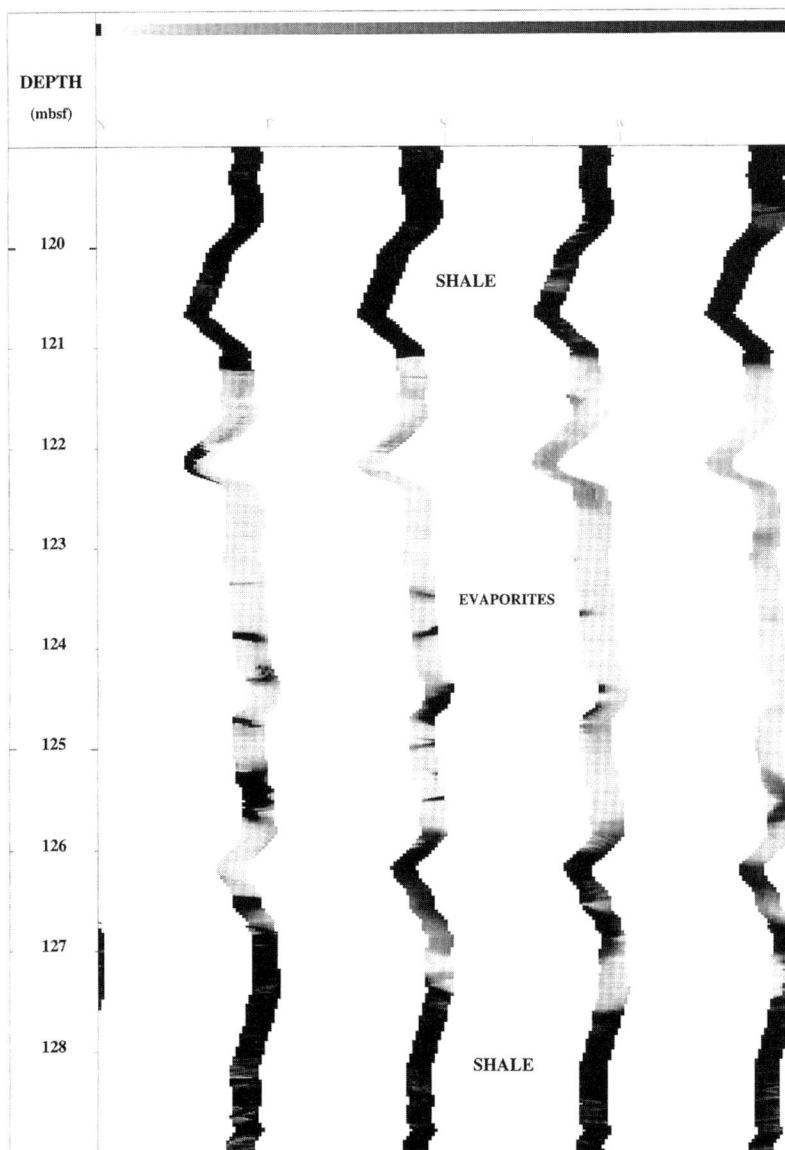

(c)

forms are mixed with reworked Eocene, mid-Miocene and, rarely, Cretaceous planktonic foraminifera. The extent of reworking decreases upwards (i.e. in part of Cores 14H, 15H and 16H). The brackish 'lago-mare' Messinian facies then pass upwards, without any observable hiatus, into fine-grained, deep-water sediments of Early Pliocene age. Reworked microfossils of Middle Eocene, Oligocene and Miocene–Early Pliocene age are present within pelagic carbonates at the

base of the Early Pliocene succession. The history of this site during the mid-Tertiary is thus debatable. One possibility is that Miocene shallow-water carbonates were deposited, as at Sites 965 and 966, and were later eroded. Another is that Miocene shallow-water carbonates were never deposited there, but that pelagic deposition took place instead, followed by removal of most of these sediments. Perhaps the northernmost Eratosthenes site was located on a carbonate slope, where only reduced pelagic deposition took place in Eocene to Early Pliocene time, combined with extensive reworking of pelagic biota.

Messinian

The Mediterranean underwent its well-known salinity crisis in the latest Miocene, as documented during DSDP Legs 13 and 42 (Hsü *et al.* 1973, 1978). Layers of evaporite, up to several kilometres thick, are inferred to exist beneath the deep Mediterranean basin, based mainly on recognition of a clear 'M' reflector on seismic profiles (Ryan *et al.* 1971; Ross & Uchupi 1977; Woodside 1977; Chaumillon & Mascle 1995). The 'M' reflector pinches out upslope on the south and southwest flanks of the Eratosthenes platform. The top of the platform was therefore above the level of the surrounding Mediterranean deep basin during the Messinian, although the flanks may have remained largely submerged. A prominent reflector, thought to be an erosion surface, was noted at the base of the inferred Pliocene sequence, based on site survey studies (Limonov *et al.* 1994).

Drilling during Leg 160 confirmed that no thick evaporites are present on the Eratosthenes platform (Site 966), or on its northern flank (Sites 965 and 967; Fig. 3). The Messinian is thin, or absent, in the Eratosthenes plateau area. However, at Site 965, several metres of reddish clays were recovered at the boundary between overlying sediments that contain Early Pliocene nannofossils and underlying shallow-water limestones. These clays contain a small fauna of ostracods of possible Messinian to Early Pliocene age, together with dolomite, aragonite and swelling clays. Also, at Site 967, near the base of an interval of muddy sediments that contains only reworked microfossils, an interval several metres thick (not recovered) was identified on geophysical borehole log records as gypsum (e.g. Fig. 6a). Evidence comes from greatly increased levels of porewater sulphate at this level and from geochemical logs. During the Messinian, the Eratosthenes platform formed a topographically raised area, marked by erosion and/or local accumulation of gypsum and ferruginous muds, possibly in small lagoons and/or lakes.

The sedimentary and tectonic history of the Cyprus margin was expected to be quite different from that of the Eratosthenes platform, as these two areas were located on opposing plates. Drilling at Site 968, near the foot of the slope of the Cyprus margin, did reveal such a contrast. Beneath a Plio-Pleistocene succession, more than 200 m of argillaceous sediments were recovered, interbedded with calcareous turbidites. The microfossils on the Cyprus margin are entirely reworked, but include

abundant brackish-water ostracods and the benthic foraminifer *Ammonia tepide*. Other clues to the depositional setting include the presence of abundant clay minerals (e.g. smectites) and kaolinite, together with scattered dolomite. Several thin layers of detrital selenitic gypsum sands were also recovered, together with thin layers (5 cm) of fine-grained alabastrine gypsum within clay. Lithoclasts of pelagic chalk and basalt and other constituents within abundant turbidites at Site 968 were derived from the Troodos ophiolite and its sedimentary cover in southern Cyprus (e.g. the Miocene Pakhna Formation; Robertson *et al.* 1991*b*). The turbidites are inferred to have accumulated in a large lake or inland sea of variable salinity. In general, lacustrine conditions are known to have prevailed widely in the late Messinian, especially in the Mediterranean (Hsü *et al.* 1978) and the circum-Black Sea region (Steininger & Rögl 1984). Site 968 was possibly located at the northern edge of a large hypersaline basin, within an existing subduction trench (*c*. 25 km wide), between Eratosthenes and Cyprus. Such trench sediments were possibly subducted in the Plio-Quaternary, explaining why there is now no evidence (based on seismic data) that thick evaporites were ever present between Cyprus and the Eratosthenes platform.

In summary, sediments of inferred Messinian age on the Eratosthenes platform (Sites 965, 967, 968) are thin or absent, whereas a thick (i.e. hundreds of metres) succession of probable Late Miocene lacustrine to hypersaline facies (i.e. the 'lago-mare'facies) is present at Site 968. This confirms a contrast between Cyprus located on the Eurasian plate and Eratosthenes on the Africa plate.

Early Pliocene

Drilling at all of the sites documents refilling of the Mediterranean at the end of the Messinian desiccation crisis (Hsü *et al.* 1978). However, conditions varied at individual locations. At Site 965, on the upper slope of the seamount, a thin (6 m) inferred Messinian interval contains rare, poorly preserved nannofossils of Late Miocene–Early Pliocene age (3.5–6 Ma). There are also possible palaeosols within an interval of reddish and brownish sticky clays. In this vicinity, the base of the Pliocene is marked by a sharply defined, flat seismic reflector, consistent with the presence of a sharp lithology contrast between Miocene lime-stone below and the weakly consolidated Pliocene fine-grained sediments above. A similar well-defined reflector is present at Site 967 to the north, where the interval between the Early Pliocene and Middle Eocene includes scattered clasts of chalk and limestone and, as noted earlier, reworked foraminifera of Eocene, Oligocene and Lower Miocene age. This is in addition to long-ranging forms that span Late Miocene, and/or Early Pliocene times. A source of both well-lithified and unlithified carbonates must thus have existed in the vicinity of Site 967.

About 65 m of interbedded nannofossil muds and matrix-supported breccias (i.e. debris flows) were recovered at the southernmost of the Eratosthenes sites (Site 966, Fig. 3) as also confirmed by Formation MicroScanner (FMS) data (Fig. 6b). In this area, the seismic character of the unconformity at the base of the Pliocene succession is much less distinct than in adjacent parts of the platform. Records were studied in four adjacent holes, allowing detailed comparisons. The matrix-supported breccias mainly formed by mass-flow processes in a tectonically unstable deep-sea setting. A similar, but less pronounced history of tectonic instability, with erosion of limestone and redeposition of clasts in a deep-marine setting, is documented at Site 965. It is clear that such tectonic instability and faulting were active at, and probably prior to, 4.5 Ma (i.e. Early Pliocene). Deposition reflects both the effects of initial tectonically triggered subsidence of the Eratosthenes platform and rapid sea-level rise at the end of the Messinian.

An entirely different Miocene–Pliocene transition is documented at the base of the Cyprus slope at Site 968, where the upward transition to the Early Pliocene is marked by an incoming of age-diagnostic nannofossils and planktonic foraminifera. Deposition of calcareous turbidites, derived from Cyprus, continued with no obvious break. Site 968 was already located below eustatic sea level during the Messinian salinity crisis; deep-marine sedimentation then ensued in the Early Pliocene with no observable break in deposition.

Pliocene–Pleistocene

At each of the four sites, the Plio-Pleistocene sequence records deep-marine accumulation. The Eratosthenes platform and its flanks were then at water depths ranging from 700 to 2900 m. The lowest part of the Early Pliocene succession already contains bathyal pelagic microfossils. Further deepening took place after the Late Pliocene, at least at Site 967, as benthic foraminifera there indicate marked upward deepening. Thin steel-grey mud turbidites in the uppermost part of the succession at the base of the Eratosthenes Seamount slope (Site 967) were deposited from dilute gravity flows containing mica, biotite and other minerals, probably derived from the Nile.

The Eratosthenes platform experienced tectonic disturbance during Late Pliocene–Pleistocene time. A hiatus of a maximum of 0.5 Ma exists within the Late Pliocene–Early Pleistocene time interval at Sites 965 and 967 on the Eratosthenes slope. At Site 966, extremely slow sedimentation rates (1–2 m Ma^{-1}) suggest that a depositional hiatus may be present in the early Pleistocene, although no micropalaeontological zones are missing. The hiatuses could reflect deformation of the Eratosthenes plateau area, as suggested by evidence of low-angle discordances and folding observable on high-resolution deep-tow seismic records (Limonov *et al.* 1994). The probable cause of this deformation is faulting of underlying pre-Pliocene–Pleistocene rocks.

Evidence of thrusting and break-up of Eratosthenes

Interpretation of seismic data indicates the presence of a small ridge at the base of the lower slope of the Eratosthenes Seamount at Site 967. This ridge is underlain by a northward-dipping fault zone (i.e. a reverse fault, or thrust). Prior to Leg 160 it was hypothesized that this ridge is in the process of tectonic detachment from the Eratosthenes Seamount (Limonov *et al.* 1994; Robertson *et al.* 1994, 1995*b*; Fig. 7). An objective of drilling at Site 967 was to test if there is any evidence of such a zone of reverse faulting. Indeed, it was hoped to drill through a décollement zone at a depth of *c.* 550 m. In the event, drilling penetrated to the maximum permitted depth (600 m), but did not reach a décollement. This was probably because previous depth estimates did not take account of the presence of thick limestones with relatively high seismic velocities, that made the reflector appear shallower than in reality. However, drilling terminated in a unit of tectonically brecciated limestones (based on core observations), suggestive of proximity to a major fault. Faults observed in the overlying cores of Eocene and Upper Cretaceous limestone exhibit mainly normal offsets as supported by FMS data (Fig. 6c). FMS log data (M. J. Jurado pers. comm. 1996) suggest that numerous zones of deformation and brecciation are present. Accordingly, it seems likely that Eratosthenes at Site 967 experienced a complex history of both extensional and compressional deformation which still requires clarification

Although a décollement was not actually penetrated, the history of thrust-disruption of the base of the northerly Eratosthenes platform slope can be inferred from seismic evidence. Faulting must be relatively recent, as Plio-Pleistocene sediments that form the uplifted ridge at the base of the slope are clearly seen to be deformed on seismic profiles. Also, the ridge has a small sedimentary

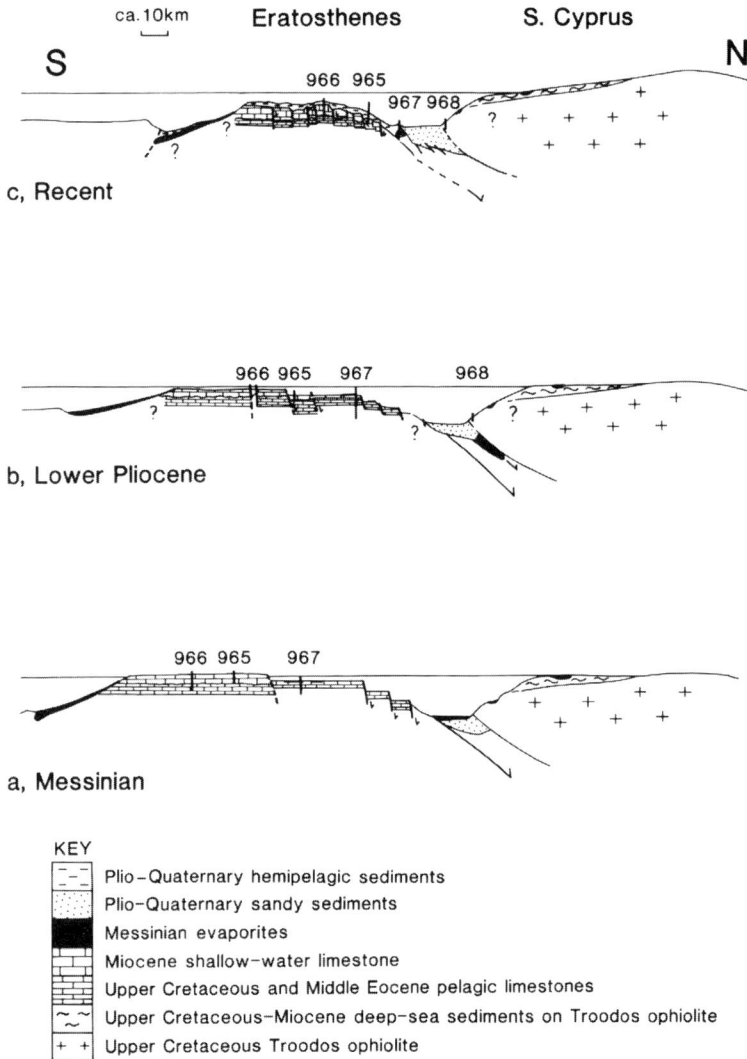

Fig. 7. Tectonic–sedimentary history of the northern Eratosthenes platform shown as a series of time slices. Note the role of extensional faulting of the platform margin and the inferred thrusting at the base of the lower slope.

basin ponded against it to the south. This basin is also well imaged on deep-tow high-resolution seismic profiles (Limonov *et al.* 1994) and can be seen to be deformed by numerous upright folds, with estimated wavelengths of only a few metres. The folds are interpreted to result from shortening of the sedimentary basin fill, associated with contemporaneous movement on the upthrust ridge to the south. Reverse faults are also imaged directly to the south, on the lower slope of the Eratosthenes Seamount.

Significance of Eratosthenes drilling for on-land accretionary processes

The Eratosthenes drilling sheds light on fundamental processes associated with the early stages of continental collision leading to the formation of mountain belts. By the time mountains begin to emerge, collisional processes are already well advanced and thus the early stages must be explored beneath the sea. Converging continental plates may not be linear and, where present, promontories and embayments may influence the timing and nature of collision in different areas (Dewey 1980). In addition, rifted continental slivers or oceanic edifices may be accreted prior to suturing of large continental plates. The arrival of an oceanic seamount at a trench has already been shown to involve subsidence and break-up of a volcanic edifice, in the case of the Daisha Seamount in the Japan trench (Le Pichon *et al.* 1987; Fig. 8). However, the Eratosthenes drilling instead investigated the collision of a carbonate platform with a convergent margin. This platform is assumed to be underlain by continental crust, based mainly on geophysical evidence (Makris *et al.* 1983). The results suggest that the platform

underwent strong flexural break-up and collapse as it approached and then entered the trench (Fig. 9). For both Daisha and Eratosthenes, break-up was largely achieved by high-angle normal faulting. Post-cruise work is continuing to model the collisional history of the Eratosthenes platform.

The drilling results also help with the general problem of preservation of continental fragments and carbonate platforms within on-land accretionary terrains (e.g. Tethys, or the Palaeozoic Calaedonian–Appalachian belt). In the case of the Daisha Seamount it appears that the bulk of the volcanic edifice is being consumed within the trench (Fig. 8). In future, little trace of its existence may remain, other than a disrupted forearc and the

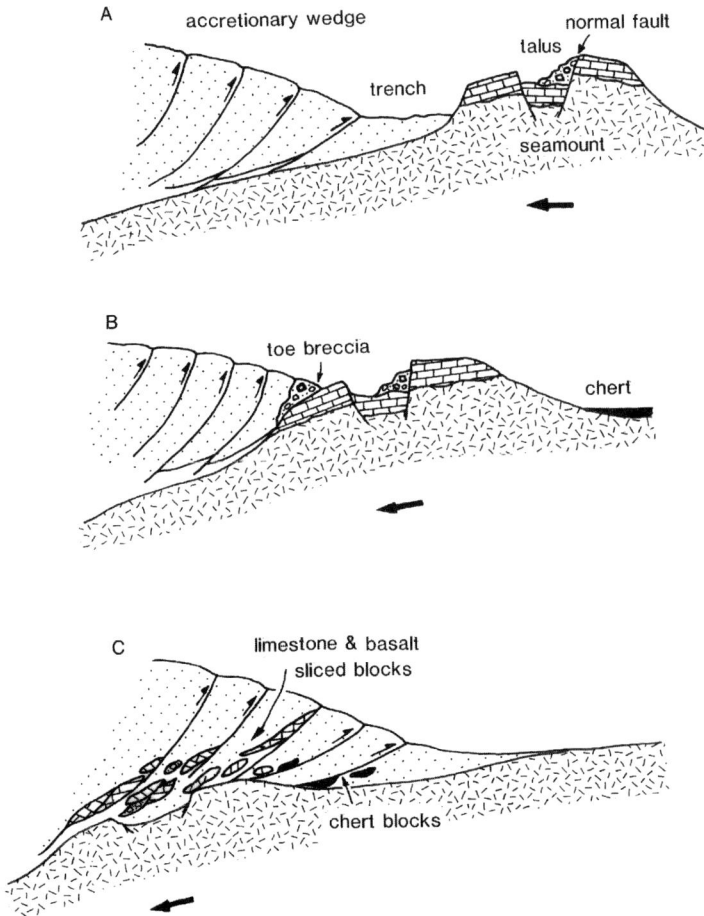

Fig. 8. Model showing processes active when an oceanic seamount arrives at a trench. The seamount undergoes large-scale normal faulting, shedding large volumes of basalt and sediment-derived clastics, often as debris flows. The bulk of the seamount is subducted while fragments are incorporated into an accretionary prism as detached blocks and debris flows (i.e. 'olistostromes'. Similarities and differences with the Eratosthenes–Cyprus trench collision are discussed in the text. From Taira *et al.* (1989); drawing courtesy of E. Pickett.

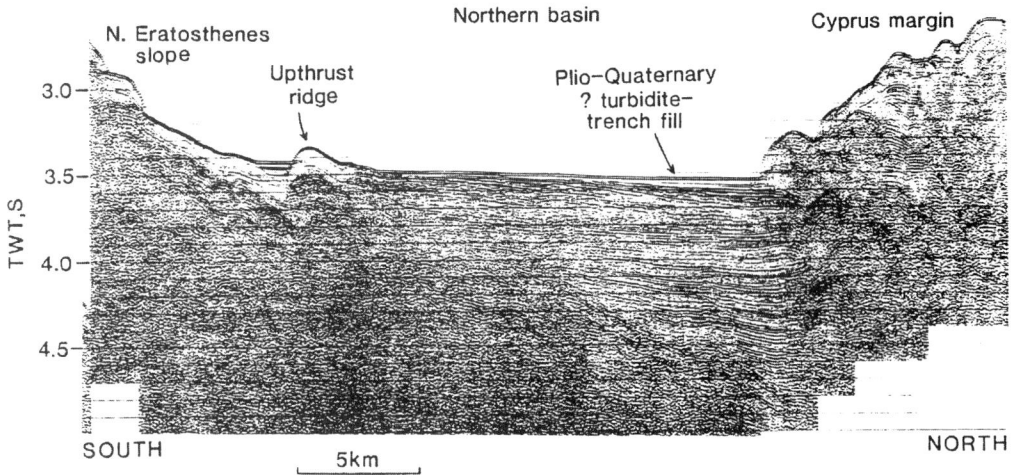

Fig. 9. Single-channel seismic profile from the northern margin of the Eratosthenes platform to the base of the southern Cyprus slope. TREDMAR Line 120. The Eratosthenes platform is breaking up and being thrust beneath southern Cyprus, with a sediment-filled trench above. Note the prominent raised ridge at the base of the northern slope. This folded Pleistocene sediments and appears to be currently active. The raised feature is interpreted as a thrust culmination above a blind thrust, related to detachment of a sliver of Eratosthenes limestones related to subduction and collision.

presence of volcanic–sedimentary debris flows and detached blocks incorporated into an accretionary prism. This debris would have a high chance of ultimate preservation in the stratigraphical record and can be regarded as recent equivalents of widespread 'olistostromes' (i.e. debris flows) found in many on-land accretionary terrains, including those in Japan. The fate of Eratosthenes would seem, similarly, to involve underthrusting and deep burial beneath the over-riding plate. The upper plate is represented by southern Cyprus, including the Troodos ophiolite. However, seismic and drill results at Site 967 from the foot of the northern slope of the Eratosthenes platform indicate that detachment of a sliver of the Eratosthenes carbonate platform above an active thrust is currently taking place (Fig. 10). If this process continues, in the future the main part of the Eratosthenes platform could be subducted, while at least one crustal sliver (*c.* 1 km thick) may be accreted and thus have a high preservation potential. Such a preserved slice (e.g. at Site 967) would comprise a basement of brecciated limestone above a thrust fault, then a *c.* 400 m thick slice of pelagic carbonates, overlain by a hemipelagic succession (with sapropels) and then terrigenous turbidites.

Similar crustal fragments have been preserved on land by processes of subduction/accretion and are widely exposed in the Tethyan orogenic belt (e.g. Robertson 1993, 1994; Figs 11 and 12). Such units

include accreted carbonate platform fragments, e.g. Upper Permian–early Mesozoic, in older (i.e. Palaeotethyan) Tethyan terrains in northern Turkey and equivalents in western Turkey. Similar units in the Greek area were accreted in the mid-Mesozoic (e.g. from the Pindos Ocean of northern Greece). Accretion of similar units also took place in the Late Cretaceous–Early Tertiary in both the Greek and southern Turkish areas (e.g. Antalya Complex).

Such accretion of small thrust slices is, however, not a mechanism to emplace large carbonate platforms on the scale of many kilometres across and kilometres thick. Examples of such carbonate platforms in the Tethyan orogenic collage include the Parnassus carbonate platform in Greece and the Bey Dağları carbonate platform in southwest Turkey (e.g. Robertson *et al.* 1991*a*). Carbonate platforms on this scale were perhaps too large to be accreted as discrete thrust slices, but were instead preserved by other means. One possibility involves oceanward relocation of the subduction zone, as in a number of SW Pacific examples (Weissel *et al.* 1982). Another possibility is suggested by site survey results for the Eratosthenes platform. The southern margin of Eratosthenes is apparently being thrust beneath the Levantine Basin to the south (Limonov *et al.* 1994; Fig. 3). Africa–Eurasia convergence is thus now beginning to be accommodated by both northward and southward underthrusting of the Eratosthenes platform. Thus, Eratosthenes may not in future be thrust far beneath

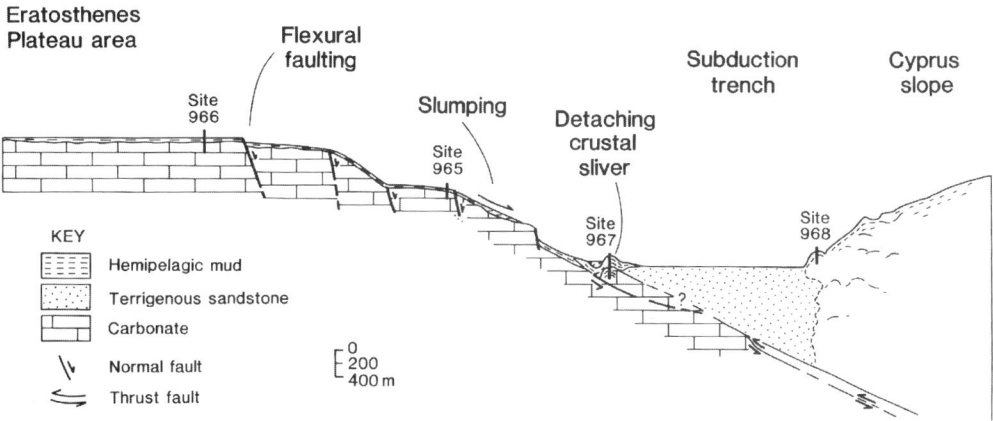

Fig. 10. Tectonic interpretation of processes affecting the northern margin of the Eratosthenes platform. These are mainly flexural normal faulting related to impingement with the subduction trench to the north, and thrust detachment of a sliver of Eratosthenes limestone at the base of the northern slope.

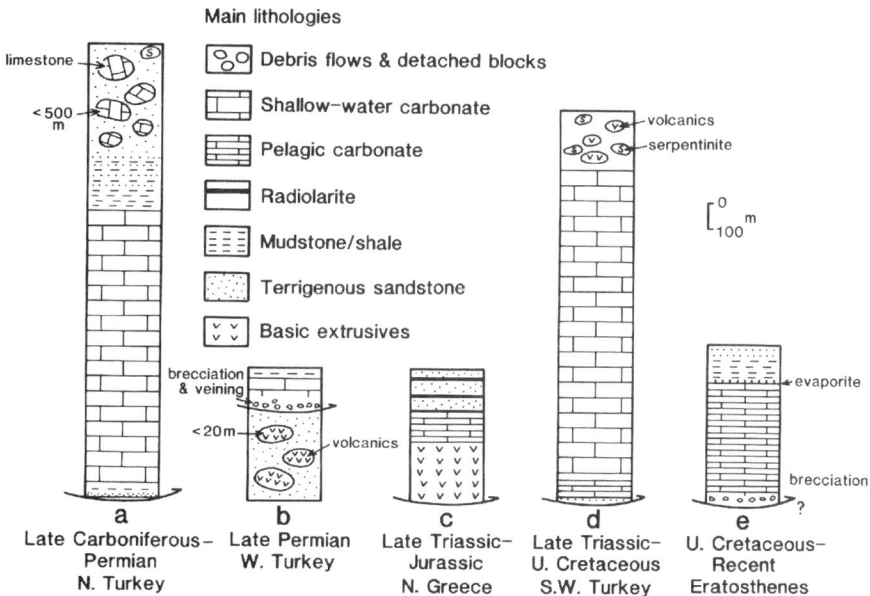

Fig. 11. Logs of limestone-dominated crust. (a–d) Slivers in melange terrains of Turkey and Greece: (**a**) Late Carboniferous–Late Permian thrust slice within an accretionary complex, Central Pontides, northern Turkey (Kargı unit); (**b**) Late Permian–Triassic thrust slice of limestone with a terrigenous basement, Çal unit, Karakaya Complex, W Turkey; (**c**) disrupted thrust slice of basalt overlain by limestones and the deep-water sediments, Alopetra unit within the Avdella Melange, Pindos Mtns, NW Greece; (**d**) large limestone thrust slice within sutured deep-sea and oceanic material, Sutçuler unit, SE part of the Antalya Complex, SW Turkey. (**e**) Eratosthenes unit at Site 967 for comparison. It is assumed that the succession is floored by a thrust fault, as in Fig. 7. All but (c) are interpreted as carbonate platforms within an open oceanic area that were emplaced by subduction and accretion; (c) represents limestone constructed on a volcanic seamount. At least (c) and (d) originated near a rifted continental margin, as inferred for Eratosthenes (e). All of these units experience obduction related to subduction–accretion processes. Data sources in Robertson *et al.* (1991, 1996).

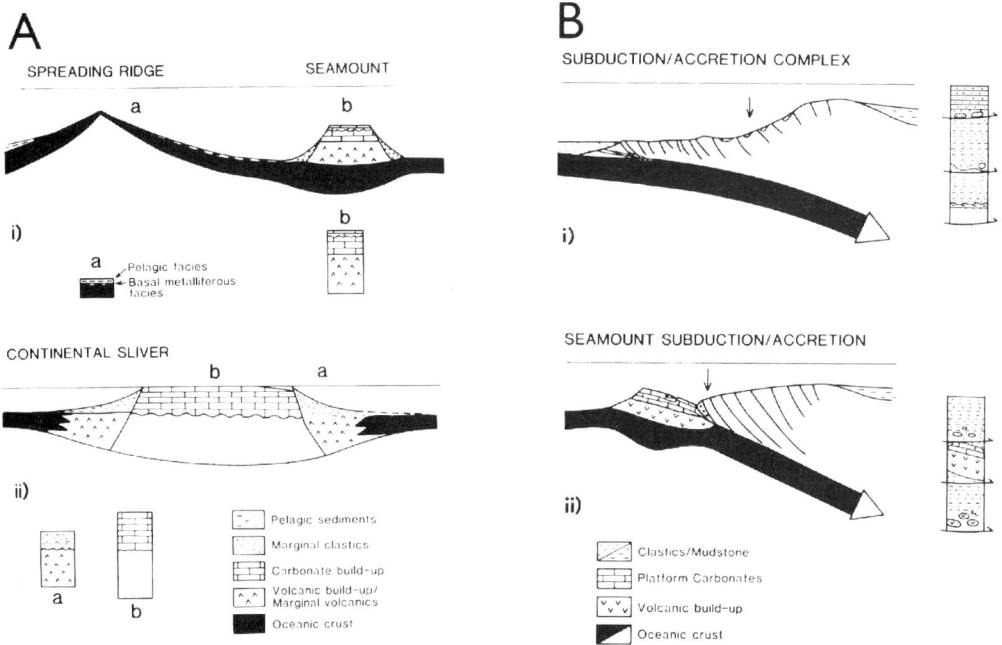

Fig. 12. Tectonic facies models: (**A**) (i) volcanic seamount of axis of spreading ridge; (ii) rifted off continental fragment; (**B**) (i) simple accretionary prism; (ii) accretion of a carbonate platform. Both continental fragments and volcanic seamounts were accreted in the Tethyan suture zone. Eratosthenes is interpreted as an accreted continental sliver.

Cyprus (and metamorphosed), but may instead remain at a high structural level between the plates, where it could later be exhumed and eroded.

Role of flexure-induced subsidence

The collision of the Eratosthenes platform can be compared with the development of foredeeps where thrust units are emplaced onto continental crust (e.g. Beaumont 1981; Stockmal *et al.* 1986). The Eratosthenes platform is interpreted as having evolved into a foredeep related to collision with the Cyprus active margin. The upward passage from shallow-water carbonates to deep-water hemi-pelagic and terrigenous sediments can be compared with that observed on land in many accreted carbonate platform slices. These include most of the examples of accreted units described earlier (Fig. 11). In addition, a similar trend from shallow- to deep-water sediments is seen in a number of Tethyan continental margins (e.g. in Oman; Robertson 1987) and in other areas (e.g. S Appalachians; Whiting & Thomas 1994), related to thrust and nappe emplacement. Preliminary calculations suggest that the Eratosthenes platform can be modelled in terms of an exponential increase

in subsidence rate with time during the Plio-Pleistocene (B.M. Whiting pers. comm. 1996). Subsidence was accompanied by important normal faulting as observed in seismic profiles (Limonov *et al.* 1994), and in cores (Emeis *et al.* 1996; Kopf & Flecker 1996). Indeed, the Eratosthenes crust did not deform according to a simple flexural model (e.g. Beaumont 1981), but underwent major normal faulting during subsidence. This coeval faulting could reflect reactivation of pre-existing zones of structural weakness within the Eratosthenes platform.

Significance for ophiolite obduction

Modes of ophiolite emplacement have long been enigmatic. One favoured model involves the collision of oceanic crust above a subduction zone with a passive continental margin. A modern example is the collision of the North Australian margin with the Banda arc (Harris 1991); another is the emplacement of the Taitao ophiolite, south Chile (Le Moigne *et al.* 1993). In the Eastern Mediterranean and Middle Eastern regions most of the large ophiolites are inferred, mainly based on geochemical evidence, to have formed above intraoceanic subduction zones (e.g. Jurassic Pindos ophiolite, N Greece; U Cretaceous Semail ophiolite, Oman (Pearce *et al.* 1984)). These ophiolites were emplaced when an above-subduction-zone ophiolite collided with a Tethyan passive margin, either the main margin of Gondwana (e.g. in Oman) or a small rifted continental fragment (e.g. Pelagonian zone, N Greece). The significance of the Eratosthenes results is that they document a Tethyan ophiolite actually in the process of being emplaced onto continental crust. Most of the other Upper Cretaceous ophiolites which formed in the southerly Neotethys (e.g. Hatay, Baer-Bassit, Oman) were emplaced onto the Arabian passive margin (i.e. Gondwana) in the latest Cretaceous and were transgressed by clastic sediments and then shallow-water carbonates. However, the Troodos ophiolite retains an unbroken sedimentary cover

from the time of genesis of the oceanic crust in the Upper Cretaceous until the Lower Miocene (Robertson 1990). The Troodos ophiolite remained within a remnant of the Neotethyan Ocean during the Tertiary. It was then strongly uplifted in the Plio-Pleistocene, with most of the uplift taking place in the Late Pliocene to mid-Pleistocene (Poole & Robertson 1992). This uplift was broadly coeval with collapse of the Eratosthenes platform and it is inferred that the two events are related. The first sign of break-up of the Eratosthenes platform is dated as Early Pliocene, suggesting the existence of a several million year time gap before evidence of strong uplift of southern Cyprus (i.e. Upper Pliocene). However, there is evidence of rapid subsidence of Eratosthenes in the Upper Pliocene to lower Pleistocene, based on microfossils (S. Spezzaferri, pers. comm., 1995), coeval with rapid

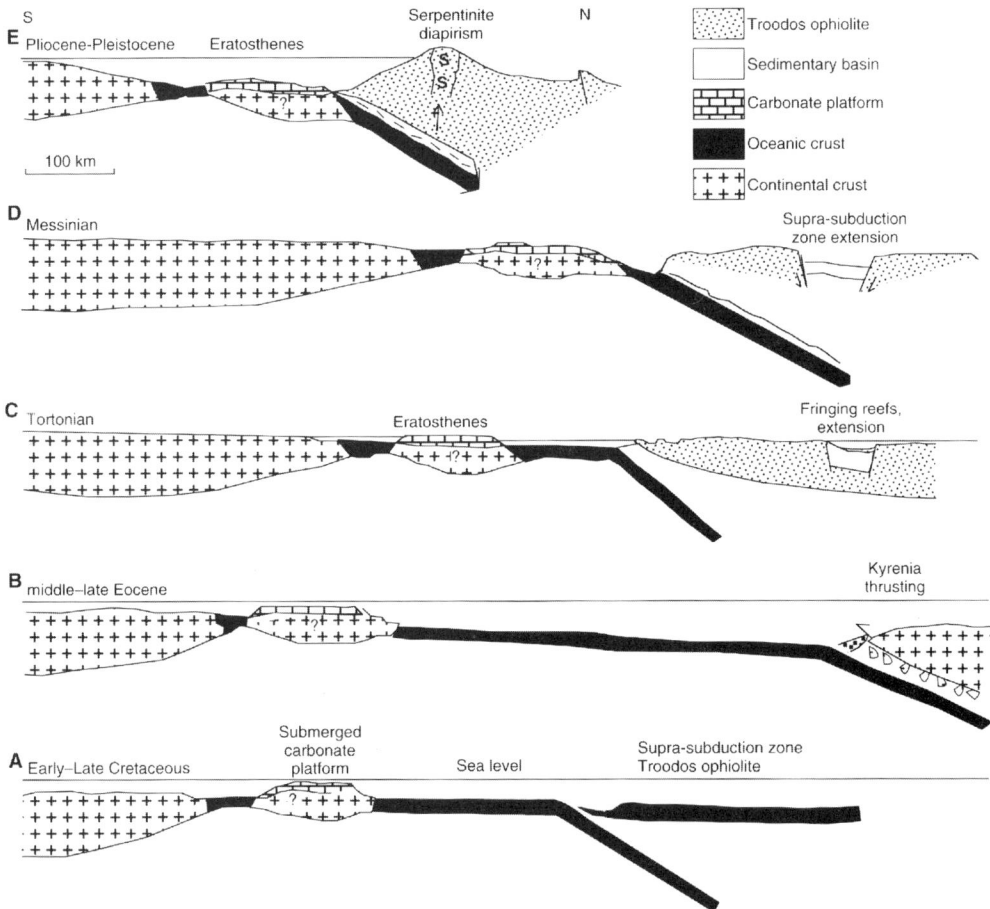

Fig. 13. Plate tectonic model for the evolution of the Eratosthenes platform in relation to Cyprus (see text for explanation).

uplift of the Troodos ophiolite. The collision of the Eratosthenes platform with southern Cyprus along the Cyprus active margin is identified as a major driving force in the emplacement of the Troodos ophiolite, with implications for other examples in the geological record.

In summary, drilling of the Eratosthenes carbonate platform has shed important light on processes of the initial stages of collision of the opposing African and Eurasian plates; it has also helped explain how carbonate platform fragments of different scales could be preserved in accretionary complexes on land, similar to examples from Mediterranean Tethyan terrains. Our working tectonic hypothesis is shown in Fig. 13.

Mud volcanism on the Mediterranean Ridge

There might seem to be little link between mud volcanism and carbonate platform collisional processes. However, early-stage collisional processes do provide a link. The structures drilled, the Milano and Napoli mud volcanoes, form part of a well-defined field of mud structures (Olympi Field) located on the northern part of the Mediterranean Ridge (Cita *et al.* 1989, 1994*a,b*; Cita & Camerlenghi 1990; Camerlenghi *et al.* 1992, 1995; Premoli-Silva *et al.* 1996; Staffini *et al.* 1993; Limanov *et al.* 1994, 1995; Fig. 14). The Mediterranean Ridge is a mud-dominated accretionary complex that was constructed in the later Tertiary, related to northward subduction of southern Neotethyan oceanic crust bordering the North African plate (Le Pichon & Angelier 1979; Le Pichon *et al.* 1982). During the Mesozoic, a

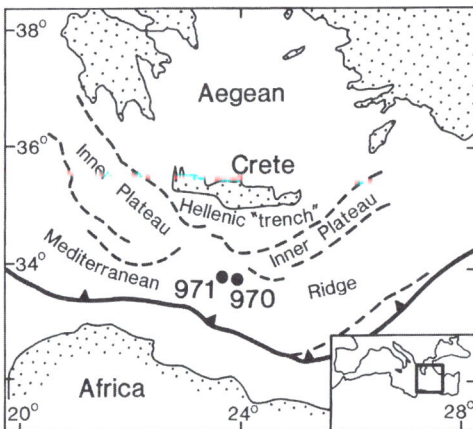

Fig. 14. Tectonic setting of the mud volcanoes drilled during Leg 160.

number of such strands of Neotethyan Ocean existed in this area, separated by microcontinents rifted from the north Gondwana margin (e.g. Adria, also know as Apulia). Together with Eratosthenes, the Mediterranean Ridge is sited within the most southerly of these Neotethyan oceanic strands. Other former oceanic basins further north in the Greek area include the Pindos and Vardar zones. Related to convergence of the African and Eurasian plates, the more northerly of these oceanic basins completely closed by Early Tertiary time (i.e. Eocene–Oligocene), such that further Africa–Eurasia convergence could be taken up only within the most southerly Neotethyan oceanic basin. Northward subduction within this basin then gave rise to the Mediterranean Ridge accretionary complex and its eastwards extension, the Cyprus active margin. Onset of subduction and the beginning of growth of the Mediterranean Ridge accretionary complex probably took place around Late Oligocene time (e.g. Kastens 1991; Meulenkamp *et al.* 1994), and subduction and accretion were well advanced by the Late Miocene (Fig. 15).

Understanding of the Mediterranean Ridge has been retarded by the existence of seismically impenetrable Messinian evaporites ('M' reflector). However, recent geophysical studies shed new light on its structure and geological history (i.e. Kastens *et al.* 1992; De Voogd *et al.* 1992; Truffert *et al.* 1993; Dickmann *et al.* 1995; Mascle *et al.* 1995; Reston *et al.* 1995; Chaumillon & Mascle 1995). It is now clear that the Mediterranean Ridge accretionary complex is unusually low and broad compared to most subduction complexes. One factor may be that during the Plio-Pleistocene the trench lay adjacent to a mainly submerged continental margin to the north (i.e. Aegean Sea), and thus only limited volumes of coarse clastic sediment reached the trench (e.g. from Crete). On the other hand, some sand probably reached the trench from North Africa, much as today. In addition, a large amount of evaporite was probably precipitated in subduction trench, ridge and associated forearc basin settings during the Messinian desiccation crisis. Evaporites within the trench were later subducted or accreted, reducing the strength of the wedge and allowing it to spread out laterally above a gently dipping subduction zone.

By the Late Pliocene or Early Pleistocene, the Mediterranean Ridge began to collide with a large promontory (Cyrenaica) of the North African plate (Fig. 14), causing the toe of the accretionary complex to be uplifted. Accretionary wedges are continuously maintained in mechanical equilibrium (Platt 1990). Maintenance of a 'critical taper' was achieved by initiating crustal thickening via

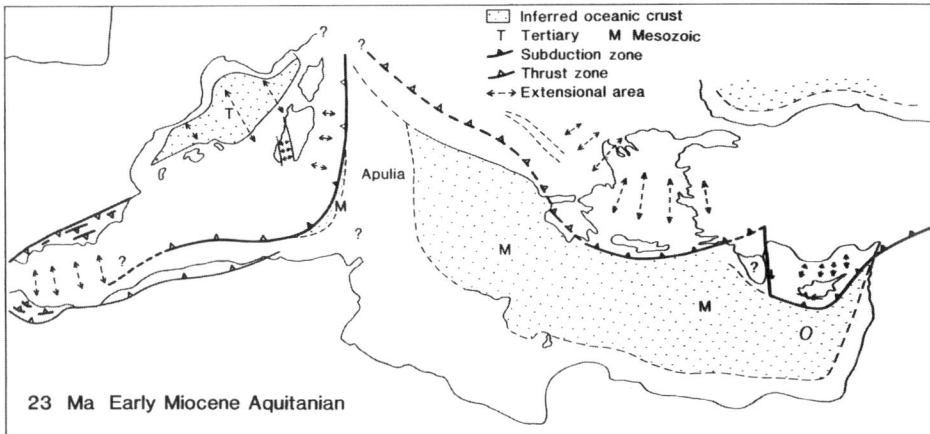

23 Ma Early Miocene Aquitanian

(a)

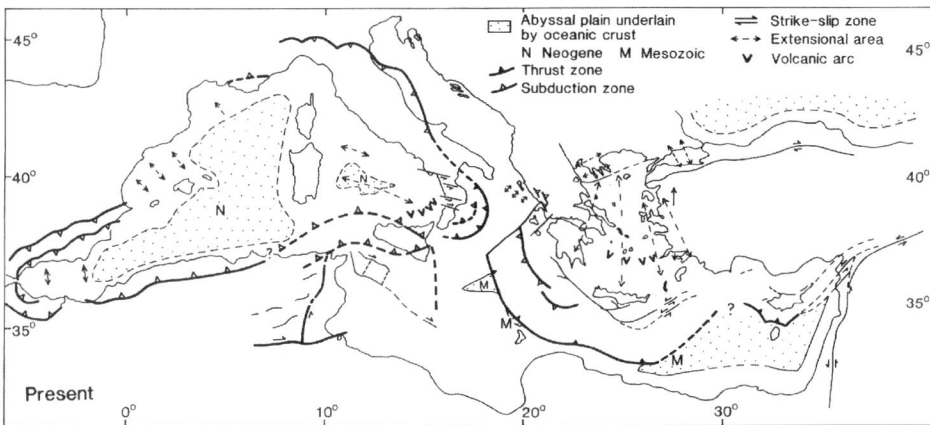

Present

(b)

Fig. 15. Plate tectonic maps showing the setting of the Mediterranean Ridge accretionary prism in: (**a**) Early Miocene; (**b**) Present. Modified from Robertson & Grasso (1995).

backthrusting of the northern margin of the Mediterranean Ridge accretionary complex. In addition, backthrusting is known to affect accretionary wedges undergoing steady-state subduction/accretion, unrelated to collision (Brown & Westbrook 1988). One probable effect of regional collision with the African margin to the south is that the northern part of the Mediterranean Ridge accretionary complex was thrust northwards over a backstop of continental crust to the north. This led to overpressuring within muddy sediments of mainly Neogene to Plio-Pleistocene age. The presence of Messinian evaporites may also have helped retain such overpressured fluids.

Drilling during Leg 160 was designed to define the internal anatomy, age and mode of mud volcanism. The two examples selected, the Milano and Napoli mud volcanoes, were already seismically well-imaged and were known to show contrasting morphologies, and possibly illustrated different stages of development, with one, Napoli, being currently active, in contrast to Milano. The drill sites aimed to constrain crest, inner flank and outer flank settings. Specifically, drilling of the crestal areas aimed to shed light on alternative extrusive (i.e. as mud-debris flows) versus intrusive (i.e. as semi-solid 'protrusions') origins (Limonov et al. 1994; Camerlenghi et al. 1995). Drilling of the flanks would reveal evidence of volcanic up-building, while it was hoped to reach pre-volcanic deep-sea sediments on the outer flanks and thus date the age of mud volcano initiation. Details of the main features of the two mud volcanoes drilled are given below (Figs 16 and 17).

Milano mud volcano

Nannofossil oozes, nannofossil clays and sapropels (i.e. organic-matter-rich muds) of early late Pleistocene age were cored at an outermost hole, beyond the mud volcano (Hole 970B). These sediments are interbedded with poorly consolidated, thin-to medium-bedded sands and silts, presumably shed from the mud volcano. Some intervals are tilted, with small normal and reverse faults. A hole in the outer flank area (Hole 970A) revealed alternations of clast-rich, mud-supported sediments ('mud breccias') and normal hemipelagic sediments. Using the FMS the lowest interval recorded there (but not cored) was identified as matrix-supported conglomerate (i.e.

mud breccia). The lowest interval actually recovered is pelagic sediment, dated 1.75 Ma. This is overlain by mud-debris flows and coarse-grained turbidites, in which the FMS clearly images numerous clasts (<5 cm to 55 cm in diameter). Above comes a thin interval of pelagic sediment, dated as *c*. 1–1.5 Ma. This, in turn, is overlain by a thick, clast-rich, mud-supported sedimentary interval with local thin sapropels (also detected with FMS) and, finally, by an uppermost pelagic interval (<1 m), dated as <0.26 Ma. By contrast, two holes on the inner flank (Hole 970C) and on the crest (Hole 970D) of the Milano mud volcano recovered gaseous muddy, silty and sandy sediments.

Napoli mud volcano

A succession on the outer flank (Hole 971A) begins with normal hemipelagic sediments and sapropels

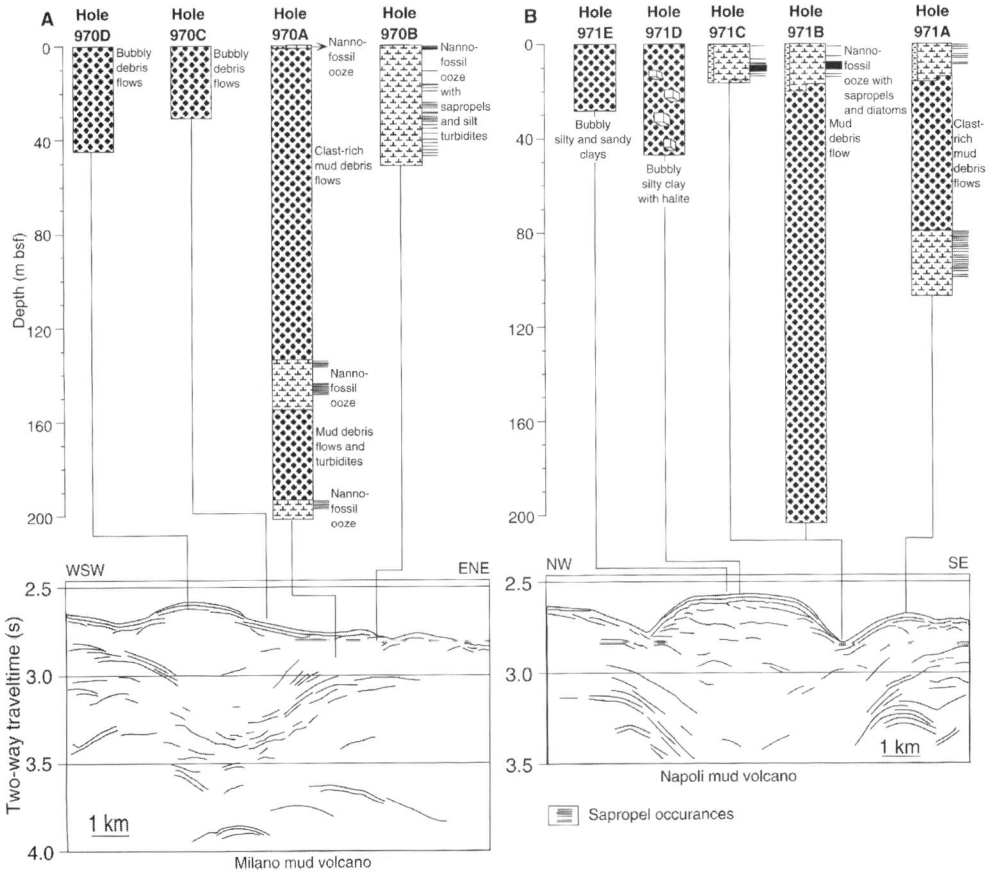

Fig. 16. Summary of the lithostratigraphy of the Milano (**A**) and Napoli (**B**) mud volcanoes drilled during Leg 160. The seismic reflectors visible within the two mud volcanoes are shown below. Note the presence of inward-dipping reflectors beneath both flanks of the Napoli and Milano structures. See text for explanation.

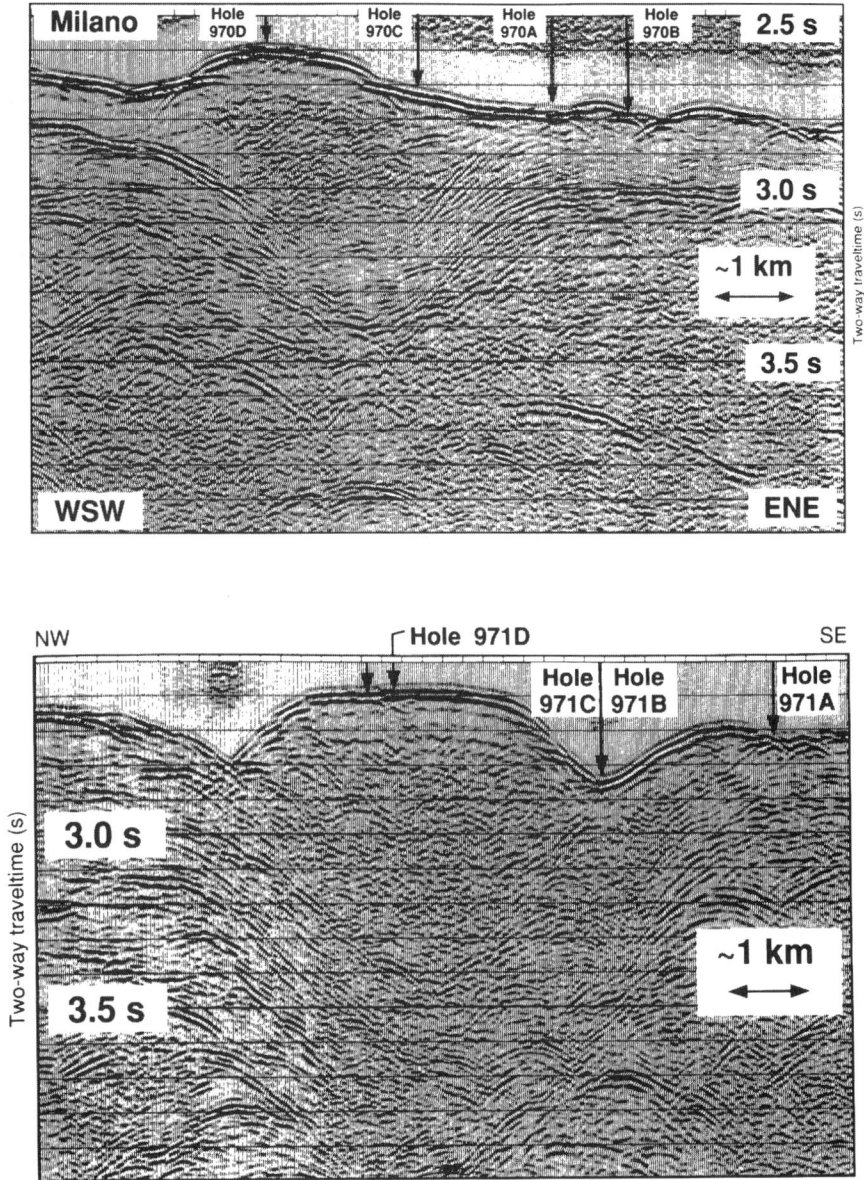

Fig. 17. Seismic profiles of the Milano and Napoli mud volcanoes based on site survey work during Leg 160. (**a**) Milano; (**b**) Napoli.

dated 0.46–1.5 Ma (i.e. within the large *Gephyrocapsa* zone). This interval is followed by clast-rich, matrix-supported sediments, in which the clasts are mainly calcareous and range from several millimetres to a few centimetres in size. The succession ends with a *c.* 20 m succession of nannofossil ooze, nannofossil clay and turbidites, dated as middle–late Pleistocene to Middle

Pliocene (<1.5 Ma). A thick unit of matrix-supported, clast-rich muddy sediments was next recovered on the inner flank (Hole 971B) and dated as between 0.26 and 0.46 Ma. Intervals with scattered clasts (<5 cm) alternate with more homogeneous silty clay. Downhole logs (especially natural gamma and resistivity) reveal layers that might correspond to relatively silty to sandy intervals. Overlying this are hemipelagic sediments, including sapropels. Hole 971C was drilled at the same location as Hole 971B, revealing an expanded sequence that probably resulted from redeposition of fine-grained sediment from the crestal area of the mud volcano into a surrounding moat-like depression. On the crest of the Napoli mud volcano (Hole 971D) gaseous silty clay was recovered, with scattered small (<5 cm) clasts of mudstone and siltstone. Angular to subrounded clasts of coarsely crystalline halite (up to 3 cm diameter) were noted within thin, silty layers, together with a few subrounded halite-cemented mudstone clasts (<5 cm diameter). The matrix is significantly finer at the crest of the Napoli mud volcano, than at Milano (Flecker & Kopf 1996). Finally, bubbly clays and silts with thin (several centimetres), more sandy layers were recovered at another crestal site (Hole 971E), together with a few small (<3 cm) clasts of mudstone and fine-grained carbonate. Rare nannofossils of Pleistocene age are present in the upper part of the section, and reworked Miocene nannofossils are common throughout.

Sediment clast and matrix types

Well-consolidated, matrix-supported, clast-rich muddy sediments dominate the flanks of both mud volcanoes (Fig. 18). These are the well-known 'mud breccias' of Cita *et al.* (1981). The matrix ranges from silty clay to rare sandy silt, and includes nannofossils, foraminifera, clay, quartz and rock fragments. The matrix-supported texture is supported by FMS data (Fig. 19). Nannofossils and planktonic foraminifer assemblages within the matrix are dominantly of Middle Miocene age. However, Eocene, Oligocene and Middle Miocene nannofossils are also present and may be reworked. In addition, Pleistocene microfossils are present in the matrix of the Napoli mud volcano. Clasts vary from mainly subangular to subrounded, and are less commonly angular or rounded. Lithologies include poorly consolidated sandstone and siltstone, weakly to well-consolidated calcareous claystone and mudstone, together with calcite- (or locally quartz-) cemented sandstone and siltstone. Sandstone clasts are mainly of plutonic igneous and metamorphic origin, and were probably mainly derived from North Africa, although a source in Crete cannot be ruled out. There are also shallow-water carbonate clasts with calcareous algae, polyzoans and reworked pelagic carbonate. The clasts in both volcanoes contain the following biota: (i) nannofossils and planktonic foraminifera of Lower–Middle Miocene (i.e. Burdigalian–Langhian) age; (ii) mixed nannofossil and planktonic foraminifer assemblages of Middle

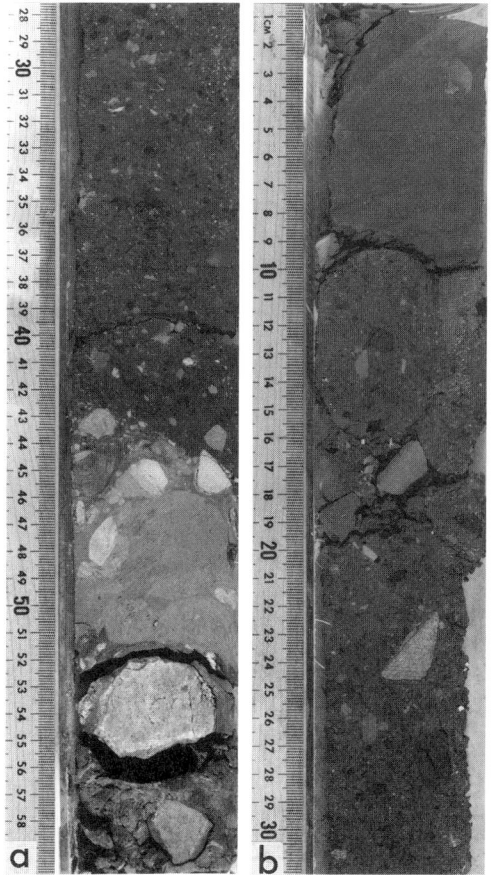

Fig. 18. Photographs of cores of the Milano mud volcano (photos: ODP). (**a**) Massive clast-rich mud. Note the small angular claystone clasts (upper) and the larger pale (sandstone) clasts (lower). Drilling disturbance has occurred in the vicinity of the large, hard clasts, 960A, 14X, 1, 30–0 cm. (**b**) Typical clast-rich mud. Clasts include sandstone (pale), micritic limestone (medium grey) and claystone (dark), Note also the hemipelagic mud (upper), 970A-15X-1, 0–30 cm. The clast-rich muds are interpreted as extrusive debris flows rather than intrusive bodies, as in some earlier interpretations. Hole 967E, Core 46R, Section 1, 0–20 cm.

Miocene, Oligocene and Eocene ages, together with Cretaceous nannofossils; and (iii) brackish-water-type ostracods of Late Messinian–Early Pliocene age in the Milano mud volcano.

Geochemical processes

The presence of clathrates (i.e. methane hydrates) was inferred in the crest of the Milano mud volcano based on the evidence of abnormally low porewater

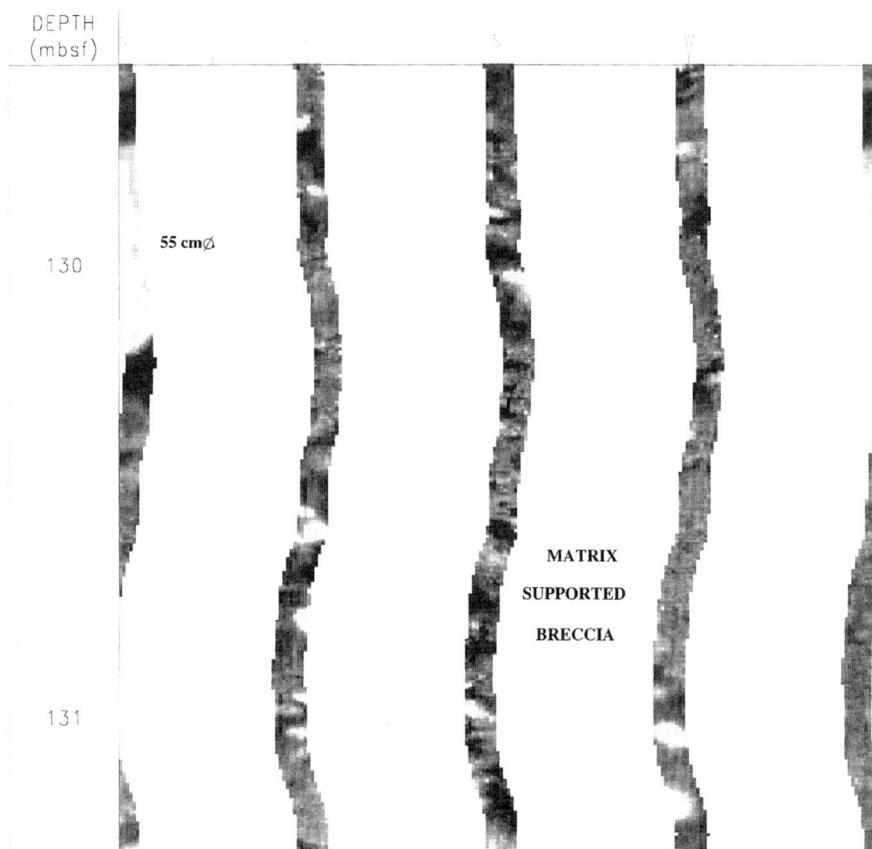

Fig. 19. Formation MicroScanner (FMS) image of typical 'mud breccias' from Hole 970A. The pale lenses correlate with lithified sandstone and limestone clasts, while the darker material is mainly matrix and small argillaceous clasts. See text for explanation (M. J. Jurado, unpublished data).

salinities (Holes 970C and 970D). These low values are probably due to the decomposition of clathrates that took place immediately after core recovery. In addition, oxygen-isotopically heavy ($\delta^{18}O$) porewaters indicate clathrate decomposition at deeper levels (G. De Lange, pers. comm. 1996). The absence of methane hydrates in the crest of the Napoli mud volcano could reflect the existence of higher pore-fluid temperatures and salinities (up to 300 g kg^{-1}) that acted to suppress clathrate formation. Low levels of sulphate in Hole 970D could relate to intense bacterial sulphate reduction in an anoxic environment. Pure methane is abundant as gas bubbles in the uppermost 30 m below the crest of the Milano mud volcano (Hole 970D), but methane concentrations drop sharply below this, consistent with the formation of methane hydrate at relatively shallow depths. In accordance with this, levels of higher hydrocarbons

relative to methane abruptly increase below the clathrate zone.

Hydrocarbon gas is abundant in the Napoli mud volcano, where methane/ethane ratios vary from 10 to 40 overall and remain constant with depth, but vary significantly from hole to hole (Holes 971A, B, D and E). In contrast to the Milano mud volcano, clathrate was not detected. The gas in the Napoli mud volcano also contains several higher hydrocarbons (up to hexane) that are currently being identified. Porewaters from the crestal holes are saturated with respect to halite. There are marked local variations in potassium content suggesting that brine of more than one source may be present (i.e. in the lower part of Hole 971A). Very high alkalinity throughout (c. 80 mmol l^{-1} in Holes 971D and 971E) probably reflects microbial consumption of methane. Sulphate decreases sharply downward in Hole 971B, probably owing to high rates of bacterial sulphate reduction. A single temperature measurement of 16.1°C was obtained in Hole 971D at 45 m below sea floor, which is 2°C above normal bottom-water temperatures. Under these conditions, hydrocarbon gases may continuously flow to the surface of the Napoli mud volcano rather than being trapped as clathrates.

Processes of mud volcanism

The lowest known interval of 'mud breccias' (i.e. seen using FMS) relates to a relatively early stage of eruption above inferred Messinian evaporites and thin (i.e. tens of metres) sediments of Pliocene? age (Fig. 20). At Milano, this was followed by deposition of a distinctive clastic interval composed of debris flows and turbidites, and then by more quiescent extrusion of large volumes of multiple debris flows that interfinger with surrounding deep-sea sediments to build up the mud volcano cone. Facies transitions, as seen at Napoli, favour non-intrusive processes of mud volcano construction.

The mud volcanism involved detachment of lithified clasts derived from an underlying sedimentary succession of at least partly Miocene age. The depth range of derivation remains poorly constrained, but preliminary petrographic and clay mineralogical evidence does not indicate very deep burial (i.e. more than anchizonal) conditions. One possibility is that the matrix was derived from near the décollement of the downgoing plate, estimated at a depth of 5–7 km (Camerlenghi *et al.* 1995). This décollement apparently cuts through Miocene sediments, based on recent high-resolution seismic data (Chaumillon & Mascle 1995). More probably, most or all of the clasts were derived from shallower depths, within the accretionary wedge (i.e. <3 km). The driving force was possibly related to backthrusting against a backstop of rigid Cretan crust to the north. Such backthrusting could have

resulted in tectonic disruption and formation of new pathways for escape of overpressured materials to the surface. Mud volcanism could have been initiated in this model when an overlying seal, where present, of impermeable Messinian evaporite was broken. Elsewhere, faulting alone may have been sufficient to trigger expulsion of overpressured fluids.

Significance of mud volcano drilling

Mud volcanoes and related mud diapirs are known in a wide variety of settings world-wide and have been the subject of numerous publications in the international literature in the last decade or so (e.g. Brown & Westbrook 1988). Many examples are unrelated to subduction (e.g. Alboran Sea in the W Mediterranean and Black Sea (Henry *et al.* 1990; Limonov *et al.* 1994)). Mud volcanoes in accretionary settings have commonly been seismically imaged, and exposed counterparts on land are often composed of muddy debris flows (Barber *et al.* 1986). However, drilling of the Milano and Napoli mud volcanoes has offered the first opportunity to characterize the internal anatomy of submarine mud volcanoes in detail and determine their age and period of activity. The discovery that both the Milano and Napoli mud volcanoes are >1 Ma old is a surprising and important result. Once initiated, mud volcanoes are capable of remaining episodically active for geologically significant periods of time. Post-cruise work and companion studies are continuing to elucidate the intriguing nature of mud volcanism on the Mediterranean Ridge. However, at present it seems likely that a number of specific factors acted in concert to promote mud volcanism (Figs 21 and 22). These were: (i) shallow northward subduction (5–7 km) giving rise to overpressured mud (Camerlenghi *et al.* 1995); (ii) structural control of

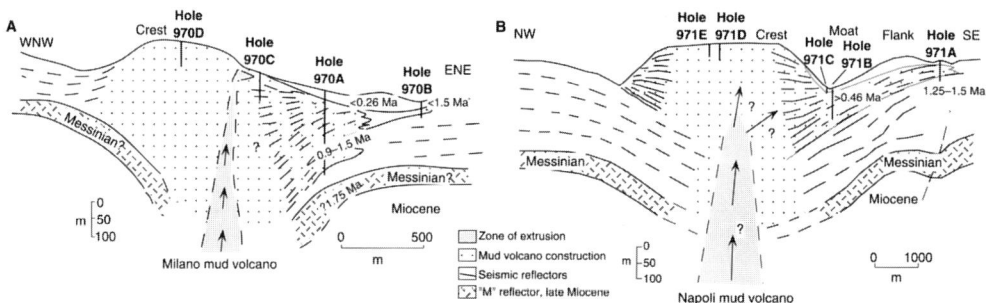

Fig. 20. Interpretation of the anatomy of (**A**) the Milano and (**B**) the Napoli mud volcanos. Clast-rich mud-debris flows were intruded from depth and were then extruded on the sea floor building up mud volcano edifices. Peripheral subsidence appears to have taken place to form a moat, with an infilling of multiple debris flows.

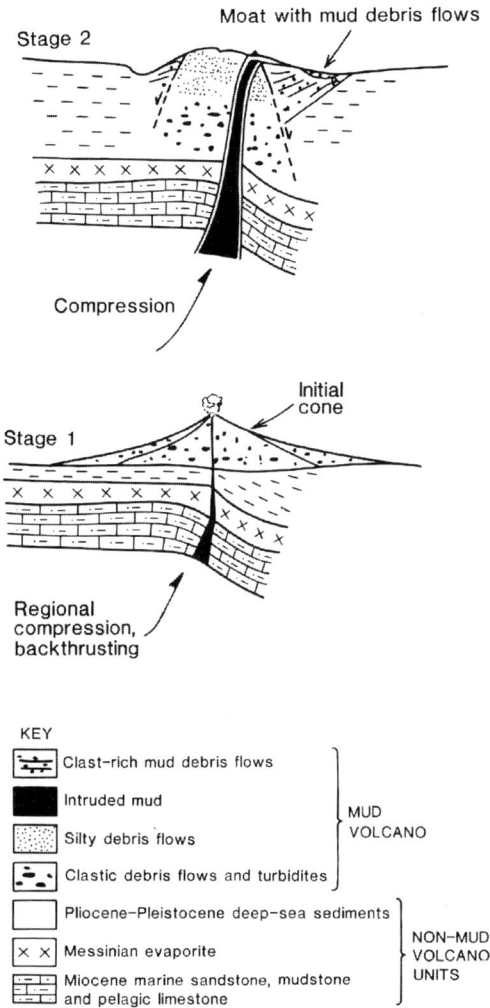

Fig. 21. Model for development of mud volcanism at the sites drilled. Mud volcanism was initiated when collision-related backthrusting increased *in situ* fluid pressure beneath an impermeable layer of Messinian evaporite. Once punctured, mud volcanism remained active for more than 1 Ma. Stage 1: early explosive phase builds up a clastic cone. Stage 2: collapse to form a moat which progressively fills with multiple debris flows. The detailed structural setting, however, remains unclear as collection of high-resolution seismic data is hampered by the presence of salt.

previously accreted material; (iii) the local presence of Messinian evaporites to help seal fluids; and (iv) backthrusting to provide zones of fluid escape.

Conclusions from Leg 160 tectonic drilling

Drilling during ODP Leg 160 in the Eastern Mediterranean has successfully shed light on

fundamental tectonic processes related to the early stages of convergence and collision of lithospheric plates.

Eratosthenes platform

The Eratosthenes platform south of Cyprus is confirmed to be in the process of incipient collision with the Cyprus active margin to the north. The platform is inferred to be a continental fragment rifted from the northern margin of the North African plate (i.e. Gondwana) in early Mesozoic time. By the mid-Cretaceous, Eratosthenes formed a shallow-water carbonate platform. Similar shallow-water carbonates are exposed in the Levant to the east. In common with the Levantine platform, the Eratosthenes carbonate platform subsided in the Late Cretaceous and was overlain by bathyal pelagic carbonate. It escaped tectonic disruption in the latest Cretaceous, associated with ophiolite emplacement in areas to the north, including Cyprus and Turkey. Eratosthenes was then still in a southerly position adjacent to the North African passive margin. Eratosthenes was tectonically uplifted by the Early Miocene to become a shallow-water carbonate platform. This uplift could relate to the initial effects of northward subduction. During the Messinian Mediterranean-wide desiccation crisis Eratosthenes was emergent, with local palaeosol development, while the flanks may have remained submerged. In the Early Pliocene it was then submerged to bathyal depths, accompanied by deposition of localized matrix-supported debris flows. The platform was later overlain by deep-water hemipelagic muds, including sapropels; subsidence then accelerated in thc Upper Pliocene–Pleistocene. This subsidence is seen as the result of collision with the Cyprus active margin to the north.

A sliver of the Eratosthenes crust is in the process of thrust detachment near the base of the northern slope of the platform. Its existence provides a mechanism for the mode of accretion of comparable platform slivers within on-land melange terrains, e.g. in Greece and Turkey. Collapse of the Eratosthenes platform in Early Pliocene–Pleistocene time was broadly coeval with the main phase of surface uplift of the Troodos ophiolite in the Late Pliocene to mid-Pleistocene. The two events are seen as intimately related to collision of the African and Eurasian plates, and associated with final emplacement of the Troodos ophiolite.

Milano and Napoli mud volcanoes

Drilling of the Milano and Napoli mud volcanoes on the Mediterranean Ridge south of Crete has shed

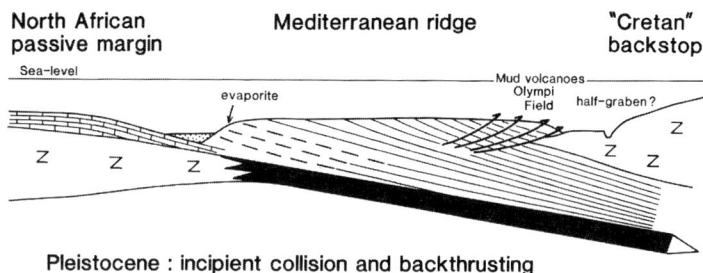

North African passive margin   **Mediterranean ridge**   **"Cretan" backstop**

Pleistocene : incipient collision and backthrusting

Fig. 22. Sketch of the inferred setting of mud volcanism on the Mediterranean Ridge accretionary complex, related to backthrusting.

light on the timing and processes of mud volcanism in accretionary settings. Both mud volcanoes are dominated by clast-rich, mud-matrix sediments, interpreted as multiple mud-debris flows. Drilling supported interpretatious made during the TREDMAR Site Survey Cruise (Limonov *et al.* 1994). The 'mud breccias' are similar to the origin of some debris flows associated with mud volcanoes on land (e.g. in Indonesia). Secondly, the presence of *in situ* hemipelagic sediments, both beneath the flanks of the mud volcano and interbedded with the mud-debris flows, shows that both mud volcanoes are >1 Ma old or older. Thus, individual mud volcanoes may remain active over geologically significant periods of time. The mud volcanoes developed above a shallow northward-dipping subduction zone in proximity to a zone of backthrusting along the northern part of the Mediterranean Ridge accretionary complex,

adjacent to a rigid backstop of Cretan continental crust to the north. Backthrusting is also known to take place within open-ocean accretionary complexes undergoing steady-state subduction (e.g. Barbados). In the case of the Mediterranean Ridge, backthrusting was a consequence of maintaining mechanical equilibrium (i.e. the critical taper), following onset of collision of the Mediterranean Ridge accretionary complex with a promontory of the North African continental margin to the south. The backthrust zone may have provided egress for overpressured materials originating from within the accretionary prism.

We thank the captain, the crew and the marine technicians for outstanding support of RV *JOIDES Resolution* during Leg 160. The drilling could not have taken place without the essential Site Survey data collected during a number of cruises over a decade or so.

References

BARBER, A. J., TJOKROSAPOETO, S. & CHARLTON, T. R. 1986. Mud volcanoes, shale diapirs, wrench faults, and melanges in accretionary complexes, Eastern Indonesia. *American Association of Petroleum Geology Bulletin*, **70**, 1729–1741.

BEN-AVAHAM, Z., SHOHAM, Y. & GINZBURG, A. 1976. Magnetic anomalies in the Eastern Mediterranean and the tectonic setting of the Eratosthenes Seamount. *Geophysical Journal of the Royal Astronomical Society*, **45**, 105–123.

BEAUMONT, C. 1981. Foreland basins. *Royal Astronomical Journal Geophysical Journal*, **65**, 291–329.

BROWN, K. & WESTBROOK, G. K. 1988. Mud diapirism and subcretion in the Barbados Ridge accretionary complex: The role of fluids in accretionary processes. *Tectonics*, **7**, 613–640.

CAMERLENGHI, A., CITA, M. B., HIEKE, W. & RICCHIUTO, T. S. 1992. Geological evidence of mud diapirism on the Mediterranean Ridge accretionary complex. *Earth and Planetary Science Letters*, **109**, 493–504.

——, ——, DELLA VEDOVA, B., FUSI, N., MIRABILE, L. & PELLIS, G. 1995. Geophysical evidence of mud

diapirism on the Mediterranean Ridge accretionary complex. *Marine Geophysical Researches,* **17**, 115–141.

CHAUMILLON, E. & MASCLE, J. 1995. Variation laterale des fronts de deformation de la Ride Mediterranéenne (Mediterranée orientale). *Bulletin de la Societé Geologique de France*, **166**(5), 463–478

CITA, M. B. & CAMERLENGHI, A. 1990. The Mediterranean Ridge as an accretionary prism in collisional context. *Societa Geologica Italiana, Memoire*, **45**, 463–480.

——, RYAN, W .F. B. & PAGGI, L. 1981. Prometheus mud breccia: an example of shale diapirism in the Western Mediterranean Ridge. *Annales Géologiques de Pays Héllenique*, **30**, 543–570.

——, ——, ERBA, E. *ET AL.* 1989. Discovery of mud diapirism in the Mediterranean Ridge – A preliminary report. *Societa Geologica Italiana, Boletin*, **108**, 537–543.

——, ERBA, E., Lucci, R. Van DER MEER, R., POTT, M. & NIETO, L. 1994a. Bottom sampling in the Olimpi mud volcano area. *In*: LIMONOV, A. F., WOODSIDE, J. M. & IVANOV, M. K. (eds) *Mud Volcanism in the Mediterranean and Black Seas and Shallow Structure of the Eratosthenes Seamount*. Initial results of the geological and geophysical investigations during the Third 'Training-through-Research' Cruise of the R/N *Gelendzhik* (June–July 1993). UNESCO Report in Marine Sciences, **64**, 89–107.

——, WOODSIDE, J. M., IVANOV, M. K. *ET AL*. 1994*b*. Fluid venting, mud volcanoes and mud diapirs in the Mediterranean Ridge. *Atti della Accademia Nazionale dei Lincei Roma, Rendoconti*, **5**, 159–169.

CLUBE, T. & ROBERTSON, A. H. F. 1986. The palaeorotation of the Troodos microplate, Cyprus, in the Late Mesozoic–early Tertiary plate tectonic framework of the Eastern Mediterranean. *Nature*, **317**, 522–525.

DERCOURT, J., RICOU, L. F. & VRIELYNCK, B. (eds) 1993. *Atlas of Tethys Palaeoenvironmental Maps*. (Gauthier-Villars), Paris.

DE VOOGD, B., TRUFFERT, C., CHAMOT-ROOKE, N., HUCHON, P., LALLEMANT, S. & LE PICHON, X. 1992. Two-ship deep seismic soundings in the Basins of the eastern Mediterranean Sea (Pasiphae cruise). *Geophysical Journal International*, **109**, 536–552.

DEWEY, J. F. 1980. Episodicity, sequence and style at convergent plate boundaries. *In*: STRANGEWAY, D. W. (ed.) *The Continental Crust and its Mineral Deposits*. Geological Society of Canada Special Paper, **20**, 553–573.

—— & ŞENGÖR, A. M. C. 1979. Aegean and surrounding areas: complex multi-plate and continuum tectonics in a convergent zone. *Geological Society of America Bulletin*, **90**, 84–92.

DICKMAN, T., RESTON, T. J. & HUENE, R. 1995. The structure of the Mediterranean Ridge near the deformation front; first results from the Imerse Project. *Terra Abstracts*, 285.

EMEIS, K.-C., ROBERTSON, A. H. F., RICHTER, D. *ET AL*. 1996. *Reports of the Ocean Drilling Program*, College Station, TX. **160**, 972.

FLECKER, R. & KOPF, A. 1996. Data Report: Clast and grain-size analysis of sediment recorded from the Napoli and Milano mud volcanoes, Eastern Mediterranean. *Proceedings of the Ocean Drilling Program Initial Reports*, College Station, TX, **160**, 529–534.

——, ROBERTSON, A. H. F., POISSON, A. & MÜLLER, C. 1995. Facies and tectonic significance of two contrasting Miocene basins in southern coastal Turkey. *In*: ROBERTSON, A. H. F. & GRASSO, M. (eds) *Thematic Set-later Tertiary–Quaternary Mediterranean Tectonics and Palaeo-environments*. *Terra Nova*, **7**(2), 221–233.

FOLLOWS, E. J. & ROBERTSON, A. H. F. 1990. Sedimentology and structural setting of Miocene reefal limestones in Cyprus. *In*: MALPAS, J. ET AL. (eds) *Ophiolites, Oceanic Crustal Analogues, Proceedings of the International Symposium*, Nicosia, Cyprus, Oct. 1987, 207–216.

——, ——, & SCOFFIN, T. P. 1996. Tectonic controls of

Miocene reefs and related carbonate facies in Cyprus. *In*: *Miocene Reefs of the Mediterranean Region*. Society of Economic Geologists Special Publication, SEPM Concepts in Sedimentology and Paleontology, **5**, 296–315.

GARFUNKEL, Z. & DERIN, B. 1984. Permian–early Mesozoic tectonism and continental margin formation in Israel and its implications for the history of the Eastern Mediterranean. *In*: DIXON, J. E. & ROBERTSON, A. H. F. (eds) *The Geological Evolution of the Eastern Mediterranean*, Geological Society, London, Special Publication, **17**, 187–201.

HARRIS, R. A. 1991. Temporal distribution of strain in the active Banda orogen: a reconciliation of rival hypotheses. *Journal of Southeast Asian Earth Science*, **6**, 373–386.

HENRY, P., LE PICHON, X., LALLEMANT, S., FOUCHER, J. P., WESTBROOK, G. & HOBART, M. 1990. Mud volcano field seaward of the Barbados accretionary complex: a deep-towed side scan sonar survey. *Journal of Geophysical Research*, **95**, 8917–8929.

HSÜ, K., CITA, M. & RYAN, W. B. F. 1973. The origin of the Mediterranean evaporites. *In*: *Initial Reports of the Deep Sea Drilling Project*, **13**. US Government Printing Office, Washington, DC, 1203–1231.

——, MONTADERT, L. *ET AL*. 1978. *Initial Reports of the Deep Sea Drilling Project*, **42A**. US Government Printing Office, Washington, DC.

KASTENS K. A. 1991. Rate of outward growth of the Mediterranean Ridge accretionary complex. *Tectonophysics*, **199**, 25–50.

——, BREEN, N. A. & CITA, M. B. 1992, Progressive deformation of an evaporite-bearing accretionary complex: SeaMARC 1, SeaBeam and piston-core observations from the Mediterranean Ridge: *Marine Geophysical Researches*, **14**, 249–298.

KEMPLER, D. 1993. *Tectonic patterns in the Eastern Mediterranean*. PhD thesis, Hebrew University of Jerusalem.

KOPF, A. & FLECKER, R. 1996. Problems associated with the interpretation of normal fault distributions in sediments recovered by Advanced Piston Coring. *Proceedings of the Ocean Drilling Program Initial Reports*, College Station, TX, **160**, 507–512.

KRASHENINNIKOV, V. A., UDINTSEV, G. B., MOURAVIOV, V. I. & HALL, T. K. 1994. Geological structure of the Eratosthenes Seamount. *In*: KRASHENINNIKOV, V. A. & HALL, J. K. (eds) *Geological Structure of the Northeastern Mediterranean* (Cruise 5 of the Research Vessel 'Akademik Nikolaj Strakhov'). Historical Productions - Hall, Jerusalem, 113–131.

LE MOIGNE, J., LAGABRIELLE, Y., BOURGOIS, J. & PALVADEAU, E. 1993. Ophiolites en contexte de dorsale en subduction: nouvelles données sur la péninsule de Taitao (Sud Chili). Compte Rendues de L'academie des Sciences, *Paris*, **317**, Series 2, 403–410.

LE PICHON, X. & ANGELIER, J. 1979. The Hellenic Arc and Trench system: A key to the neotectonic evolution of the eastern Mediterranean area. *Tectonophysics*, **60**, 1–42.

——, LYBERIS, N., ANGELIER, J. & RENARD, V. 1982. Strain distribution over the East Mediterranean

Ridge: a synthesis incorporating new sea-beam data. *Tectonophysics,* **86**, 243–274.

——, IIYAMA, T., CHAMLEY, H. *ET AL.* l987. Nankai Trough and Zenisu Ridge: a deep-submersible survey. *Earth and Planetary Science Letters,* **83**, 199–213.

LIMONOV, A. F., WOODSIDE, J. M. & IVANOV, M. K. (eds) 1994. *Mud Volcanism in the Mediterranean and Black Seas and Shallow Structure of the Eratosthenes Seamount.* Initial results of the geological and geophysical investigations during the third 'Training through-Research' Cruise of the R/V *Glendzhik* (June–July, 1993). UNESCO Reports in Marine Sciences, **64**, 173 pp.

——, KENYON, N. Y., IVANOV, M. K. & WOODSIDE, J. M. (eds) 1995. *Deep-sea depositional systems of the Western Mediterranean and mud volcanism on the Mediterranean Ridge.* UNESCO Report in Marine Sciences, **67**, 172.

LIVERMORE, R. A. & SMITH, A. G. 1984. Some boundary conditions for the evolution of the Mediterranean region. *In:* STANLEY, D. J. & WEZEL, F.-C. (eds). *Geological evolution of the Mediterranean Basin.* Springer-Verlag, Berlin, 83–100.

MAKRIS J., BEN-AVRAHAM & EELEFTHERIOU, S. 1983. Seismic refraction profiles between Cyprus and Israel and their interpretation. *Geophysical Journal of the Royal Astronomical Society*, **75**, 575–591.

MART, Y., ROBERTSON, A. H. F. & WOODSIDE, J. AND ODP LEG 160 SHIPBOARD SCIENTIFIC PARTY, 0000. *Cretaceous tectonic setting of the Eratosthenes Seamount in the Eastern Mediterranean Neotethys*: Initial results of ODP Leg 10. *Compte Rendues de l' Academie des Sciences,* Paris, **324**, serie 11a, 127–134..

MASCLE, J., CHAUMILLION, E. & PEDERSEN, H. l995. Structural variations and deformational styles of the Mediterranean Ridge: new constraints from the PRISMED MCS survey. *International Earth Sciences Colloquium on the Aegean Region 1995,* Izmir and Güllük, Turkey, Abstract, 38.

MEULENKAMP, J. E., WORTEL, M. J. R., VAN WARNE., W. A., SPAKMAN, W., & HOOGERDUIJN STRATING, E. 1994. On the Hellenic subduction zone and the geodynamic evolution of Crete since the late middle Miocene. *Tectonophysics,* **146**, 203–215.

MORRIS, A., CREER, K. M. C. & ROBERTSON, A. H. F. l990. Palaeomanetic evidence for clockwise rotations related to dextral shear along the Southern Troodos Transform Fault, Cyprus. *Earth and Planetary Science Letters,* **99**, 250–262.

PEARCE, J. A., LIPPARD, S. J. & ROBERTS, S. 1984. Characteristics and tectonic significance of supra-subduction zone ophiolites. *In:* KOKELAAR, B. P. & HOWELLS, M. F. (eds) *Marginal Basin Geology.* Geological Society, London, Special Publications, **16**, 77–89.

PLATT, J. P. 1990. Thrust mechanics in highly overpressured accretionary wedges. *Journal of Geophysical Research,* **95**(B6), 9025–9034.

POOLE, A. J. & ROBERTSON, A. H. F. 1992. Quaternary uplift and sea-level change at an active plate boundary, Cyprus. *Journal of the Geological Society, London,* **148**, 909–921.

PREMOLI SILVA, I., ERBA, E., SPEZZAFERRI, S. & CITA, M.

B. 1996. Variation in age of the diapiric mud breccia along and across the axis of the Mediterranean Ridge accretionary complex, *Marine Geology,* **132**, 175-202.

RESTON, T. J., VON HUENE, R., FRUEHN, J., DICKMAN, T. l995. A section across the Mediterranean Ridge – results of the IMERSE project. *EOS, American Geophysical Union Fall Meeting Supplement,* Abstract, F625.

ROBERTSON, A. H. F. 1987. The transition of a passive margin to an Upper Cretaceous foreland basin related to ophiolite emplacement in the Oman Mountains. *Geological Society of America Bulletin,* **99**, 633–653.

—— 1990. Tectonic evolution of Cyprus: Ophiolites and oceanic lithosphere. *In:* MALPAS, J., MOORES, E. M., PANAYIOTOU, A. & XENOPHONTOS, C. (eds) *Proceeding of International Symposium*, Troodos 1987, Geological Survey Department, Nicosia, Cyprus, 235–252.

—— 1993. Mesozoic–Tertiary sedimentary and tectonic evolution of Neotethyan carbonate platforms, margins and small ocean basins in the Antalya complex, S.W. Turkey. *In:* FROSTICK, L. E. & STEEL, R. (eds) *Tectonic Controls and Signatures in Sedimentary Successions.* International Association of Sedimentologists, Special Publication, **20**, 415–465.

—— 1994. Role of the tectonic facies concept in orogenic analysis and its application to Tethys in the Eastern Mediterranean region. *Earth-Science Reviews,* **37**, 139–213.

—— & DIXON, J. E. l984. Introduction: aspects of the geological evolution of the Eastern Mediterranean. *In:* DIXON, J. E. & ROBERTSON A. H. F. (eds) *The Geological Evolution of the Eastern Mediterranean.* Geological Society, London, Special Publications, **17**, 1–74.

—— & GRASSO, M. 1995. Overview of the Late Tertiary tectonic and palaeo-environmental development of the Mediterranean region. *Terra Nova,* **7**, 114–127.

—— & XENOPHONTOS, C. 1993. Development of concepts concerning the Troodos ophiolite and adjacent units in Cyprus. *In:* PRICHARD, H. M., ALABASTER, T., HARRIS, N. B. & NEARY, C. R. (eds) *Magmatic Processes and Plate Tectonics.* Geological Society, London, Special Publications, **70**, 85–120.

——, CLIFT, P. D., DEGNAN, P. & JONES, G. l991a. Palaeogeographic and palaeotectonic evolution of the Eastern Mediterranean Neotethys. *Palaeogeography, Palaeoclimatology, Palaeoecology,* **87**, 289–344.

——, EATON, S., FOLLOWS, E. J. & MCCALLUM, J. E. 1991b. The role of local tectonics versus global sea-level change in the Neogene evolution of the Cyprus active margin. *In:* MACDONALD, D. I. M. (ed.) *Sedimentation, Tectonics and Eustasy.* International Associations of Sedimentologists, Special Publication, **12**, 331–369.

——, KIDD, R. B., IVANOV, M. K., LIMONOV, A. F., WOODSIDE, J. M., GALINDO-ZALDIVAR, J. & NIETO, L. 1994. Probing continental collision in the Mediterranean Sea. *EOS,* **75**(21), 233, 238.

—— EMEIS, K.-C., RICHTER, C. *ET AL.* l995a. Evidence of

collisional processes associated with ophiolite obduction in the Eastern Mediterranean: Results of Ocean Drilling Program Leg 160. *Geology Today*, **5** (11), 213, 219–221.

——, ——, —— ET AL. 1995*b*. Eratosthenes Seamount, easternmost Mediterranean: evidence of active collapse and thrusting beneath Cyprus. *Terra Nova*, **7**(2), 254–265.

——, DIXON, J. E., BROWN, S., COLLINS, A., MORRIS, A., PICKETT, E., SHARP, I. & USTAÖMER, T. 1996. Alternative tectonic models for the Late Palaeozoic–Early Tertiary development of Tethys in the Eastern Mediterranean region. *In:* MORRIS, A. & TARLING, D. H. (eds) *Palaeomagnetism and Tectonics of the Mediterranean Region.* Geological Society, London, Special Publications, **105**, 39–263.

ROSS, D. A. & UCHUPI, E. 1977. Structure and sedimentary history of southeastern Mediterranean Sea–Nile Cone. *American Association of Petroleum Geologists Bulletin*, **61**, 872–902.

RYAN, W. B. F., STANLEY, D. J., HERSEY, J. B., FAHLQUIST, J. B. & ALLEN, T. D. 1971. The tectonics and geology of the Mediterranean Sea. *In:* MAXWELL, A. (ed.) *The Sea.* John Wiley, New York, **4**, 387–492.

——, HSÜ, K. J. ET AL. 1973. *Initial Reports of the Deep Sea Drilling Program.* US Government Printing Office, Washington, **13** (Pt. 1).

STAFFINI, F., SPEZZAFERRI, S. & AGHIB, F. 1993. Mud diapirs of the Mediterranean Ridge: sedimentological and micropalaeontological study of the mud breccia. *Rivista Italiano Paleontologie Stratiraphie,* **99**, 225–254.

STEININGER, F. F. & RÖGL, F. 1984. Paleogeography and palinspastic reconstruction of the Neogene of the Mediterranean and Paratethys. *In:* DIXON,. J. E. &

ROBERTSON, A. H. F. (eds) *The Geological Evolution of the Eastern Mediterranean.* Geological Society, London, Special Publications, **17**, 659–668.

STOCKMAL, G. S., BEAUMONT, C. & BOUTILIER, B. 1986. Geodynamic models of convergent margin tectonics. Transition from a rifted margin to overthrust belt and consequences for foreland basin development. *American Association of Petroleum Geologists Bulletin*, **70**, 727–730.

TAIRA, A., TOKUYAMA, H. & SOH, W. 1989. Acccretion tectonics and evolution of Japan. *In:* BEN-AVRAHAM, Z. (ed.) *The Evolution of the Pacific Ocean Margins.* Oxford University Press, 100–123.

TRUFFERT, C., CHAMOT-ROOKE, N., LALLEMANT, S., DE VOOGD, B., HUCHON, P. & LE-PICHON, X. 1993. The crust of the Western Mediterranean Ridge from deep seismic data and gravity modelling. *Geophysical Journal International,* **114**, 360–372.

WEISSEL, J. K., TAYLOR, B. & KARNER, G. D. 1982. The opening of the Woodlark basin subduction of the Woodlark spreading system, and the evolution of northern Melanesia since mid-Pliocene time. *Tectonophysics*, **87**, 253–277.

WHITING, B. M. & THOMAS, W. A. 1994 Three-dimensional controls on the subsidence of a foreland basin associated with a thrust belt recess, Black Warrior basin, Alabama and Mississippi. *Geology*, **22**, 727–730.

WOODSIDE, J. M. 1977. Tectonic elements and crust of the eastern Mediterranean Sea. *Marine Geophysical Research*, **3**, 317–354.

—— 1991. Disruption of the African plate margin in the Eastern Mediterranean. *In:* SALEM, M. J. (ed.) *The Geology of Libya.* Elsevier, Oxford, **6**, 2319–2329.

Rare earth element anomalies in the Nankai accretionary prism, Japan

JANE L. ALEXANDER

*Department of Geological Sciences, University College London, Gower Street,
London WC1E 6BT, UK*

*Address for correspondence: Department of Mineralogy, The Natural History Museum,
Cromwell Road, London SW7 5BD, UK*

Abstract: Rare earth element (REE) patterns are often used as provenance indicators, but there is growing evidence that this signal may be distorted by REE mobility. This study focuses on turbiditic and hemipelagic muds from Ocean Drilling Program (ODP) Site 808, drilled in the toe of the Nankai accretionary prism. Most of these muds have a typical shale REE composition, but several samples are enriched in REE. Four of these are associated with hydrothermal deposits, rich in manganese (up to 18% MnO). A further two have minor mineral phases, zircon and florencite, with high REE concentrations. There is also a heavy REE enrichment in one sample from the décollement zone. This has not yet been explained, but it does not appear to have a mineralogical control.

Rare earth element (REE) patterns play a key role in understanding the provenance of mudrocks (McLennan 1989). However, there is uncertainty about the extent to which REE are mobile in muds. Mobility may be due to a combination of pore-fluid complexation, mineral precipitation and adsorption of REE to mineral surfaces. Such processes vary depending on temperature, pressure and chemical conditions, and may lead to a fractionation between heavy and light REE. This study focuses on REE associated with turbiditic and hemipelagic muds from the Ocean Drilling Program (ODP) Leg 131, Site 808, drilled in the toe of the Nankai accretionary prism (Fig. 1; Shipboard Scientific Party 1991a).

The Nankai accretionary prism marks the boundary between the Philippine Sea Plate and the Eurasia Plate. The Shikoku Basin, part of the Philippine Sea Plate, is being subducted beneath the Southwest Japan Arc (Fig. 1). This basin formed as a backarc basin behind the Izu–Bonin Arc during the Oligocene to middle Miocene (Shipboard Scientific Party 1991b), and the relict spreading centre is currently being subducted close to Site 808. Sediments in the central part of the basin were mainly hemipelagic, with some input of hydrothermal material close to the spreading centre. This can be seen in the form of umbers (beige-coloured muds enriched in Mn, Ca and trace elements) and other metalliferous muds.

Initial geochemical studies suggested that there are several anomalies in the REE patterns, parti-cularly at the décollement zone and in the ancient umbers (Pickering *et al.* 1993). However, REE and major element analyses, and mineralogical studies, were all carried out on different groups of samples (Shipboard Scientific Party 1991a; Underwood *et al.* 1993), therefore the relationships between these components could not be determined accurately, as their concentrations change over a few centimetres of core. Recent resampling of the core has allowed a more detailed study of the décollement (945.0–964.3 metres below sea floor (mbsf)) and other sediments enriched in REE.

The hemipelagic muds all have similar major element and REE compositions, and are composed mainly of quartz, feldspar, and mixed-layer clays (this work; Shipboard Scientific Party 1991a). Darker layers, noted during sampling, also contain pyrite (FeS_2). This has decomposed in some fractured samples to form jarosite ($KFe_3(OH)_6(SO_4)_2$) giving them a distinctive yellow colour. Décollement samples have REE, trace and major element concentrations similar to the other hemipelagic muds, despite being brecciated. These four types of mud are referred to as background muds, dark muds, yellow muds and décollement muds, respectively. Two distinctly lighter layers are referred to as umbers, because of their major element concentrations. Two layers that are slightly lighter than the surrounding muds, and are enriched in manganese and iron, are referred to as metalliferous muds.

The main aim of this study is to investigate the

ALEXANDER, J. L. 1998. Rare earth element anomalies in the Nankai accretionary prism, Japan. *In*: CRAMP, A., MACLEOD, C. J., LEE, S. V. & JONES, E. J. W. (eds) *Geological Evolution of Ocean Basins: Results from the Ocean Drilling Program*. Geological Society, London, Special Publications, **131**, 273–285.

273

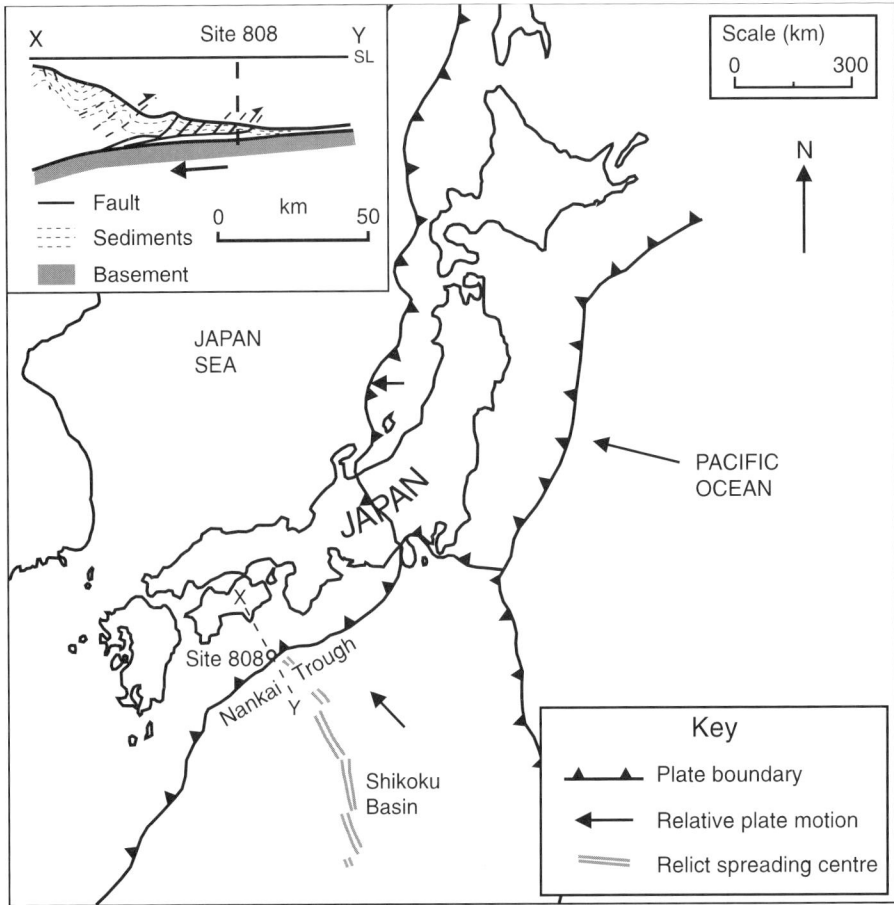

Fig. 1. Location of ODP Site 808 (modified after Shipboard Scientific Party 1991*b*).

relationships between major element chemistry, mineralogy and REE anomalies. A knowledge of the correlation between the REE concentrations and mineralogy within each sample will provide a very good indication of whether the REE signature is a result of provenance or a product of fluid–mineral interactions.

Method

Samples were selected after inspecting the core from ODP Site 808C. They include 29 samples taken at frequent intervals in the décollement zone, 30 background muds from above and below the décollement, two bands which were distinctly lighter in colour, and 18 other samples with subtly different colour or fracturing characteristics to the background muds (Table 1).

The samples were dried for one week at 50°C. At higher temperatures, there is a risk of altering the surface structure of clay minerals and fixing exchangeable ions (Lewis & McConchie 1994). Preliminary crushing using a fly press with hardened steel plates reduced the sample grain size to less than 5 mm. Contamination is minimal with this technique (Fairchild *et al.* 1988). Each sample was then crushed for 30 s in a Tema® disk mill with an agate barrel, resulting in a fine powder (grain size *c.* 40 µm). Finally, the powders were dried for a further 24 h at 50°C.

For each sample, 0.2 g of powder was digested in an open PTFE beaker using the following method. Concentrated HNO_3 (1.5 ml) was added to each sample, and evaporated to dryness on a sand-bath. This was followed by fuming the sample with 12 ml HF (40%) and 2 ml $HClO_4$ and then digesting the residue in 3 ml concentrated HNO_3

and 18 ml deionized water. Samples were then filtered using Whatman No. 42 filter papers, made up to exactly 100 ml and stored in HDPE bottles.

ICP–AES analysis

All samples were analysed for major and minor elements using an ARL 3410 Minitorch inductively coupled plasma atomic emission spectrometer (ICP–AES) at the Natural History Museum, London. Drift samples containing all the elements of interest were run after every fourth sample to correct for drift during each run. Several acid blanks and reference standards (SGR-1 and NBS1633a) were also run. The blanks were below the detection limits and reference standards were in good agreement with published data (Govindaraju 1989).

ICP–MS analysis

All samples were analysed for the rare earth elements, uranium and thorium using a V G Elemental PlasmaQuad inductively coupled plasma mass spectrometer (ICP–MS) at the University of Bristol. Each sample was spiked with a 100 ppb Re and Ru internal standard. Several acid blanks and reference standards (SGR-1 and NBS1633a) were also run. The blanks showed no contamination with REE, U or Th, and reference standards were in good agreement with published data (Govindaraju 1989).

XRD and electron microprobe analysis

A small selection of powders were analysed using a position-sensitive detector X-ray diffractometer (XRD), to investigate their mineralogy and compare it with the results of bulk chemical analysis. They were chosen to include samples with high REE concentrations, as well as several background muds for comparison. These samples were also analysed using a Hitachi S-2500 scanning electron microscope, to examine minor mineral phases and their influence on the anomalous REE levels in some muds.

Estimation of errors

Systematic errors in the results from both the ICP–AES and the ICP–MS were eradicated by using either drift correction or internal standards. This correction was confirmed by checking with reference samples (SGR-1 and NBS1633a). The effects of random errors are represented by the error bars in the figures. They are a combination of errors in analysis, sample mass and sample volume. Analysis errors were calculated from the standard deviations of repeat runs for ICP–MS analysis, and detection limits for ICP–AES analysis.

Results

The major element chemistry of the hemipelagic muds varies little with depth, although several samples have anomalously high or low concentrations of one or more elements (Fig. 2). Most of the mudrocks have a composition of between 15% and 20% Al_2O_3, approximately 2.5% MgO, 2% Na_2O and 4% K_2O, with less than 1% MnO. Minor and trace element concentrations are also relatively constant with depth, with only a few samples having anomalously high or low concentrations (Fig. 2). XRD results indicate that the major mineral phases are quartz, feldspar and mixed-layer clays, which is consistent with the major element chemistry. The concentrations of iron, sulphur and calcium are more variable than most other elements, and are related to the pyrite and calcite content of the mudrocks. Titanium and zircon are present at lower levels than reported in other shales (< 4000 ppm Ti and 100 ppm Zr; cf. Gromet et al. 1984; Condie 1991), suggesting that there are very few detrital minerals. REE concentrations are almost constant with depth, but are enriched at the same depths as the major and minor element anomalies (Fig. 3, Table 1). The REE concentrations and chondrite-normalized patterns for the four types of hemipelagic muds are similar to those for average shales (Fig. 4; Taylor and McLennan 1985).

Anomalies

There are seven distinctly anomalous samples out of the 79 analysed. They are enriched in some or all of the REE, and also have anomalous major and minor element chemical signatures. The two samples with the highest REE concentrations (up to 187 ppm La) correspond with the two distinctly lighter-coloured layers, observed at 1060.08 mbsf and 1098.96 mbsf. They are also enriched in Mn, Ca, S, Sr and Ba, and depleted in Al, Mg, Na, K and Ti. A further two samples, at 1025.11 mbsf and 1032.05 mbsf, are enriched in all REE, Mn and Fe, but to a lesser extent (up to 47 ppm La). These samples were slightly lighter in colour than the background muds. Two samples are enriched in REE, at 963.73 mbsf and 1054.62 mbsf (up to 41 ppm La). They are both enriched in Ca, but were not visibly different to the background muds. The final REE anomaly (up to 43 ppm La) is in one of the yellow muds (1052.58 mbsf), which is also enriched in Zr.

Table 1. *REE concentrations in samples from Site 808C*

Core	Section	Interval	Depth	Type	La	Ce	Pr	Nd	Sm	Eu	Gd	Tb	Dy	Ho	Er	Tm	Yb	Lu
64R	04	99-101	910.99	BM	31.44	69.27	6.68	23.63	4.55	1.36	7.45	0.64	3.45	0.60	1.94	0.30	1.86	0.31
65R	01	50-52	915.60	BM	29.81	68.78	6.52	23.87	4.15	1.31	7.07	0.61	3.16	0.58	1.55	0.26	1.40	0.24
65R	02	102-104	917.62	BM	30.71	67.92	6.70	22.88	4.42	1.26	6.49	0.59	3.12	0.56	1.59	0.23	1.35	0.25
66R	01	140-142	926.20	YM	36.45	84.00	7.27	25.83	4.64	1.56	9.27	0.79	3.64	0.56	1.96	0.19	1.66	0.27
66R	03	60-62	928.40	BM	26.79	63.07	5.92	22.34	4.07	1.33	6.40	0.62	3.15	0.61	1.55	0.24	1.52	0.25
67R	01	46-48	934.96	BM	31.87	71.97	6.84	23.25	4.09	1.20	5.96	0.59	3.28	0.61	1.86	0.26	1.53	0.26
67R	01	126-128	935.76	YM	31.39	71.95	6.49	22.01	3.89	1.39	7.20	0.63	3.84	0.56	1.84	0.18	1.63	0.27
67R	02	16-18	936.16	BM	28.47	67.60	6.53	23.41	4.29	1.35	6.90	0.65	3.59	0.65	1.88	0.29	1.78	0.30
68R	01	54-56	944.74	D	31.59	74.43	6.70	23.54	4.12	1.42	8.73	0.75	3.62	0.58	1.92	0.18	1.64	0.27
68R	01	80-82	945.00	D	34.25	77.70	6.50	22.97	4.14	1.31	7.03	0.64	3.50	0.59	1.88	0.21	1.87	0.30
68R	01	104-106	945.24	D	29.33	69.49	6.99	24.26	4.63	1.42	8.36	0.72	3.83	0.61	1.87	0.26	1.76	0.29
68R	01	125-127	945.45	D	27.87	65.90	6.12	22.53	3.86	1.26	6.33	0.67	3.32	0.58	1.73	0.25	1.61	0.28
68R	02	4-6	945.74	D	28.45	67.62	6.51	23.89	4.58	1.51	7.87	0.76	4.01	0.68	2.10	0.30	1.84	0.29
68R	02	30-32	946.00	D	28.85	68.43	6.52	23.12	4.06	1.13	7.06	0.57	3.35	0.58	1.89	0.29	1.62	0.31
69R	01	20-22	954.00	D	26.09	59.47	5.80	21.45	4.04	1.24	6.96	0.57	3.43	0.55	1.83	0.25	1.65	0.27
69R	01	36-38	954.16	D	22.08	53.18	5.15	18.59	3.54	1.21	6.18	0.54	3.00	0.46	1.59	0.22	1.41	0.24
69R	01	57-59	954.37	D	35.21	79.17	7.13	25.28	4.41	1.39	9.11	0.75	3.81	0.60	1.98	0.20	1.81	0.28
69R	01	99-101	954.79	D	29.90	69.66	7.00	24.92	4.55	1.55	9.28	0.76	4.27	0.69	1.97	0.32	1.96	0.36
69R	01	120-122	955.00	D	30.19	66.87	6.56	22.68	4.32	1.16	6.26	0.59	3.33	0.67	1.94	0.27	1.76	0.29
69R	02	27-29	955.57	D	34.01	76.70	6.85	24.46	4.34	1.61	7.84	0.75	4.06	0.67	2.08	0.24	1.93	0.32
69R	02	60-62	955.90	D	28.34	66.53	6.21	22.29	4.01	1.24	6.27	0.60	3.19	0.58	1.75	0.24	1.60	0.28
69R	02	92-94	956.22	D	35.83	79.59	7.06	25.23	4.36	1.49	7.68	0.67	3.78	0.61	1.93	0.19	1.71	0.28
69R	02	106-108	956.36	D	26.07	59.13	5.74	19.22	3.18	1.11	5.53	0.47	2.40	0.45	1.55	0.21	1.64	0.27
69R	02	125-127	956.55	D	33.55	73.78	6.54	23.49	4.00	1.41	8.17	0.66	3.80	0.62	2.00	0.19	1.98	0.28
69R	03	10-12	956.90	D	28.20	63.64	6.56	24.66	4.31	1.47	7.37	0.69	3.72	0.66	2.00	0.28	1.86	0.31
69R	03	34-36	957.14	D	31.45	72.55	6.95	25.09	4.58	1.78	10.12	0.79	4.26	0.68	2.20	0.24	1.81	0.29
69R	03	75-77	957.55	D	41.61	78.13	7.08	25.27	4.23	1.60	8.08	0.64	3.66	0.60	1.94	0.16	1.69	0.31
69R	03	105-107	957.85	D	32.17	73.65	6.95	25.47	4.58	1.63	9.29	0.76	3.82	0.63	1.98	0.21	1.75	0.24
69R	04	20-22	958.50	D	34.27	75.23	6.96	23.73	4.35	1.49	8.27	0.69	3.40	0.57	1.90	0.17	1.73	0.27
69R	04	44-46	958.74	D	31.63	69.99	6.67	23.27	4.07	1.53	8.87	0.78	3.81	0.60	1.94	0.20	1.76	0.24
69R	04	91-93	959.21	D	26.21	56.75	5.79	20.47	3.94	1.29	6.58	0.59	3.07	0.53	1.50	0.23	1.52	0.26
69R	04	110-112	959.40	D	36.37	80.93	7.88	26.70	4.96	1.27	7.10	0.64	3.71	0.58	1.91	0.24	1.67	0.30
69R	04	125-127	959.55	D	30.44	68.79	6.92	24.30	4.74	1.50	8.20	0.70	3.68	0.67	2.12	0.28	1.84	0.32
69R	05	2-4	959.82	D	36.42	80.73	7.44	25.72	4.54	1.30	6.86	0.67	3.57	0.61	1.75	0.24	1.59	0.26
69R	05	57-59	960.37	D	32.20	71.84	6.72	22.89	4.07	1.48	7.98	0.66	3.46	0.56	1.85	0.20	1.64	0.27
70R	01	8-10	963.48	DA	29.21	63.31	6.51	24.26	5.13	1.56	8.64	0.86	4.87	0.96	2.97	0.45	2.96	0.48
70R	01	33-35	963.73	D	31.96	72.40	6.95	25.35	4.33	1.45	7.74	0.73	4.09	0.68	2.14	0.29	1.83	0.35
70R	01	62-64	964.02	BM	28.41	65.04	6.12	20.98	3.81	1.51	8.36	0.71	3.60	0.56	1.74	0.18	1.57	0.22
70R	03	46-48	966.96	BM	27.53	62.58	6.20	22.61	4.19	1.39	7.26	0.61	3.29	0.57	1.82	0.26	1.64	0.28
71R	01	74-76	973.84	BM	35.92	79.14	6.72	23.98	4.03	1.33	7.26	0.61	3.50	0.56	1.94	0.18	1.68	0.26
71R	04	47-49	978.07	BM	26.23	58.72	5.90	20.68	4.28	1.35	6.57	0.56	3.06	0.52	1.72	0.21	1.47	0.24

Core	Sect.	Interval	Type	Depth														
72R	02	26-28	YM	984.56	28.78	64.01	5.56	18.47	2.93	1.27	6.37	0.48	2.81	0.40	1.51	0.12	1.36	0.22
72R	02	84-86	BM	985.14	30.77	70.04	6.60	24.10	3.92	1.32	6.53	0.64	3.49	0.63	1.82	0.25	1.62	0.26
72R	04	71-73	BM	987.98	28.27	67.09	6.45	23.04	4.03	1.25	7.37	0.64	3.24	0.57	1.77	0.25	1.59	0.26
76R	03	62-64	DM	1024.12	24.41	58.94	5.26	18.85	3.20	1.48	8.63	0.65	3.33	0.51	1.69	0.18	1.45	0.23
76R	04	11-13	MM	1025.11	47.31	135.02	12.88	53.24	11.91	3.60	18.70	1.91	10.42	1.75	4.32	0.50	2.88	0.49
77R	01	33-35	DM	1030.23	28.14	70.10	6.01	20.41	3.88	1.55	8.13	0.72	3.28	0.49	1.81	0.16	1.53	0.22
77R	01	72-74	BM	1030.62	32.61	78.67	6.90	23.65	3.98	1.68	8.70	0.73	3.75	0.59	1.97	0.21	1.84	0.26
77R	02	65-67	MM	1032.05	32.11	81.33	7.69	29.83	5.75	1.89	11.65	0.99	5.32	0.89	2.63	0.35	2.41	0.45
77R	04	78-80	DM	1035.18	24.68	59.25	5.55	19.62	3.53	1.17	6.08	0.50	2.77	0.51	1.58	0.22	1.46	0.24
77R	04	96-98	BM	1035.36	30.77	72.62	6.74	23.97	4.12	1.34	7.19	0.61	3.49	0.59	1.87	0.30	1.77	0.32
78R	01	30-32	DM	1039.40	33.90	77.42	6.84	22.84	3.94	1.19	6.13	0.51	3.05	0.59	1.66	0.25	1.42	0.24
78R	01	61-63	BM	1039.71	34.20	79.52	6.93	23.51	4.16	1.65	9.66	0.68	4.02	0.64	2.24	0.21	1.92	0.32
78R	05	51-53	BM	1045.61	32.27	76.67	6.47	22.12	4.07	1.56	7.55	0.63	3.49	0.54	1.70	0.13	1.57	0.26
78R	05	102-104	DM	1046.12	28.56	68.74	6.46	23.38	4.34	1.46	7.53	0.65	3.28	0.59	1.79	0.27	1.71	0.26
79R	02	110-112	YM	1050.90	15.89	40.11	3.91	14.57	3.34	1.56	4.82	0.49	2.62	0.53	1.44	0.21	1.43	0.23
79R	03	128-130	YMA	1052.58	43.43	119.86	11.61	42.52	8.78	2.61	13.32	1.32	6.95	1.19	2.97	0.35	2.22	0.35
79R	04	78-80	BM	1053.58	29.27	71.49	6.84	24.39	4.15	1.47	9.42	0.64	3.59	0.58	1.86	0.28	1.69	0.26
79R	05	32-34	BMA	1054.62	40.93	89.54	8.11	29.52	5.70	2.21	13.85	1.17	5.79	0.97	2.89	0.37	2.45	0.41
79R	05	47-49	DM	1054.77	31.18	69.48	6.43	22.47	3.76	1.71	9.16	0.66	3.45	0.49	1.59	0.21	1.56	0.25
80R	01	19-21	YM	1057.99	29.69	69.16	6.05	20.11	3.18	1.08	5.40	0.48	2.53	0.46	1.40	0.24	1.30	0.23
80R	01	93-95	BM	1058.73	34.12	80.71	6.90	22.40	3.92	1.39	7.06	0.57	3.11	0.50	1.88	0.14	1.52	0.26
80R	02	7-9	YM	1059.37	23.12	58.00	5.49	18.82	3.60	1.27	6.30	0.54	3.01	0.46	1.45	0.20	1.24	0.21
80R	02	78-80	U	1060.08	186.74	301.92	23.40	86.64	14.38	5.73	45.78	4.10	24.27	5.61	15.82	2.04	11.49	1.83
80R	02	105-107	DM	1060.35	31.78	72.98	6.50	22.38	3.98	1.58	7.71	0.62	3.22	0.50	1.72	0.14	1.43	0.24
80R	03	78-80	BM	1061.58	30.82	70.80	6.96	25.19	4.30	1.79	8.51	0.73	3.75	0.59	1.86	0.25	1.68	0.28
80R	04	16-18	BM	1062.46	28.10	68.46	6.11	21.97	3.74	1.51	9.03	0.67	3.39	0.52	1.78	0.17	1.52	0.28
81R	01	30-32	YM	1067.40	28.58	68.16	6.18	20.88	3.86	1.16	6.74	0.51	2.96	0.49	1.60	0.23	1.43	0.23
81R	01	96-98	BM	1068.06	28.49	67.70	6.27	22.34	4.18	1.74	10.47	0.78	3.71	0.57	1.82	0.25	1.56	0.28
81R	02	60-62	BM	1069.20	33.44	75.33	6.92	24.51	4.10	1.33	6.50	0.60	3.09	0.53	1.62	0.25	1.47	0.25
82R	02	49-51	BM	1078.49	36.99	84.94	7.09	24.02	4.11	1.30	7.29	0.61	3.31	0.60	2.12	0.18	1.75	0.27
82R	02	82-84	BM	1078.82	32.98	77.77	6.45	21.95	3.76	1.29	6.70	0.56	3.05	0.47	1.73	0.16	1.43	0.23
83R	03	66-68	YM	1087.96	31.46	71.86	5.96	19.86	3.23	1.19	6.82	0.49	2.70	0.40	1.55	0.12	1.50	0.23
83R	03	30-32	BM	1089.10	27.83	63.22	5.61	18.91	3.10	1.15	6.28	0.51	2.78	0.47	1.52	0.14	1.36	0.25
84R	02	77-79	BM	1094.27	28.26	64.93	6.24	22.13	4.05	1.18	5.00	0.58	3.17	0.56	1.78	0.28	1.74	0.24
85R	01	66-68	U	1098.96	168.39	349.32	29.66	121.32	22.69	21.20	52.48	4.98	29.19	5.91	15.67	1.99	11.66	1.95
85R	01	120-122	BM	1099.50	34.31	81.46	7.55	25.23	4.27	1.28	6.54	0.65	3.44	0.59	1.90	0.24	1.68	0.27
85R	03	37-39	BM	1101.67	35.65	80.43	6.79	23.02	3.68	1.28	6.94	0.56	3.11	0.53	1.76	0.12	1.62	0.25
85R	03	76-78	BM	1102.06	30.73	74.08	6.88	24.79	4.26	1.48	8.42	0.77	3.39	0.59	1.90	0.30	1.75	0.26

Samples are numbered following the ODP standard method.
Depths are in metres below sea floor (mbsf) and REE concentrations in ppm.
Sample types are: BM, background muds; D, décollement; DM, dark muds; YM, yellow muds; U, umber; MM, metalliferous muds; BMA, background mud anomaly; DA, décollement anomaly; YMA, yellow mud anomaly.

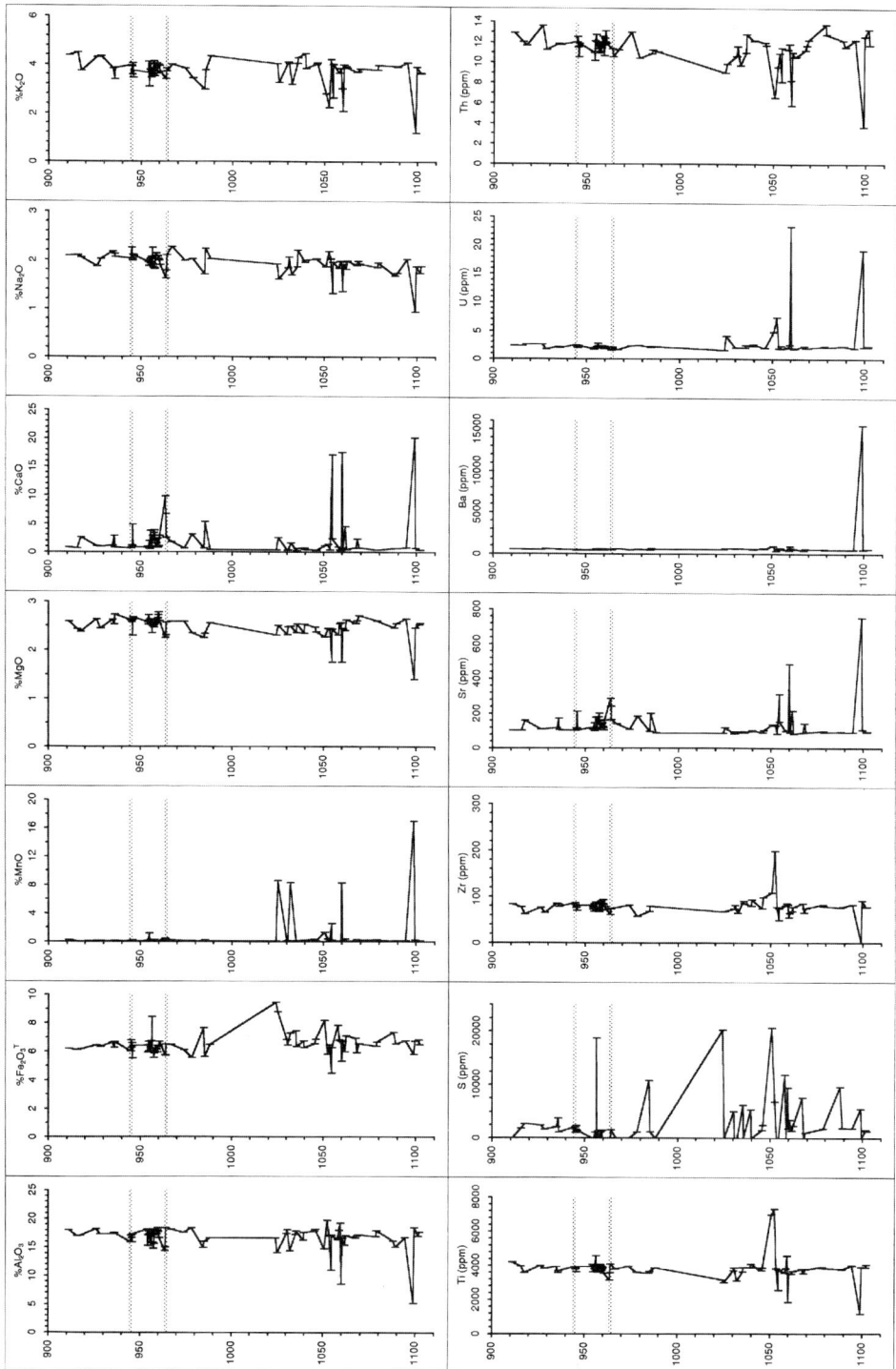

Fig. 2. Major and minor element concentrations versus depth (mbsf). Dotted lines represent the top and bottom of the décollement zone.

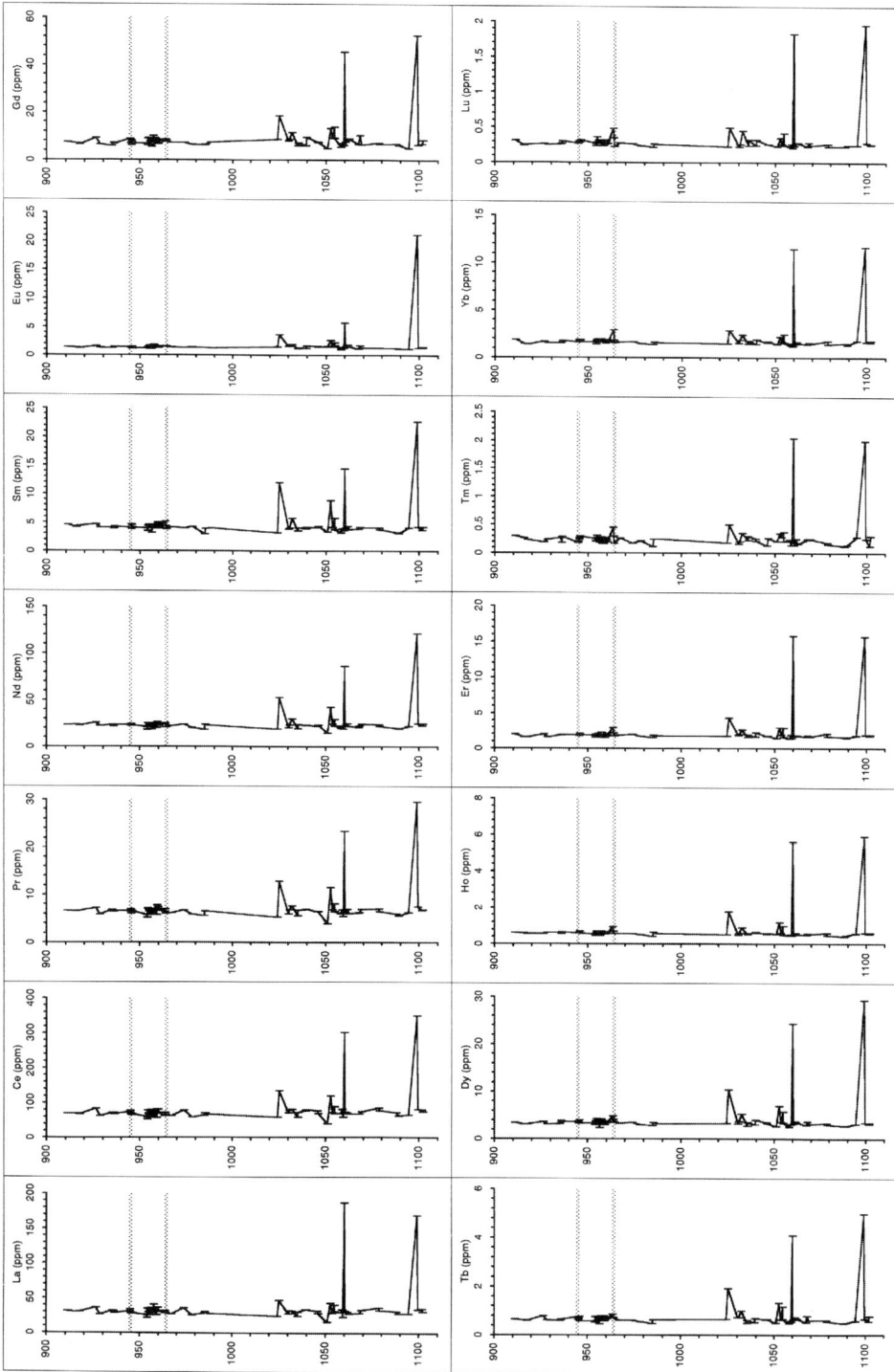

Fig. 3. REE concentrations versus depth (mbsf). Dotted lines represent the top and bottom of the décollement zone.

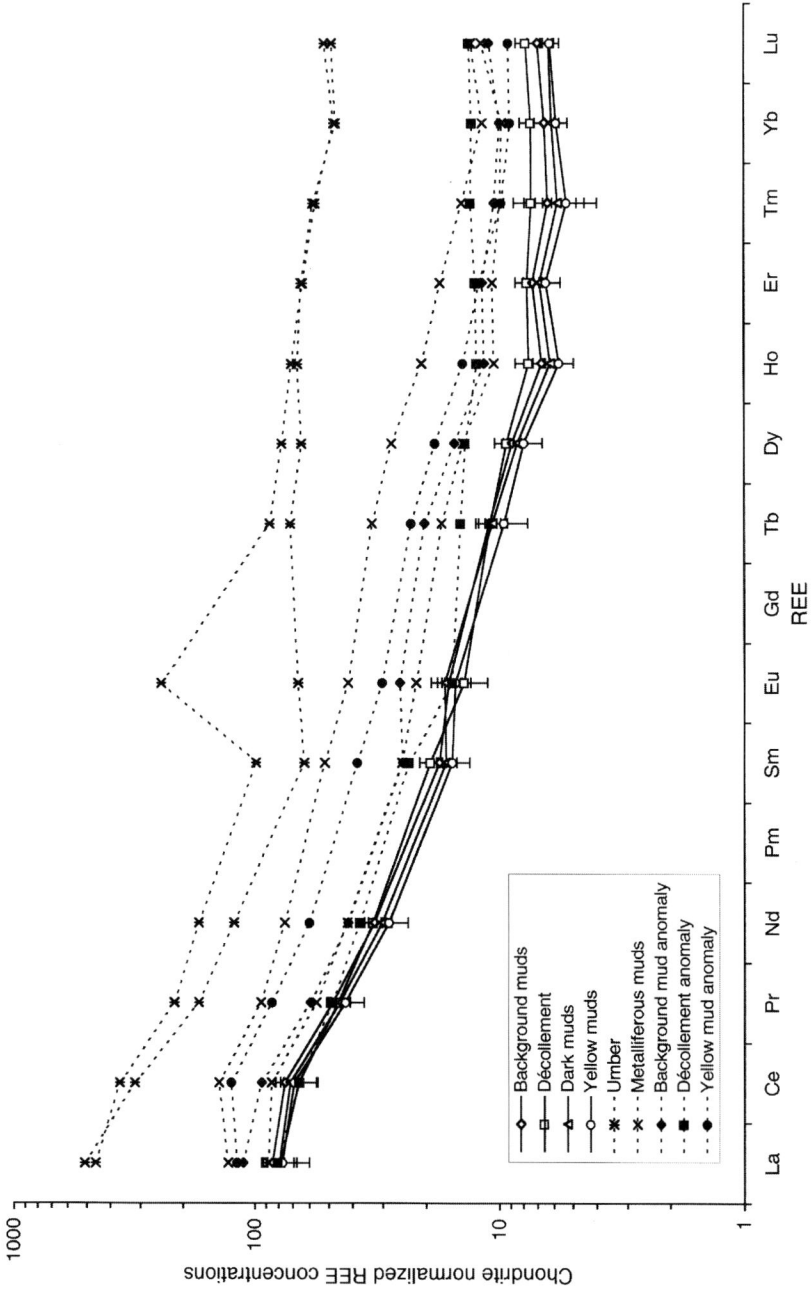

Fig. 4. Chondrite-normalized REE concentrations. Solid lines represent averages for background muds, with error bars showing the standard deviations. Dotted lines represent individual anomalous samples.

Discussion

Major element chemistry and XRD results indicate that the predominant minerals in the background muds are quartz, feldspar and illite/smectite mixed-layer clays. Most of the REE in these samples will be adsorbed onto clay mineral surfaces. Previous work on mudrocks has shown that clays have high total REE concentrations, with similar patterns to average shale, and quartz and feldspar have low total REE concentrations (Taylor & McLennan 1985). There are very few detrital minerals, such as zircon, monazite, titanite and allanite, as these muds are mainly hemipelagic. This was confirmed by the low levels of zirconium measured, and by examining selected samples using an electron microscope. Possible heavy minerals, which appear as small, bright grains, were then analysed using the electron microprobe. These detrital minerals would normally contribute significantly to the REE levels, particularly the heavy REE (Taylor & McLennan 1985; McLennan 1989; Götze & Lewis 1994), but this is not observed in these samples.

The muds in the décollement zone are chemically and mineralogically the same as the background muds. Trace element and REE compositions fall within the same ranges as the background muds (Figs 2 and 3), despite the much greater fracturing and brecciation, which would allow greater fluid flow along this zone. The dark-coloured muds have a very similar major element composition and mineralogy to the background muds, but are enriched in both iron and sulphur. There is a strong correlation between the levels of iron and sulphur (Fig. 5). The equation of the regression line through the dark muds implies that the ratio of Fe : S is 1 : 1.578, after converting mass to number of atoms. Pyrite (FeS_2) was visible in some samples under an optical microscope, but another more iron-rich phase must also be present. All of the yellow muds lie below the regression line for dark muds, suggesting that iron has been leached as the pyrite decomposed. These yellow samples are all fractured, providing numerous paths for pore fluid flow. Jarosite ($KFe_3(OH)_6(SO_4)_2$), a common product of pyrite decomposition (e.g. Bladh 1982; Michel & van Everdingen 1987), has been identified by XRD analysis.

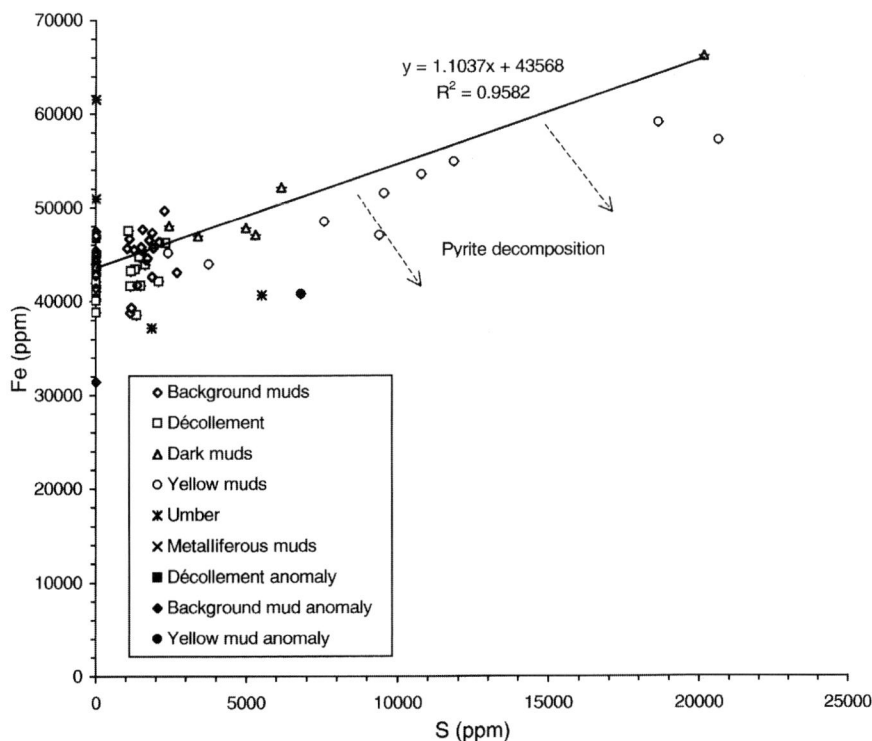

Fig. 5. Plot of iron concentrations versus sulphur concentrations, showing relationships with mineralogy. The regression line is fitted through the data for the dark muds, and all of the yellow muds lie below this line.

Umbers

The two distinctly lighter samples from 1060.08 mbsf and 1098.96 mbsf have a much lower clay content than the background muds, as indicated by their low aluminium concentrations (Fig. 6). Pickering *et al.* (1993) identified similar samples from this core as umbers, based upon their major element chemistry. They have high manganese concentrations (up to 18% MnO), similar to ophiolitic umbers (Robertson 1975), and it is thought that the samples analysed here are also umbers. In contrast to such ophiolitic umbers, they have high levels of calcium (Fig. 2), although this has been reported in some modern-day deposits (Hodkinson *et al.* 1994). The umbers found here are contained within the sediments, rather than directly overlying basalt. This may be due to the high rate of sediment input, or they may be a more distal product of a hydrothermal vent.

Previous workers (Underwood *et al.* 1993) found a general region with elevated levels of Ca, Mg, Fe and Mn between 1087 and 1111 mbsf. However, the more detailed sampling used here shows that these increased levels are present in discrete layers, whose chemistry changes with their stratigraphic position. The lower two samples are chemically similar to umbers and oxide sediments found above the axial lavas at Suhaylah, Oman (Robertson & Boyle 1983; Robertson & Fleet 1986), while the stratigraphically higher metalliferous muds are similar to oxide sediments higher in the sequence at Suhaylah. It is not possible to consider the three-dimensional structure of these hydrothermal deposits, as data are only available from this single core.

The REE concentrations are enriched in these two umber samples, similar to other ferro-manganese oxyhydroxide crusts (Fleet 1984; Robertson & Fleet 1986; De Carlo & McMurtry 1992). The deeper umber, which was probably closer to the ridge, is more enriched in REE than the shallower umber. It has a chondrite normalized REE pattern that is very similar in shape to that of modern vent fluids (Fig. 4; James *et al.* 1995). The positive europium anomaly is typical of an oceanic hydrothermal system (McLennan 1989). In both these samples and in the metalliferous muds, there is a strong correlation between calcium and the REE, which is not present in samples with much lower manganese concentrations (Fig. 7).

Metalliferous muds

The two metalliferous muds, at 1025.11 mbsf and 1032.05 mbsf, are enriched in manganese and iron, and have a lower percentage of clay minerals than the background muds (Figs 2, 5 and 6). Unlike the umbers they are not enriched in other elements associated with hydrothermal deposits, such as sulphur, barium and strontium. This, along with their stratigraphically higher position, suggests that they were deposited further from the spreading centre, but were probably still associated with the hydrothermal system. These metalliferous muds are enriched in REE, but to a lesser extent than the umbers, and they do not exhibit the same oceanic hydrothermal system patterns. However, they do display the same correlation with calcium (Fig. 7). Again, the REE are probably associated with the ferromanganese oxyhydroxide mineralization, and

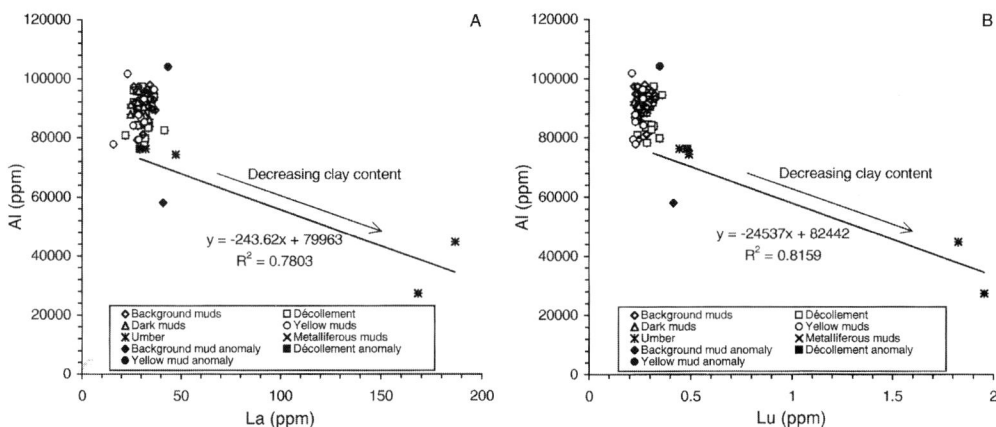

Fig. 6. Aluminium concentrations versus (**A**) lanthanum and (**B**) lutetium concentrations. Samples with anomalously high REE have a lower clay content than the background muds. The regression lines are fitted through the umbers, metalliferous muds, and anomalous background and décollement muds, all of which contain low levels of clay minerals.

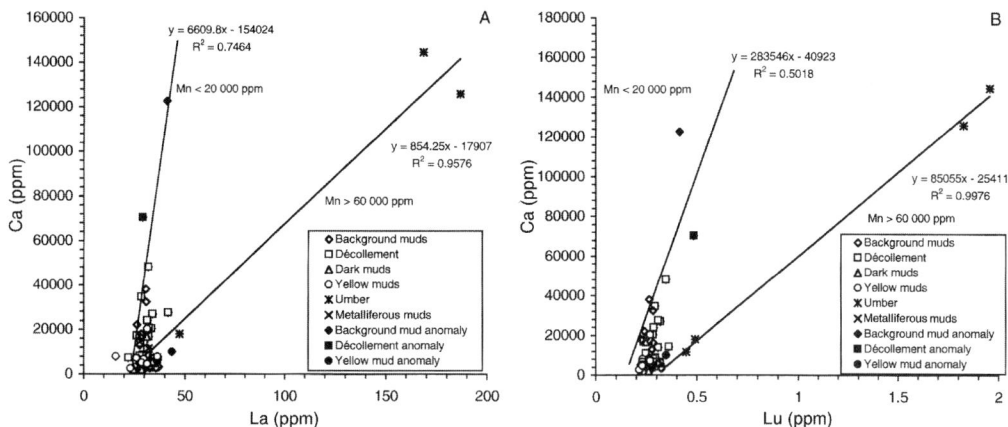

Fig. 7. Calcium concentrations versus (**A**) lanthanum and (**B**) lutetium concentrations. There are correlations between the REE and calcium, which are dependent on manganese content. The regression lines for samples with >60 000 ppm manganese are fitted through the umbers and metalliferous muds. The regression lines for samples with <20 000 ppm manganese are fitted through all muds with >20 000 ppm calcium (except the umbers and metalliferous muds).

there is a slight positive cerium anomaly, similar to that described by Piper (1974) in other such oxyhydroxides. Major element chemistry and electron microprobe studies indicate that the REE in these two samples are not present in accessory mineral phases.

Background mud anomaly

The anomalous background mud, at 1054.62 mbsf, is enriched in calcium, but is otherwise chemically similar to the other background muds (Fig. 2). There may be a correlation between calcium and REE, but it is much poorer than that of the high manganese samples (Fig. 7). XRD studies indicate that the calcium is present in calcite, which normally has much lower REE concentrations than clays (McLennan 1989). If the REE were contained in major mineral phases, the sample should be depleted in REE rather than enriched, due to the dilution effect of the calcite. As expected, electron microprobe results for this sample indicate that the REE are present in an accessory mineral phase. It is composed of aluminium, phosphorus and calcium, as well as the REE, and is probably a florencite $(REEAl_3(PO_4)_2(OH)_6)$ – crandallite $(CaAl_3(PO_4)_2(OH)_5H_2O)$ solid solution (Burt 1989).

Décollement mud anomaly

The décollement mud, at 963.73 mbsf, initially appears similar to the background mud anomaly. It is enriched in calcium, in the form of calcite, and would therefore be expected to be depleted rather than enriched in REE. However, the light REE

concentrations are similar to the background muds, and the enrichment becomes greater towards the heavy REE. In the original geochemical work on the same core by Pickering et al. (1993), three out of the four samples from the décollement showed this anomaly. In the more detailed sampling described here, covering the same depth range, only one out of the 29 samples is enriched in heavy REE. These four anomalous samples are not adjacent, suggesting that the feature causing the anomaly must only be present at discrete levels within the décollement zone.

Several possible causes for the heavy REE anomaly have been explored, but none have yet successfully explained it. Unlike the anomalous background mud, electron microprobe work on this sample and the three anomalous samples of Pickering et al. (1993) has not identified any accessory mineral phase containing high REE concentrations. The accessory minerals present are the same as those found in the background muds, and are therefore very unlikely to be responsible for the anomaly. Differences in clay mineralogy could potentially account for differences in adsorbed REE. Ca-montmorillonite can be enriched in the heavy REE, compared to smectite (Maza-Rodriguez et al. 1992), and this was investigated as a possible source of the anomaly. However, XRD results show that there is very little difference in clay mineralogy between the anomalous and background samples. The anomalous samples appear to have a slightly higher ratio of smectite to illite than the background, the opposite of what would be required to explain the anomaly. It is possible that fluids enriched in heavy REE flowed through the décollement at discrete levels, and that these heavy

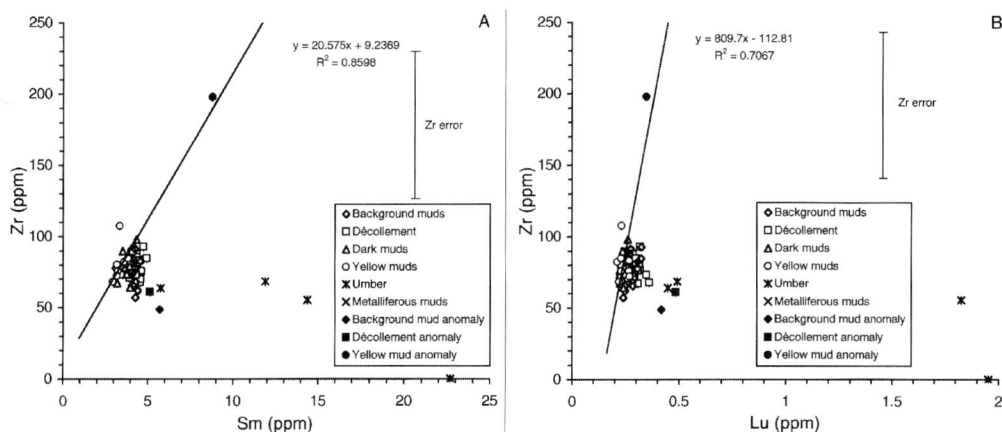

Fig. 8. Zirconium concentrations versus (**A**) samarium and (**B**) lutetium concentrations. The yellow mud anomaly is probably due to the presence of zircon. Other anomalous samples have similar zircon levels to the background muds. The regression lines are fitted through all of the yellow muds.

REE were adsorbed to mineral surfaces. However, there is no evidence for this at present.

Further work is planned, to selectively extract different mineral phases and examine where the REE are actually located. It should be possible to determine the proportion of REE in carbonates (e.g. calcite), sulphides (e.g. pyrite), oxides (e.g. Fe and Mn oxides) and adsorbed to mineral surfaces, for each sample (Lewis & McConchie 1994). This should help to determine the differences between anomalous and background samples.

Yellow mud anomaly

The anomalous yellow mud, at 1052.58 mbsf, is enriched in REE, especially the middle and heavy REE. Its chemistry is similar to the other yellow muds, except that it is enriched in zirconium (Fig. 8). This suggests the presence of zircon ($ZrSiO_4$), which can contain high levels of REE (Taylor & McLennan 1985; McLennan 1989; Götze & Lewis 1994). The background muds and other anomalous samples all have much lower zirconium concentrations.

Conclusions

The hemipelagic muds from ODP Site 808 are composed mainly of quartz, feldspar and mixed-layer clays, and have a typical shale REE com-

position. Results from this study are in agreement with previous work (Shipboard Scientific Party 1991a; Pickering et al. 1993; Underwood et al. 1993), but have additionally allowed the relationships between REE, major and minor element concentrations and mineralogy to be investigated.

The samples with the highest REE concentrations have a hydrothermal origin, and the REE are probably located in ferromanganese minerals. Other REE anomalies result from the minor mineral phases of the sediments, such as zircon and florencite. The anomaly in the décollement zone has not yet been explained, although several possible reasons, such as REE-enriched accessory minerals or changes in clay mineralogy, have been ruled out. Future work, such as quantitative electron microprobe analysis and sequential leaching experiments, should confirm the explanations for the anomalies described here, and allow interpretation of the décollement anomaly.

This work was funded by a NERC studentship, reference no. GT4/94/214/G, and forms part of the author's PhD studies. Samples were supplied through the assistance of the international Ocean Drilling Program. XRD analyses were carried out by M. Batchelder at the Natural History Museum. My supervisors, E. H. Bailey and K. T. Pickering provided helpful discussions and comments throughout the course of this work. I would also like to thank A. H. F. Robertson and A. D. Saunders for helpful and constructive reviews of the first draft.

References

BLADH, K. W. 1982. The formation of goethite, jarosite, and alunite during the weathering of sulfide-bearing felsic rocks. *Economic Geology*, **77**, 176–184.

BURT, D. M. 1989. Compositional and phase relations among rare earth element minerals. *In*: LIPIN, B. R. & McKAY, G. A. (eds) *Geochemistry and Mineralogy of Rare Earth Elements*. Reviews in Mineralogy, **21**, 259–307.

CONDIE, K. C. 1991. Another look at rare earth elements in shales. *Geochimica et Cosmochimica Acta*, **55**, 2527–2531.

DE CARLO, E. H. & McMURTRY, G. M. 1992. Rare-earth element geochemistry of ferromanganese crusts from the Hawaiian Archipelago, central Pacific. *Chemical Geology*, **95**, 235–250.

FAIRCHILD, I., HENDRY, G., QUEST, M. & TUCKER M. 1988. Chemical analysis of sedimentary rocks. *In*: TUCKER, M. (ed.) *Techniques in Sedimentology*. Blackwell Scientific Publications, Oxford, 274–354.

FLEET, A. J. 1984. Aqueous and sedimentary geochemistry of the rare earth elements. *In*: HENDERSON, P. (ed.) *Rare Earth Element Geochemistry*. Elsevier, Oxford, 343–373.

GÖTZE, J. & LEWIS, R. 1994. Distribution of REE and trace elements in size and mineral fractions of high-purity quartz sands. *Chemical Geology*, **114**, 43–57.

GOVINDARAJU, K. 1989. *Geostandards Newsletter, Special Issues*, **13**.

GROMET, L. P., DYMEK, R. F., HASKIN, L. A. & KOROTEV, R. L. 1984. The "North American shale composite": Its compilation, major and trace element characteristics. *Geochimica et Cosmochimica Acta*, **48**, 2469–2482.

HODKINSON, R. A., STOFFERS, P., SCHOLTEN, J., CRONAN, D. S., JESCHKE, G. & ROGERS, T. D. S. 1994. Geochemistry of hydrothermal manganese deposits from the Pitcairn Island hotspot, southeastern Pacific, *Geochimica et Cosmochimica Acta*, **58**, 5011–5029.

JAMES, R. H., ELDERFIELD, H. & PALMER, M. R. 1995. The chemistry of hydrothermal fluids from the Broken Spur site, 29°N Mid-Atlantic Ridge. *Geochimica et Cosmochimica Acta*, **59**, 651–659.

LEWIS, D. W. & McCONCHIE, D. 1994. *Analytical Sedimentology*. Chapman & Hall, London.

McLENNAN, S. M. 1989. Rare earth elements in sedimentary rocks: Influence of provenance and sedimentary processes. *In*: LIPIN, B. R. & McKAY, G. A. (eds) *Geochemistry and Mineralogy of Rare Earth Elements*. Reviews in Mineralogy, **21**, 169–200.

MAZA-RODRIGUEZ, J., OLIVERA-PASTOR, P., BRUQUE, S. & JIMENEZ-LOPEZ, A. 1992. Exchange selectivity of lanthanide ions in montmorillonite. *Clay Minerals*, **27**, 81–89.

MICHEL, F. A. & VAN EVERDINGEN, R. O. 1987. Formation of a jarosite deposit on Cretaceous shales in the Fort Norman area, Northwest Territories. *Canadian Mineralogist*, **25**, 221–226.

PICKERING, K. T., MARSH N. G. & DICKIE B. 1993. Data Report: Inorganic major, trace and rare earth element analysis of the muds and mudstones from site 808. *In*: HILL, I. A., TAIRA, A., FIRTH, J.V. *ET AL.* (eds) *Proceedings of the Ocean Drilling Program, Scientific Results*, **131**, 427–450.

PIPER, D. Z. 1974. Rare earth elements in the sedimentary cycle: a summary. *Chemical Geology*, **14**, 285–304.

ROBERTSON, A. H. F. 1975. Cyprus umbers: Basalt–sediment relationships on a Mesozoic ocean ridge. *Journal of the Geological Society, London*, **131**, 511–531.

—— & BOYLE, J. F. 1983. Tectonic setting and origin of metalliferous sediments in the Mesozoic Tethys. *In*: RONA P. A. *ET AL.* (eds) *Hydrothermal Processes at Seafloor Spreading Centres*. Nato Conference Series, 595–663.

—— & FLEET, A. J. 1986. Geochemistry and palaeo-oceanography of metalliferous and pelagic sediments from the Late Cretaceous Oman ophiolite. *Marine and Petroleum Geology*, **3**, 315–337.

SHIPBOARD SCIENTIFIC PARTY. 1991a. Site 808. *In*: TAIRA, A., HILL, I., FIRTH, J. V. *ET AL.* (eds) *Proceedings of the Ocean Drilling Program, Initial Reports*, **131**, 71–269.

—— 1991b. Geological background and objectives. *In*: TAIRA, A., HILL, I., FIRTH, J. V. *ET AL.* (eds) *Proceedings of the Ocean Drilling Program, Initial Reports*, **131**, 5–14.

TAYLOR, S. R. & McLENNAN, S. M. 1985. *The Continental Crust: its composition and evolution*. Blackwell Scientific Publications, Oxford.

UNDERWOOD, M. B., PICKERING, K. T., GIESKES, J. M., KASTNER, M. & ORR, R. 1993. Sediment geochemistry, clay mineralogy, and diagenesis: A synthesis of data from Leg 131, Nankai Trough. *In*: HILL, I. A., TAIRA, A., FIRTH, J. V. *ET AL.* (eds) *Proceedings of the Ocean Drilling Program, Scientific Results*, **131**, 343–363.

Application of FMS images in the Ocean Drilling Program: an overview

M. A. LOVELL[1], P. K. HARVEY[1], T. S. BREWER[1], C. WILLIAMS[1],
P. D. JACKSON[2] & G. WILLIAMSON[1]

[1]*Leicester University Borehole Research, Geology Department, Leicester University
LE1 7RH, UK*
[2]*British Geological Survey, Keyworth, Nottingham NG12 5GG, UK*

Abstract: Piecing together the evolution of the ocean basins increasingly relies on the integration of data from both recovered core and downhole measurements. This task is often complicated by the limited amount of core recovered by the Ocean Drilling Program (ODP) and the lack of understanding of the downhole data. The availability of downhole electrically based images since ODP Leg 126 in 1989 provides scientists with the visual means of examining the nature of the subsurface, and for tying disparate core to the continuous downhole data. These Formation MicroScanner (FMS) images are unfortunately based on a relatively crude resistivity measurement which provides the interpreter with only an estimate of the resistivity of the rock but, where there are variations in resistivity which correspond to variations in fabric or structure, the measurement response is often sufficient to provide a detailed visual record. Scientists participating in the ODP have explored the use of these images in tackling a wide range of problems from volcanic and sediment stratigraphy to structure and tectonic applications. The determination of core orientation and the mapping of intervals where core recovery is incomplete in particular provide the geologist with a means of carrying out field studies based on borehole and core observations which were previously unthinkable. This paper aims to provide a brief introduction to this subject, and in reviewing some of the principal results to date, illustrates the use of downhole FMS images in the ODP.

The scientific study of the ocean crust relies on a variety of techniques from large-scale remote geophysical measurements through to detailed laboratory studies on recovered material. The Ocean Drilling Program (ODP), together with its predecessors International Phase of Ocean Drilling (IPOD) and Deep Sea Drilling Project (DSDP), has emphasized the recovery of material through drilling. Unfortunately, whilst this approach provides real groundtruth, in the sense of hands-on material for examination in the laboratory, it frequently yields incomplete information due to the disparate and discontinuous nature of the core recovery. An alternative approach, and one which may be seen as complementary, is the use of down-hole measurements to provide *in situ*, continuous characterization of the ocean crust. Traditionally these wireline log measurements have provided simple logs versus depth of different physical and chemical parameters. Often the measured parameters may not be of direct interest and thus some effort at relating measured parameters to other physical or chemical properties is attempted.

Examples from the ODP are used here to illustrate the application of Formation Micro-Scanner (FMS) images in a sedimentological context for evaluating bedding, fabrics and facies, and in a volcanic context for identifying contacts and constructing lithostratigraphies. The use of the images for mapping fractures, determining stress distributions, and orientating core are also documented. In addition, an outline of the measurement basis, the nature of the images, and the limitations of the data acquired by the ODP compared with that routinely acquired by industry are detailed. Finally, we present results from numerical modelling which support our observations within the ODP. Overall the aim is to introduce the many ways in which electrical images can contribute to geological interpretation.

Electrical borehole image measurement and display

In the past decade, logging tool developments have enabled the simultaneous acquisition of multiple closely spaced microresistivity logs which may be presented as visual images (Fig. 1) reflecting variations in the electrical conductivity of the rock, on

LOVELL, M. A., HARVEY, P. K., BREWER, T. S., WILLIAMS, C., JACKSON, P. D. & WILLIAMSON, G. 1998.
Application of FMS images in the Ocean Drilling Program: an overview. *In*: CRAMP, A., MACLEOD, C. J.,
LEE, S. V. & JONES, E. J. W. (eds) *Geological Evolution of Ocean Basins: Results from the Ocean
Drilling Program*. Geological Society, London, Special Publications, **131**, 287–303.

the millimetre scale (Lloyd *et al.* 1986; Ekstrom *et al.* 1987; Boyeldieu & Jeffreys 1988; Bourke *et al.* 1989; Luthi & Souhaite 1990; Bourke 1992). These electrical images, whilst not equivalent to optical images, provide the geologist with an opportunity to view the subsurface formations in their complete state, and in effect 'to carry out field mapping of the ocean crust' (Bloomer 1995 pers. comm.).

In the ODP the downhole tool (Formation MicroScanner™ Schlumberger; Fig. 1) has four arms which are opened downhole and force four pads against the borehole wall (Pezard *et al.* 1990). Each of the four pads contains two overlapping rows of eight button-electrodes. Due to operational constraints (logging tools are run through drillpipe) the FMS tool deployed routinely is a slimline version which, when combined with the large hole size, provides only limited coverage of the borehole wall.

The electrical images are acquired using pad-based measurements (Ekstrom *et al.* 1987) in which multiple button-electrodes are held at a constant potential relative to a return electrode higher up the tool. Current is thus injected into the formation and the current emanating from each electrode is monitored continuously. The lower tool body and pad faces adjacent to the electrodes are held at the same potential as the electrodes relative to the return electrode. This arrangement constitutes passive focusing where, in principle, the constant potential of the tool body ensures that no current flows axially along the borehole; current therefore flows from the button-electrodes into the formation in a direction normal to the axis of the borehole.

Being at a fixed potential relative to the current return electrode, the current emanating from each button varies in response to the formation resistivity immediately adjacent to it. The conductance of each button is measured downhole (as the current passed divided by the potential relative to the current return electrode) and transmitted to the surface for processing. Given a constant potential difference the measured values can be seen to be directly proportional to the current flowing from each electrode.

Electrical images are thus based on a two-electrode measurement which will contain unknown proportions of the measured value due to electrode polarization and contact impedances. Electrode polarization is a non-linear, frequency-dependent phenomenon, while contact impedances are also unpredictable as they are controlled by electrochemical processes in addition to simple ohmic effects. Four-electrode methods, where currents do not emanate from electrodes used to measure potential differences, were developed to overcome these effects and are used in standard downhole resistivity logging tools such as the dual laterolog. Consequently, the conductance–depth traces from adjacent electrodes, while displaying similar character, often have different offsets and amplitudes. In order to display geological features the conductance–depth traces for each button are individually corrected for offset and amplitude. Thus while these measurements of the current flowing from each button-electrode are used to produce images depicting geological structure, they do not represent quantitative estimations of formation resistivity.

The depth of investigation of the FMS is ill-defined due to the combined effect of a number of factors, in particular the nature of the two-electrode measurements (as described above) and the resistivity structure of the formation. It is clear that the current flows a finite distance into the formation (illustrated in models by Williams *et al.* (1995*b*)) as it travels over a vertical distance of approximately 10 m from the pads to the return electrode, but the measurement is influenced primarily by variations in the near-pad region where current density is highest and the measurement is thus controlled primarily by the near-surface features of the formation. Although often referred to in the literature as a few centimetres (e.g. Molinie & Ogg 1992*b*), for a homogeneous formation with a smooth borehole the depth of investigation may be comparable with that of a shallow lateral log measurement (0.25 m or more; Bourke *et al.* 1989) but in normal logging environments is very shallow. Numerical modelling by Williams *et al.* (1995*a,b*; Williams 1996) quantifies depth of investigation for conductive fractures to be typically 2–5 cm and for localized conductive or resistive anomalies to be around 1–2 cm, but this is dependent on resistivity contrasts and the geometry of anomalous features.

By offsetting the two rows of buttons it is possible to increase the sampling around the borehole wall (with measurements made every 2.5 mm) and improve the quality of the image generated when the current intensities at each button are mapped into greyscales. In addition, the tool includes inclinometry and accelerometry measurements to enable its orientation and speed to be measured. Speed corrections are necessary to ensure accurate portrayal of the images, whilst inclinometry provides reference for orientation of the images.

Figure 1 demonstrates schematically the FMS tool in the borehole and how a single pad provides a series of current intensity versus depth curves which are then converted into a strip-like image of the borehole wall. The current intensity is converted to variable intensity greyscale or colour images through a series of processing steps (Serra 1989; Harker *et al.* 1990). These correct for

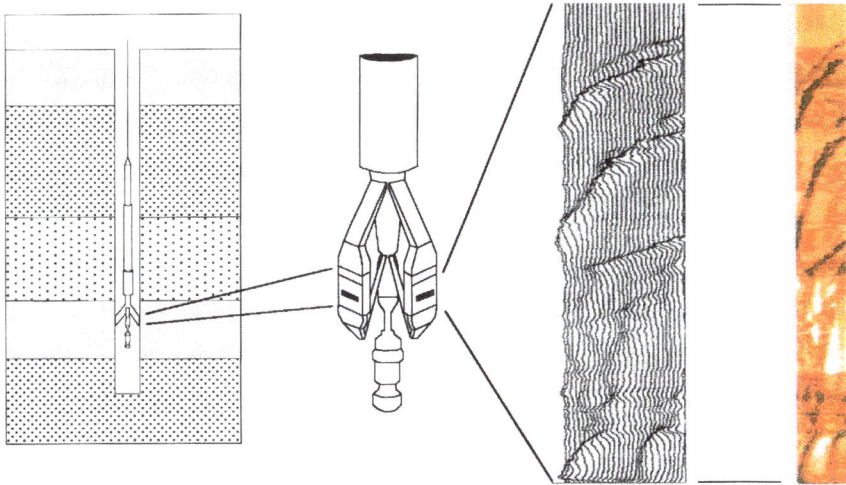

Fig. 1. Schematic diagram illustrating the acquisition of Formation MicroScanner (FMS) image data.

variations in the focusing current, speed of tool movement up the hole, and localized differences in response between electrode buttons. The resulting image is then statically normalized such that each greylevel is represented by an equal area on the final image. This produces a large-scale visualization of the data which unfortunately may mask important fine details. This may be overcome by the use of dynamic normalization which attempts to enhance such localized features through contrast magnification within a sliding window. The choice of window size is critical here in that it must be larger than the smallest feature of interest but smaller than the thickness of the unit being investigated. An alternative means of enhancing the image is through histogram equalization in which the entire greyscale is used within a sliding window: this works well when there is no electronic noise in the data (Serra 1989; Molinie & Ogg 1992b).

Processed images may be presented as an intact borehole image on a workstation or as an unrolled view of the (flattened) borehole. In the latter case the borehole image is effectively cut open along its length (usually along the north axis, 0°) and presented as four separate strip images; each is orientated with respect to north based on inclinometry measurements made during the tool's ascent up the borehole. Physical distortion of the images will occur unless the horizontal and vertical scales are equal. This is an important aspect to consider when interpretation of detailed sedimentology or textural-based features is required. Thus the flattened images are displayed as (in the ODP) four separate strip-like images (see Fig. 2). In assessing the images, convention dictates that in greyscale

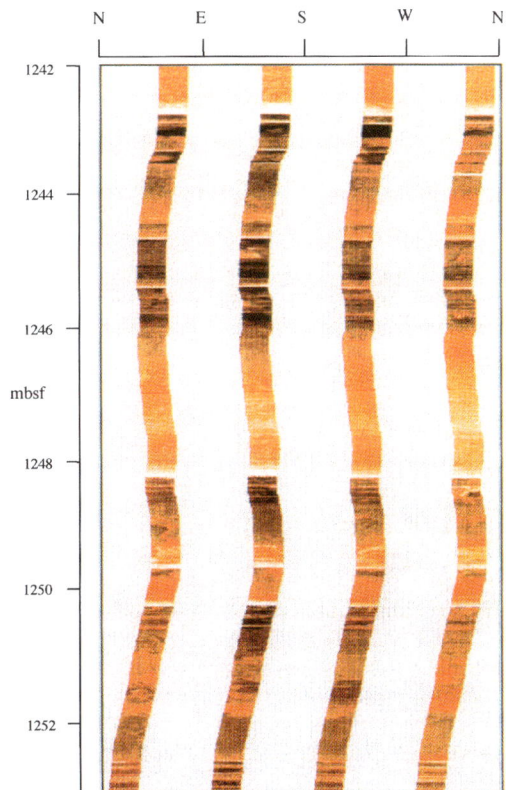

Fig. 2. Downhole electrical images of turbidite sequences in the Izu-Bonin arc (ODP Hole 792E, Leg 126); white/yellow indicate coarser-grained, more resistive sediments, orange/brown finer-grained, more conductive sediments.

images black represents more conductive, with white more resistive, which translates to brown/black (conductive) and white/yellow (resistive) for most colour image schemes.

Bourke (1989) documented various image artefacts which can complicate the interpretation of real geological features if they are not first recognized and diagnosed. He identified four categories of artefact from acquisition, borehole, processing and derived images: acquisition effects arise due to tool sticking, tool rotation and dead electrodes, all of which can affect the ODP images; borehole effects generally involve mudcake effects which are usually absent in the ODP because only seawater is normally circulated; processing routines may create images which lack detail due to strong resistivity contrasts and poorly chosen dynamic normalization windows, in addition to effects due to inadequate speed correction; derived images may confuse the field geologist because they have a different appearance to optical images of the borehole wall. Since this last category includes perception of the image in terms of the subsurface geology it could be extended to include the parallel difficulty of visualizing three dimensional geology from borehole wall images; this can be relevant in any situation, but in an ODP context, for example, attempts may be made to identify features such as individual basalt pillow rims where they are intersected by the borehole.

Refinements in the processing of FMS images within the ODP were discussed by Pratson et al. (1995). In detailing three separate stages to the depth-shifting of the data resulting from operational constraints, tool sticking, and the need for reference between multiple tool runs, they noted that the alteration of processing parameters can noticeably affect the quality of the FMS image and proposed using the double integration of the downhole accelerometer data constrained by surface-measured cable depth to determine the motion of the tool.

Constructing lithostratigraphies

Bed boundaries/unit contacts

The first FMS images obtained by the ODP (Pezard et al. 1992; Hiscott et al. 1992, 1993) show excellent near-horizontal bed boundaries (Fig. 2) within a volcaniclastic turbidite succession. The borehole wall images are presented as an unrolled borehole which has been cut open along its vertical length. Note that the images are orientated with respect to north, but that because of the relatively large size of the ODP hole the image coverage is typically only 20–30% at maximum (Table 1). In enlarged or washed-out zones the coverage is proportionately less. The images presented in Fig. 2

Table 1. *Percentage of borehole wall imaged as borehole diameter changes for four pad slimline FMS used in the ODP*

Borehole diameter (in)	% of borehole wall imaged
10.0	22.6
11.0	20.5
12.0	18.8
13.0	17.4
14.0	16.1
15.0	15.0

Width of single FMS pad is 1.8 inches
Maximum extension of caliper arms is 15 inches

illustrate the sharp contrast at the base of each bed which then grades upwards into darker shades, corresponding to a change from coarser-grained to finer-grained sediment. The coarser sands exhibit a higher resistivity than the finer clays; within ocean crust electrical resistivity in most rocks depends on both the amount of pore space and the proportion of clays present, decreasing as each increases. The only exceptions to this general trend are limited to the presence of electronic conductors such as pyrite which effectively short-circuit the electrical paths through the pore space (Lovell 1985; Pezard et al. 1988; Lovell & Pezard 1990). Using FMS images it is thus possible to construct a continuous lithostratigraphy for the borehole even where there is disparate core recovery (see for examples: Hiscott et al. 1993; Cooper et al. 1995; Demanay 1995; Lincoln et al. 1995).

The use of downhole logs for determining palaeoclimate is documented within the ODP (deMenocal et al. 1992; deMenocal 1994). The FMS provides high-resolution images which, where sedimentation rates are high, can provide sufficiently detailed palaeoclimate data. deMenocal et al. (1992) note the upper limit on temporal resolution set by the vertical resolution of logging tools and suggest a need to address both tool resolution and log response in often subtly changing lithologies, whilst deMenocal (1994) discusses the use of FMS images for matching core to logs through ash horizons.

Demanay et al. (1995) used FMS images in conjunction with the gamma ray log to develop a precise lithostratigraphy of the Palaeocene volcanic succession at the Greenland margin. Using onboard processing at sea to produce colour images at the 1/200 scale it was possible to identify the base and vesiculated top of a lava flow. The gamma ray log provided additional constraints where the composition of successive lava flows is different. Later shore-based processing enabled definition of features less than 1 cm and the identification of volcanological structures such as the degree of vesiculation, flow banding, brecciated flow tops and bases.

Meredith & Tada (1992) attributed alternatively low- and high-resistivity beds identified from FMS images as interbedding of chert and porcellanite layers. In particular they proposed a reinterpretation of shipboard observations which were based on poor core recovery concerning the extent and correlation of units.

Sometimes the changes in sedimentation do not yield distinct changes in FMS images. Thus Molinie & Ogg (1992a) matched Milankovitch eccentricity cycles to features identified on FMS images and found some degree of matching with chert bands or concentrations of increased silicification of the sediment, whilst the gamma ray log also suggested clay enrichment, but this was not discernible from the poorly cemented radiolarite on the FMS images.

ODP Leg 128 was devoted to drilling part of the Japan Sea, which is a prime example of a backarc basin. During Leg 128 three sites were drilled in the northern Yamato Basin (Sites 799, 794 and 798), of which one hole (Hole 794D) penetrated the basement. In Hole 794D the basement lies at 540 metres below sea floor (mbsf) and is composed of a series basalts, dolerites and minor intercalated sediments. Core recovery in the basement was poor, averaging 21.7%, and the 34.85 m of core is heterogeneously distributed within the 20 core barrels (Ingle *et al.* 1990). Consequently the identification of the different rock types was difficult, which was further hampered by the moderate to high degrees of alteration of the igneous rocks, as was the location of lithological boundaries. Ultimately the lithological boundaries were located with better precision by use of the logging results than could be achieved from the core barrels (Ingle *et al.* 1990). The basement section was divided into eight igneous units, separated by either sedimentary or altered volcanic material. These eight units were interpreted as four dolerite sills intruded into soft sediments at shallow depths and four dolerite basalts which were either erupted onto the sea floor or intruded into soft sediments near to the sea floor. All of the igneous lithologies contain veins which are either shallow (25–40°) or steeply (*c.* 60°) dipping, infilled by calcite, chlorite + talc or talc + chlorite + calcite (Ingle *et al.* 1990). FMS images were only obtained over a short interval in Hole 794D (560–595 mbsf), but the quality of these images is extremely good (Fig. 3). The FMS images come from the lower part of a 47 m thick plagioclase–pyroxene phyric leuco-dolerite sill (lopolith; Ingle *et al.* 1990), from which 9.9 m of core was recovered, the majority of which was

obtained from the lower part of the sill. The shipboard scientists interpreted the numerous resistivity gradations on the FMS images as reflecting cumulate layering within the sill (Fig. 3),

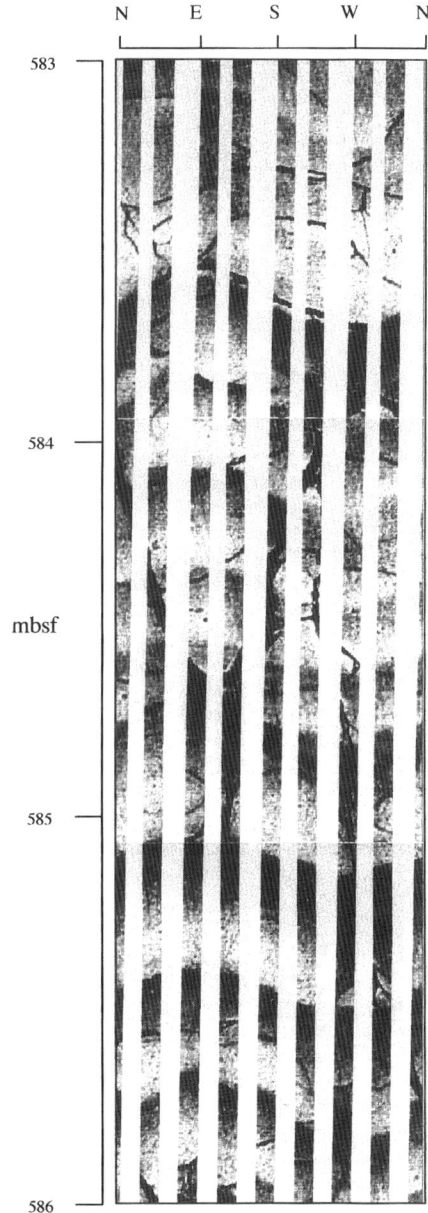

Fig. 3. Cumulate layering in crystalline rock revealed by downhole electrical images (ODP Hole 794D, Leg 128). Ingle *et al.* (1990).

whereas the infilled veins and fractures were easily identified. Although this interpretation may be perfectly valid, there are certain problems.

(i) The FMS images are from close to the base of the sill (592 mbsf). Since the sill is interpreted as intruded into soft wet sediments, the margins of such intrusion commonly develop pillow textures, which could give this type of FMS image.

(ii) Close to the margins of such a high-level sill it is unlikely that such well-developed cumulate layering would be found.

(iii) The cumulate layering would be defined by very distinct mineralogical variations which should be apparent in other downhole logging measurements. The sonic, dual lateral log and gamma ray log have very uniform responses over this section of hole, suggesting that the mineralogy of the rock is more uniform and not recording rapid variations due to cumulate layering.

Therefore the FMS image (Fig. 3) displays a very pronounced textural pattern, a feature not observed in any of the core, which may represent cumulate layering (Ingle *et al.* 1990), and illustrates the care necessary when integrating the known geology with the interpretation of downhole images.

Sedimentological structures

Luthi (1990) considers the nature of visual image texture as defined by Hall (1979) as a repetitive arrangement of a basic pattern with some degree of randomness. Furthermore, in applying texture recognition to FMS images Luthi notes that borehole image texture is a macroscopic expression of spatial changes in rock texture and (for sedimentary rocks) is directly related to sedimentary structures in a broad sense. Below, this definition is used to extract lithological information about different formations in both the sediment section and basement section, but here we consider detailed applications to the identification of sedimentary structures.

Hiscott *et al.* (1992) used FMS images to interpret beds with apparent cross-lamination and infer ripple migration directions in the Izu-Bonin Forearc

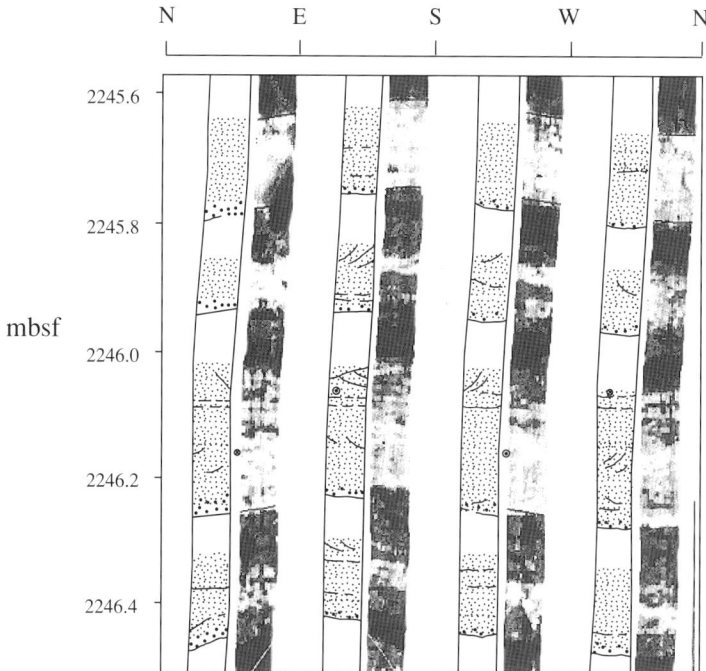

Fig. 4. Cross-lamination identified on FMS images (ODP Hole 792E, Leg 126) taken from Hole 792E in the Izu-Bonin Forearc Basin. Here the FMS images are interpreted as beds with apparent cross-lamination and the sketches are correctly positioned relative to compass directions (whereas for this diagram the FMS images have been moved relative to these). Circled dots and circled crosses represent, respectively, the heads and tails of arrows aligned with the inferred ripple migration directions; black indicates conductive layers (finer grained), white more resistive (coarser grained). After Hiscott *et al.* (1992).

Basin (Fig. 4). This example shows the combination of a qualitative approach to the interpretation, similar to making initial sedimentological observations on core, combined with using the orientation data to correctly position those observations in space.

The beds in Fig. 4 appear to be offset vertically with respect to each other. This is due to the beds dipping shallowly towards the northwest. Figure 5 illustrates how a planar dipping layer (e.g. a bedding plane or fracture) intersecting a vertical borehole would generate a sinusoidal feature around the unwrapped borehole wall. Note that the position of the maximum and minimum of the sinusoid enables the direction of dip to be determined, whilst the amplitude of the sinusoid (together with the borehole size) can be used to determine the magnitude of the dip:

$$\text{Dip angle} = \tan^{-1}\left(\frac{H}{D}\right)$$

$$\text{Strike} = \text{Azimuth @ Maximum} \pm 90°$$

where H is the amplitude of the sinusoid and D is the diameter of the hole.

Most software packages for working with FMS data allow sinusoids to be picked either manually or automatically so that the orientation of planar features can be quantified. The accuracy of such measurements obviously depends on the clarity and definition of the sinusoidal feature in the image, but also on the angle of dip. There is a fundamental problem in geology which is well known in that as the dip approaches the horizontal the strike of a planar feature becomes more difficult to define; consequently the errors associated with such measurements can be substantial. In addition, with FMS data there is a bias against, and potential difficulty in measuring, steeply dipping features. These problems arise from the difficulty of following near-vertical features in ODP boreholes, often with only 20% or less of the borehole actually present in the image.

In a study of the Amazon Fan, Pirmez *et al.* (1994) identified channel-levee and mass-flow deposits and noted that sedimentary structures often poorly preserved in the cores could be identified on the FMS images. They also used groundtruthing between intermediate scale logs, FMS images and core to extend interpretation to zones of no recovery.

Lithological textures

Within the context of FMS images the notion of visual zonation is subjective. Whilst it is possible to define threshold conductivity and hence greylevels or colours in determining sand–shale sequences (e.g. Harker *et al.* 1990; Luthi 1990), there is an additional aspect to textural analysis which compares differences in visual patterns between different formations. This has been used tentatively in a qualitative and semi-quantitative manner in carbonate sequences for identifying different patterns effected by dolomitization (Jackson *et al.* 1993; Cooper *et al.* 1995; Ogg *et al.* 1995a) and in the volcanic section of the ocean crust for quantifying stratigraphic proportions of pillows, flows and breccias (Brewer *et al.* 1995). Whilst at this stage there are limited calibration data available in the sense of documented examples of textural variations, and the results presented here are qualitative – and hence subjective comparisons of texture – there is clearly a case for using borehole image texture as a stratigraphic tool. When used in conjunction with other core and log data this approach is strengthened. Work in progress to quantify texture includes using neural networks to extract heterogeneity indices (e.g. Hall *et al.* 1996; Dolfus *et al.* in press) and extend the range of applications of borehole images.

ODP Leg 139 drilled four sites in the Middle Valley Rift, located at the northern end of the Juan de Fuca Ridge (Davis *et al.* 1992) as part of an ongoing project detailing sedimented ridges. The location is a few hundred kilometres off the coast of western North America between the Blanco and Sovanco fracture zones. The Middle Valley was specifically chosen since the spreading centre is completely covered by Pleistocene turbidites and sites at which a series of holes were drilled cover a range of hydrogeological environments. Downhole logging was carried out at Sites 856, 857 and 858, and FMS images vary in quality between the individual holes. Hole 856H drilled a sulphide

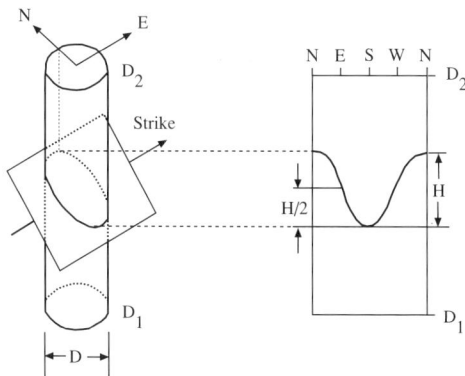

Fig. 5. Schematic illustration of the intersection of a planar dipping layer with a vertical borehole in an unwrapped borehole image display.

complex and the FMS images are particularly poor in the rugose sections of the hole, while the remainder of the images show considerable variation, displaying a mottled appearance which reflects the marked anisotropic lithological structure of the sulphide complex (Davis *et al.* 1992). Sites 857 and 858 drilled buried basement, and logging results are available from Holes 857D and 858F. In both holes useful FMS images were obtained within the sediments and the basement and also successful images of the sediment/basement interface (Fig. 6). In both holes the sediments display much darker greyscale images and have frequent banding or layering.

FMS images of basalts near to the sediment/basement interface are characterized by (Davis *et al.* 1992):

(i)   numerous highly conductive thin features, infilled and open fractures: these features have variable dips, the steeper sets have been interpreted as a set of thermal stress fractures, whereas the remainder are probably radial cooling fractures related to the formation of the pillow lavas;

(ii)   relatively large areas of generally more uniform low resistivity, which are bounded by curved surfaces: these images are interpreted in this hole as pillow lavas;

(iii)   zones of high resistivity bounded by the curved pillow surfaces: this part of the image is probably recording the interpillow debris, which is composed of spalled and altered basalt and a variety of secondary minerals (e.g. clays).

The extremely prominent image contrast between the basalts and sediments produces a very sharp sediment–basalt interface (Fig. 6), which is an important constraint in constructing the lithostratigraphy.

Another example of textural differences is presented in Fig. 7, taken from Leg 133 of the ODP, where a short section of Hole 816B is displayed. Jackson *et al.* (1993) used a novel concept of velocity–resistivity ratio, determined from conventional logs, in attempting to identify indicators of different styles of cementation in carbonate-reef environments where diagenesis is a major factor. The ratio as defined is dimensionless, being itself a ratio of two dimensionless parameters (see Jackson *et al.* (1993) for precise details). Furthermore, the individual velocity and resistivity parameters appear not to be related to depth or compaction trends but rather to be controlled by spatially variable diagenesis. In Fig. 7 the FMS images are presented for a zone above and below a layer of very high-velocity resistivity ratio. Jackson *et al.* (1993) interpret the FMS image above and below as

being consistent with a pore morphology similar to a dolomitized rhodolith-bearing mudstone described onboard ship as having substantial connected porosity. The zone between these two (220.7 mbsf) is a thin layer with low resistivity, suggesting a conduit of even higher connectivity than is observed already for the surrounding formations; but the conventional resistivity logs have a depth of investigation of at least a metre and hence the zone is probably a major connection in the formation. Drawing on supportive results from numerical models of pore channels, the authors produce a convincing case for the interpretation of the feature as a well-cemented zone of high, well-connected porosity which extends at least 1 to 2 m into the formation; furthermore they suggest this to be the product of dissolution during exposure and fabric destruction during diagenesis. On the left of Fig. 7 between 220.8 and 221.4 mbsf the image is saturated due to the static normalization of the data over the whole interval. Here a single greyscale level (or colour) may be compared at different depths and approximated as being equal. Note the enhancement of the conductive feature on the right using dynamic normalization where each greyscale level is independent of any other outside the moving window. Jackson *et al.* (1993) suggested this feature might represent a significant pathway for fluid movement.

Cooper *et al.* (1995) used a combination of core descriptions, conventional log responses and FMS images to establish detailed stratigraphic columns for the ODP sites in Guyot carbonate platforms. Initially core descriptions were compared with log responses to provide a broad stratigraphy; the units thus defined often contained several major carbonate facies. Next, FMS images were added to provide 'type' facies log responses which could then be used to interpret FMS images where no core recovery was achieved. Cooper *et al.* (1995) note that it was immediately apparent that core recovery in the two sites studied was highly preferential. This raises a cautionary note with respect to the integration of core and log data where the latter are effectively continuous yet the former are often disparate and biased. Obviously care is needed and Cooper *et al.* (1995) astutely identify the use of the quantitative response of conventional logs in supporting any interpretation based primarily on FMS images. Furthermore, they note that the greatest source of error in the identification of facies from FMS images concerns hole size. For both sites hole diameters were often larger than the maximum opening diameter of the FMS calipers. Consequently the data for specific short depth intervals were often of poor quality with poor pad contact degrading the quality of the images, and on some images only one or two pads made sufficient

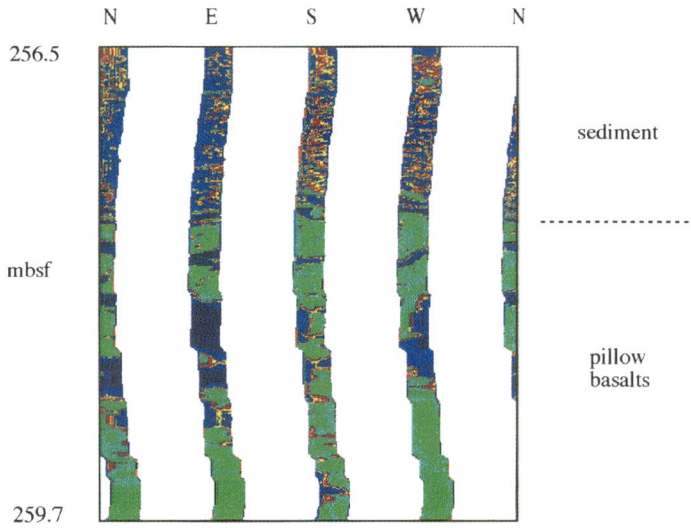

Fig. 6. Textural variations seen on FMS images at sediment–basalt interface in ODP Hole 858F, Leg 139. Green is more resistive, blue less resistive. Davis *et al.* (1992).

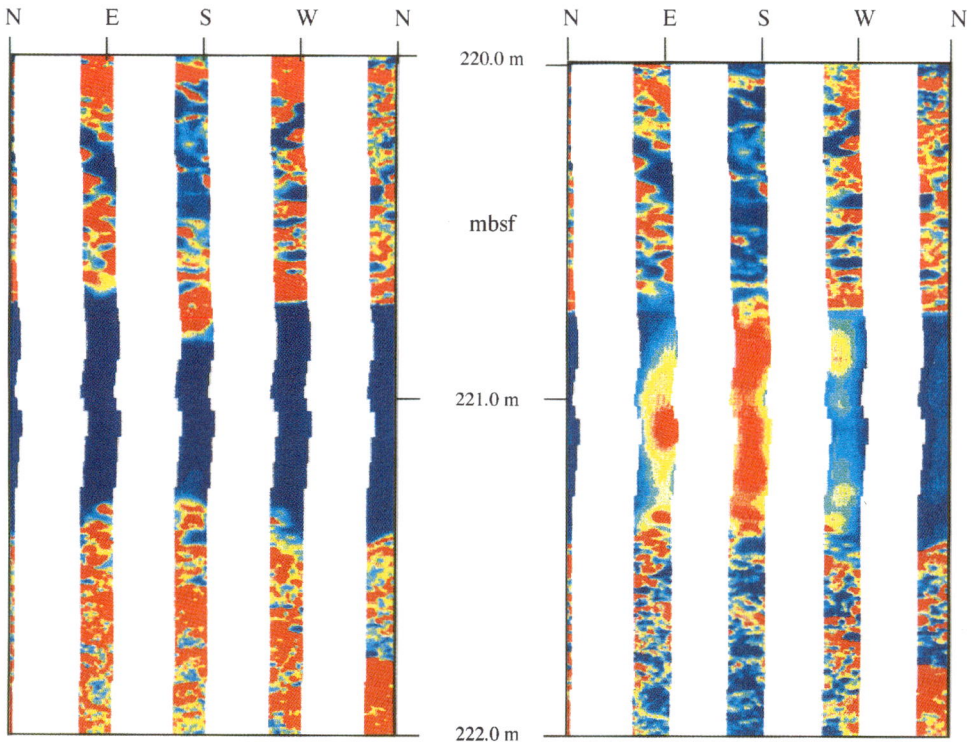

Fig. 7. FMS images of carbonates showing how static normalization (left) aids extended depth comparison even over a short interval, whereas dynamic normalization aids the investigation of a thin layer at the expense of detail either side of that layer. Red is more resistive, blue less resistive. Example from ODP Hole 816C, Leg 133; after Jackson *et al.* (1993).

contact with the borehole wall. In their inter-pretation, Cooper *et al.* (1995) used resistivity, gamma ray, density and caliper logs in addition to FMS images and core descriptions.

Ogg *et al.* (1995*a*) use independent observations from cored sediments (typically less than 5% recovery), resistivity and gamma ray logs, and vertical successions and detailed textures derived from FMS images to construct carbonate facies. Then, using these facies they are able to reconstruct the depositional history of the carbonate platform. To achieve this they assume that the FMS does not give absolute resistivity values but instead refer to relative high and low values, equivalent to differ-ences in greylevel intensities, which are then compared to the highest resolution resistivity log available, the unaveraged spherically focused log (SFLU). In this manner they demonstrate that it is possible to use the FMS images to determine the homogeneity of carbonate grains and cementation patterns at the centimetre scale. This approach is similar to that of Cooper *et al.* (1995) but realizes a difference in detailed nomenclature and implied environmental differences. Consequently Ogg *et al.* (1995*a*) note the importance of terminology in calibrating FMS images to core: whilst Cooper *et al.* describe a 'mudstone with moldic porosity', Ogg *et al.* (1995*a*, *b*) refer to a similar FMS texture as a 'bioclastic wackestone–packstone', the precise nature of this terminology referring to post- and pre-diagenetic alteration respectively.

An important aspect here is the use of static-processed images which enable comparison of different greylevels as relative variations in resistive response of the sediments. Ogg *et al.* (1995*a*) used static normalization over the entire 296 m section to produce 16 greylevels linearly normalized between the highest and lowest value; in this way a particular shade of grey approximates to the same relative resistivity. Unfortunately, because such images do not show much detail, dynamically normalized images were additionally used to identify sedimentary features such as bedding and structures. The same approach is used for similar sites by Ogg (1995) and Ogg *et al.* (1995*b*)

In a similar manner, Salimullah & Stow (1995) have used qualitative variations in texture to describe ichnofacies within turbidites and hemitur-bidites. They recognized six different types of FMS bioturbation patterns which could be correlated with the characteristic trace or pattern and possible ichnofacies in recovered core pieces. They noted the importance of identifying image patchiness caused by conglomerates/debrites, image artefacts (physical or processing) and bioturbation, and qualified their initial interpretation with the need for more certain identification using improved FMS

resolution and better controlled core/image calibration. Furthermore, they identified the need for software development capable of handling fine-scale textural discrimination and subtle pattern recognition, much along the lines being advanced by Hall *et al.* (1996).

Structure, tectonics and stress

Fractures and tectonic features

Fractures are often visible on FMS images where a contrast in resistivity occurs between the fracture and the host rock. In the ODP an open fracture will typically be filled with seawater and hence will appear conductive against the resistive rock background. Often, however, the fracture is partially or completely filled with secondary mineralization. Where this infill is composed of clays the fracture may again appear as a conductive feature. Where the infill is composed of say quartz or calcite, the fracture may be more or less resistive than the host rock depending on the extent to which the fracture is filled; this feature will again be visible to the FMS if a resistivity contrast exists between the fracture and the rock, but may not be apparent if the contrast is insufficient.

Chabernaud (1994) used FMS images to study the effect of the collision of the d'Entrecasteaux Ridge with the central New Hebrides Island Arc. Structural features such as bedding, faults, fractures, folds and shear zones, automatically detected and interactively mapped from the images, were interpreted to document the deformation processes caused by the collision. This combination of using automatic techniques for detecting features, then applying interactive methods for effectively interpreting and labelling such events, works well and provides an opportunity for rejecting unreliable results. Chabernaud (1994) provides examples of interpreting different con-ductive features in the FMS images as, respectively, a thrust fault zone, small faults and individual fractures. In shear zones the borehole is often enlarged and the FMS images are poor, but the FMS data allow the orientation and thickness of the zone to be determined. Using the geometry (dips and strikes) of planar features mapped interactively from FMS images at Site 832, an angular uncon-formity is also realized. At Site 829, in the accretionary complex of the subduction zone, FMS images helped to locate and orientate the imbricated thrust sheets off-scraped from the ridge and arc.

Structural details of the Cote D'Ivoire–Ghana transform margin were extracted from FMS images and integrated with geochemical logs and core observations by Goncalves & Ewert (1995).

Similar work by Bristow *et al.* (1995) was extended to include seismic data in studying the evolution of Gran Canaria; high-resolution FMS data augmented core-based studies of ash abundance and thickness in zones of poor recovery and provided dip and azimuth information for the tectonic evolution of the islands.

Brewer *et al.* (1995) describe an example from ODP Hole 896A (Fig. 8) in which the basement section (195.1 to 469 mbsf) is composed of pillow

recovered core   FMS image

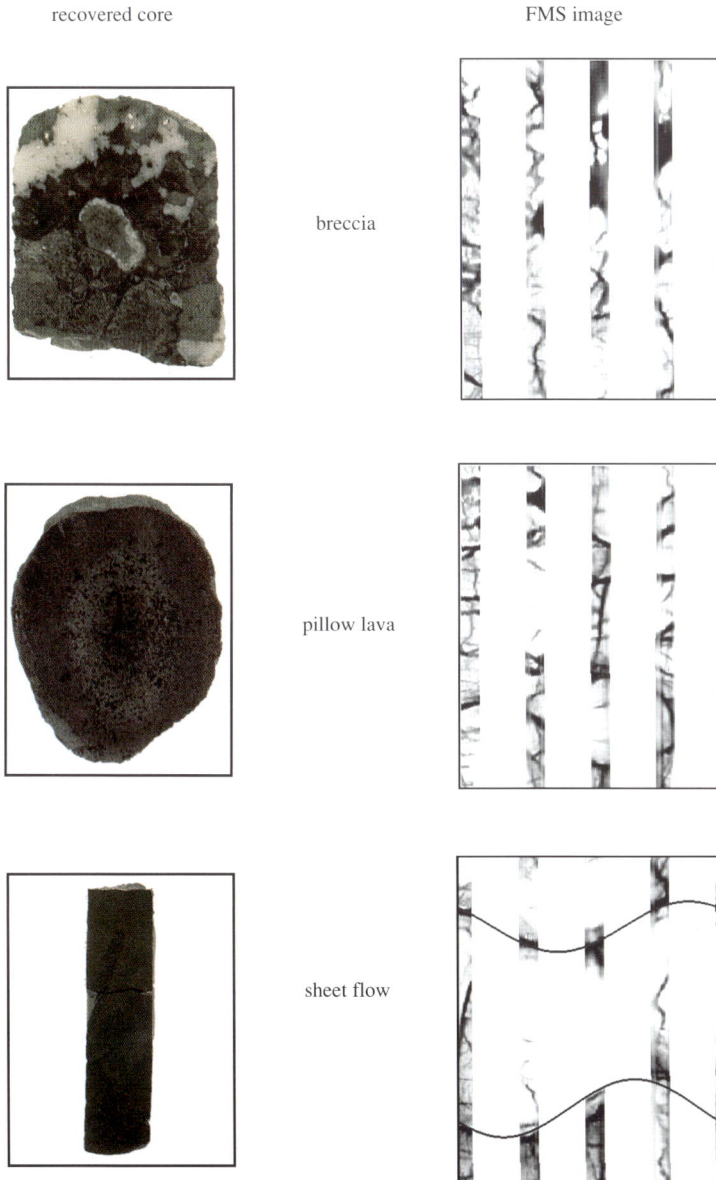

breccia

pillow lava

sheet flow

Fig. 8. Core photographs of the different lithologies recovered in the drilling of ODP Hole 896A. Adjacent to each rock type is an example of its typical FMS image. In the case of the sheet flow the resistivity is relatively uniform and the sinusoids are the position of the flow top and base. In the pillow image the green represents pillow interiors, which are transected by variably dipping fractures and surrounding interpillow material (blue). Throughout the hole the breccias are characterized by mottled images, which reflect the localized variations in rock types (and physical properties). White zones are more resistive, black less resistive.

lavas, massive flows and breccias (Alt *et al.* 1993). With the exception of pillow rims, the majority of the rocks are slightly altered (<10%) and variably veined (Alt *et al.* 1993). Pervasive background-reducing alteration coupled with saponite and minor pyrite replacement of olivine account for the grey colour of the core. Oxidative alteration is manifested by dark grey to yellow and red alteration halos which commonly occur around smectite veins (Alt *et al.* 1993). In the pillow lavas and massive flows veins are usually <1 mm in thickness and commonly infilled by dark and light green saponite and aragonite. Other vein minerals include analcite, fibrous zeolite and pyrite. The orientation of the various types of veins is problematic, since many of the core pieces were small and so were unorientated with respect to the azimuth. Where the core was orientated, it was evident that in the massive units at least one set of fractures is relatively steeply dipping probably cooling joints, whereas in the pillow lavas the distribution of fractures is more random and relates to radially orientated cooling joints. Since all of the fractures are infilled with clay minerals, there exists a localized resistivity anisotropy within the core. This anisotropy is evident in the FMS images, where part of the image texture is produced by the resistivity contrast of the fractures relative to the basalt. Thus, pillow lavas are characterized by anastomosing networks of fractures, whereas the massive units have a more ordered, steeply dipping fracture network.

Chabernaud (1994) also notes that whilst the FMS is an efficient means of recording *in situ* tectonic features and provides a continuous record of most structures even where core recovery is poor, it fails to resolve very small-scale features seen on cores.

More recently a concerted effort on Hole 504B has yielded a large data set relating to the state of the ocean crust at that site. Nearly 7700 features were mapped over 225 m of the hole, averaging 34.3 per metre, by deLarouziere *et al.* (1994). These were described in terms of geometry and aperture and provided a very detailed description of the upper basement structures in terms of geometry and volume. Pezard *et al.* (1995) report similar use of FMS images in mapping 34 000 fractures over 1600 m of basement. Dilek *et al.* (1996) used FMS images and core to study the structure of the sheeted dyke complex in Hole 504B. They found that fracturing is the main deformation process during the evolution of the crust, but whilst there is intense microfracturing in the core at specific depths, which may be related to faults or local deformation zones, open subhorizontal fractures observed in the core are generally absent from the FMS images of the borehole walls, implying that

they are drilling induced. Specific intervals were identified on the FMS images where marked zones of subhorizontal and vertical fractures are present but the interpretation of these as fault zones is considered highly subjective given the limited recovery of fault rocks (Dilek *et al.* 1996). The authors end by suggesting that the dyke–gabbro transition may be just beneath and that detailed structural interpretation of this important reference section remains critical. In the same study Ayadi *et al.* (1996) and Pezard *et al.* (1996) mapped 4500 traces of fractures from FMS images over the lowermost 167 m of the hole. They record that steep to vertical features dominate but again note that many might be related to drilling. They further suggest that borehole enlargements and near-vertical fracturing originate from specific cooling of the hole.

Core orientation and stress measurements

In addition to providing visual, high-resolution, *in situ* images of the subsurface formations, the FMS also provides important orientation information concerning the position of the four arms which collect the raw data for image creation. These arms also provide four separate positions around the borehole wall from which two orthogonal diameters can be calculated (Plumb & Hickman 1984). These two diameters often form the major and minor axes of the elliptical shape of the borehole due to the local stress field, thus allowing the regional stress field to be evaluated from analysis of multiple boreholes (Bell 1990; Evans & Brereton 1990; Caillet *et al.* 1994; Chabernaud 1994). Pezard *et al.* (1992) provide an example of this approach in deducing the horizontal stress field orientation in the Izu-Bonin Arc, combining results from both FMS and borehole televiewer data (Fig. 9).

Orientated FMS images can also be used to reorientate core. MacLeod *et al.* (1994) describe the use of the FMS (and borehole televiewer) in structural and palaeomagnetic studies through the matching of distinctive inclined planar features.

The effect of borehole enlargement on the measurement of elongation direction by four-arm caliper tools such as the FMS is discussed in the literature (Evans & Brereton 1990; Celerier *et al.* 1996). Under certain conditions, one or both of the caliper directions may be saturated and the constraint of the direction and amplitude of the elongation becomes less well constrained. The use of data redundancy techniques to compensate for low-quality data in bad hole conditions where washouts are prevalent was accomplished during ODP Leg 147 to the Hess Deep (Celerier *et al.* 1996; MacLeod *et al.* 1995). By merging data from

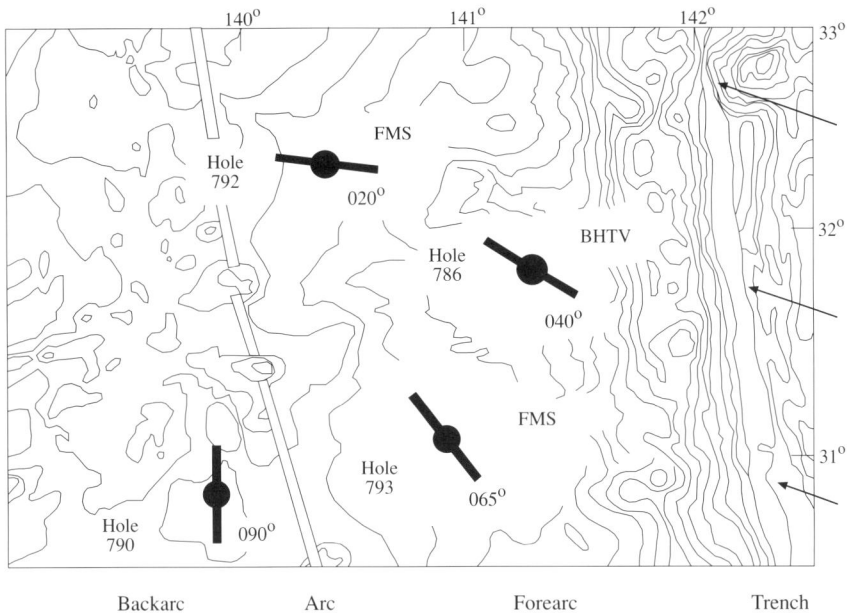

Fig. 9. Direction of horizontal stress field orientation in the Izu-Bonin Arc determined from observations of borehole ellipticity with acoustic televiewer images (Hole 786B), FMS images (Holes 792E and 793B) in the forearc, and geological evidence from opening of the Sumisu Rift in the southernmost part of the backarc region. The relative velocity vector between the Pacific plate and the Philippine Sea plate is indicated to the east of the trench. After Pezard *et al.* (1992).

multiple passes of the FMS logging tool, the confidence with which interpretations could be made was increased and often differed from initial conclusions based on single-pass data. They note, however, that under such conditions the complete 360 degree coverage of an acoustic televiewer tool would better constrain the detailed shape of the borehole. Consequently they were able to extract quantitative information from the FMS images, constrain the fracture orientations and suggest directions of preferential enlargement.

Numerical modelling of the FMS

The direct analysis and interpretation of FMS images is complemented by quantitative modelling of the tool response and current flow (Ekstrom *et al.* 1987; Bourke *et al.* 1989; Luthi & Souhaite 1990; Williams *et al.* 1995*a*, *b*; Williams 1996).

A three-dimensional numerical simulation of an electrical imaging tool is developed by Williams *et al.* (1995*a*) and applied to a model based on data from the ODP. Resistivity and FMS data for a 10 m interval from Leg 126, Hole 793B, which corresponds to a graded bed from a turbidite sequence in the Izu-Bonin Forearc, is shown in Fig. 10a. The resistivity of the bed decreases from bottom to top

due to variations in particle size and clay content (as described earlier in the 'Bed boundaries/unit contacts' section); this is clearly indicated by both the conventional resistivity logs (Fig. 10a, left) and the FMS image (Fig. 10a, right) which varies upwards from paler (corresponding to a more resistive section) to darker (more conductive) grey-scales. The focused, deep and shallow resistivity logs correspond closely, in response to relatively thick, homogeneous horizontal layers. These well-constrained downhole data are used by Williams *et al.* (1995*a*) to provide resistivity values for a model which consists of a series of horizontal layers (Fig. 10b, left). The model borehole diameter is set to match that of the caliper and the borehole fluid resistivity corresponds to that of seawater. This model is the input into the imaging tool simulation which generates a synthetic tool response (Fig. 10b, centre). This in turn is used to create a simulated image (Fig. 10b, right) which matches that of the actual FMS-measured images closely, confirming the validity of the images in this case (as would be expected for this comparatively homogeneous formation).

Other models are used to investigate the tool's depth of investigation and possible image artefacts. Simulations of conductive fractures of different

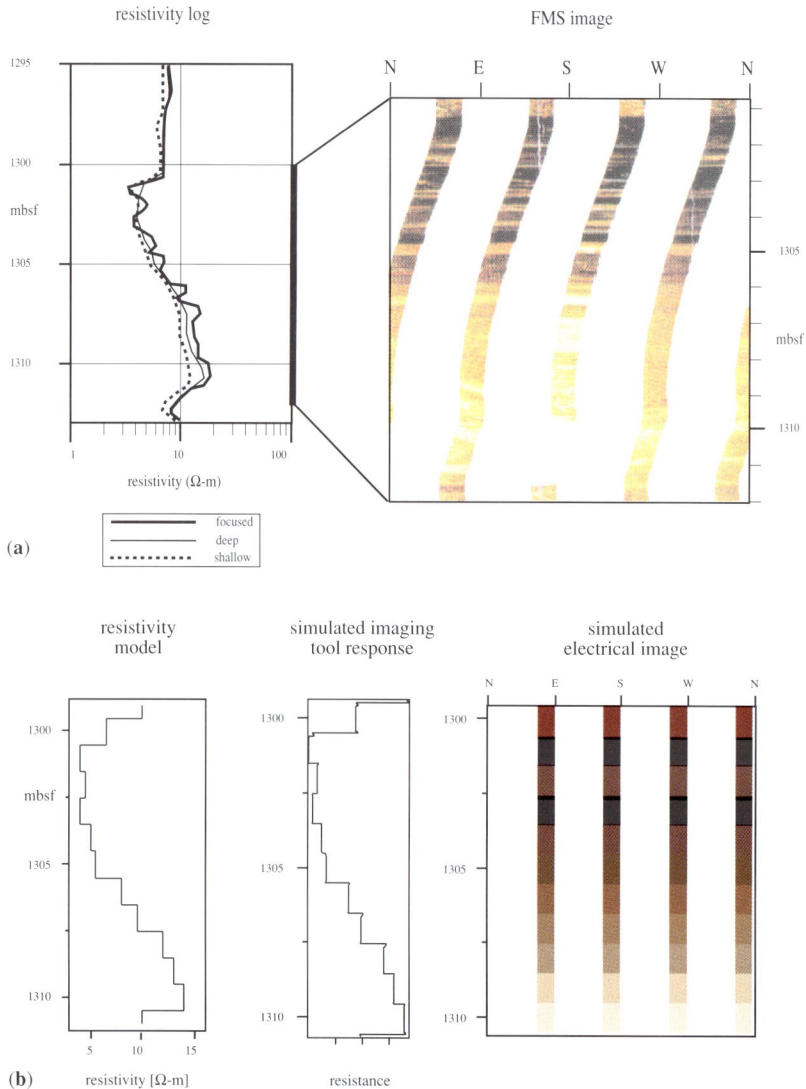

Fig. 10. Comparison of real FMS images of turbidites from the Izu-Bonin Arc with numerically simulated images. After Williams *et al.* (1995*a*).

depths (Williams *et al.* 1995*a, b*) quantify the typical tool depth of investigation as 2–5 cm, but also indicate that anomalous pale fringes may be observed around shallow features due to passive focusing effects. This is used to identify anomalous conductive features in FMS from Hole 835B, thought to be caused by poor pad contact (Williams 1996). Models of isolated resistive and conductive anomalies (Williams *et al.* 1995*b*; Williams 1996) demonstrate the likelihood of the current from the FMS travelling a finite distance into the formation, and hence the possibility of the tool 'seeing' the

effect of an isolated pore space or feature which does not intersect the borehole wall. Again, this is used to aid interpretation of a section of FMS from Hole 835B which is characterized by isolated resistive spots that are thought to be caused by clusters of foraminifera in an otherwise homogeneous section of nanofossil ooze (Williams 1996).

This type of approach, when developed in parallel with direct interpretation of data, is useful in helping to understand how real geological features are imaged by the FMS as well as serving as a reminder of the limitations of the tool and the potential for generation of artefacts.

Conclusion

The development of a slimline FMS for deployment by the Ocean Drilling Program heralded a new phase for the *in situ* study of the ocean crust. Unfortunately, because of hole conditions, operational constraints and the broad range of lithologies encountered, the data acquired in the ODP are often of an inferior standard to those acquired by industry. Through careful calibration to recovered core, however, scientists have successfully tackled relevant issues to enable the incorporation of FMS images into their interpretation toolbox. The range of applications is broad in terms of both geological environment and scientific questions. Whilst the images provide the opportunity for seeing the state of the crust at depth at high resolution, the quantitative nature of the information has wide-reaching implications for future studies across the breadth of the ODP.

The FMS images provide geologists with *in situ* images of the borehole at a scale which is directly comparable to optical observations on the core. These images allow direct comparison between core and log data and create a route to core–log integration and correlation which encompasses the multidimensional nature of the rock. Thus the geometry and spatial heterogeneity of the core may be considered during the integration process. Improvements in desk-top computing combined with the continual reduction in cost should enable easier access by scientists to both FMS images, FMS interpretation capabilities, and the integration of core and log data.

Within ODP there are proposals to deploy a logging-while-drilling measurement (Resistivity At Bit: RAB) which provides lower resolution but full borehole (360°) coverage. These lower-resolution images would complement the existing FMS images and, together, their combination may provide a better total image of the subsurface geology.

The authors acknowledge support from NERC (GST 102/684) and Z & S group (software and support). PDJ acknowledges the Director of the British Geological Survey for permission to publish this paper.

References

ALT, J. C., KINOSHITA, H., STOKKING, L. B. *ET AL.* 1993. *Proceedings of the Ocean Drilling Program, Initial Reports*, **148**. College Station, TX (Ocean Drilling Program).

AYADI, M., PEZARD, P .A. & DE LAROUZIERE, F. D. 1996. Fracture distribution from downhole electrical images at the base of the sheeted dike complex in Hole 504B. *In*: ALT, J. C., KINOSHITA, H., STOKKING, L. B. & MICHAEL, P. J. (eds) 1996. *Proceedings of the Ocean Drilling Program, Scientific Results*, **148**. College Station, TX, 307–315.

BELL, J. S. 1990. Investigating stress regimes in sedimentary basins using information from oil industry wireline logs and drilling records. *In*: HURST, A., LOVELL, M. A. & MORTON, A. C. (eds) *Geological Applications of Wireline Logs*. Geological Society, London, Special Publications, **48**, 305–325.

BOURKE, L. T. 1989. Recognising artefact images of the Formation Microscanner. *In*: *30th Annual Logging Symposium Transactions.* Society of Professional Well Log Analysts, 25 pp. (Later reprinted: Bourke, L. T. 1990. *In Borehole Imaging Reprint Volume.* Society of Professional Well Log Analysts, 191–215.)

—— 1992. Sedimentological borehole image analysis in clastic rocks: a systematic approach to interpretation. *In*: HURST, A., GRIFFITHS, C. M. & WORTHINGTON, P. F. (eds) *Geological Applications of Wireline Logs II*. Geological Society, London, Special Publications, **65**, 31–42.

——, DELFINER, P., TROUILLER, J-C, FETT, T., GRACE, M.,

LUTHI, S., SERRA, O. & STANDEN, E. 1989. Using Formation Microscanner images. *The Technical Review*, **37**(1), 16–40.

BOYELDIEU, C. & JEFFREYS, P. 1988. Formation Microscanner: new developments. *Society of Professional Well Log Analysts 11th European Formation Evaluation Symposium.* Oslo, Norway, paper WW.

BREWER, T. S., LOVELL, M. A., HARVEY, P. K. & WILLIAMSON, G. 1995. Stratigraphy of the ocean crust in ODP Hole 896A from FMS images. *Scientific Drilling*, **5**, 87–92.

BRISTOW, J. F., HARVEY, P. K. & LOVELL, M. A. 1995. Chemical and structural evolution of Gran Canaria – ODP Leg 157; evidence from the integration of core, downhole logging, and seismic. *EOS*, **76**(46), F325.

CAILLET, G., DEBOAISENE, R., MATHIS, B. & ROUX, C. 1994. The present day stress regime in some deep structures of quadrant-25, Offshore Norway. *Bulletin des centres de recherches exploration-production Elf Aquitaine*, **18**(2), 381–390.

CELERIER, B., MACLEOD, C. J. & HARVEY P. K. 1996. Constraints on the geometry and fracturing of Hole 894G, Hess Deep, from Formation Microscanner logging data. *In*: MEVEL, C., GILLIS, K. M., ALLAN, J. F. & MEYER, P. S. (eds) *Proceedings of the Ocean Drilling Program, Scientific Results*, **144**. College Station, TX, 329–342.

CHABERNAUD, T. J. 1994. High resolution electrical imaging in the New Hebrides Island Arc: Structural analysis and stress studies. *In*: GREENE, H. G.,

COLLOT, J.-Y., STOKKING, L.B. *ET AL. Proceedings of the Ocean Drilling Program, Scientific Results,* **134**. College Station, TX, 591–606.

COOPER, P., ARNAUD, H. M. & FLOOD, P. G. 1995. Formation Microscanner logging responses to lithology in Guyot carbonate platforms and their implications: Sites 865 and 866. *In*: WINTERER, E. L., SAGER, W. W., FIRTH, J. V. & SINTON, J. M. (eds) *Proceedings of the Ocean Drilling Program, Scientific Results,* **143**. College Station, TX, 329–372.

DAVIS, E. E., MOTTL, M. J., FISHER, A. T. *ET AL.* 1992. *Proceedings of the Ocean Drilling Program Initial Reports,* **139**. College Station, TX (Ocean Drilling Program).

deLAROUZIERE, F. D., PEZARD, P .A. & AYADI, M. 1994. Downhole measurements and electrical images in ODP Hole 896A, Costa Rica Rift. *EOS*, **75**(44), 317.

DEMANAY, A., CAMBRAY, H. & VANDAMME, D. 1995. Lithostratigraphy of the volcanic sequences at Hole 917A, Leg 152, S.E. Greenland Margin. *Journal of the Geological Society*, **152**(6), 943–946.

deMENOCAL, P. 1994. Downhole logs as Palaeoclimate tools: a case study from ODP Leg 128, Sea of Japan. *EOS*, **75**(44), 309.

——, BRISTOW, J. F. & STEIN, R. 1992. Paleoclimate applications of downhole logs: Pliocene–Pleistocene results from Hole 798B, Sea of Japan. *In*: PISCIOTTO, K. A., INGLE, J. C., von BREYMANN, M. T. *ET AL. Proceedings of the Ocean Drilling Program, Scientific Results*, **127/128**, Part 1, 393–407.

DILEK, Y., HARPER, G. D., PEZARD, P. A. & TARTAROTTI, P. 1996. Structure of the sheeted dike complex in Hole 504B (Leg 148). *In*: ALT, J. C., KINOSHITA, H., STOKKING, L. B. & MICHAEL, P. J. (eds) *Proceedings of the Ocean Drilling Program, Scientific Results,* **148**. College Station, TX, 229–243.

DOLFUS, D., CAMBRAY, H. & PEZARD, P. In press. Identification of lithological features from FMS images using neural networks. *Scientific Drilling*.

EKSTROM, M. P., DAHAN, C., CHEN, M-Y., LLOYD, P. & ROSSI, D. J. 1987. Formation imaging with microelectrical scanning arrays. *The Log Analyst*, **28**, 294–306.

EVANS, C. J. & BRERETON, N. R. 1990. Insitu crustal stress in the United Kingdom from borehole breakouts. *In*: HURST, A., LOVELL, M. A. & MORTON, A. C. (eds) *Geological Applications of Wireline Logs*. Geological Society, London, Special Publications, **48**, 327–338.

GONCALVES, C. A. & EWERT, L. 1995. Sedimentary and structural relationship of the CoteD'Ivoire-Chana Transform Margin – ODP Leg 159: evidence from downhole log measurements. *EOS*, **76**(46), F597

HALL, E. L. 1979. *Computer Image Processing and Recognition*. Academic, New York.

HALL, J., PONZI, M., GONFALINI, M. & MALETTI, G. 1996. Automatic extraction and characterisation of geological features. *Transactions of the Society of Professional Well Log Analysts,* New Orleans.

HARKER, S. D., MCGANN, G. J., BOURKE, L. T. & ADAMS, J. T. 1990. Methodology of Formation Microscanner image interpretation in Claymore and Scapa Fields (North Sea. *In*: HURST, A., LOVELL, M. A. & MORTON, A. C. (eds) *Geological Applications of Wireline Logs*. Geological Society, London, Special Publications, **48**, 11–25.

HISCOTT, R. N., COLELLA, A., PEZARD, P., LOVELL, M. A. & MALINVERNO, A. 1992. Sedimentology of deep-water volcaniclastics, Oligocene Izu-Bonin Forearc Basin, based on Formation Microscanner images. *In*: TAYLOR, B., FUJIOKA, K. *ET AL. Proceedings of the Ocean Drilling Program, Scientific Results,* **126**, 75–96.

——, ——, ——, —— & —— 1993. Basin plain turbidite succession of the Oligocene Izu-Bonin intraoceanic forearc basin. *Marine and Petroleum Geology*, **10**, 450–466.

INGLE, J. C. JR, SUYEHIRO, K., von BREYMANN, M. T. *ET AL.* 1990. *Proceedings of the Ocean Drilling Program, Initial Reports*, **128**. College Station, TX (Ocean Drilling Program).

JACKSON, P. D., JARRARD, R. D., PIGRAM, C. J. & PEARCE, J. M. 1993. Resistivity/porosity/velocity relationships from downhole logs: an aid for evaluating pore morphology. *In*: MCKENZIE, J. A., DAVIES, P. J., PALMER-JULSON, A. *ET AL. Proceedings of the Ocean Drilling Program, Scientific Results,* **133**, 661–686.

LINCOLN, J. M., ENOS, P. & OGG, J. G. 1995. Stratigraphy and diagenesis of the carbonate platform at Site 873, Wodejebato Guyot. *In*: HAGGERTY, J. A., PREMOLI SILVA, L., RACK, F. & McNUTT, M. K. (eds) *Proceedings of the Ocean Drilling Program, Scientific Results*, **144**. College Station, TX, 255–269.

LLOYD, P. M., DAHAN, C. & HUTIN, R. 1986. Formation imaging with electrical scanning arrays. A new generation of stratigraphic high resolution dipmeter tool. *Transactions of the Society of Professional Well Log Analysts, 10th European Formation Evaluation Symposium*, Paper L.

LOVELL, M. A. 1985. Thermal conductivity and permeability assessment by electrical resistivity measurements in marine sediments. *Marine Geotechnology*, **6**, 205–240.

—— & PEZARD, P. A. 1990. Electrical properties of basalts from DSDP Hole 504B: a key to the evaluation of pore space morphology. *In*: HURST, A., LOVELL, M. A. & MORTON, A. C. (eds) *Geological Applications of Wireline Logs*. Geological Society, London, Special Publications, **48**, 339–345.

LUTHI, S. M. 1990. Sedimentary structures of clastic rocks identified from electrical borehole images. *In*: HURST, A., LOVELL, M. A. & MORTON, A. C. (eds) *Geological Applications of Wireline Logs*. Geological Society, London, Special Publications, **48**, 3–10.

—— & SOUHAITE, P. 1990. Fracture apertures from electrical borehole scans. *Geophysics*, **55**(7), 821–833.

MacLEOD, C. J., PARSON, L. M. & SAGER, W. W. 1994. Reorientation of cores using FMS and BHTV: application to structural and palaeomagnetic studies with the Ocean Drilling Program. *In*: HAWKINS, J., PARSON, L., ALLAN, J. *ET AL.* (eds) *Proceedings of the Ocean Drilling Program, Scientific Results*, **135**, 301–311.

——, CELERIER, B. & HARVEY, P. K. 1995. Further techniques for core reorientation by core-log integration: application to structural studies of lower crust in Hess Deep, Eastern Pacific. *Scientific Drilling*, **5**(2), 77–86.

MEREDITH, J. A. & TADA, R. 1992. Evidence for Late Miocene cyclicity and broad-scale uniformity of sedimentation in the Yamato Basin, Sea of Japan, from Formation Microscanner data. *In*: TAMAKI, K., SUYEHIRO, K., ALLAN, J., MCWILLIAMS, M. ET AL. *Proceedings of the Ocean Drilling Program, Scientific Results*, **127/128**, Part 2. College Station, TX, 1037–1046.

MOLINIE, A. J. & OGG, J. G. 1992a. Milankovitch cycles in Upper Jurassic and Lower Cretaceous radiolarites of the Equatorial Pacific: spectral analysis and sedimentation rate curves. *In*: LARSON, R. L., LANCELOT, Y. ET AL. *Proceedings of the Ocean Drilling Program, Scientific Results*, **129**. College Station, TX, 529–547.

—— &—— 1992b. Data report: Formation Miroscanner imagery of Lower Cretaceous and Jurassic sediments from the Western Pacific (Site 801). *In*: LARSON, R. L., LANCELOT, Y. ET AL. *Proceedings of the Ocean Drilling Program, Scientific Results*, **129**. College Station, TX, 671–691.

OGG, J. G. 1995. MIT Guyot: Depositional history of the carbonate platform from downhole logs at Site 878 (Lagoon). *In*: HAGGERTY, J. A., PREMOLI SILVA, L., RACK, F. & MCNUTT, M. K. (eds) *Proceedings of the Ocean Drilling Program, Scientific Results*, **144**. College Station, TX, 337–359.

——, CAMOIN, G. F. & JANSA, L. 1995a. Takuyo-Daisan Guyot: Depositional history of the carbonate platform from downhole logs at Site 879 (Outer Rim). *In*: HAGGERTY, J. A., PREMOLI SILVA, L., RACK, F. & MCNUTT, M. K. (eds) *Proceedings of the Ocean Drilling Program, Scientific Results*, **144**. College Station, TX, 361–380.

——, —— & VANNEAU, A. A. 1995b. Limalok Guyot: depositional history of the carbonate platform from downhole logs at Site 871 (Lagoon). *In*: HAGGERTY, J. A., PREMOLI SILVA, L., RACK, F. & MCNUTT, M. K. (eds) *Proceedings of the Ocean Drilling Program, Scientific Results*, **144**. College Station, TX, 233–253.

PEZARD, P. A. HOWARD, J. J. & LOVELL, M. A. 1988. The influence of clays on the electrical conductivity of basalts from Hole 504B and changes in the structure of the pore geometry due to hydrothermal alteration of the crust. *In*: BECKER, K., SAKAI, H. ET AL. *Proceedings of the Ocean Drilling Program, Scientific Results*, **111**.

——, LOVELL, M. A. & ODP LEG 126 SHIPBOARD SCIENTIFIC PARTY 1990. Downhole images: electrical scanning reveals the nature of the ocean crust. *EOS*, 709.

——, HISCOTT, R. N., LOVELL, M. A., COLLELA, A. & MALINVERNO, A. 1992. Evolution of the Izu-Bonin intraoceanic forearc basin, western Pacific, from cores and FMS images. *In*: HURST, A., GRIFFITHS, C. M. & WORTHINGTON, P. F. (eds) *Geological Applications of Wireline Logs II*. Geological Society, London, Special Publications, **65**, 43–69.

——, CORROTTI, P. & AADI, M. 1995. Fractures, faults and tectonic stresses in the upper oceanic crust from ODP core and downhole measurements. *EOS*, **76**(46), F325.

——, BECKER, K., REVIL, A., AYADI, M. & HARVEY, P. K. 1996. Fractures, porosity, and stress in the dolerites of Hole 504B, Costa Rica Rift. *In*: ALT, J. C., KINOSHITA, H., STOKKING, L. B. & MICHAEL, P. J. (eds) *Proceedings of the Ocean Drilling Program, Scientific Results*, **148**. College Station, TX, 317–329.

PIRMEZ, C., KRONEN, J. D., THIBAL, J. & ODP LEG 155 SHIPBOARD SCIENTIFIC PARTY. 1994. Preliminary results of core-log-seismic data integration, ODP Leg 155, Amazon Fan. *EOS*, **75**(44), 315.

PLUMB, R. A. & HICKMAN, S. H. 1984. Stress-induced borehole elongation: a comparison between the four-arm dipmeter and the borehole televiewer in the Auburn geothermal well. *Journal of Geophysical Research*, **90**, 5513–5521

PRATSON, E. L., PRIMEZ, C. & GOLDBERG, D. 1995. A refinement of FMS depth shifting in the Ocean Drilling Program, *EOS*, **76**(46), F326.

SALIMULLAH, A. R. M. & STOW, D. A. V. 1995. Ichnofacies recognition in turbidites/hemiturbidites using enhanced FMS images: examples from ODP Leg 129. *The Log Analyst*, **36**(4), 38–49.

SERRA, O. 1989. *Formation Microscanner Image Interpretation*. Schlumberger Educational Services, Houston, Texas.

WILLIAMS, C. G. 1996. *Assessment of electrical resistivity properties through development of three-dimensional numerical models*. PhD thesis, Leicester University.

WILLIAMS, C., JACKSON, P., LOVELL, M., HARVEY, P. & REECE, G. 1995a. Numerical simulation of electrical imaging tools. *4th International Conference of the Brazilian Geophysical Society – 1st Latin American Geophysical Conference*, Rio de Janeiro, Volume II, 744–746.

——, ——, ——, —— & ——. 1995b. Numerical simulation of downhole electrical images. *Scientific Drilling*, **5**, 93–98.

A statistical study of hydraulic piston coring, ODP Legs 101–149

YIR-DER E. LEE & T. J. G. FRANCIS

Ocean Drilling Program, Texas A&M University, 1000 Discovery Drive, College Station, Texas 77845, USA

Abstract: The operational performance of the Advanced Piston Corer over the first 49 legs of the Ocean Drilling Program (ODP) is compared to the physical properties of the cores recovered. Piston coring performance in different sediment types is summarized. The pullout force measured on extracting the piston corer from the sediment is shown to correlate with the shear strength of the sediment cored. Pullout force thus provides a direct measurement of the *in situ* shear strength. Hydraulic piston coring could be used to determine the tensional load-bearing capabilities of sediments where Tension Leg Platforms may be tethered.

Since its first use on Deep Sea Drilling Project (DSDP) Leg 64 in 1978, hydraulic piston coring has grown to occupy a more and more important place in scientific ocean drilling. Like all mechanical devices, the coring tool itself has been modified and upgraded over the years, so that the Advanced Piston Corer (APC) used on the *JOIDES Resolution* in 1995 is vastly superior to the 1978 prototype. Similarly, the strategies for using this coring technique have evolved. With increasing emphasis on studies of climate change, triple APC holes are now often made at drill sites to ensure complete recovery of the soft sedimentary section before penetration is pursued with rotary drilling to greater depths. Consequently, most of the core acquired and curated by the Ocean Drilling Program (ODP) has been recovered by hydraulic piston coring.

The vast majority of studies relying on hydraulic piston coring are naturally concerned with what can be learned from examination of the core recovered. This study focuses on the operation of the coring tool. Data acquired during APC coring over a large number of ODP legs have been studied to try to understand how the performance of the APC is affected by the physical properties of the sediments sampled and by other operational parameters. This paper summarizes only the highlights of this study, focusing in particular on the relationship of pullout forces observed in APC coring to the shear strength of the sediment recovered.

Evolution of hydraulic piston coring tools

The prototype Hydraulic Piston Corer used on DSDP Leg 64 in 1978 was referred to as HPC-15, since it was designed to recover a 15 ft (4.6 m) core

(Storms *et al.* 1983). This was followed in 1981 by the Variable Length Hydraulic Piston Corer (VLHPC) which was used on DSDP Leg 80 (Poag & Foss 1985). The VLHPC could be fitted with core barrels ranging in length from 3.5 to 9.5 m. These early hydraulic piston corers tended to be pulled apart on being extracted from stiff sediments, so development was continued to make a more robust and reliable tool. The result of this development was the Advanced Piston Corer (APC) which was first used on DSDP Leg 94 in 1983 (Huey *et al.* 1984; Ruddiman *et al.* 1987).

The Ocean Drilling Program began sea-going operations in 1985 with the 1983 version of the APC. Pullout was limited to 60 kips (60000 lb force or 27.2 tonnes). It was found that the APC was able to penetrate sediments which were strong enough to exert even higher forces on pullout than 60 kips. Further development therefore ensued, slightly reducing the penetration capability of the tool but increasing the pullout force which could be applied. This modification, referred to as the APC-129, was first used on ODP Leg 130 in 1990 (Kroenke *et al.* 1991). The pullout limitation for the APC-129 is 100 kips.

The next modification, which was introduced in 1992, was the 'washover' facility. This allowed a tightly stuck APC to be recovered by drilling over it using circulation and rotation. The advent of this development allowed APC coring to be used at the unprecedented depth of 458 m below the sea floor (mbsf) in Hole 883B on ODP Leg 145 (Rea *et al.* 1993).

A schematic representation of the APC assembly is shown in Fig. 1. The nominal maximum length of core recovered by APC coring during ODP has always been 9.5 m. However, when comparing

LEE, Y-D. E. & FRANCIS, T. J. G. 1998. A statistical study of hydraulic piston coring, ODP Legs 101–149. *In*: CRAMP, A., MACLEOD, C. J., LEE, S. V. & JONES, E. J. W. (eds) *Geological Evolution of Ocean Basins: Results from the Ocean Drilling Program*. Geological Society, London, Special Publications, **131**, 305–316.

305

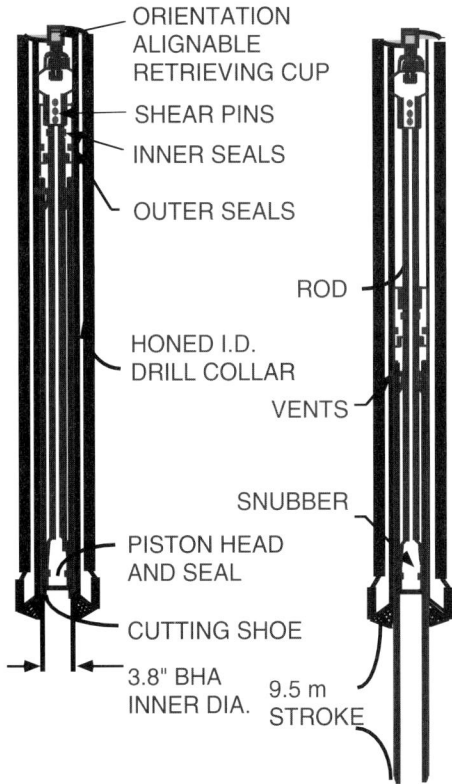

ORIENTATION
ALIGNABLE
RETRIEVING CUP

SHEAR PINS

INNER SEALS

OUTER SEALS

ROD

HONED I.D.
DRILL COLLAR

VENTS

SNUBBER

PISTON HEAD
AND SEAL

CUTTING SHOE

3.8" BHA
INNER DIA.   9.5 m
                  STROKE

Fig. 1. Schematic representation of the Advanced Piston Corer (ODP file drawing).

APC performance across a large number of legs, it is important to remember that the tool itself has been changed a number of times, as indicated above, in order to improve the efficiency of its operation.

The maximum penetration of APC coring

A total of 389 APC holes were cored on ODP Legs 101–149. At sites where APC coring could be conducted, more than one APC hole was generally cored. The reasons for stopping APC coring at these 389 holes are summarized in Fig. 2. Most APC holes were terminated because of a scientific or an operational decision. For example, the target depth for a particular hole might have been reached; or the first APC core at a site had 100% recovery and consequently could not be used to establish the mudline (the sediment/water interface). In that situation the drill string was pulled clear of the sea floor, the ship offset, and another hole started to establish the mudline.

In order to study the geological limits to hydraulic piston coring, it was necessary to exclude all holes which were terminated for non-geological reasons. Sometimes the data were insufficient to allow the reason for stopping an APC hole to be established. In order to be sure that APC coring was terminated because of the physical properties of the sediments being cored, it was decided to eliminate all holes except those where the APC was followed by Extended Core Barrel (XCB) coring. Hence 158 holes were determined to have been terminated for

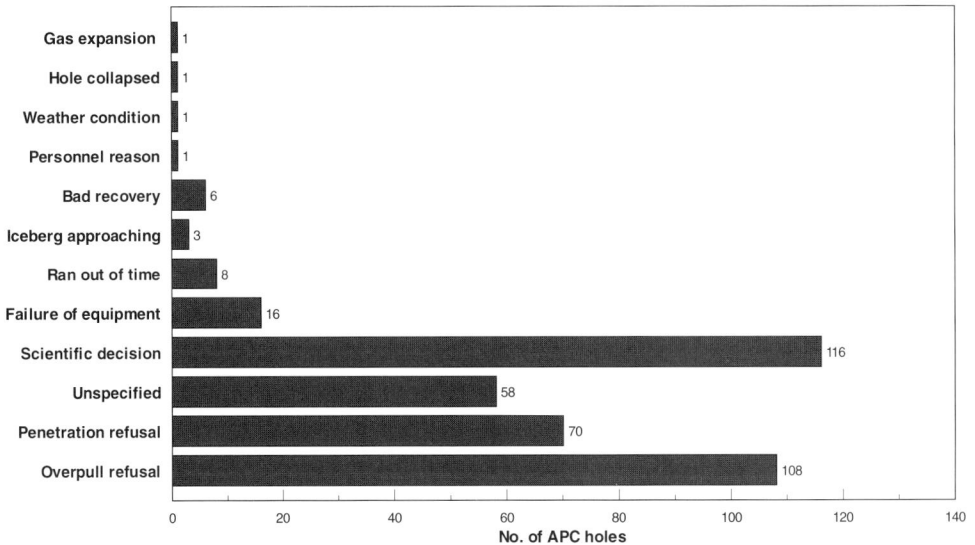

Fig. 2. Reasons for stopping coring at the 389 APC holes drilled on ODP Legs 101–149.

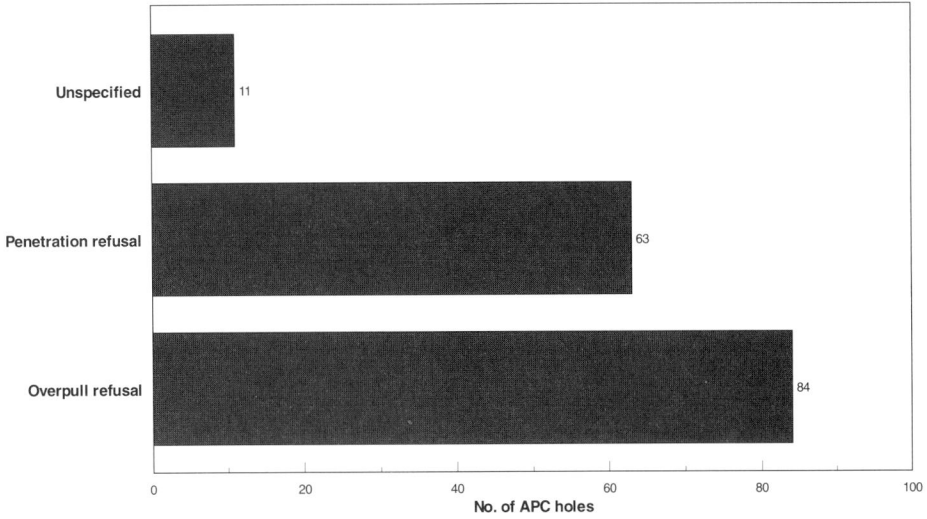

Fig. 3. Reasons for stopping APC coring at the 158 holes where XCB coring followed.

geological reasons. Figure 3 summarizes the reasons for terminating APC coring at these holes.

If the APC fails to penetrate a formation because it is too stiff or hard, which would be indicated by the corer failing to achieve a full stroke, coring is terminated due to 'penetration refusal'. If full penetration is achieved, but the core barrel cannot be extracted from the sediment without approaching the pullout limitation of the corer, the coring is terminated due to 'overpull refusal'. The force involved in withdrawing the 9.5 m long core barrel from the sediment is generally too great to be achieved by the wireline, so pullout is normally achieved by raising the whole drill string with the draw-works.

Operational data and the physical properties of the deepest core in each hole were assembled for 158 APC holes from three sources:

(a) *ODP Initial Reports* volumes;
(b) ODP's computerized database;
(c) Coring Technician data sheets.

A histogram of the maximum penetration depths achieved in these 158 holes is shown in Fig. 4. The

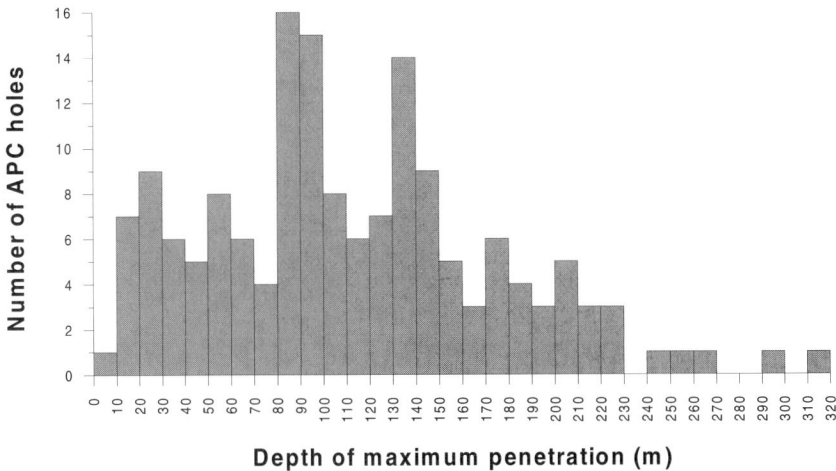

Fig. 4. Distribution of the maximum penetration achieved in 158 APC holes.

Depth of maximum penetration (m)

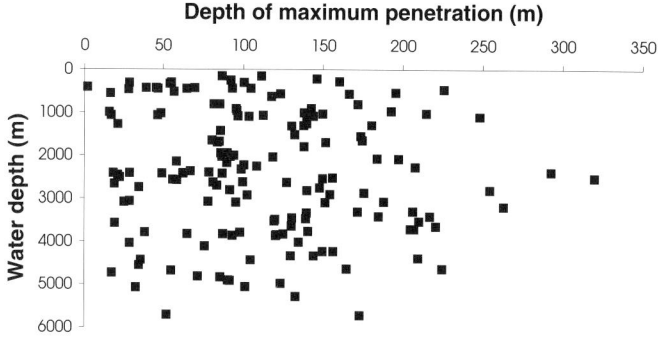

Fig. 5. Plot of maximum penetration versus water depth for 158 APC holes.

mean depth of maximum penetration is 112 m whilst the median depth is 100.3 m. These figures are much less than would now be expected for the depths of APC coring on dedicated palaeoceanographic legs because the latter are biased towards targets with predominantly pelagic sediments. The greatest penetration achieved in continuous APC coring in this data set of 158 holes was 320 m (Leg 130, Hole 806B: nanofossil ooze with foraminifers), whilst the least was 1.9 m (Leg 119, Hole 739B: diamicton). Note that the record APC penetration of 458 mbsf in Hole 883B mentioned earlier was achieved after XCB coring had been tried higher up in the hole.

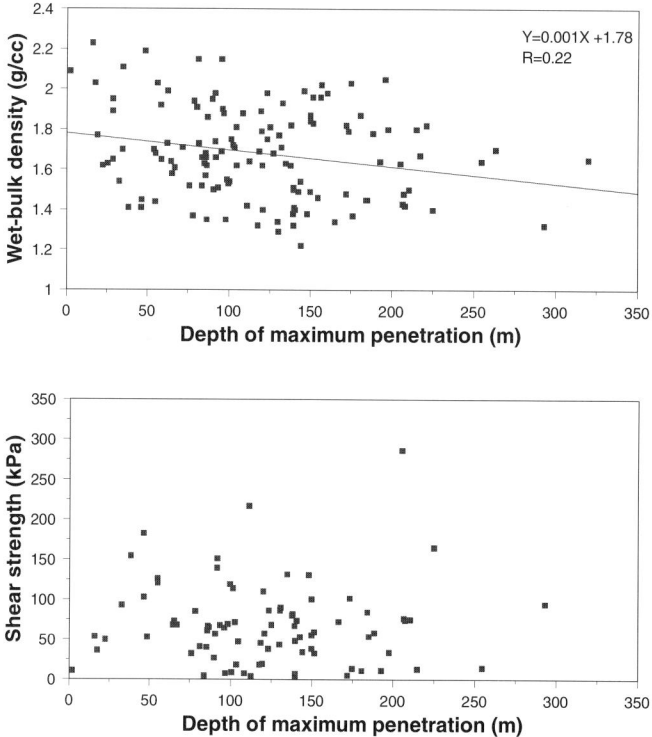

$$Y = 0.001X + 1.78$$
$$R = 0.22$$

Fig. 6. Wet-bulk density and undrained shear strength of the deepest cores plotted against depth of maximum penetration, all data.

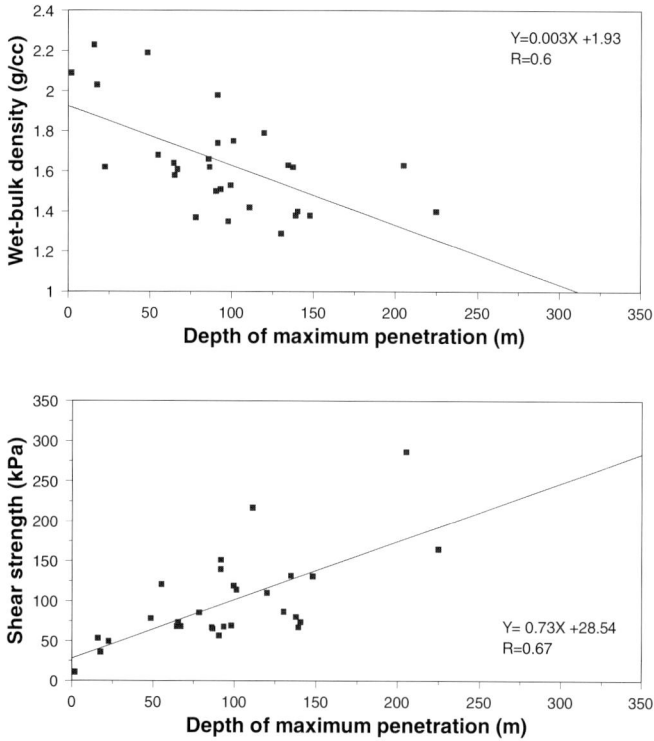

Fig. 7. As for Fig. 6 but including only the data for normally consolidated sediments.

It might be thought that the maximum penetration in APC holes would be related to the water depth for two reasons:

(a) the APC core barrel is driven into the sediment by the hydraulic energy stored in the drill string. With a longer drill string more hydraulic energy is available for the operation;

(b) coarser-grained sediments might be more likely to be present at the shallow water sites.

Maximum penetration is plotted against water depth for the 158 APC holes in Fig. 5. It is apparent that there is no correlation at all between these two parameters.

Relationship of sediment properties to maximum APC penetration

In an attempt to relate the maximum penetration achieved in APC coring to the physical properties of the sediments being cored, the bulk density and shear strength measured in the deepest core at each hole were plotted against maximum penetration (Fig. 6). Whilst one can be certain that the maximum penetration was consistently and accurately recorded throughout the eight years of ODP operations represented by Legs 101–149, measurements of the physical properties of the sediments in the shipboard laboratory were not so consistently

Table 1. *The average depth of APC holes and the mean properties of the deepest cores for the two types of APC refusal*

Refusal type	No. of holes	Mean penetration depth (m)	Mean water content (%)	Mean wet-bulk density (g cm^{-3})	Mean porosity (%)	Mean shear strength (kPa)	Mean clay mineral (%)
Overpull refusal	84	122.3	44.09	1.62	66.93	72.7	26.7
Penetration refusal	63	95.8	33.56	1.85	57.07	49.0	34.0
Unspecified	11						
Total APC holes	158	112.0					

Table 2. *The average depth of APC holes and the mean properties of the deepest cores as a function of sediment type*

Sediment type major	subgroup	No. of holes	Mean penetration depth (m)	Mean water content (%)	Mean wet-bulk density (g cm^{-3})	Mean porosity (%)	Meann shear strength (kPa)
Calcareous sediment	Pelagic calcareous	44	134.6	39.94	1.67	63.62	43.2
	Non-pelagic carbonate	6	114.3	26.11	1.97	50.38	7.7
	Calcareous mixed (non-/pelagic)	7	132.1	27.15	1.87	49.97	28.1
	Calcareous with siliciclastic	11	93.9	41.34	1.68	65.48	95.6
	Total	68	126.0	37.28	1.72	61.35	47.0
Siliceous sediment	Siliceous ooze	15	164.6	59.12	1.38	80.85	78.6
	Siliceous with siliciclastic	14	61.8	54.87	1.46	76.78	122.6
	Total	29	114.1	57.00	1.42	78.82	100.5
Biogenic mixed sediment		8	150.4	53.51	1.44	75.06	89.6
Siliciclastic sediment		34	79.9	30.44	1.88	54.50	83.8
Volcaniclastic sediment		14	103.2	36.58	1.82	58.23	54.1
No recovery		5					
Total APC holes		158	112.0				

made. The latter measurements were made by many different shipboard scientists and the laboratory equipment changed in this period, so the measurements available for the deepest cores vary in quantity and quality. For example, some cores have no shear strength measurements, while others have several which were averaged to represent the complete core.

It is clear from Fig. 6 that the maximum penetration achieved in an APC hole correlates only weakly with the sediment bulk density and not at all with the shear strength. This is hardly surprising, because the data include many different sediment types, deposited at different rates, in different water depths and at different times, with different consolidation histories and so on. However, for 83 of the deepest cores the data were available to allow the consolidation state of the sediment to be calculated. Following Skempton (1954), 49 of the deepest cores were categorized as underconsolidated, 29 as normally consolidated and five as overconsolidated. When the plot shown in Fig. 6 is restricted to normally consolidated sediments, both the bulk density and shear strength of the deepest cores are found to correlate significantly with the depth of maximum penetration (Fig. 7). This result indicates that APC coring in normally consolidated sediments tends to be terminated at shallow depth by high-density sediments whilst at greater depth termination was due to higher shear strength.

A similar result for the whole data set is shown in Table 1. This summarizes the physical properties of the sediment according to the two types of refusal. The average penetration of APC holes terminated by overpull refusal is significantly greater than for those terminated by penetration refusal. Furthermore, the mean index properties of the deepest APC cores also differ significantly between the two types of refusal. Holes with penetration refusal terminate in sediments which are denser, less porous and weaker than the sediments in which overpull refusal occurs.

One would expect that sediment type would have an important effect on the performance of APC coring. To determine this, the lithologies of the deepest cores were classified into five major types following the classification scheme of Mazzullo *et al.* (1987) and their mean properties tabulated (Table 2). Calcareous and siliceous sediments were further subdivided. The following broad conclusions can be drawn from these data, all statistically significant:

(a) significantly greater APC penetrations are achieved in biogenic (calcareous, siliceous, mixed calcareous and siliceous) than in non-biogenic sediments;

(b) the best APC performance is obtained in siliceous oozes, followed by biogenic mixed sediments and pelagic calcareous sediments;

(c) the presence of a siliciclastic component in the sediment significantly reduces the depth to which APC coring can penetrate.

There is nothing particularly new in these conclusions. However, it is comforting that this statistical analysis of APC coring confirms the subjective experience of ODP operational staff involved in APC coring on a regular basis.

Pullout forces in APC coring

The pullout forces experienced in APC coring are routinely recorded in the data sheets compiled by the Coring Technicians on the drill ship. Data sheets going back to Leg 101 are kept on file in the Engineering and Drilling Operations department at ODP, Texas A&M University. The measurement of

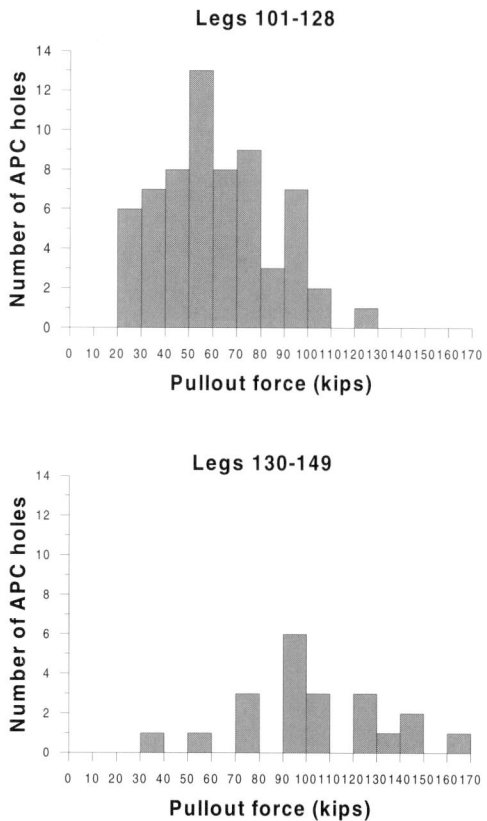

Fig. 8. Distribution of pullout forces for the last cores in APC holes terminated by overpull refusal. The 1983 APC was in use for Legs 101–128; APC-129 was used for Legs 130–149.

Fig. 10. Shipboard measurements of undrained shear strength down four APC holes. Measurements made on the deepest core in each hole are shown by open circles.

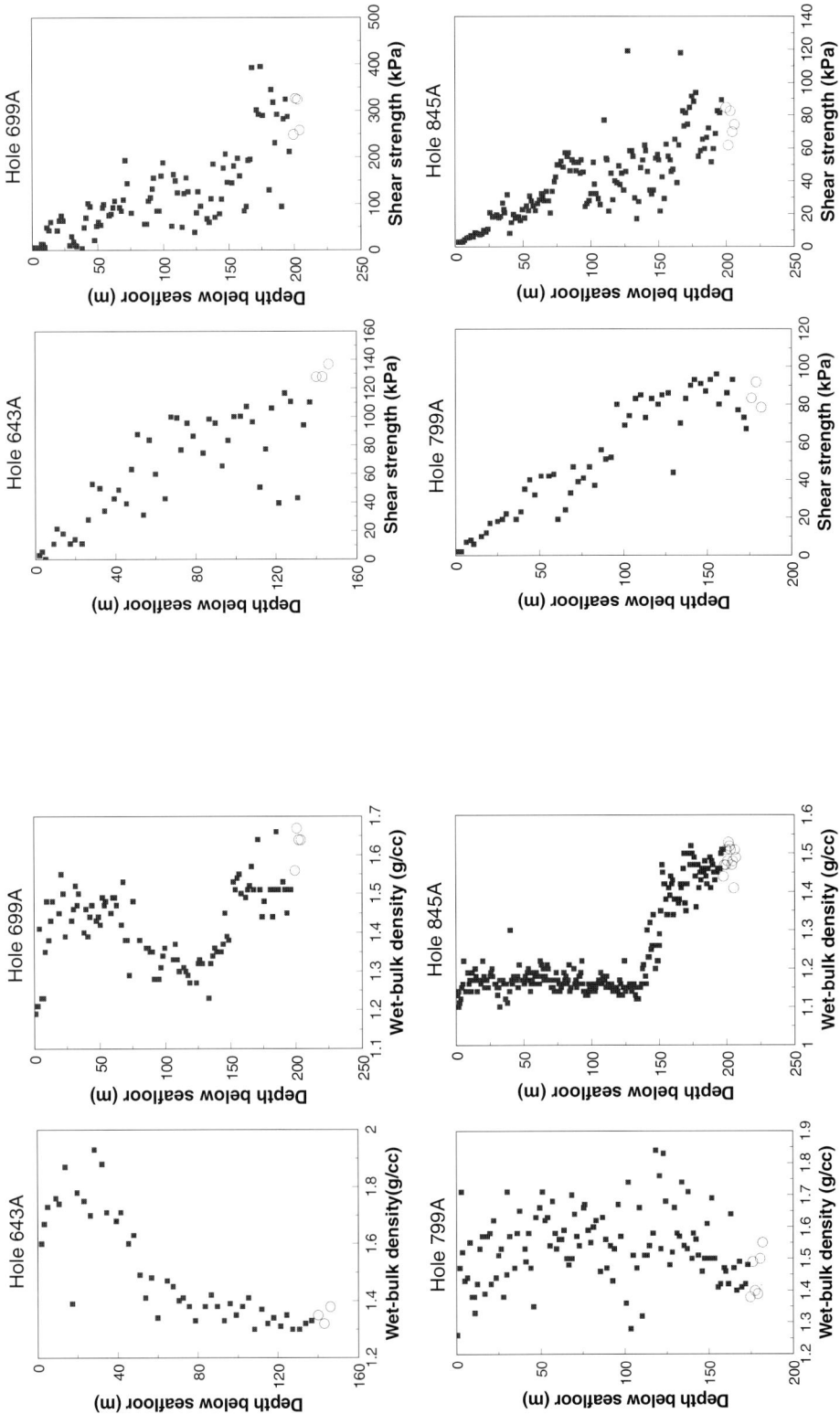

Fig. 9. Laboratory measurements of wet-bulk density down four APC holes. Measurements made on the deepest core in each hole are shown by open circles.

Fig. 12. Pullout force plotted against the mean wet-bulk density of each core for four APC holes.

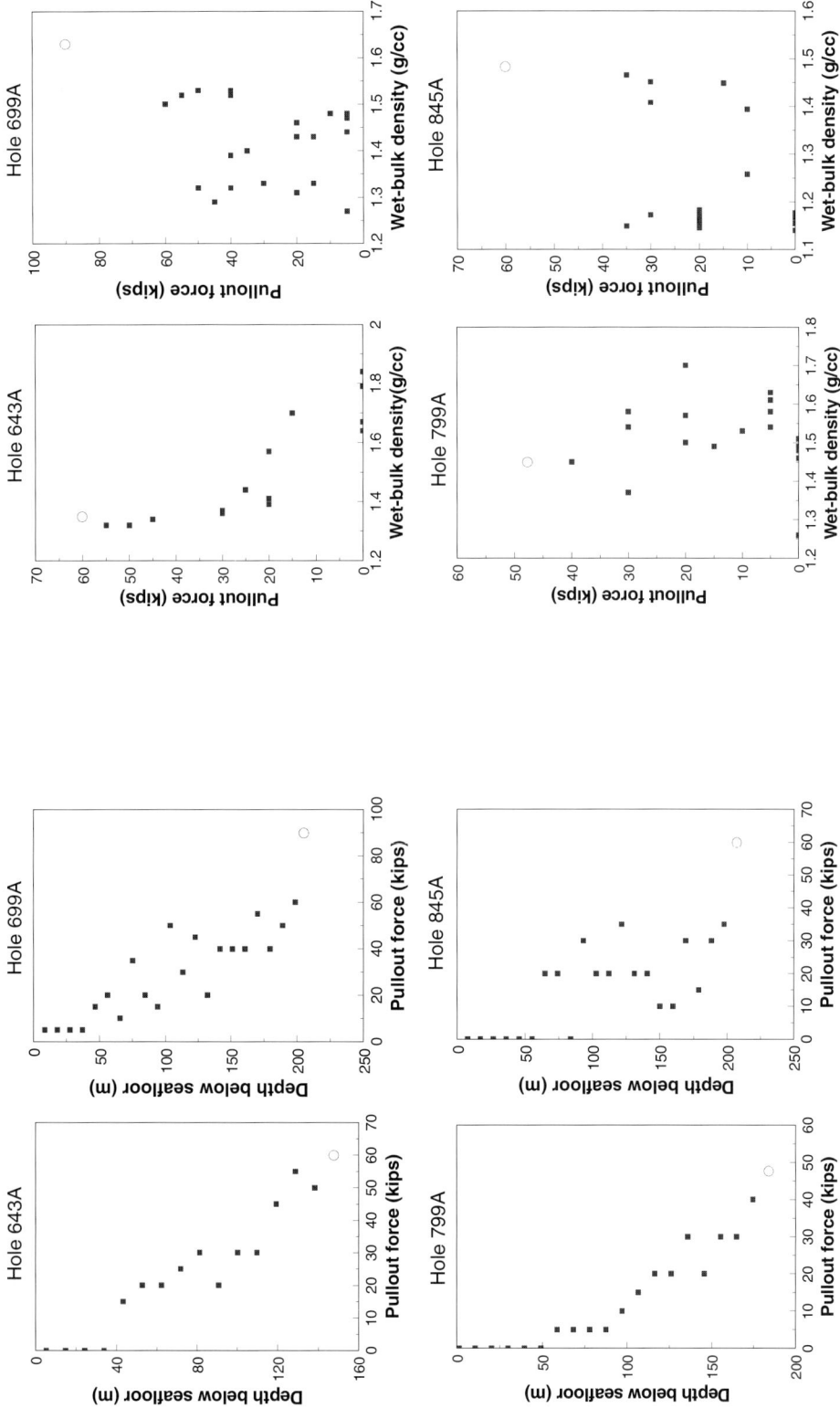

Fig. 11. Pullout force plotted against depth below sea floor for four APC holes.

pullout force is obtained from the driller's weight indicator which shows the total weight suspended from the derrick. Since the latter includes the weight of the drill string and of the travelling gear in the derrick, the pullout force cannot be measured very accurately, as it represents only a small percentage of the total load. Furthermore the load on the derrick fluctuates with the heave of the ship. Hence the pullout force is recorded by the Coring Technician to a precision of only 5 or 10 kips. In very weak sediments it may not be possible to measure the pullout force at all.

Histograms of the pullout forces measured for the deepest APC core in holes which were terminated by overpull refusal are shown in Fig. 8. On Legs 101–128, the 1983 model APC was in use, for which the pullout force was limited to 60 kips. On Legs 130–149 the more robust APC-129 was in use, with an operational limit to the pullout force of 100 kips. It is apparent in both cases that the measured pullout force often exceeded the operational limit. This reflects the difficulty the Coring Technicians have in predicting what the pullout force for the next core will be. On the other hand, it also demonstrates that ODP is using the APC aggressively in response to the high scientific demand for this very efficient coring tool.

Relationship of the pullout force to the physical properties of the sediment

In spite of the difficulty in measuring the pullout force accurately, it is worth asking whether there is any information about the *in situ* physical properties of the sediment being cored in the measurements being made. To answer this question, four APC holes for which detailed physical property data were available were selected, as detailed in Table 3. Detailed descriptions of the sediments recovered in these holes are given in the relevant *Initial Reports* volume (Eldholm *et al.* 1987;

Ciesielski *et al.* 1988; Ingle *et al.* 1990; Mayer *et al.* 1992).

The wet-bulk-density measurements made on the cores from these four holes are shown in Fig. 9, and the undrained shear strengths in Fig. 10. The pullout force measured on extracting each core is plotted against depth below the sea floor for the same four holes in Fig. 11. With some scatter, the pullout force increases with depth below the sea floor at all four holes.

In Fig. 12 the pullout force measured for each core is plotted against the mean wet-bulk density of the sediment recovered in that core. Clearly, there is no correlation between these two parameters.

In Fig. 13 the pullout force measured for each core is plotted against the effective overburden pressure for that core. Since the effective overburden pressure increases monotonically with depth and the pullout force has been observed to increase with depth below the sea floor, one would expect some correlation between these two parameters. If the withdrawal of the APC core barrel were simply a frictional process, which did not involve any failure of the sediment, this correlation would be good. But this is not thought to be a realistic model of the coring process. The sediment must fail as the core barrel is thrust into it. It then 'heals' (i.e. recovers its strength) depending on the time the corer remains at rest within it, then is again subjected to failure as the corer is extracted.

The pullout force measured for each core is plotted against its average shear strength in Fig. 14. Quite good correlation is observed between pullout force and shear strength at all four holes. Thus the pullout force is providing a direct measurement, albeit not very accurately, of the *in situ* shear strength of the sediment being cored.

With the growth of the offshore oil industry over the past few decades a considerable body of geotechnical knowledge has been developed concerning the loads which piles in marine sediments can bear (API 1993). With the develop-

Table 3. *The four holes selected to study relationship of pullout force to sediment physical properties*

Leg	Hole	Water depth (m)	Deepest APC Core (m)	Location
104	643A	2753	147.8	Norwegian Sea
114	699A	3705	205.1	Sub-Antarctic South Atlantic
128	799A	2073	184.1	Japan Sea
138	845A	3704	207.1	Eastern Equatorial Pacific

Hole 643A

Hole 699A

Hole 799A

Hole 845A

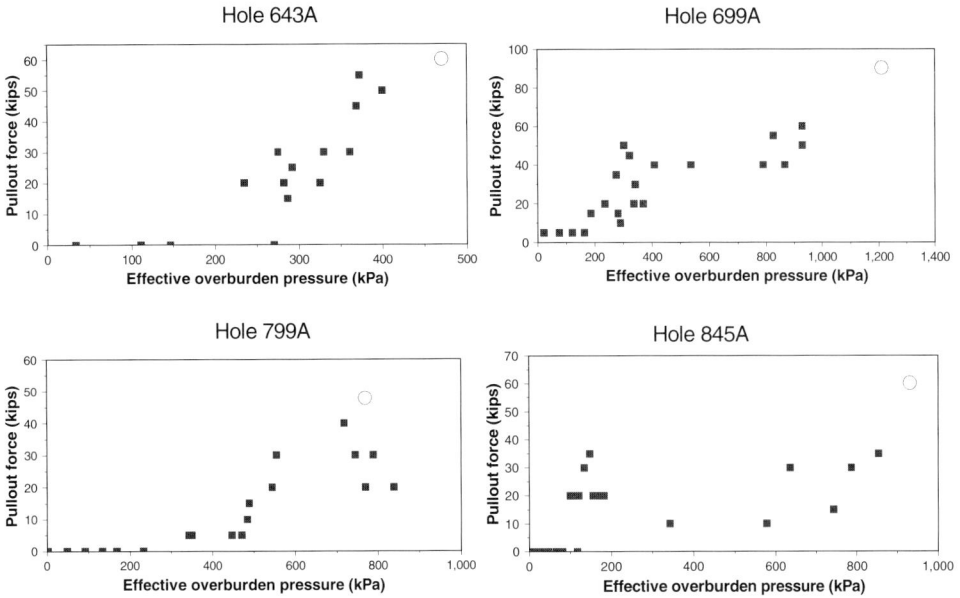

Fig. 13. Pullout force plotted against the effective overburden pressure of each core for four APC holes.

ment of tension leg platforms (TLPs), axial pullout forces have to be calculated in order to determine factors of safety for the large cylindrical piles which anchor them to the seabed. This problem is exactly analogous to that of determining the pullout

force for an APC. Indeed, a single TLP pile is very similar to an APC core barrel scaled up by a factor of about ten. Extracting pullout force and shear strength data from ODP's database may therefore be of value to the offshore oil industry. Alter-

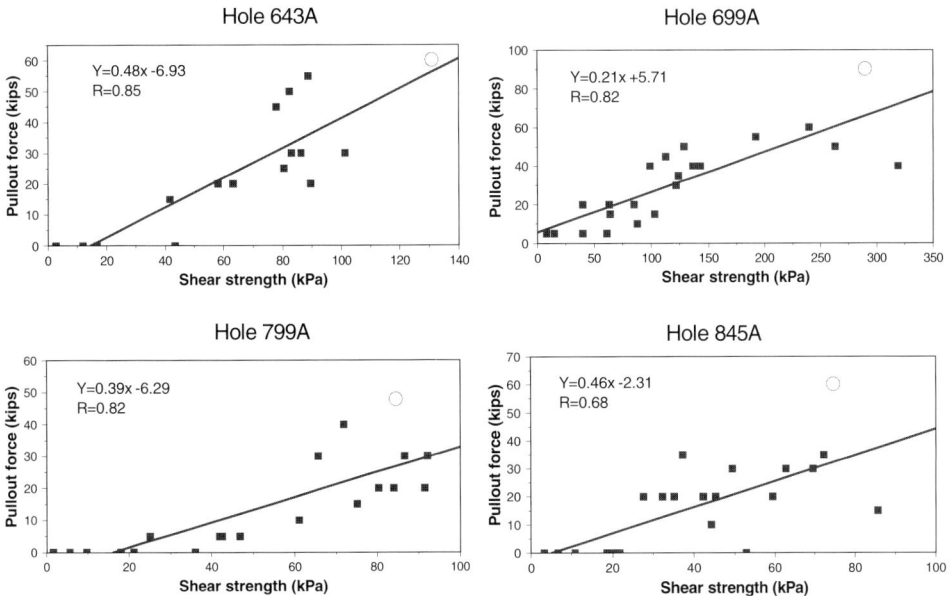

Hole 643A

Y=0.48x -6.93
R=0.85

Hole 699A

Y=0.21x +5.71
R=0.82

Hole 799A

Y=0.39x -6.29
R=0.82

Hole 845A

Y=0.46x -2.31
R=0.68

Fig. 14. Pullout force plotted against the average shear strength of each core for four APC holes.

natively, hydraulic piston coring could be carried out at the site where a TLP is to be installed in order to determine the tensional load-bearing capabilities of the sediment directly.

We thank J.-L. Briaud, G. Foss, R. Grout, C. Richter and M. Storms for discussions. J. Perry and R. Lawrence kindly assisted us in compiling data from various ODP databases.

References

AMERICAN PETROLEUM INSTITUTE. 1993. *Recommended Practice for Planning, Designing and Constructing Fixed Offshore Platforms.* API RP2A-WSD.

CIESIELSKI, P. F., KRISTOFFERSEN, Y. ET AL. 1988. *Proceedings of the Ocean Drilling Program, Initial Reports*, **114**. Ocean Drilling Program, College Station, TX.

ELDHOLM, O., THIEDE, J., TAYLOR, E. ET AL. 1987. *Proceedings of the Ocean Drilling Program, Initial Reports*, **104**. Ocean Drilling Program, College Station, TX.

HUEY, D. P., STORMS, M. A. & CAMERON, D. H. 1984. Design and Operation of a Hydraulic Piston Corer. DSDP Technical Report No. 21, National Technical Information Service, Springfield, Virginia.

INGLE, J. C., SUYEHIRO, K., VON BREYMANN, M. T. ET AL. 1990. *Proceedings of the Ocean Drilling Program, Initial Reports*, **128**. Ocean Drilling Program, College Station, TX.

KROENKE, L. W., BERGER, W. H., JANACEK, T. R. ET AL. 1991. *Proceedings of the Ocean Drilling Program, Initial Reports*, Leg 130. Ocean Drilling Program, College Station, TX.

MAYER, L., PISIAS, N., JANACEK, T. ET AL. 1992. *Proceedings of the Ocean Drilling Program, Initial Reports*, **138**. Ocean Drilling Program, College Station, TX.

MAZZULLO, J. M., MEYER, A. & KIDD, R. 1987. New sediment classification scheme for the Ocean Drilling Program. *In*: MAZZULLO, J. M. & GRAHAM, A. G. (eds) *Handbook for Shipboard Sedimentologists.* Ocean Drilling Program Technical Note, **8**, 45–67.

POAG, C. W. & FOSS, G. N. 1985. Explanatory notes. *In*: PRELL, W. L., GARDNER, J. V. ET AL. *Initial Reports of the Deep Sea Drilling Project*. US Government Printing Office, Washington DC, **68**, 5–13.

REA, D. K., BASOV, I. A., JANECEK, T. R. ET AL. 1993. *Proceedings of the Ocean Drilling Program, Initial Reports*, **145**. Ocean Drilling Program, College Station, TX.

RUDDIMAN, W. F., KIDD, R. B., THOMAS, E. ET AL. 1987. Introduction, background and explanatory notes. *In*: RUDDIMAN, W. F., KIDD, R. B., THOMAS, E. ET AL. (eds) *Initial Reports of the Deep Sea Drilling Project*. US Government Printing Office, Washington DC, **94**, 5–17.

SKEMPTON, A. W. 1954. The structure of inorganic soil (discussion). *Proceedings of the American Society of Civil Engineers.* **80**, Report **478**, 19–22.

STORMS, M. A., NUGENT, W. & CAMERON, D. H. 1983. *Design and Operation of the Hydraulic Piston Corer.* DSDP Technical Report No. 12, National Technical Information Service, Springfield, Virginia.

Index

Page numbers in *italics* refer to Tables or Figures

accelerator mass spectrometry (AMS) and ^{14}C 115
accretionary terrains 254–8
accretionary wedge, Mediterranean 260–2
advanced piston corer (APC)
    development of 305–6
    penetration of 306–9
        relation to sediment type 309–11
    pullout force 311–14
        relation to sediment type 314–16
Aghulas Basin 32, 45, 46
Aghulas Platform 29
Amazon Fan
    evolution 111–13
    sediment analysis 113–14
        dating 114–17
        mass transport studies 130
            methods 130–3
            results 133–41
            results discussed 141–7
        stable isotope analysis
            methods 114
            results 118–22
            results discussed 122–5
Amirante Channel 45, 46
Angola Basin 29, 32, 48
Angulobracchia crassa 76
anoxia 29
Antalya Complex 245
Antarctic Basin 32
    AABW flows 45, 48
Antarctic bottom water (AABW)
    Miocene circulation study
        method 45–6
        results 46–52
Antarctic–Australian Discordance (AAD) 45, 46
areas of potential hiatus formation (APHiD) 43
Argentine Basin 45
Atlantic Ocean 5–6, 29–30, 31–2, 142
    Cretaceous radiolarian zonation
        results 72–80
        sites 71–2
    Eemian planktonic foraminiferal studies
        isotope analyses
            methods 92
            results 94–6
            results discussed 97–8
    Miocene benthic foraminiferal studies
        isotopes 60, 64–5
        species 57–8, 63–4
    Quaternary–Recent circulation 112–13, 124–5
    see also Amazon Fan

Australia, Cretaceous plate motion 29
Australian Basin 45

Baer-Bassit 245
Bellinghausen Basin 32, 45, 49
Bengal Fan 151–3
    geochemical and mineralogical analysis
        methods 155–6
        results 156–60
        results discussed 161–8
    lithofacies description 153–5
    provenance 168–70
benthic foraminiferal analysis
    methods 57, 133
    results 57–60, 140, *142*
    results discussed 63–4, 143
Bermuda Rise 72, *75*
Bey Daglari carbonate platform 256
Blake event 1130
Bond cycles 103
Bouma sequences 213
Brazil Basin 45
Broken Ridge 30
bulk density, relation to piston coring depth *308, 309,*
    311, *312*

δ^{13}C
    Eemian planktonic foraminiferal analysis
        methods 92
        results 95–6
        results discussed 97–8
    Miocene benthic foraminiferal analysis
        methods 57
        results 60
    Pleistocene planktonic foraminiferal analysis
        methods 114
        results 118–22
        results discussed 122–5
    relation to radiolarian extinction 80
^{14}C dating 115
California, Gulf of
    greyvalue time series analysis
        methods 103
        results 103–5
        results discussed 105–9
Campbell Basin 32
Campbell Plateau 30, 31
carbonate compensation depth (CCD) 142

317

carbonate platforms
  Eratosthenes Platform (Seamount) 244–6
    accretion potential 254–8
    break-up evidence 253–4
    flexure-induced subsidence 258
    role in ophiolite emplacement 258–60
    stratigraphy
      Cretaceous–Tertiary 246–52
      Messinian evidence 252
      Plio-Pleistocene 252–3
Caribbean Current 112
Central Indian Basin 30
  AABW flows 48
Central Indian Ridge 31
Central Lau spreading centre 213, 214, 232
Chagos-Laccadive Ridge 31
chemical index of alteration 162
Chile Ridge 32
Cibicides wuellerstorfi isotope record 133, 141
  Eemian
    methods of analysis 92
    results 94–6
    results discussed 97–8
  Miocene
    methods of analysis 56–7
    results 57–60
    results discussed 63–5
clathrates 264–6
Costa Rica Dorric *84*
Costa Rica Rift ocean crust study
  drill holes compared 190–1
  drill holes described
    alteration 184–5
    fracture patterns 182–4, 188–90
    seismic wave velocity 185–7
    stratigraphy 181–2, 187–8
  tectonic synthesis 191–4
Cretaceous
  palaeobathymetry *8–15*
  Quiet Zone 5
  radiolarian zonation
    results 72–80
    sites 71–2
  stratigraphy of Eratosthenes Platform 246
Crolanium triagulare 76
Crozet Basin 32
Crozet hotspot 30, 32
Crucella cachensis 80
Cycladophora davisiana Ehrenberg
  low latitude stratigraphy
    methods of preparation 85
    results 86–7
    results discussed 87–8
Cyprus 259–60

Daisha Seamount 255
Dansgaard-Oeschger events 103
deep carbonate units, Amazon Fan 129
deep tow profiler records 201–2
deglaciation effects on ^{13}C record 122–3
Dorypyle anisa 77
DSDP leg 13 244
DSDP leg 42 244

DSDP leg 47B, site 398 72
DSDP leg 64 305
DSDP leg 69, site 504
  alteration 184–5
  fracture pattern 182–4
  seismic wave velocity 185–7
  stratigraphy 181–2
DSDP leg 70 *200*
DSDP leg 79, site 545 72
DSDP leg 93, site 603 72
DSDP leg 94 305

East Pacific Rise 30, 31, 32
Eastern Lau spreading centre 213–14, 232
Eemian
  pollen records 96–7
  stable isotope record
    methods of analysis 92
    results 94–6
    results discussed 97–8
El Niño Southern Oscillation (ENSO) 101, 104
electrical borehole images *see* formation microscanner
Emiliania huxleyi 131, 135
Equatorial Front *84*
Eratosthenes Platform (Seamount)
  accretion potential 254–8
  break-up evidence 253–4
  flexure-induced subsidence 258
  role in ophiolite emplacement 258–60
  setting 244–6
  stratigraphy
    Cretaceous–Tertiary 246–52
    Messinian evidence 252
    Plio-Pleistocene 252–3

Falklands Plateau 29
Florida Current 112
foraminifera
  Eemian planktonic studies
    methods of analysis 92
    results 95–6
    results discussed 97–8
  Miocene benthic studies
    methods of analysis 57
    results 60
  Pleistocene planktonic studies
    methods of analysis 114
    results 118–22
    results discussed 122–5
  Quaternary benthic and planktonic studies
    methods of analysis 133
    results 137–41
    results discussed 143
Formation microscanner (FMS)
  image numerical modelling 299–301
  mode of operation 288–90
  use
    bed boundary recognition 290–2
    fracture pattern identification 296–8
    sedimentary structure identification 292–3
    stress analysis 298–9
    texture recognition 293–6

fracture patterns
  in boreholes 296–8
  in ocean crust 182–4

Galapagos spreading centre
  sedimentary rate study
    measurement 201–2
    relation to latitude 205–6
    thickness variation 202–5
    use in seafloor dating 206–7
gateways 29, 31–2
geochemistry
  Bengal Fan study
    methods of analysis 155–6
    results 157–60
    results discussed 161–2
  Lau Basin study *220*, 222, 233
    methods of analysis 234
    results 234–6
    results discussed 236–40
  Nankai accretionary prism
    methods of analysis 274–5
    results 275–80
    results discussed 281–4
  ocean crust 185
Gephyrocapsa 132, 133
glaciation effects on ^{13}C record 123–5
Globigerina inflata isotope record
  methods of analysis 92
  results 94–6
  results discussed 97–8
Globigerinoides ruber 114, 118–22, 133, 140
Globigerinoides sacculifer 114, 118–22, 133, 140, 141
Globigerinoides trilobus 114, 118–22, 133, 140
Globorotalia hexagonus 137
Globorotalia tosaensis 132
Globorotalia truncata 114, 118–22, 133, 140
Globorotalia tumida flexuosa 137
Gondwana fragmentation 7–10
Greenland
  ice core 92
  volcanic lithology interpretation 290–1
greyvalue time series analysis
  methods 103
  results 103–5
  results discussed 105–9
Guaymas Basin 101, *102*
  greyvalue time series analysis
    methods 103
    results 103–5
    results discussed 105–9
Gulf Stream 112–13
Guyana Current 112

Halesium crassum 76
Hatteras Formation 71
Heinrich events 123–5
hemipelagites, cf. turbidites 213
Holocene compared with Eemian *see* Eemian

Holocene stratigraphy
  comparison with Eemian
    pollen records 96–7
    stable isotope record
      methods of analysis 92
      results 94–6
      results discussed 97–8
Lau Basin
  description 215–19
  interpretation 219–26
hotspots
  depth anomalies 6–7
  South Atlantic 29, 30
hydraulic piston corer (HPC)
  development of 305–6
  first use 305
hydrothermal alteration 185
hydrothermal flux
  Lau Basin study 233
    methods of analysis 234
    results 234–6
    results discussed 236–40

Indian Ocean
  AABW flows 46–8, 49
  Cretaceous plate setting 5–6
  Miocene benthic foraminifera
    isotopes 60, 64–5
    species 58–60, 63–4
  Plio-Pleistocene radiolarian stratigraphy 87–8
    84, 85–6
inductively coupled plasma atomic emission
  spectrometry (ICP-AES) 234, 275
inductively coupled plasma mass spectrometry (ICP-MS)
  275
instrumental neutron activation analysis (INAA) 156
interglacials, dating of 141–3
International Phase of Ocean Drilling (IPOD) 287
iron accumulation 239–40
isochrons use in palaeobathymetry 6
isopach map use in palaeobathymetry 6

Japan
  Nankai accretionary prism 273
    composition 273–4
    geochemical analysis
      method 274–5
      results 275–80
      results discussed 281–4
Japan Sea drilling programme analysis 291–2

Kerguelen Plateau 29, 48

Lake Mungo excursion 130
latitude effect on sedimentation rate 205–6

Lau Basin 213–15, 231
  drill sites 232–3
  evolution 231–2
  hydrothermal flux 233
  Miocene–Holocene stratigraphy
    description 215–19
    interpretation 219–26
  sediment geochemistry study
    methods 234
    results 234–6
    results discussed 236–40
  tectonic history 226
Lau Ridge 213
lithostratigraphy, use of formation microscanner 290–2

M reflector 260
Madagascar Basin 30, 45, 46
Madagascar Ridge 46
Madingley Rise 84
magnetostratigraphy 114–15, 130
Mamonia Complex 245
manganese accumulation 239–40
Mascarene Plateau 31
mass accumulation rate (MAR)
  measurement
    method 201–2
    results 202
  relation to latitude 205–6
  thickness variation
    characterization 202–3
    scaling 203–5
  use in seafloor dating 206–7
mass transport deposits (MTD)
  Amazon Fan occurrence 129–30
    sources 144–7
    timing 144
Maud Rise 29, 32
Mazagan Plateau 72, 74
Mediterranean Ridge
  mud volcanoes 261–2
    age 266–7
    formation 266
    geochemical processes 264–6
    Milano 262
    Napoli 262–4
    sediment types 264
  tectonic setting 260–1
Mediterranean Sea
  Eratosthenes Platform 244–6
    accretion potential 254–8
    break-up evidence 253–4
    flexure-induced subsidence 258
    role in ophiolite emplacement 258–60
    stratigraphy
      Cretaceous–Tertiary 246–52
      Messinian evidence 252
      Plio-Pleistocene 252–3
  Miocene benthic foraminifera 60, 63
  mud volcanoes
    age 266–7
    geochemical processes 264–6
    Milano 262
    Napoli 262–4

    sediment types 264
    tectonic setting 243–6
  Messinian evaporites 260
  Messinian salinity crisis 252
  metamorphic grade, ocean crust 184–5, 193
  methane hydrates 264–6
  microfractures in ocean crust 182–4
  microresistivity logs 287
  Milankovitch cycle recognition by FMS 291
  Milano mud volcano structure 243, 262
  mineralogy of Bengal Fan
    methods 155–6
    results 156–7
    results discussed 161–2
  Miocene
    AABW circulation study
      methods 45–6
      results 46–52
    benthic foraminifera analysis
      isotopes 60, 64–5
      species 57–8, 63–4
  Mita gracilis 76
  Mita spoletoensis 76
  Mozambique Basin 46
  Mozambique Channel 29, 45, 46
  Mozambique Ridge 29
  mud volcanoes
    formation 266
    Mediterranean Ridge
      age 266–7
      geochemical processes 264–6
      Milano 262
      Napoli 262–4
      sediment types 264

Nankai accretionary prism 273
  composition 273–4
  geochemical analysis
    method 274–5
    results 275–80
    results discussed 281–4
nannofossils in Pleistocene stratigraphy
  methods of analysis 130–2
  results 133–7
Napoli mud volcano structure 243, 262–4
Nd isotope ratios in Bengal Fan 166–8
Neogene stratigraphy
  Bengal Fan 152–3
  Lau Basin
    description 215–19
    interpretation 219–26
Neogloboquadrina dutertrei 114, 118–22
Nicobar Fan 48
Ninetyeast Ridge 30, 48
North Brazil Coastal Current (NBCC) 112, *113*, 124
North Equatorial Countercurrent (NECC) *84*, *113*
North Equatorial Current (NEC) *84*, *113*
Northern Component Water (NCW) 56, 66
North Atlantic Deep Water (NADW) 56, 67
Nuttallides umboniferus
  Miocene
    Atlantic Ocean 57–8
    Indian Ocean 58–60

$\delta^{18}O$ from foraminiferal analysis
  Eemian
    methods 92
    results 94–5
    results discussed 97–8
  Miocene
    methods 57
    results 60
  Pleistocene planktonic studies
    methods 114
    results 118–22
    results discussed 122–5
  Quaternary benthic and planktonic studies
    methods 133
    results 139–41
    results discussed 143
obduction 258–60
ocean anoxic events (OAE), Cretaceous *73*, *74*, *75*,
  80
ocean crust *see* Costa Rica Rift
ocean crust
  ages 33, *34–5*
  dating methods 202–3, 206–7
  drill penetration *33*
  structure in Costa Rica Rift
    drill holes compared 190–1
    drill holes described
      alteration 184–5
      fracture patterns 182–4, 188–90
      seismic wave velocity 185–7
      stratigraphy 181–2, 187–8
      tectonic synthesis 191–4
ODP leg 101 to 149, survey of piston coring
  penetration data 306–9
    relation to sediment 309–11
    pullout forces 311–14
    relation to sediment 314–16
ODP leg 103, sites 638/641 72
ODP leg 111, site 677
  description 84, 85
  radiolarian stratigraphy 87–8
ODP leg 116, sites 171 to 179 152, 153
ODP leg 128, site 794 291–2
ODP leg 130 305
ODP leg 131, site 808 273
ODP leg 133, site 816 294, *295*
ODP leg 135, sites 834 to 839 214–15
  description 215–19
  interpretation 219–26
  site 836 233, *235*, *237*, 238, *240*
  site 837 233, *235*, *237*, 238, *240*
  site 838 233, *235*, *238*, 239, *240*
  site 839 233, *235*, 238–9, *240*
ODP leg 138
  site 709
    description 84, 85–6
    radiolarian stratigraphy 87–8
  site 847
    description 84, 86
    radiolarian stratigraphy 87–8
  site 850 84, 86
  site 851
    description 85, 86
    radiolarian stratigraphy 87–8

ODP leg 139, sites 856 to 858 293–4
ODP leg 145 305
ODP leg 148, site 896
  fracture pattern 188–9
  stratigraphy 187–8
ODP leg 155
  sites 931/935/942/944/946 *136*
  site 932 113
    methods of analysis 114–17
    results 118–19
    results discussed 122–5
  site 933 113, *136*, *137*, *139*, *140*
    methods of analysis 114–17
    results 119–22
    results discussed 122–5
  site 936 *136*, *138*, *141*, *142*
ODP leg 160 244
ODP site 658 92
  methods of analysis 92
  results 94–6
  results discussed 97–8
Olduvai time slab radiolarian stratigraphy
  method of preparation 85
  results 86–7
  results discussed 87–8
Olympi Field 260
ophiolite emplacement 258–60
Orcadas Basin 29
overburden pressure in relation to piston coring 314,
  315

Pacific Ocean
  AABW flows 45, 49
  Costa Rica Rift study
  Costa Rica Rift ocean crust study
    drill holes compared 190–1
    drill holes described
      alteration 184–5
      fracture patterns 182–4, 188–90
      seismic wave velocity 185–7
      stratigraphy 181–2, 187–8
    tectonic synthesis 191–4
  Lau Basin study 213–15, 231
    drill sites 232–3
    evolution 231–2
    hydrothermal flux 233
    Miocene–Holocene stratigraphy
      description 215–19
      interpretation 219–26
    sediment geochemistry study
      methods 234
      results 234–6
      results discussed 236–40
    tectonic history 226
  radiolarian abundance study
    methods of preparation 85
    results 86–7
    results discussed 87–8
  Santa Barbara Basin study
    methods of analysis 274–5
    results 275–80
    results discussed 281–4

palaeobathymetry
    methods in reconstruction 3–5
    southern oceans study
        equations 7
        methods of analysis 5–7
        results 7–32
        results discussed 32–7
Palaeocene volcanism 290–1
palaeoclimate signal recognition by FMS 290
palaeomagnetic dating 114–15
PALIOS 3
PALSOS 3
Panama Basin Galapagos sedimentation study
    measurement 201–2
    relation to latitude 205–6
    thickness variation 202–5
    use in seafloor dating 206–7
Parnassus carbonate platform 256
permeability of ocean crust 186, 188
Peru Current *84*
Pindos ophiolite 259
piston corer
    development of 305–6
    penetration 306–9
        relation to sediment 309–11
    pullout forces 311–14
        relation to sediment 314–16
planktonic foraminifera in stable isotope analysis
    record for Eemian
        methods 92
        results 94–6
        results discussed 97–8
    record for glacial–interglacial cycle
        methods of analysis 114
        methods of dating 114–17
        results 118–22
        results discussed 122–5
    record for Quaternary debris flows
        methods 132–3
        results 137–41
        results discussed 143
Pleistocene
    stratigraphy of Amazon Fan
        methods of analysis 130–3
        results 133–41
        results discussed 141–7
    stratigraphy of Lau Basin
        description 215–19
        interpretation 219–26
Plio-Pleistocene
    radiolarian stratigraphy
        method of preparation 85
        results 86–7
        results discussed 87–8
    tectonics of Mediterranean 260–1
Podocapsa 77
pollen records, Eemian 96–7
porosity of ocean crust 186, 188
power spectra 104, *106–7*
Pseudodictyomitra lodogaensi 76
Pseudodictyomitra nakesekoi 80
Pseudodictyomitra pseudomacrocephelata 77
Pseudoemiliania lacunosa 131–2, 133
Pulleniatina obliquiloculata 114, 118–22, 133, 139, 140

pullout force
    in advanced piston coring 311–14
    relation to sediment type 314–16
pyrite in FMS logs 290

Quaternary *see* Pleistocene

radiolaria
    Cretaceous zonation
        results 72–80
        sites 71–2
    low latitude stratigraphy
        methods of preparation 85
        results 86–7
        results discussed 87–8
REE in accretionary prism muds
    methods of analysis 274–5
    results 275–80
    results discussed 281–4
residual depth anomalies 6–7
Réunion hotspot 30, 31
ridge jump 5–6
Rio Grande Rise 29
Ross Sea and AABW flows 45–6, 49

Santa Barbara Basin 101, *102*
    greyvalue time series analysis
        methods 103
        results 103–5
        results discussed 105–9
Sao Paulo Rise 29
Sayade Malha Bank 31
Scotia Sea 31
sea surface temperature (SST) 94
sea-floor dating 202–3, 206–7
sediment thickness use in palaeobathymetry 6
sediment-free bathymetry 36–7, *38*
sedimentation rates, Lau Basin *224*
seismic wave velocity in ocean crust 185
Semail ophiolite 259
shear strength and piston coring depth *308, 309,* 311,
    312
solar activity cycle 104
Somali Basin 45, 46, 52
South Atlantic Ocean AABW flows 48–9
South Equatorial Current *84*
Southeast Indian Ridge 45, 52
southern oceans
    Cretaceous–Recent palaeobathymetry
        equations 7
        methods of analysis 5–7
        results 7–32
        results discussed 32–7
Southwest Indian Ridge 32, 45
spectral analysis 104, *106–7*
Sr isotope ratios in Bengal Fan 168
stable isotope analysis *see* ^{13}C *also* ^{18}O
stress analysis in FMS images 298–301
subduction history, Mediterranean 243–4
sunspot cycles 104

Taitao ophiolite 258
Tasman Fracture Zone 31
Tasman Rise 31
Tasman Sea 30
    AABW flows 45, 49
Tauride carbonate platform 245, 246
tectonic setting analysis by FMS 296–8
tension leg platforms 315–16
Tertiary
    palaeobathymetry *16–32*, 29–31
    stratigraphy of Eratosthenes Platform 246–52
    see also Miocene *also* Palaeocene
Tethyan Outflow Water (TOW) 63–6
texture analysis in FMS images 293–6
Theoconus coronatus 77
Tofua Arc 213
Tonga Ridge 213
Torculum coronatum 76–7
trace element analysis of Bengal Fan
    methods 156
    results 159–60
    results discussed 162–8
Troodos ophiolite 259–60
turbidites 211–12
    cf. hemipelagites 213

umbers 273, 282
upwelling 94
Uvigerina sp. 133

Valu Fa Ridge 214
variable length hydraulic piston corer (VLHPC) 305
Vema Channel 45
Vigo Seamount 72, *73*
Vostok Deuterium ice core 92

Walvis Ridge 29, 45
weathering analysis in Bengal Fan 161
Weddell Sea 29
    AABW flows 45, 49
West African Shelf *see* ODP site 658
Wharton Basin 29, 30, 45, 48
Wharton Ridge 30, 32

X-ray diffraction (XRD) analysis 155, 275
X-ray fluorescence (XRF) analysis 155

Yamato Basin drilling programme analysis 291–2